SOCIETÀ ITALIANA DI FISICA

RENDICONTI
DELLA
SCUOLA INTERNAZIONALE DI FISICA
«ENRICO FERMI»

CXXIII Corso
a cura di B. MARAVIGLIA
Direttore del Corso
VARENNA SUL LAGO DI COMO
VILLA MONASTERO
13-21 Ottobre 1992

Doppia risonanza magnetica nucleare

1993

SOCIETÀ ITALIANA DI FISICA
BOLOGNA-ITALY

ITALIAN PHYSICAL SOCIETY

PROCEEDINGS
OF THE
INTERNATIONAL SCHOOL OF PHYSICS
«ENRICO FERMI»

Course CXXIII
edited by B. MARAVIGLIA
Director of the Course
VARENNA ON LAKE COMO
VILLA MONASTERO
13-21 October 1992

Nuclear Magnetic Double Resonance

1993

NORTH-HOLLAND
AMSTERDAM · OXFORD · NEW YORK · TOKYO

Copyright ©, 1993, by Società Italiana di Fisica

All rights reserved. No part of this publication may be reproduced, stored in a retrieval system, or transmitted, in any form or by any means, electronic, mechanical, photocopying, recording or otherwise, without the prior permission of the copyright owner.

PUBLISHED BY
North-Holland
Elsevier Science Publishers B.V.
P.O. Box 211
1000 AE Amsterdam
The Netherlands

SOLE DISTRIBUTORS FOR THE USA AND CANADA:
Elsevier Science Publishing Company, Inc.
655 Avenue of the Americas
New York, N.Y. 10010
U.S.A.

Technical Editor
P. PAPALI

Library of Congress Cataloging-in-Publication Data

International School of Physics "Enrico Fermi" (1992 : Varenna, Italy)
 Nuclear magnetic double resonance : Varenna on Lake Como, Villa Monastero, 13-21 October 1992 / edited by B. Maraviglia.
  p. cm. -- (Proceedings of the International School of Physics "Enrico Fermi" ; course 123)
  Title on added t.p.: Doppia risonanza magnetica nucleare.
  At head of title: Italian Physical Society.
  ISBN 0-444-81823-5
  1. Electron nuclear double resonance--Congresses. I. Maraviglia, B. II. Società italiana di fisica. III. Title. IV. Title: Doppia risonanza magnetica nucleare. V. Series: International School of Physics "Enrico Fermi." Proceedings of the International School of Physics "Enrico Fermi" ; course 123.
QC762.6.E45I58 1992
538'.362--dc20   93-48564
                                                                  CIP

Proprietà Letteraria Riservata
Printed in Italy

This Course
is under the auspices
of the
Groupement AMPERE,
which promoted
and
scientifically supported it.

INDICE

B. MARAVIGLIA – Preface .... pag. XV

Gruppo fotografico dei partecipanti al Corso   fuori testo

M. GOLDMAN – Introduction to some basic aspects of NMR.

1.	Nuclear interactions...	pag.	1
	1˙1.	The Zeeman interaction	»	1
	1˙2.	Electron-nucleus interactions..	»	3
		1˙2.1. Magnetic interactions.....................................	»	4
		1˙2.2. Electric interactions..	»	7
		1˙2.3. The case of metals	»	9
	1˙3.	Dipole-dipole interactions...	»	9
	1˙4.	Exchange interactions	»	11
	1˙5.	Appendix: transformation properties of 2nd-rank tensors upon rotation	»	12
2.	Spin-lattice relaxation revisited..	»	14
	2˙1.	Derivation of the master equation	»	15
		2˙1.1. Evolution in the interaction representation	»	18
		2˙1.2. Evolution in the Schrödinger representation	»	20
		2˙1.3. Relation between both approaches.................	»	21
		2˙1.4. Conditions of validity of the theory	»	22
		2˙1.5. Extension to solids	»	23
	2˙2.	Drawbacks of alternative approaches. A selection	»	24
		2˙2.1. Use of an interaction representation that does not remove all of $\mathscr{H}_0$	»	25
		2˙2.2. Premature assumption that the average of a product equals the product of the averages	»	26

	2˙2.3.	Calculation of the time evolution of all elements of the density matrix	pag.	27
	2˙2.4.	Guesses about the form of the density matrix	»	28
	2˙2.5.	Use of the memory function formalism	»	29
3.	Spin temperature		»	31
3˙1.	Spin temperature in zero field		»	31
3˙2.	Spin temperature in low field		»	34
3˙3.	Spin temperature in high field		»	35
3˙4.	R.f. irradiation in high field		»	38
3˙5.	Value and evolution of the spin temperature		»	39
	3˙5.1.	Sudden change of the Hamiltonian	»	40
	3˙5.2.	Adiabatic variation of the Hamiltonian	»	40
3˙6.	What gives rise to a r.f. signal?		»	42
3˙7.	Verification of spin temperature theory		»	44
	3˙7.1.	General	»	44
	3˙7.2.	The case of effective Hamiltonians	»	45
4.	The case of a periodic Hamiltonian		»	52
4˙1.	Introduction		»	52
4˙2.	Fast modulation		»	55
	4˙2.1.	The basic equations	»	55
	4˙2.2.	The next-order approximation	»	56
	4˙2.3.	The general expansion	»	57
4˙3.	Large anisotropic chemical shifts		»	63
	4˙3.1.	Spin-spin relaxation	»	65
	4˙3.2.	Application of the theory. Generalities	»	66

CH. P. SLICHTER – Double resonance.

1.	What is double resonance, and why do it?	»	69
2.	Discovery of the Pound-Overhauser family	»	70
3.	A model system	»	71
4.	The electron-nuclear case: the Overhauser effect	»	71
5.	Cross-relaxation double resonance	»	78
6.	Spin coherence double resonance—S-flip-only echoes	»	82
7.	Spin coherence double resonance—coherence transfer	»	85
8.	Spin echo double resonance (SEDOR)	»	88
9.	SEDOR signals when there are several S spins	»	92
10.	Examples of SEDOR studies	»	94
Explanatory note		»	101
Polarization of nuclear spins in metals		»	101
Measurement of the spin and gyromagnetic ratio of ^{13}C by the collapse of spin-spin splitting		»	104
Effects of perturbing radiofrequency fields on nuclear-spin coupling		»	106
Method of polarizing nuclei in paramagnetic substances		»	117
Polarization of phosphorous nuclei in silicon		»	120

L. EMSLEY and A. PINES – Lectures on pulsed NMR (2nd edition).

Introduction		»	123
1.	Multiple-quantum NMR	»	123
1˙1.	Dipolar couplings and molecular structure	»	123
1˙2.	Onset of spectral complexity	»	124

	1˙3.	Simplification by multiple-quantum transitions	pag.	127
	1˙4.	Analogy to chemical isotopic labeling	»	128
	1˙5.	Obtaining multiple-quantum spectra	»	129
	1˙6.	Theory of multiple-quantum NMR: preliminaries	»	130
	1˙7.	Multiple-quantum signal	»	132
	1˙8.	Special case: one-quantum FID point by point	»	132
	1˙9.	General case: multiple-quantum FID	»	133
	1˙10.	Time-reversal (conjugate) detection	»	134
	1˙11.	Effect of phase shifts	»	135
	1˙12.	Time-proportional phase incrementation (TPPI)	»	137
	1˙13.	Double-quantum NMR in solids	»	137
	1˙14.	Double-quantum spin locking	»	138
	1˙15.	Molecular structure by multiple-quantum NMR	»	141
	1˙16.	Selective n-quantum excitation	»	145
	1˙17.	Multiple-quantum NMR in solids	»	147
	1˙18.	Selection rules in multiple-quantum dynamics	»	151
	1˙19.	Total coherence transfer	»	151
	1˙20.	Bilinear rotation pulses	»	153
2.	Coherent-averaging theory		»	153
	2˙1.	Introduction	»	153
	2˙2.	Multiple-pulse line narrowing	»	155
	2˙3.	Magic-angle spinning	»	157
	2˙4.	Double rotation	»	160
3.	Spin decoupling		»	165
	3˙1.	Spin $I = 1/2$ pair	»	165
	3˙2.	One-quantum offset and r.f. amplitude dependence	»	166
	3˙3.	Double-quantum decoupling	»	168
	3˙4.	Double-quantum offset and r.f. amplitude dependence	»	168
	3˙5.	Comment on the relationship between spin decoupling an multiple-quantum excitation	»	171
4.	Interaction of radiation and matter		»	171
	4˙1.	Two-level system in a quantized field	»	171
	4˙2.	Fictitious spin $I = 1/2$ operators	»	173
	4˙3.	Evolution of the two-level system	»	174
	4˙4.	Evolution off resonance	»	175
	4˙5.	Adiabatic rapid passage	»	175
	4˙6.	Three-level system in a quantized field	»	177
	4˙7.	Fictitious spin $I = 1$ operators	»	178
	4˙8.	Double-quantum (two-photon) Hamiltonian	»	178
	4˙9.	Evolution of the system	»	179
	4˙10.	Double-quantum adiabatic rapid passage	»	180
5.	Group theory and dynamics		»	182
	5˙1.	Molecular motion	»	183
	5˙2.	Group theory and exchange	»	185
	5˙3.	Relevant representations	»	186
	5˙4.	Summary of symmetry considerations	»	187
	5˙5.	Example: dynamics of solid benzene	»	188
	5˙6.	Macroscopic motional averaging	»	190
	5˙7.	Sample spinning	»	190
	5˙8.	Group theory of motional averaging	»	191
	5˙9.	Averaging under cubic and icosahedral symmetry	»	193

5˙10.	Dynamic-angle spinning	pag.	195
5˙11.	Isotropic-anisotropic correlation spectra	»	199
5˙12.	Variable-angle correlation spectroscopy	»	201
6. Cross-polarization		»	203
6˙1.	Spin temperature	»	203
6˙2.	Methods for obtaining thermal contact	»	204
6˙3.	Statistical picture of cross-polarization	»	206
6˙4.	Hartmann-Hahn mismatch	»	208
6˙5.	Thermodynamics of heteronuclear cross-polarization	»	210
6˙6.	Resolved heteronuclear coupling	»	213
7. Unitary bounds on spin dynamics		»	216
7˙1.	Unitary evolution	»	216
7˙2.	The entropy limit	»	217
7˙3.	Bounds on unitary evolution	»	217
7˙4.	Redfield's description of polarization transfer	»	218
7˙5.	Bounds in $I_N S$ systems	»	219
7˙6.	The thermodynamic limit	»	220
7˙7.	Two-dimensional bounds	»	221
7˙8.	Transfer of basis operators	»	223
7˙9.	Nonunitary evolution	»	223
7˙10.	Cross-polarization echoes	»	225
8. Zero-field NMR		»	226
8˙1.	Zero-field NQR of deuterium	»	228
8˙2.	Two-dimensional zero-field NMR	»	230
8˙3.	Zero-field pulses	»	233
8˙4.	Calculation of the zero-field spectrum	»	236
8˙5.	Average over orientational distribution	»	237
8˙6.	Dipolar coupled spin $I = 1/2$ pair or quadrupolar spin $I = 1$	»	237
8˙7.	Effects of motion	»	240
8˙8.	Magnetic resonance with a SQUID detector	»	241
8˙9.	Comment on relationship of spatially selective pulses to zero-field NMR	»	242
9. Geometric phases		»	243
9˙1.	Context for geometric phases	»	243
9˙2.	Classical holonomy	»	244
9˙3.	Quantum holonomy	»	245
9˙4.	Equations for the geometric phase	»	248
9˙5.	Explicit calculation for spin $I = 1/2$	»	249
9˙6.	The Aharonov-Bohm effect	»	251
9˙7.	Geometric phase in NMR interferometry	»	252
9˙8.	Fractional quantum numbers	»	254
9˙9.	Nonunitary behavior, quantum projection	»	256
9˙10.	Geometry of light	»	257
9˙11.	Rotation of cats	»	258

M. MEHRING – A guided tour through double-resonance phenomena.

1.	Introduction	»	267
2.	Double resonance in three- and multi-level systems	»	269
2˙1.	Population trasfer	»	270

			2˙1.1. Saturation of the 2-3 transition	pag.	271
			2˙1.2. Inversion of the 2-3 transition	»	272
		2˙2.	Spin alignment and pulsed ENDOR experiments	»	272
		2˙3.	Spin echo double resonance (SEDOR)	»	276
			2˙3.1. The 1-2 spin echo	»	277
			2˙3.2. SEDOR effect	»	277
			2˙3.3. Spinor experiments	»	278
		2˙4.	Cross-polarization in the rotating frame	»	280
		2˙5.	Coherence transfer	»	281
	3.	Many-spin systems		»	282
		3˙1.	Spin temperature and spin density matrix	»	283
		3˙2.	Spin calorimetry	»	284
		3˙3.	From Hilbert space to Liouville space	»	285
		3˙4.	Spin calorimetry in Liouville space	»	286
			3˙4.1. Energy conservation	»	287
			3˙4.2. Isentropic mixing	»	288
		3˙5.	SEDOR	»	289
		3˙6.	Double-resonance spin dynamics	»	291
	4.	Unconventional double-resonance experiments		»	294
		4˙1.	Rb-Xe double resonance and rotating the frame	»	294
		4˙2.	Hyperfine spectroscopy with correlation to an electron spin (HYSCORE)	»	295
		4˙3.	Optical Zeeman double resonance with a laser diode	»	296
	5.	Summary		»	298

B. H. MEIER – Polarization transfer experiments.

1.	Introduction	»	301
	1˙1. A simple application: the two-spin system	»	305
2.	Application to the MOIST experiment	»	309
3.	Larger spin systems	»	314
4.	Time-dependent Hamiltonians	»	317
5.	Spin diffusion	»	324
6.	Polarization echoes	»	330

R. FREEMAN – Introduction to two-dimensional NMR in liquids.

1.	Introduction	»	335
2.	*Pseudo*-two-dimensional spectroscopy	»	337
3.	Correlation spectroscopy (COSY)	»	340
4.	Total-correlation spectroscopy	»	345
5.	Chemical-exchange spectroscopy	»	346
6.	Nuclear Overhauser effect	»	346
7.	Nuclear Overhauser spectroscopy (NOESY)	»	349
8.	Forbidden transitions	»	349
9.	«INADEQUATE»	»	350
10.	Spin echoes	»	353
11.	«Broad-band decoupled» proton spectra	»	356
12.	Discussion	»	359

R. FREEMAN – Selective excitation in high-resolution NMR.

1. Introduction... pag. 363
2. The DANTE sequence .... » 364
3. Shaped pulses .... » 365
4. Phase gradients .... » 366
5. «Spin pinging».. » 367
6. Practical implementation .... » 368
7. Resolution enhancement.. » 368
8. Reduction of dimensionality .... » 370
9. The Hartmann-Hahn experiment.. » 373
10. Stepwise coherence transfer (DAISY) .... » 374

R. FREEMAN – Fine structure in two-dimensional spectra.

1. Introduction.. » 381
2. The structure of COSY cross-peaks .... » 381
3. Extraction of coupling constants .... » 386
4. J-extension .... » 387
5. J-deconvolution .... » 389
6. J-doubling... » 393

C. ZWAHLEN, S. J. F. VINCENT and G. BODENHAUSEN – Selective double resonance and coherence transfer.

1. Introduction.. » 397
2. Doubly selective irradiation .... » 398
3. Matrix representations .... » 401
4. Evolution, coherence transfer and detection .... » 404
5. Conclusions... » 410

R. CAMPANELLA, S. CAPUANI, F. DE LUCA and B. MARAVIGLIA – Double-resonance J-coupling imaging.

1. Introduction.. » 413
2. Double resonances to increase the signal-to-noise ratio.. » 413
3. J-coupling imaging .... » 416
4. Results and conclusions .... » 418

E. W. RANDALL – Some double-resonance methods in imaging experiments.

Introduction.. » 423
1. Some thoughts on definition .... » 424
2. Classification .... » 426
3. Introduction to double-resonance imaging .... » 434
   3`1. Nitrogen .... » 435
   3`2. Imaging ^{14}N .... » 436
   3`3. Imaging ^{15}N .... » 439
      3`3.1. Direct method... » 439
      3`3.2. ^{15}N{^{1}H} Overhauser method .... » 439
      3`3.3. Indirect methods, ^{1}H{^{15}N} .... » 444

N. Lugeri, F. De Luca, B. C. De Simone and B. Maraviglia – Rotating-frame spectroscopy and imaging under radio- and audio-frequency excitation.

1. Introduction .... pag. 449
2. General features of RCF experiments .... » 450
3. Theoretical backgrounds .... » 450
4. Magic-angle rotating-frame experiments .... » 455
    4'1. Interaction representation .... » 455
    4'2. TRCF Hamiltonians .... » 459
5. Applications of MARF experiments .... » 462
    5'1. MARF imaging .... » 462
    5'2. Relaxation studies .... » 466

M. Bloom – A proposed Stern-Gerlach experiment on individual spins in solids.

1. Introduction .... » 473
2. Review of the Stern-Gerlach experiment .... » 474
    2'1. Classical description .... » 474
    2'2. Quantum description .... » 475
3. The transverse Stern-Gerlach experiment .... » 475
    3'1. Resonance in the TS-G experiment—quantization of I_{x_e}. » 476
4. The folded Stern-Gerlach experiment .... » 476
    4'1. Analysis of the FS-G experiment for a frictionless oscillator » 477
    4'2. Inclusion of oscillator damping .... » 478
    4'3. Some practical considerations concerning mechanical oscillators .... » 478
    4'4. Criterion for detectability of a single proton spin by FS-G resonance .... » 479
5. Basic differences between Stern-Gerlach experiments and NMR ... » 480
6. Stern-Gerlach experiments on macroscopic spin systems .... » 481
    6'1. Polarized samples .... » 482
    6'2. Unpolarized samples .... » 482
Note added (January, 1993) .... » 483

D. J. Lurie and I. Nicholson – Proton-electron double-resonance imaging of exogenous and endogenous free radicals *in vivo*.

1. Introduction .... » 485
2. Summary of relevant results from dynamic nuclear polarization theory .... » 486
    2'1. The DNP enhancement .... » 487
    2'2. The coupling factor .... » 487
    2'3. The leakage factor .... » 487
    2'4. The saturation factor .... » 487
    2'5. The maximum theoretical enhancement .... » 488
    2'6. EPR irradiation power .... » 488
3. Implementation of PEDRI .... » 489
    3'1. Magnetic-field strength considerations .... » 489
    3'2. Basic PEDRI pulse sequence .... » 490

	3˙3.	Interleaved PEDRI pulse sequence	pag.	490
	3˙4.	Field-cycled PEDRI	»	491
	3˙5.	Snapshot PEDRI	»	492
4.	Hardware for PEDRI		»	492
	4˙1.	The magnet	»	492
	4˙2.	The double-resonance r.f. coil assembly	»	494
5.	Applications of PEDRI		»	496
	5˙1.	Imaging exogenous free radicals *in vivo*	»	496
		5˙1.1. Desirable features of a PEDRI contrast agent	»	496
		5˙1.2. Nitroxide free radicals	»	496
		5˙1.3. *In vivo* PEDRI studies	»	498
	5˙2.	Imaging endogenous free radicals *in vitro* and *in vivo*	»	498
		5˙2.1. Spin trapping	»	499
		5˙2.2. PEDRI with spin trapping	»	500
6.	Summary and conclusions		»	501

Preface.

The evolution of the magnetic-resonance area keeps expanding with respect both to the theoretical foundations and to the applied research in more and more fields. Magnetic double resonances were introduced some decades ago with immediate impact on several subjects. The example of the Overhauser effect with its theoretical and practical consequences is sufficient to stress the role of double resonances, which have spread to almost every section of magnetic resonances.

The booming field of space-resolved NMR, both with NMR imaging and *in vivo* spectroscopy, has not taken yet advantage of all double-resonance methods used in spectroscopy. At the same time the great interest in investigating all the possible ways to develop more advanced methodologies, in order to increase the signal-to-noise ratio of nuclei occurring in low concentration or having a low gyromagnetic ratio, drew me to the conclusion that it was time to gather some of the best NMR people at the Enrico Fermi School in Varenna.

After some exchange of opinions with the invited lecturers, the coordinated plan of the course was decided. The Enrico Fermi School of Varenna is intended as an advanced school, therefore all the selected students were already active in NMR research. For this reason this book of proceedings is at an advanced level, although a smooth graduality is always present. Pound-Overhauser double resonance, cross-relaxation and spin coherence double resonance together with spin decoupling dynamics and other fundamental concepts are carefully dealt in this book. Equal attention is reserved to several applications to liquids and solids, but special care is devoted to the first applications of double resonances to space-resolved double NMR like heteronuclear imaging, proton-electron double-resonance imaging, etc.

Thus this book is of great interest for physicists, chemists and engineers who research new lines of advancement of NMR spectroscopy and space-resolved NMR. The excellent correlation among the lecturers unifies the whole sequel of written monographic reports into a real single textbook.

As occurred for the 1986 NMR course of Varenna, this new important contribution on nuclear magnetic double resonance was a relevant scientific success, which is widely and thoroughly reported in this book.

B. MARAVIGLIA

SOCIETÀ ITALIANA DI FISICA
SCUOLA INTERNAZIONALE DI FISICA «E. FERMI»
CXXIII CORSO - VARENNA SUL LAGO DI COMO - VILLA MONASTERO - 13-21 Ottobre 1992

Introduction to Some Basic Aspects of NMR.

M. GOLDMAN

Service de Physique de l'Etat Condensé, Centre d'Etudes de Saclay
91191 Gif-sur-Yvette Cedex, France

1. – Nuclear interactions.

We review briefly the principal interactions experienced by nuclear spins that make magnetic resonance feasible, and that disturb it in a way that gives access to the properties of bulk matter. These can be called the «basic bricks» of NMR.

General references are [1, 2].

1˙1. *The Zeeman interaction.* – The interaction energy of a magnetic dipole M with a magnetic field B is classically of the form

$$(1.1) \qquad E = -M \cdot B.$$

When M is the magnetic moment of an electron or of an atomic nucleus, this is called the Zeeman interaction.

It is a well-established fact that nuclei that possess a magnetic moment M also possess an angular momentum $\mathscr{S}$, and that both are collinear:

$$(1.2) \qquad M = \gamma \mathscr{S},$$

where γ, the gyromagnetic ratio, is characteristic of each nuclear species.

A consequence of eqs. (1.1) and (1.2) is the celebrated Larmor theorem, which is the law of dynamic evolution of the magnetic moment:

$$(1.3) \qquad \frac{\mathrm{d}}{\mathrm{d}t} M = \boldsymbol{\omega} \wedge M,$$

where the frequency ω is equal to

$$(1.4) \qquad \boldsymbol{\omega} = -\gamma \boldsymbol{B}.$$

When the field B is constant in magnitude and orientation, say B_0 along the axis Oz, the evolution of M, when not parallel to Oz, is a rotation around Oz at constant mutual angle, with frequency

$$\omega_0 = -\gamma B_0,$$

called the Larmor frequency. This kind of motion bears the name of precession, and it is called the Larmor precession.

The very same result (1.3)-(1.4) can be obtained either by a classical or a quantum-mechanical treatment.

The classical chain of implications is the following:

i) It results from the form (1.1) of the energy that the magnetic moment $\boldsymbol{M}$ experiences a torque

(1.5) $$\boldsymbol{C} = \boldsymbol{M} \wedge \boldsymbol{B}.$$

ii) The torque is equal to the time derivative of the angular momentum:

(1.5a) $$\boldsymbol{C} = \frac{d}{dt}\mathscr{S}.$$

iii) The Larmor theorem (eqs. (1.3) and (1.4)) is then a direct consequence of the relation (1.2) between $\boldsymbol{M}$ and $\mathscr{S}$.

The quantum-mechanical development goes as follows:

i) To the energy (1.1) there corresponds a Hamiltonian (in frequency units)

(1.6) $$\mathscr{H} = -\hbar^{-1}\boldsymbol{M}\cdot\boldsymbol{B},$$

where $\boldsymbol{M}$ is a quantum-mechanical operator. (In practice it is the sum of the magnetic-moment operators $\boldsymbol{\mu}_i$ of a large number of identical nuclei.)

ii) The state of the system is described by a density matrix σ, whose evolution is described by the Liouville-von Neumann equation:

(1.7) $$\frac{d\sigma}{dt} = -i[\mathscr{H}, \sigma].$$

iii) The expectation value of the magnetic moment is

(1.8) $$\langle \boldsymbol{M} \rangle = \text{Tr}(\boldsymbol{M}\sigma).$$

iv) The angular momentum $\mathscr{S}_i$ of an individual nucleus is

(1.9) $$\mathscr{S}_i = \hbar \boldsymbol{I}_i,$$

where $\boldsymbol{I}_i$ is the vectorial spin operator, whose components are the generators of rotations in the spin space of the nucleus, which spans an irreducible representation $\mathscr{D}_I$ of the rotation group. We have $\mathscr{S} = \Sigma \mathscr{S}_i$.

The Larmor equations (1.3) and (1.4), where M has to be replaced by its expectation value $\langle M \rangle$, are deduced from eqs. (1.6) to (1.9) through the commutation relations between spin operator components.

The identity of the classical and quantum-mechanical equations of motion has an important consequence: it is possible to describe NMR in classical terms. This, however, is restricted to pure precession, and it ceases to be true when dealing with the evolution of spins experiencing mutual interactions, or the polarization transfer between such spins. Approximate pictorial representations have been elaborated to describe such situations, which some people find efficient and manipulate with mastery. I must confess that this is not my case.

Remark I. It is wrong to say that a nucleus has a well-defined spin. A nucleus has a ground state and a number of excited states, each of which is a manifold with a given spin different from state to state. The «spin» of the nucleus is that of the state on which NMR is performed, usually the ground state.

Remark II. The collinearity between spin and magnetic moment is of quantum-mechanical origin. The magnetic-moment operator has matrix elements within each nuclear level as well as between nuclear levels. In practice, the Zeeman interactions are so much smaller than the energy separations between nuclear levels that the Zeeman matrix elements linking distinct states can be ignored in a perturbation treatment. This amounts to limiting the magnetic moment to its projection onto the Hilbert space of the nuclear state under study. The proportionality of this truncated magnetic moment to the spin of that state is a consequence of the Wigner-Eckart theorem of group theory: within an irreducible representation of the rotation group, all tensorial operators of a given rank are proportional to each other. The proportionality factor between μ_i and I_i, namely $\gamma\hbar$, may differ from one nuclear state to another.

Remark III. The practice in NMR is to distinguish, in the field B experienced by a given nucleus, that part which arises from the electrons of the atom, or ion, or molecule to which the nucleus belongs. This contribution is at the origin of the chemical shift (and the indirect interactions) as seen in the next subsection. The remaining part of the field (the main part) is called the external field B_0. Since it includes the contribution from the magnetic moments of the nearby molecules, it can vary locally. This is at the origin of a serious theoretical complexity that will not be analysed here.

1`2. *Electron-nucleus interactions.* – Two kinds of interactions need to be considered: magnetic interactions giving rise to chemical shifts and indirect interactions, and electric interactions giving rise to nuclear quadrupole interactions. They are analysed in turn.

1`2.1. Magnetic interactions.

Schematic description. We consider the case of diamagnetic molecules, whose electronic ground state is nonmagnetic.

It acquires a small magnetization under the perturbing effect of a magnetic field. This magnetization is proportional to the field and is nonuniform spatially. It creates an extra field at the site of the nearby nuclei, which depends on the electronic wave function.

The perturbing field has two origins:

i) the external field $\boldsymbol{B}_0$, which creates an extra nuclear field of the rough form $-\sigma \boldsymbol{B}_0$: it is the chemical shift;

ii) the field produced by a nuclear magnetic dipole $\boldsymbol{\mu}_1$, which creates an extra field at the site of a nearby nuclear moment $\boldsymbol{\mu}_2$. This results in an interaction which is bilinear in the two magnetic moments, called indirect interaction because it is mediated by the electrons.

The elements of a theoretical description. Consider a system containing n electrons (index k) and N nuclei (index q). We assume that the nuclei are fixed (Born-Oppenheimer approximation). The relevant part of the Hamiltonian of the electrons for the present problem is

$$(1.10) \qquad \mathcal{H} = \sum_{k=1}^{n} \left\{ \frac{1}{2m} \left(\boldsymbol{p}_k + \frac{e}{c} \boldsymbol{A}_k \right)^2 + 2\beta \boldsymbol{s}_k \cdot \mathrm{curl}\, \boldsymbol{A}_k \right\},$$

where m is the electron mass, $\boldsymbol{p}_k = (\hbar/i)\nabla_k$ its momentum, $\beta = e\hbar/2mc$ the Bohr magneton, $\boldsymbol{s}_k$ the electronic spin and $\boldsymbol{A}_k$ the vector potential. The relevant part of the latter is

$$(1.11) \qquad \boldsymbol{A}_k = \boldsymbol{A}_k^0 + \sum_{q=1}^{N} \boldsymbol{A}_k^q,$$

where

$$(1.12) \qquad \boldsymbol{A}_k^0 = \frac{1}{2} (\boldsymbol{B}_0 \wedge \boldsymbol{r}_k)$$

is the contribution from the external field $\boldsymbol{B}_0$, and

$$(1.13) \qquad \boldsymbol{A}_k^q = \mathrm{curl}\, (\boldsymbol{\mu}_q / r_{kq}) = (\boldsymbol{\mu}_q \wedge \boldsymbol{r}_{kq})/r_{kq}^3$$

is the contribution from the nuclear magnetic moment:

$$\boldsymbol{\mu}_q = \gamma \hbar \boldsymbol{I}_q.$$

The terms in the Hamiltonian (1.10) that yield the chemical shift of the nucleus q are those linear in B_0 and I_q, for instance,

$$(A_k^0 A_k^q + A_k^q A_k^0).$$

Its nuclear form is obtained by taking its average over the electronic ground-state wave function:

$$\langle 0 | (\ldots) | 0 \rangle.$$

Other terms have vanishing average in the ground state, and their contribution is obtained by second-order perturbation.

It turns out from calculation that the chemical shift originates mostly from the orbital part of the electronic Hamiltonian, and the indirect interactions mostly from its spin part.

Form of the interactions. The chemical-shift interaction is of the form

(1.14) $$\mathcal{H}_{CS} = \gamma \boldsymbol{B}_0 \cdot \boldsymbol{\sigma} \cdot \boldsymbol{I},$$

where $\boldsymbol{\sigma}$ is a second-rank tensor called the nuclear shielding tensor.

It has long been assumed that this tensor was symmetrical. In that case the coupling (1.14) can be written

(1.15) $$\mathcal{H}_{CS} = \gamma \sigma_0 \boldsymbol{B}_0 \cdot \boldsymbol{I} + \gamma \boldsymbol{B}_0 \cdot \boldsymbol{\sigma}' \cdot \boldsymbol{I},$$

where the first term is the isotropic-chemical-shift interaction, and the second one is the anisotropic-chemical-shift interaction: $\boldsymbol{\sigma}'$ is a traceless second-rank tensor. Equation (1.15) can also be written

(1.16) $$\mathcal{H}_{CS} = \gamma (B_{0X} \sigma_X I_X + B_{0Y} \sigma_Y I_Y + B_{0Z} \sigma_Z I_Z),$$

where the principal axes OX, OY, OZ are orthogonal, and σ_X, σ_Y, σ_Z are the principal values of the nuclear shielding tensor.

If expressed in a different frame $Oxyz$ (where Oz is usually parallel to the external field), there will appear in this Hamiltonian second-order spherical harmonics, as shown later.

In fact, the tensor $\boldsymbol{\sigma}$ may contain an antisymmetric part, leading to a contribution to the Hamiltonian of the form

(1.17) $$\mathcal{H}_{CS}^{anti} = (\boldsymbol{B}_0 \wedge \boldsymbol{\Sigma}) \cdot \boldsymbol{I},$$

where $\boldsymbol{\Sigma}$ is a vector.

The chemical-shift interaction is always much smaller than the Zeeman interaction (that is $|\boldsymbol{B} - \boldsymbol{B}_0| \ll |\boldsymbol{B}_0|$), and it can be treated as a perturbation.

In a solid, one can, therefore, retain only the part of $\mathcal{H}_{CS}$ that commutes with I_z, that is,

$$\mathcal{H}'_{CS} = \gamma(\lambda_X^2 \sigma_X + \lambda_Y^2 \sigma_Y + \lambda_Z^2 \sigma_Z) B_0 I_z = \tag{1.18}$$

$$= \gamma \left[\sigma_0 + \frac{3\lambda_Z^2 - 1}{2} \sigma'_Z + \frac{\lambda_X^2 - \lambda_Y^2}{2} (\sigma'_X - \sigma'_Y) \right] B_0 I_z ,$$

where

$$\sigma_0 = \frac{1}{3}(\sigma_X + \sigma_Y + \sigma_Z), \tag{1.19}$$

$$\sigma'_X = \sigma_X - \sigma_0, \quad \text{etc.} \tag{1.20}$$

and $\lambda_X, \lambda_Y, \lambda_Z$ are the cosines of Oz with respect to the axes $OXYZ$.

With the standard polar angles

$$\begin{cases} \lambda_Z = \cos\theta, \\ \lambda_X = \sin\theta\cos\varphi, \\ \lambda_Y = \sin\theta\sin\varphi, \end{cases} \tag{1.21}$$

we obtain

$$\mathcal{H}'_{CS} = \gamma\sigma_0 B_0 I_z + \gamma B_0 I_z \left[\frac{3\cos^2\theta - 1}{2} \sigma'_Z + \frac{1}{4}\sin^2\theta\cos 2\varphi(\sigma'_X - \sigma'_Y) \right]. \tag{1.22}$$

The resonance frequency is

$$\omega_0 = -\gamma B_0 (1 - \sigma_0 - [\ldots]),$$

where $[\ldots]$ is the term in square brackets on the right-hand side of eq. (1.22).

In liquids, the anisotropic part of $\mathcal{H}_{CS}$ vanishes on the average. Its only effect is to contribute to relaxation. The static part of $\mathcal{H}_{CS}$ reduces to its isotropic part.

The antisymmetric part of $\mathcal{H}_{CS}$ is off-diagonal with respect to the Zeeman interaction, and does not influence the resonance spectrum. It is usually very small. It is only recently that such an interaction has been detected, through its contribution to relaxation [3].

In a similar fashion, the indirect interaction between two spins 1 and 2 is of the form

$$\mathcal{H}_{\text{ind}1,2} = \boldsymbol{I}_1 \cdot \boldsymbol{J} \cdot \boldsymbol{I}_2 , \tag{1.23}$$

where $\boldsymbol{J}$ is a second-rank tensor. Its scalar part is

$$\mathcal{H}'_{\text{ind}1,2} = J\boldsymbol{I}_1 \cdot \boldsymbol{I}_2 , \tag{1.24}$$

where the indirect coupling constant J is the angular average of the tensor $\boldsymbol{J}$.

1´2.2. *Electric interactions.* Consider a nucleus with charge density $\rho_n(r_n)$, surrounded by electrons with charge density $\rho_e(r_e)$. The electrostatic energy between the nucleus and the electrons is

$$(1.25) \qquad W_E = \int \rho_e(r_e) \rho_n(r_n) |r_e - r_n|^{-1} \, dr_e \, dr_n .$$

One uses the standard expansion

$$(1.26) \qquad (r_e - r_n)^{-1} = 4\pi \sum_{l=0}^{\infty} \sum_{m=-l}^{l} \frac{1}{2l+1} \frac{r_<^l}{r_>^{l+1}} Y_l^{m*}(\theta_n, \varphi_n) Y_l^m(\theta_e, \varphi_e),$$

where $r_<$ and $r_>$ are the smallest and the largest of $|r_n|$ and $|r_e|$. In practice, the nucleus is so small that $r_< = |r_n|$ and $r_> = |r_e|$.

It can be proved on general grounds that only even values of l are allowed in W_E. The largest part on the right-hand side of eq. (1.26) corresponds to $l = 2$ (the term $l = 0$ is a constant) and it is the only one we consider in the following.

One obtains for W_E a product of electronic and nuclear tensorial operator components. For the electronic part, one takes its average over the electronic ground-state wave functions. As far as the nuclear part is concerned, one keeps only its matrix elements within the nuclear ground state. We can then use the Wigner-Eckart theorem, that is write that the nuclear tensorial operator is proportional to that built from the nuclear spin operators. For $l = 2$, one obtains the quadrupole interaction in the form

$$(1.27) \qquad \mathcal{H}_Q = \sum_{j,k} \left(\frac{\partial^2 V}{\partial x_j \partial x_k} \right) Q_{jk} \qquad (j, k = x, y, z),$$

where V is the electric potential, so that the double derivatives are components of the electric-field gradient, a classical quantity that is the ground-state average of the electronic $l = 2$ operators in W_E, and

$$(1.28) \qquad Q_{jk} = \frac{eQ}{6I(2I-1)} \left\{ \frac{3}{2}(I_j I_k + I_k I_j) - \delta_{jk} I(I+1) \right\}.$$

Q is called the nuclear quadrupole moment. It results from general selection rules that a quadrupole moment can exist only for spins $I > 1/2$.

Let V_{ZZ}, V_{XX} and V_{YY} be the principal values of the electric-field gradient. We have then

$$(1.29) \qquad \mathcal{H}_Q = \frac{eQ}{6I(2I-1)} \{V_{ZZ} Q_{ZZ} + V_{XX} Q_{XX} + V_{YY} Q_{YY}\}.$$

The Laplace equation reads

$$(1.30) \qquad V_{XX} + V_{YY} + V_{ZZ} = 0.$$

For a nucleus in a site of cubic symmetry, all principal values must be equal, i.e. they vanish according to eq. (1.30): quadrupole interactions exist only at sites of symmetry less than cubic.

Using the standard notation

(1.31) $$\begin{cases} V_{ZZ} = eq, \\ (V_{XX} - V_{YY})/V_{ZZ} = \eta, \end{cases}$$

we obtain

(1.32) $$\mathscr{H}_Q = \frac{e^2 qQ}{4I(2I-1)} \left\{ 3I_Z^2 - I(I+1) + \frac{1}{2} \eta (I_+^2 + I_-^2) \right\},$$

where

$$I_\pm = I_X \pm iI_Y.$$

The asymmetry parameter η vanishes when the field gradient is axially symmetric. $\mathscr{H}_Q$ can be expressed in a frame $Oxyz$, which introduces second-order spherical harmonics.

Remark I. Only internal, but not external (*i.e.* man-made) field gradients are strong enough to have a significant effect. The field gradients from atoms external to a given nucleus also affect the electrons surrounding the nuclear spin. They in turn affect the field gradient at the nuclear site by a factor that can be huge (in excess of 100, in some cases). This so-called antishielding, or Sternheimer, factor cannot be calculated with accuracy. It is then difficult to predict the value of eq in eq. (1.32).

Remark II. Thermal motions in solids (phonons, or librations in molecular crystals) modulate $\mathscr{H}_Q$ through the modulation of the principal values and/or the principal-axes orientations of the field gradient. The resonance spectrum is determined by the average value of $\mathscr{H}_Q$. The results are a decrease of the quadrupole splitting and a possible variation of the asymmetry parameter η.

A particularly remarkable case is that of molecules containing different nuclear isotopes besides the quadrupolar nuclei under study. (It is a recent fashion to call «quadrupolar» nuclei with a quadrupole moment.) Molecules of different weights have different moments of inertia, different motion amplitudes, and, therefore, different average $\mathscr{H}_Q$'s. This shows up by the observation of a temperature-dependent resonance splitting into several lines [4].

Half-integer spins in high field. In such a high field that the Zeeman interaction is much larger than the quadrupole interaction, the latter can be limited to its secular part, *i.e.* the part that commutes with I_z, as a result of first-order perturbation. It is of the form

(1.33) $$\mathscr{H}'_Q \propto 3I_z^2 - I(I+1).$$

It has equal diagonal elements in the states $|+1/2\rangle$ and $|-1/2\rangle$ and, to this order of approximation, it does not modify the resonance frequency between these states. If $\mathscr{H}_Q$ is not too small, this is the only transition that is observable in a powder sample.

However, in a second-order perturbation, the energy shifts experienced by these two states are different:

$$\Delta E_{\pm 1/2} = \sum_n \frac{\langle \pm 1/2 | \mathscr{H}_Q | n \rangle \langle n | \mathscr{H}_Q | \pm 1/2 \rangle}{E_{\pm 1/2} - E_n}.$$

This results in an orientation-dependent frequency splitting between these states, of the form

(1.34) $$\Delta\omega(1/2 \leftrightarrow -1/2) = \lambda(\theta, \varphi) \frac{\Omega^2}{\omega_0},$$

where ω_0 is the Larmor frequency and Ω is the quadrupole frequency (something proportional to $e^2 qQ$). As a consequence, the $+1/2 \leftrightarrow -1/2$ resonance is broadened in a powder sample.

It results from calculation that $\lambda(\theta, \varphi)$ is a sum of spherical harmonics of orders 0 (constant), 2 and 4. This causes a problem for MAS experiments with quadrupolar nuclei. MAS consists in the fast spinning of the sample around an axis at the «magic» angle $\theta = \cos^{-1}(1/\sqrt{3})$ from the axis Oz, which results in a vanishing average value for the spherical harmonics Y_2, associated with anisotropic chemical shifts and dipolar interactions, and yields narrow resonance lines. However, this rotation *does not* average to zero 4th-order harmonics, and, with half-integer quadrupolar nuclei, the second-order splitting of the $+1/2 \leftrightarrow -1/2$ transition shows up by a residual broadening of the individual lines of MAS spectra in powders. The cures to this desease are described in other lectures of this School.

1˙2.3. *The case of metals.* The interaction of nuclear spins with the conduction electrons of metals, although it can be described through the same formalism as in paragraph 1˙2.1 above, has several distinctive features.

The shift of resonance frequency, proportional to the electronic magnetization, is known as the Knight shift. It is mostly due to the scalar so-called contact interaction, and proportional to the electron density $|\psi_e(0)|^2$ at the nuclear site. As for the indirect interactions mediated by the conduction electrons, they are of relative long range and of oscillatory character. They are known as the RKKY interactions (for RUDERMAN, KITTEL, KOSIYA and YOSIDA).

1˙3. *Dipole-dipole interactions.* – The classical interaction energy between two magnetic-dipole moments μ_1 and μ_2 at a distance r is

(1.35) $$E_{12} = -\boldsymbol{\mu}_1 \boldsymbol{B}_{2 \to 1} = -\boldsymbol{\mu}_2 \cdot \boldsymbol{B}_{1 \to 2},$$

where $\boldsymbol{B}_{i \to j}$ is the field created by moment μ_i at the site of moment μ_j. From standard magnetostatics, this is equal to

$$(1.36) \qquad E_{12} = \frac{1}{r^3} \{\boldsymbol{\mu}_1 \cdot \boldsymbol{\mu}_2 - 3(\boldsymbol{\mu}_1 \cdot \boldsymbol{n})(\boldsymbol{\mu}_2 \cdot \boldsymbol{n})\},$$

where $\boldsymbol{n} = \boldsymbol{r}/r$ is the unit vector in the direction joining the moments.

In quantum mechanics, the energy E corresponds to $\hbar \mathcal{H}$, where $\mathcal{H}$ is the Hamiltonian in frequency units, and the magnetic moment is proportional to the spin (we limit ourselves for simplicity to nuclear spins):

$$(1.37) \qquad \boldsymbol{\mu}_i = \gamma_i \hbar \boldsymbol{I}_i.$$

The Hamiltonian corresponding to eq. (1.36) is, therefore,

$$(1.38) \qquad \mathcal{H}_{12} = \frac{\gamma_1 \gamma_2 \hbar}{r^3} \{\boldsymbol{I}_1 \cdot \boldsymbol{I}_2 - 3(\boldsymbol{I}_1 \cdot \boldsymbol{n})(\boldsymbol{I}_2 \cdot \boldsymbol{n})\}.$$

In a frame with OZ parallel to $\boldsymbol{n}$, this yields

$$(1.39) \qquad \mathcal{H}_{12} = \frac{\gamma_1 \gamma_2 \hbar}{r^3} \{\boldsymbol{I}_1 \cdot \boldsymbol{I}_2 - 3 I_{1Z} I_{2Z}\} = \frac{\gamma_1 \gamma_2 \hbar}{r^3} \{I_{1X} I_{2X} + I_{1Y} I_{2Y} - 2 I_{1Z} I_{2Z}\}.$$

This is the dipole-dipole (or dipolar) Hamiltonian between these two spins.

Let us use a frame $Oxyz$, and let θ and φ be the polar angles of OZ with respect to this frame. Equation (1.39) can then be written under the following form (see, e.g., ref. [5], p. 3):

$$(1.40) \qquad \mathcal{H}_{12} = -\frac{\gamma_1 \gamma_2 \hbar}{r^3} \sum_{q=-2}^{+2} \mathcal{F}_q^*(\theta, \varphi) T_q(I_1, I_2),$$

where the $\mathcal{F}_q$ and T_q are normalized tensorial components whose form is given below.

On the one hand,

$$(1.41) \qquad \begin{cases} \mathcal{F}_0 = \sqrt{\frac{3}{2}} (3\cos^2 \theta - 1), \\ \mathcal{F}_{\pm 1} = \mp 3 \sin\theta \cos\theta \exp[\pm i\varphi], \\ \mathcal{F}_{\pm 2} = \frac{3}{2} \sin^2 \theta \exp[\pm 2i\varphi]. \end{cases}$$

It can be verified that the $\mathcal{F}_q$ are proportional to 2nd-rank spherical harmonics:

$$(1.42) \qquad \mathcal{F}_q = 2 \sqrt{\frac{6\pi}{5}} Y_2^q;$$

and we have, for the angular averages,

$$(1.43) \qquad \overline{\mathcal{F}_q} = 0,$$

$$(1.44) \qquad \overline{\mathcal{F}_q \mathcal{F}_{q'}^*} = (6/5) \delta_{qq'}.$$

Why choose such a normalization that the right-hand side of (1.44) contains the factor 6/5? The reason is merely to make eqs. (1.40) and (1.41) look simpler. This reason is not compelling.

We have, on the other hand,

(1.45)
$$\begin{cases} T_0 = \dfrac{1}{\sqrt{6}}(3I_{1z}I_{2z} - \mathbf{I}_1 \cdot \mathbf{I}_2), \\ T_{\pm 1} = \mp \dfrac{1}{2}(I_{1\pm}I_{2z} + I_{1z}I_{2\pm}) = -T^{\dagger}_{\mp 1}, \\ T_{\pm 2} = \dfrac{1}{2}I_{1\pm}I_{2\pm} = T^{\dagger}_{\mp 2}. \end{cases}$$

These operators have the following properties:

(1.46)
$$\begin{cases} [I_z, T_m] = mT_m, \\ [I_\pm, T_m] = \sqrt{2(2+1) - m(m \pm 1)}\, T_{m\pm 1}, \end{cases}$$

where $I_\alpha = I_{1\alpha} + I_{2\alpha}$,

(1.47)
$$\mathrm{Tr}(T_m) = 0,$$

(1.48)
$$\mathrm{Tr}(T_m T^{\dagger}_{m'}) = \delta_{mm'}\, \frac{I_1(I_1+1)}{3} \cdot \frac{I_2(I_2+1)}{3}.$$

It should be noted that, as is most clearly seen from its form (1.39), the dipolar interaction bears some similarity with the forms (1.16) and (1.29) for the chemical shift and quadrupole interactions. The same is true for the indirect interactions. It is because of this analogy that expressing any one of these interactions in a different frame leads to the appearance of 2nd-order spherical harmonics. This has been proved so far only for the dipolar interactions, which are somewhat special in that they correspond to an axially symmetric tensorial interaction, whereas this is not necessarily so for the other bilinear interactions.

For the convenience of the reader, the transformation laws of 2nd-rank tensor components will be given in the appendix of the present section.

1`4. *Exchange interactions.* – Exchange interactions are most commonly met with electrons, and they play a fundamental role in electron magnetic ordering. They are much less common with nuclei.

The origin of exchange is best explained for a system of two identical spins 1/2. It is known from the Pauli principle that their wave function must be antisymmetric. The wave function for two particles can be written as the product of an orbital part and a spin part. Two spins 1/2 can be combined into an antisymmetric singlet (total spin 0) associated with a symmetric orbital function and a symmetric triplet (total spin 1) associated with an antisymmetric orbital function. If the two particles experience an orbital interaction (for instance, an electric interaction), its expectation value will be different for different symmetries of the orbital wave function, that is also for different symmetries of the spin function. This difference in orbital energy can be described phenomenologically by a Hamiltonian term which depends only on the spin variables, whose eigenmanifolds are the triplet and singlet subspaces, with eigenvalues that depend only on the spin number.

An obvious candidate is the operator $\boldsymbol{I}_1 \cdot \boldsymbol{I}_2$, whose eigenvalues are $1/4$ and $-3/4$ in the triplet and the singlet state, respectively, and the exchange interaction between the two spins is of the form

$$\mathscr{H}_{\text{ex}} = J \boldsymbol{I}_1 \cdot \boldsymbol{I}_2 \,. \tag{1.49}$$

These considerations can be extended to a system of many identical particles.

If the basis used for the orbital part of the total wave function is built from wave functions of individual particles, it can be shown that the condition for exchange to exist is that there is some overlap between these functions. Nuclear dimensions are typically in the 10^{-13} cm range, whereas internuclear distances are of the order of 10^{-8} cm, so that the overlap between the wave functions of distinct nuclei is as a rule negligible, and so is their exchange interaction.

A noticeable exception is that of solid ^{3}He which experiences exchange interactions much larger than the dipolar interactions. The overlap between neighbouring ^{3}He nuclei is brought about by their large zero-point motion, due to their light weight. The question of exchange in solid ^{3}He is in fact much more complicated than sketched above (see, *e.g.*, ref.[5], Chap. 3).

It has long be thought that solid ^{3}He was the only example of nuclear exchange. However, exchange interactions have recently been observed in some metallic trihydrides in the liquid state around room temperature [6].

1˙5. *Appendix: transformation properties of 2nd-rank tensors upon rotation.* – By an extension of eqs. (1.45), we can construct five components of a 2nd-rank tensor from those of two vectors of unit length $\boldsymbol{u}$ and $\boldsymbol{v}$:

$$\begin{cases} T_0 = \dfrac{1}{\sqrt{6}} (3 u_z v_z - \boldsymbol{u} \cdot \boldsymbol{v})\,, \\[4pt] T_{\pm 1} = \mp \dfrac{1}{2} (u_\pm v_z + u_z v_\pm)\,, \\[4pt] T_{\pm 2} = \dfrac{1}{2} u_\pm v_\pm \,, \end{cases} \tag{1.50}$$

where $u_\pm = u_x \pm i u_y$.

A change of axes, as used in the preceding subsections, is equivalent to a rotation, so that it is sufficient to know how these components are transformed by a rotation. It is well known from group theory that this transformation is described by a Wigner matrix: T_q is transformed into $\tilde{T}_q$, with

$$\tag{1.51} \tilde{T}_q = \sum_{q'} T_{q'} \mathscr{D}_{q'q}^{(2)} .$$

We derive the explicit formulae by elementary means.

First of all, a rotation can be characterized by its Euler angles α, β, γ. It consists of the succession of

    a rotation of angle γ around Oz,
    a rotation of angle β around Oy,
    a rotation of angle α around Oz.

(*Note.* The usual definition is sligthly different. The fact that it amounts to the one given above is proved, *e.g.*, in ref.[7], Chap. 3.)

It is then sufficient to know the effect of rotations around Oz or Oy.

Rotation of angle φ around Oz. It yields

$$\tag{1.52} \tilde{T}_q = \exp[-iq\varphi] T_q .$$

Rotation of angle β around Oy. With the notations

$$\cos\beta = c, \quad \sin\beta = s,$$

the components x, y, z of a vector are transformed as follows:

$$\tag{1.53} \begin{cases} x \to cx - sz, \\ z \to cz + sx, \\ y \to y . \end{cases}$$

According to (1.51) it is elementary to find that we have (see ref.[5], p. 3)

$$\tag{1.54a} \tilde{T}_0 = \frac{3c^2 - 1}{2} T_0 - \sqrt{\frac{3}{2}} cs(T_1 - T_{-1}) + \sqrt{\frac{3}{8}} s^2 (T_2 + T_{-2}),$$

$$\tag{1.54b} \begin{cases} \tilde{T}_1 = \sqrt{\dfrac{3}{2}} cs\, T_0 + \dfrac{1}{2}(2c^2 + c - 1) T_1 - \dfrac{1}{2}(2c^2 - c - 1) T_{-1} \\ \qquad\qquad\qquad\qquad\qquad - \dfrac{1}{2} s(1+c) T_2 + \dfrac{1}{2} s(1-c) T_{-2}, \\ \tilde{T}_{-1} = -\tilde{T}_1^\dagger, \end{cases}$$

$$\tag{1.54c} \begin{cases} \tilde{T}_2 = \sqrt{\dfrac{3}{8}} s^2 T_0 + \dfrac{1}{2} s(1+c) T_1 + \dfrac{1}{2} s(1-c) T_{-1} \\ \qquad\qquad\qquad\qquad + \dfrac{1}{4}(1+c)^2 T_2 + \dfrac{1}{4}(1-c)^2 T_{-2}, \\ \tilde{T}_{-2} = \tilde{T}_2^\dagger . \end{cases}$$

It can be seen, according to eqs. (1.41) and (1.51), that it is only the Wigner matrix elements $\mathscr{D}_{q'0}^{(2)}$ that are proportional to 2nd-order spherical harmonics for all q'.

However, if not axially symmetric, the chemical shift and quadrupolar interactions, expressed in the frame of their principal axes (eqs. (1.22) and (1.32)), contain both components T_0 and $T_{\pm 2}$.

This case can be analysed as follows. Consider a tensorial expression

$$(1.55) \qquad O = \boldsymbol{u} \cdot \boldsymbol{K} \cdot \boldsymbol{v},$$

where $\boldsymbol{K}$ is a 2nd-rank symmetric traceless tensor, of principal values a, b and $-(a+b)$ along the principal axes OZ, OX and OY. Equation (1.55) reads

$$(1.56) \qquad O = a u_Z v_Z + b u_X v_X - (a+b) u_Y v_Y,$$

or else after some trivial manipulations

$$(1.57) \qquad O = A(3 u_Z v_Z - \boldsymbol{u} \cdot \boldsymbol{v}) + B(3 u_X v_X - \boldsymbol{u} \cdot \boldsymbol{v}) = \sqrt{6}\,(A T_{0Z} + B T_{0X})$$

with

$$(1.58) \qquad \begin{cases} A = \dfrac{1}{3}(2a+b), \\ B = \dfrac{1}{3}(2b+a). \end{cases}$$

By analogy with the passage from eqs. (1.39) to (1.40) we have, therefore, in the reference frame $Oxyz$

$$(1.59) \qquad O = \sum_{q=-2}^{+2} [A \mathscr{F}_q^*(\theta_Z, \varphi_Z) + B \mathscr{F}_q^*(\theta_X, \varphi_X)] T_q(\boldsymbol{u}, \boldsymbol{v}),$$

where θ_Z, φ_Z and θ_X, φ_X are the polar angles of the principal axes OZ and OX, respectively.

The coefficients of the T_q's are proportional to 2nd-order spherical harmonics as anticipated.

2. – Spin-lattice relaxation revisited.

Spin-lattice relaxation is the phenomenon of evolution of a spin system towards thermal equilibrium with the lattice, defined as the degrees of freedom of the medium in which the spins are embedded, other than those of the spins. It was realized shortly after the beginning of NMR (and EPR, by the fact) that this evolution is determined by the modulation of the spin-lattice interactions as a result of the lattice dynamics, whence the importance of spin-lattice relaxation, as a witness to this dynamics.

A word about language: we call spin-lattice relaxation the evolution brought about by randomly modulated interactions, let it be that of the longitudinal or transverse components of the magnetizations. It is only when it is due to *static* spin-spin interactions that the decay of transverse magnetization is referred to as spin-spin relaxation.

The basic formalisms of spin-lattice relaxation have been developed from 1948 to about 1960, essentially in the following references:

BLOEMBERGEN, PURCELL and POUND (BPP)[8],
WANGSNESS and BLOCH[9] and BLOCH[10],
SOLOMON[11],
ABRAGAM (ref.[1], Chap. VIII),
REDFIELD[12].

The formalisms that emerged from these developments are
good,
efficient,
well understood (at least by their authors).

However, some of them contain minor blemishes in the form of lack of rigour of presentation, which may cause (and have caused) some perplexity among beginner students of NMR. It is the purpose of this lecture to present a formal derivation of relaxation theory that avoids these pitfalls as much as possible. We will, therefore, not analyse specific relaxation mechanisms, but we will only be concerned with the form of the relaxation equations and their justification.

One must also mention the numerous later developments of relaxation theory, which fall broadly into two categories, those that are pointless or of no practical incidence, and those that are completely wrong.

The derivation given below conforms closely to the formalism of Abragam (ref.[1], Chap. VIII). It is followed by a section where some of the approaches or variants that are erroneous or likely to be misused are listed and analysed.

2˙1. *Derivation of the master equation.* – We consider a nuclear-spin system whose Hamiltonian consists of a main, time-independent Hamiltonian $\mathcal{H}_0$, plus a randomly varying term $\mathcal{H}_1(t)$, of vanishing average value, the so-called spin-lattice coupling:

(2.1) $$\mathcal{H} = \mathcal{H}_0 + \mathcal{H}_1(t).$$

We assume for the time being that $\mathcal{H}_0$ has only discrete levels (*i.e.* it is a Zeeman or quadrupole interaction, or a combination of both).

The evolution of the density matrix σ of the spin system is described by the

Liouville–von Neumann equation

(2.2) $$\frac{d\sigma}{dt} = -i[\mathcal{H}, \sigma].$$

The first step consists in isolating the effect of the spin-lattice coupling by using an interaction representation which removes *all* of the static Hamiltonian $\mathcal{H}_0$. When $\mathcal{H}_0$ is purely Zeeman, this amounts to using a rotating frame.

Any operator Q of the laboratory frame is replaced in this representation by the operator

(2.3) $$\tilde{Q}(t) = \exp[i\mathcal{H}_0 t] Q \exp[-i\mathcal{H}_0 t].$$

The evolution equation for the density matrix in this representation is

(2.4) $$\frac{d}{dt}\tilde{\sigma} = -i[(\tilde{\mathcal{H}} - \mathcal{H}_0), \tilde{\sigma}],$$

or else, according to eq. (2.1),

(2.5) $$\frac{d}{dt}\tilde{\sigma}(t) = -i[\tilde{\mathcal{H}}_1(t), \tilde{\sigma}(t)].$$

This last equation can be formally integrated to yield

(2.6) $$\tilde{\sigma}(t) = \tilde{\sigma}(0) - i\int_0^t [\tilde{\mathcal{H}}_1(t'), \tilde{\sigma}(t')]\,dt'.$$

By inserting this form into the right-hand side of eq. (2.5), we obtain

(2.7) $$\frac{d}{dt}\tilde{\sigma}(t) = -i[\tilde{\mathcal{H}}_1(t), \tilde{\sigma}(0)] - \int_0^t [\tilde{\mathcal{H}}_1(t), [\tilde{\mathcal{H}}_1(t'), \tilde{\sigma}(t')]]\,dt'.$$

Two modifications must be brought to this equation. Firstly, one must take an ensemble average of all terms. This may seem strange, since the very concept of density matrix is intimately linked to statistical average. That a further averaging is necessary stems from the fact that the operator $\mathcal{H}_1(t)$ is a random function of time; in different parts of the system, otherwise identical, simulated by different members of a Gibbs ensemble, it may have a different history, resulting in a different density matrix $\tilde{\sigma}$. One consequence of taking this average is that, since by definition $\mathcal{H}_1(t)$ has a vanishing average value, so does the first term on the right-hand side of eq. (2.7).

The second modification to eq. (2.7) is that $\tilde{\sigma}$ must be replaced by

(2.8) $$\tilde{\sigma}(t) \to \tilde{\sigma}(t) - \tilde{\sigma}_{eq},$$

where σ_{eq} is the thermal-equilibrium form of the density matrix. This is a consequence of the finite lattice temperature, as seen from a quantum treatment of the lattice [9, 10]. The remarkably simple form of eq. (2.8) is valid only in the high-temperature limit, that pertains to most of NMR studies.

We then obtain, in place of eq. (2.7),

$$(2.9) \qquad \frac{d}{dt} \bar{\tilde{\sigma}}(t) = - \int_0^t \overline{[\tilde{\mathcal{H}}_1(t), [\tilde{\mathcal{H}}_1(t'), (\tilde{\sigma}(t') - \tilde{\sigma}_{eq})]]} \, dt',$$

where the overbars mean an ensemble average.

Equation (2.9) is *rigorous*. The difference with the standard treatment [1, 7] is that we have not used a second-order Taylor expansion of $\tilde{\sigma}(t)$, which would have yielded $\tilde{\sigma}(0)$ instead of $\tilde{\sigma}(t')$ on the right-hand side of eq. (2.9).

Now, the Hamiltonian $\mathcal{H}_1(t)$ can be decomposed into a sum of terms, as follows:

$$(2.10) \qquad \mathcal{H}_1(t) = \sum_\alpha V_\alpha F_\alpha(t) = \sum_\alpha V_\alpha^\dagger F_\alpha^*(t),$$

where the V_α are spin operators and the $F_\alpha(t)$ are random functions of time. The equality of the two forms simply expresses the fact that $\mathcal{H}_1$ is a Hermitian operator, as befits a Hamiltonian term.

The decomposition in eq. (2.10) is made in such a way that

$$(2.11) \qquad \tilde{V}_\alpha(t) = \exp[i\mathcal{H}_0 t] V_\alpha \exp[-i\mathcal{H}_0 t] = \exp[i\omega_\alpha t] V_\alpha.$$

Such a decomposition is always possible. For instance, V_α may be limited to having only one nonvanishing matrix element $\langle i|V_\alpha|j\rangle$ in a basis of eigenstates of $\mathcal{H}_0$, with

$$\langle i|\mathcal{H}|i\rangle - \langle j|\mathcal{H}|j\rangle = \omega_\alpha.$$

This is not the only possible choice.

We can then write eq. (2.9) in the form

$$(2.12) \qquad \frac{d}{dt} \bar{\tilde{\sigma}}(t) = - \sum_{\alpha,\beta} \int_0^t \overline{[\tilde{V}_\alpha(t), [\tilde{V}_\beta^\dagger(t'), (\tilde{\sigma}(t') - \tilde{\sigma}_{eq})]] F_\alpha(t) F_\beta^*(t')} \, dt'.$$

As noted earlier, different histories of the random spin-lattice coupling, that is of the random functions $F_\alpha(t)$, $F_\beta^*(t')$ in different members of the ensemble (*i.e.* different molecules or different remote parts of the system), result in different density matrices $\tilde{\sigma}(t')$, and the average on the right-hand side of eq. (2.12) is a joint average over the spin part (*i.e.* $\tilde{\sigma}(t')$) and the orbital part (*i.e.* $F_\alpha(t) F_\beta^*(t')$).

In order to proceed, we make a physical hypothesis. Let τ_c be the correlation time of the random Hamiltonian, that we define loosely for the moment as the time scale for the fluctuation of the random functions $F(t)$: a substantial decay of the product $F_\alpha(t) F_\beta^*(t')$ takes place on the average as soon as

$$t - t' \sim \tau_c,$$

and it becomes negligible for $t - t'$ much larger than τ_c.

We *assume* that the variation of $\tilde{\sigma}$ is slow on the time scale τ_c (conditions for this to be true will be derived *a posteriori*) and we choose $t \gg \tau_c$ in eq. (2.12). This has two consequences (plus a third one to be discussed later). Firstly, it is a good approximation to replace $\tilde{\sigma}(t')$ by $\tilde{\sigma}(t)$ on the right-hand side of eq. (2.12), since only values of t' close to t (to within a few τ_c) contribute significantly to the integral. Secondly, since each member of the ensemble has experienced a random perturbation for many times τ_c, the effect of the different histories of their fluctuations for different members average out, so that the individual $\tilde{\sigma}(t)$ are all very nearly the same, that is all equal to $\overline{\tilde{\sigma}}(t)$, and the overbar may be omitted. Under these assumptions, eq. (2.12) becomes

$$(2.13) \quad \frac{d}{dt} \tilde{\sigma}(t) = - \sum_{\alpha, \beta} \int_0^t [\tilde{V}_\alpha(t), [\tilde{V}_\beta^\dagger(t'), (\tilde{\sigma}(t) - \tilde{\sigma}_{eq})]] \times \overline{F_\alpha(t) F_\beta^*(t')} \, dt'.$$

The second consequence of short τ_c is, therefore, that the average of the product of spin and orbital parts, on the right-hand side of eq. (2.12), can be replaced by the product of the averages of the spin and orbital parts.

We limit ourselves in the following to stationary random functions (a realistic assumption in most practical cases), *i.e.* such that

$$(2.14) \quad \overline{F_\alpha(t) F_\beta^*(t')} = G_{\alpha\beta}(|t - t'|).$$

We have reached the exact point of departure between the standard relaxation theory and the «revisited» one. Logic would require that we go ahead along the proper path without even mentioning past imperfections. However, the large availability and wide acceptance of the standard treatment make it desirable to begin by analysing briefly the problems which it is confronted with, if only to put their minor character and easy cure into perspective.

2˙1.1. *Evolution in the interaction representation*. With the help of eqs. (2.11) and (2.14), eq. (2.13) becomes

$$(2.15) \quad \frac{d}{dt} \tilde{\sigma}(t) = - \sum_{\alpha, \beta} [V_\alpha, [V_\beta^\dagger, (\tilde{\sigma}(t) - \tilde{\sigma}_{eq})]] \times$$

$$\times \int_0^t \exp[i\omega_\alpha t - i\omega_\beta t'] G_{\alpha\beta}(t - t') \, dt'.$$

The integral on the right-hand side is equal to

$$\text{(2.16)} \quad \int_0^t \ldots dt' = \exp[i(\omega_\alpha - \omega_\beta)t] \int_0^t \exp[i\omega_\beta(t-t')] G_{\alpha\beta}(t-t')\,dt',$$

and, since the decay time τ_c of the correlation function $G_{\alpha\beta}(\tau)$ is much smaller than t, the integral on the right-hand side can be extended to infinity. (This is the third consequence of short τ_c.) We then obtain, in place in eq. (2.15),

$$\text{(2.17)} \quad \frac{d}{dt}\tilde{\sigma}(t) = -\sum_{\alpha,\beta}[V_\alpha,[V_\beta^\dagger,(\tilde{\sigma}(t)-\tilde{\sigma}_{eq})]]\times\exp[i(\omega_\alpha-\omega_\beta)t]J_{\alpha\beta}(\omega_\beta),$$

where

$$\text{(2.18)} \quad J_{\alpha\beta}(\omega) = \int_0^\infty G_{\alpha\beta}(\tau)\exp[i\omega\tau]\,d\tau$$

is the spectral density of the correlation function $G_{\alpha\beta}(\tau)$.

If one looks at the evolution of a quantity of operator Q, as seen in the interaction representation (*i.e.* most of the time in the rotating frame),

$$\text{(2.19)} \quad \langle Q(t)\rangle_r = \text{Tr}\{Q\tilde{\sigma}(t)\},$$

its derivative is a sum of smooth terms, those for which $\omega_\alpha = \omega_\beta$, and of oscillatory terms. If the oscillation frequencies $\omega_\alpha - \omega_\beta$ are large, compared with the average decay rate of $\langle Q\rangle_r(t)$, their contribution to its evolution consists of oscillations of *small* amplitude that can be ignored. This amounts to discarding the oscillatory terms on the right-hand side of eq. (2.17), *i.e.* to limit oneself to those terms with $\omega_\alpha = \omega_\beta$. This is known as the adiabatic approximation. As an example, it is fully justified for the transverse relaxation in liquids of unlike spins, that is such that the difference of their Larmor frequencies is much larger than their linewidth.

There are cases, however, when the frequencies of the oscillatory terms are not large, in the sense given above. This is the case, for instance, for like nuclear spins experiencing small chemical-shift differences. Many other examples are found in EPR, Mössbauer spectroscopy and μSR. It is sometimes claimed that the adiabatic approximation is an inherent feature of the standard relaxation treatment, which is, therefore, not adapted to dealing with cases where this approximation breaks down. These cases are usually analysed through a much more complicated approach.

These criticisms are not completely founded, and it is possible, for instance, to treat the transverse nuclear relaxation of nearly like spins through a slight modification of the preceding formalism (see, *e.g.*, ref.[7], p. 251). This is, however, not very satisfactory in that the limit between fast and slow oscillation is highly subjective. It would be much better to have a unified treatment valid for all cases. This is precisely the purpose of the «revisited» version presented here.

2˙1.2. Evolution in the Schrödinger representation. Starting from eq. (2.13) for the evolution of the density matrix as seen from the interaction-representation point of view, we go back to the laboratory frame, that is to the Schrödinger representation. From

(2.20) $$\sigma(t) = \exp[-i\mathcal{H}_0 t]\tilde{\sigma}(t)\exp[i\mathcal{H}_0 t],$$

a corollary of eq. (2.3), we obtain

(2.21) $$\frac{d}{dt}\sigma(t) = -i[\mathcal{H}_0, \sigma(t)] + \exp[-i\mathcal{H}_0 t]\frac{d}{dt}\tilde{\sigma}(t)\exp[i\mathcal{H}_0 t].$$

From the property

$$UABU^\dagger = UAU^\dagger UBU^\dagger,$$

where U is a unitary operator, and according to eqs. (2.13) and (2.14), we obtain

(2.22) $$\frac{d}{dt}\sigma(t) = -i[\mathcal{H}_0, \sigma(t)] -$$

$$-\sum_{\alpha,\beta}\int_0^t [V_\alpha, [\tilde{V}_\beta^\dagger(t'-t), (\sigma(t)-\sigma_{eq})]]G_{\alpha\beta}(t-t')\,dt'.$$

The term under the integral depends on t' only through $t - t' = \tau$. We may as above extend the integral over τ to infinity and we obtain, according to eq. (2.11) and in analogy with eq. (2.17),

(2.23) $$\frac{d}{dt}\sigma(t) = -i[\mathcal{H}_0, \sigma(t)] - \sum_{\alpha,\beta}[V_\alpha, [V_\beta^\dagger, (\sigma(t)-\sigma_{eq})]]J_{\alpha\beta}(\omega_\beta).$$

The main (and fundamental) difference with eq. (2.17) is that all oscillatory terms have disappeared, together with the problems that they raise.

Using the well-known property

(2.24) $$\mathrm{Tr}\{A[B,C]\} = \mathrm{Tr}\{[A,B]C\},$$

the expectation value $\langle Q \rangle = \mathrm{Tr}\{Q\sigma\}$ in the laboratory frame of an operator Q has an evolution governed by the equation

(2.25) $$\frac{d}{dt}\langle Q \rangle = \langle -i[Q, \mathcal{H}_0]\rangle(t) - \sum_{\alpha,\beta}J_{\alpha\beta}(\omega_\beta)\cdot$$

$$\cdot\{\langle[[Q, V_\alpha], V_\beta^\dagger]\rangle(t) - \langle[[Q, V_\alpha], V_\beta^\dagger]\rangle_{eq}\}.$$

This is an equation relating expectation values, and there is no need to make assumptions as to the form of $\sigma(t)$. The double commutators $[[Q, V_\alpha], V_\beta^\dagger]$ will often yield operators differing from Q, and will, therefore, yield cross-relaxation effects. Let us then consider a proper set of operators Q_i, one of which at least being of practical interest, chosen in such a way that we have

(2.26) $$[Q_i, \mathscr{H}_0] = -\Omega_i Q_i .$$

Equation (2.25) then yields a system of coupled equations of the form

(2.27) $$\frac{d}{dt}\langle Q_i\rangle(t) = i\Omega_i \langle Q_i\rangle(t) - \sum_j \lambda_{ij}\{\langle Q_j\rangle(t) - \langle Q_j\rangle_{eq}\}$$

whose solution is straightforward.

2˙1.3. *Relation between both approaches.* Let us compare these equations with those corresponding to $\langle Q_i\rangle_r(t)$, as viewed in the interaction representation. We have

(2.28) $$\langle Q_i\rangle_r(t) = \text{Tr}\{Q_i \tilde{\sigma}(t)\} = \text{Tr}\{Q_i \exp[i\mathscr{H}_0 t]\sigma(t)\exp[-i\mathscr{H}_0 t]\} =$$
$$= \text{Tr}\{\exp[-i\mathscr{H}_0 t] Q_i \exp[i\mathscr{H}_0 t]\sigma(t)\} =$$
$$= \exp[-i\Omega_i t]\text{Tr}\{Q_i \sigma(t)\} = \exp[-i\Omega_i t]\langle Q_i\rangle(t),$$

whence, according to eq. (2.27),

(2.29) $$\frac{d}{dt}\langle Q_i\rangle_r(t) = -\sum_j \exp[i(\Omega_j - \Omega_i)t]\lambda_{ij}\{\langle Q_j\rangle_r(t) - \langle Q_j\rangle_{r\,eq}\} .$$

A comparison with eq. (2.17) makes it possible to pinpoint the origin of the oscillatory terms in the evolution equation of $\langle Q_i\rangle_r(t)$: they originate from cross-relaxation with other operators Q_j, whose oscillation frequency under the effect of the static Hamiltonian $\mathscr{H}_0$ is different from that of Q_i. The great practical advantage of the laboratory frame evolution equations (2.27) is to allow a rigorous account of the effect of these cross-relaxations.

Let us illustrate this by the case of two coupled quantities Q_1 and Q_2, corresponding, for instance, to the transverse operators I_+ and S_+ of two spins with different Larmor frequencies, in which case we have

$$\langle Q_1\rangle_{eq} = \langle Q_2\rangle_{eq} = 0 .$$

The evolution equations are

(2.30) $$\begin{cases}\frac{d}{dt}\langle Q_1\rangle = i\Omega_1\langle Q_1\rangle - \lambda_{11}\langle Q_1\rangle - \lambda_{12}\langle Q_2\rangle, \\ \frac{d}{dt}\langle Q_2\rangle = i\Omega_2\langle Q_2\rangle - \lambda_{22}\langle Q_2\rangle - \lambda_{21}\langle Q_1\rangle.\end{cases}$$

The simplest method for solving this system is through the use of Laplace transforms. Let $A_1(\mathscr{Z})$ and $A_2(\mathscr{Z})$ be the Laplace transforms of $\langle Q_1\rangle$ and $\langle Q_2\rangle$, respectively. The system (2.30) yields

(2.31) $$\begin{cases} \mathscr{Z} A_1 - \langle Q_1 \rangle(0) = i\Omega_1 A_1 - \lambda_{11} A_1 - \lambda_{12} A_2 , \\ \mathscr{Z} A_2 - \langle Q_2 \rangle(0) = i\Omega_2 A_2 - \lambda_{22} A_2 - \lambda_{21} A_1 . \end{cases}$$

In the limit when

$$|\Omega_1 - \Omega_2| \gg \lambda_{ij} ,$$

it is easily found that we have

(2.32) $$A_1(\mathscr{Z}) \simeq \frac{\langle Q_1 \rangle(0)}{\mathscr{Z} - i\Omega_1 + \lambda_{11}}$$

with an error of order of $\lambda_{12}/|\Omega_1 - \Omega_2|$, and similar results for $A_2(\mathscr{Z})$. Equation (2.32) corresponds to

(2.33) $$\langle Q_1 \rangle(t) = \langle Q_1 \rangle(0) \exp\left[(i\Omega_1 - \lambda_{11})t\right]$$

and, similarly,

(2.34) $$\langle Q_2 \rangle(t) = \langle Q_2 \rangle(0) \exp\left[(i\Omega_2 - \lambda_{22})t\right].$$

These solutions are identical to those obtained by discarding the cross-relaxation terms $\lambda_{12}\langle Q_2 \rangle$ and $\lambda_{21}\langle Q_1 \rangle$ from eqs. (2.30). This amounts to the adiabatic approximation, that is to the neglect of the terms oscillating at the frequencies $\pm(\Omega_1 - \Omega_2)$ in the evolution equations in the interaction representation.

More generally, it can be shown from the theory of linear equations that if, in the system (2.27), one has $|\Omega_i - \Omega_j| \gg \lambda_{ij}, \lambda_{ji}$, one can discard the cross-relaxation terms proportional to λ_{ij} and λ_{ji}, in accordance with the adiabatic approximation.

2˙1.4. Conditions of validity of the theory. The main condition for the validity of the theory is that the evolution of the density matrix be slow on the time scale τ_c of the random perturbation correlation functions. In fact, if we calculate the variation of the expectation value $\langle Q \rangle_r(t)$ from eq. (2.12), it is easily proved that the condition enabling one to use eq. (2.25) for calculating $\langle Q \rangle(t)$ in the laboratory frame is that the evolution of $\langle Q \rangle_r(t)$ be slow on the time scale τ_c. The conditions of validity of the theory, therefore, depend on the quantity Q whose relaxation is studied.

Consider, for instance, longitudinal and transverse relaxations due to dipolar couplings in liquids (see, e.g., ref. [7], Chap. 9). These relaxations are exponential, with relaxation times T_1 and T_2, respectively. Let ω be the Larmor frequency, $(\Delta\omega)^2$ the average square of the dipolar interactions and τ_c the correlation time of the exponential correlation functions. Transverse relaxation can be produced by the longitudinal components of the dipolar field, for which $\omega_\beta = 0$.

We then have, as an order of magnitude,

(2.35) $$\frac{1}{T_2} \sim (\Delta\omega)^2 \tau_c.$$

The condition of validity of the theory is

(2.36) $$\tau_c/T_2 \ll 1,$$

that is,

(2.37) $$\Delta\omega \tau_c \ll 1.$$

Longitudinal relaxation involves only transverse dipolar fields, with $\omega_\beta = \omega$ or 2ω. We then have

(2.38) $$\frac{1}{T_1} \sim \frac{(\Delta\omega)^2 \tau_c}{1 + \omega^2 \tau_c^2}.$$

The condition for the validity of the theory

(2.39) $$\frac{\tau_c}{T_1} \sim \frac{(\Delta\omega \tau_c)^2}{1 + \omega^2 \tau_c^2} \ll 1$$

is much less stringent than for transverse relaxation. It is either eq. (2.37) or

(2.40) $$\frac{\Delta\omega}{\omega} \ll 1$$

in the case when $\omega\tau_c \gg 1$.

2.1.5. Extension to solids. In NMR, a solid is characterized by the existence of static secular spin-spin interactions, for instance, as most often, the dipolar interactions $\mathcal{H}_D'$. (In that respect, solids where $\mathcal{H}_D'$ is modulated through motion at a rate satisfying eq. (2.37), through vacancy jumps, for instance, behave qualitatively as liquids.)

The decay of the transverse magnetization is then governed by $\mathcal{H}_D'$ and is nonexponential. Spin-lattice relaxation in that case concerns other variables such as the longitudinal magnetization or the dipolar energy.

The main Hamiltonian $\mathcal{H}_0$ is then, say, a sum of Zeeman and secular dipolar interactions of the form

(2.41) $$\mathcal{H}_0 = \mathcal{Z} + \mathcal{H}_D'.$$

This is a case where the spectrum of $\mathcal{H}_0$ *does not* consist of discrete levels: the discrete levels of the Zeeman interaction term $\mathcal{Z}$ are quasi-continuously broadened by the dipolar interactions.

The random perturbation is still of the form (2.10). We choose the V_α in such a way that

(2.42) $$[\mathcal{Z}, V_\alpha] = \omega_\alpha V_\alpha,$$

so that

(2.43) $$\tilde{V}_\beta^\dagger(t) = \exp[i\mathcal{H}_0 t] V_\beta^\dagger \exp[-i\mathcal{H}_0 t] =$$
$$= \exp[i\mathcal{H}_D' t]\exp[i\mathcal{Z} t] V_\beta^\dagger \exp[-i\mathcal{Z} t]\exp[-i\mathcal{H}_D' t] = \exp[i\omega_\beta t] V_\beta^\dagger(t)$$

with

(2.44) $$V_\beta^\dagger(t) = \exp[i\mathcal{H}_D' t] V_\beta^\dagger \exp[-i\mathcal{H}_D' t].$$

When calculating the evolution of a variable $\langle Q_i \rangle$, we are led, according to eq. (2.22), to calculate, *i.e.*, an integral of the form

$$\int_0^\infty \langle [[Q_i, V_\alpha], V_\beta^\dagger(-\tau)] \rangle(t) \exp[i\omega_\beta \tau] G_{\alpha\beta}(\tau) \, \mathrm{d}\tau.$$

If, for instance, the correlation function $G_{\alpha\beta}$ is exponential:

$$G_{\alpha\beta}(\tau) \propto \exp[-\tau/\tau_c],$$

this corresponds to an integral of the form

$$\int_0^\infty K(\tau; t) \exp\left[\left(i\omega_\beta - \frac{1}{\tau_c}\right)\tau\right] \mathrm{d}\tau.$$

The dependence of the function K on τ is determined by the dipolar interactions $\mathcal{H}_D'$ and is, therefore, complicated. The above integral is, however, easily handled provided that the decay of K as a function of τ is slow on the time scale $|i\omega_\beta - 1/\tau_c|^{-1}$. As easily shown by an integration by parts [13], this integral is then, to a good approximation, equal to

$$K(0; t) \times \frac{\tau_c}{1 - i\omega_\beta \tau_c},$$

i.e. it is the same as if the dipolar term $\mathcal{H}_D'$ were absent.

We will not discuss the case when τ_c is long, which is also tractable, but through a very different approach [14] (see, however, ref. [13]).

2'2. *Drawbacks of alternative approaches. A selection.* – We analyse briefly some developments of the theory different from that given above, that are found in the literature. Some are simply erroneous; others, while excellent from the theoretical point of view, may lead to erroneous results, when handled without enough care. The cases listed are but a short selection of what exists.

2·2.1. Use of an interaction representation that does not remove all of $\mathcal{H}_0$. It is a representation defined through

(2.45) $$Q \to \tilde{Q} = \exp[i\mathcal{H}t] Q \exp[-i\mathcal{H}t]$$

with $\mathcal{H} \neq \mathcal{H}_0$.

The evolution of the density matrix in this representation is

(2.46) $$\frac{d}{dt}\tilde{\sigma}(t) = -i[(\tilde{\mathcal{H}}_0 - \mathcal{H} + \tilde{\mathcal{H}}_1(t)), \tilde{\sigma}(t)],$$

i.e. it involves the nonrandom term $\tilde{\mathcal{H}}_0 - \mathcal{H}$ in addition to the random one. This procedure does not allow one to derive cleanly an expression for the evolution of $\tilde{\sigma}(t)$ that is workable, that is of practical usefulness for the analysis of real problems, so much so when $\tilde{\mathcal{H}}_0$ is time-dependent. However, there are cases where it seems intuitively justified to use such a procedure, as follows:

1) Derive as above an evolution equation for $\tilde{\sigma}(t)$ as if the static Hamiltonian were $\mathcal{H}$, and the term $\tilde{\mathcal{H}}_0 - \mathcal{H}$ were absent; the corresponding derivative is called $(d\tilde{\sigma}/dt)_{\text{rel}}$ and is considered as the contribution of relaxation.

2) Add an independent contribution from the residual term $\tilde{\mathcal{H}}_0 - \mathcal{H}$ (chosen to be time-independent).

The evolution equation for $\tilde{\sigma}$ is then claimed to be

(2.47) $$\frac{d\tilde{\sigma}}{dt} = i[(\tilde{\mathcal{H}}_0 - \mathcal{H}), \tilde{\sigma}] + \left(\frac{d\tilde{\sigma}}{dt}\right)_{\text{rel}}.$$

This is the case, for instance, for two like spins I and S with two very close Larmor frequencies $\omega_I = \omega_0 + \delta$ and $\omega_S = \omega_0 - \delta$, or spins experiencing an indirect interaction $JI \cdot S$. The Hamiltonian $\mathcal{H}$ is then taken equal to the average Zeeman interaction $\omega_0(I_z + S_z)$ in the former case, and to the Zeeman interaction in the latter.

We show briefly for the first example how the correct treatment provides a justification of eq. (2.47).

We consider the evolution equations (2.30) for the case when

$$Q_1 = I_+, \quad Q_2 = S_+,$$
$$\Omega_1 = \omega_I = \omega_0 + \delta, \quad \Omega_2 = \omega_S = \omega_0 - \delta.$$

These are evolution equations in the laboratory frame.

Let us now use a frame rotating at the *average* Larmor frequency ω_0 with respect to both spins, and let us call $\langle I_+ \rangle'$ and $\langle S_+ \rangle'$ the average values in this frame.

We have

(2.48) $$\begin{cases} \langle I_+ \rangle' = \exp[-i\omega_0 t]\langle I_+ \rangle, \\ \langle S_+ \rangle' = \exp[-i\omega_0 t]\langle S_+ \rangle, \end{cases}$$

and we obtain from eqs. (2.30)

(2.49) $$\begin{cases} \dfrac{d}{dt}\langle I_+ \rangle' = i\delta\langle I_+ \rangle' - \lambda_{11}\langle I_+ \rangle' - \lambda_{12}\langle S_+ \rangle', \\ \dfrac{d}{dt}\langle S_+ \rangle' = -i\delta\langle S_+ \rangle' - \lambda_{22}\langle S_+ \rangle' - \lambda_{21}\langle I_+ \rangle'. \end{cases}$$

These equations look exactly as if they had been derived from eq. (2.47), with a slight difference: the frequencies of the correlation function power spectra $J_{\alpha\beta}(\omega_\beta)$ entering the rates λ_{ij} are, in the correct treatment, the actual Larmor frequencies ω_I and ω_S, whereas, in the approximate treatment leading to eq. (2.47), these frequencies are equal to the average frequency ω_0. It is easily shown that the difference is negligible when one has

(2.50) $$\delta\tau_c \ll 1,$$

which is often the case for homonuclear spins in liquids. Therefore, unless in the presence of pretty long correlation times, it is justified to use eq. (2.47) for spin-lattice relaxation, where $\mathcal{H}_0$ is the sum of the average Zeeman interactions of each homonuclear species. It is, however, better to derive this justification from a correct theory than from a grossly approximate one.

2·2.2. *Premature assumption that the average of a product equals the product of the averages.* We refer explicitly to the assumption that could be made about the average on the right-hand side of eq. (2.12):

(2.51) $$\overline{\text{spin} \times \text{lattice}} = \overline{\text{spin}} \times \overline{\text{lattice}}.$$

This would lead, for $d\langle Q \rangle_r(t)/dt$ in, say, the adiabatic approximation ($\omega_\alpha = \omega_\beta$), to terms of the form

$$\int_0^t \mathrm{Tr}\{[[Q, V_\alpha], V_\beta^\dagger]\tilde{\sigma}(t')\}\exp[i\omega_\beta(t-t')]G_{\alpha\beta}(t-t')\,dt',$$

i.e. integrals of the form

(2.52) $$I(t) = \int_0^t C(t')D(t-t')\,dt'.$$

Such integrals are most easily exploited through their Laplace transforms. Let $\mathcal{I}(\mathcal{Z})$, $\mathcal{C}(\mathcal{Z})$ and $\mathcal{D}(\mathcal{Z})$ be the Laplace transforms of $I(t)$, $C(t)$ and $D(t)$, respectively, that is, for instance,

$$\mathscr{S}(\mathscr{Z}) = \int_0^\infty I(t)\exp[-\mathscr{Z}t]\,dt\,. \tag{2.53}$$

Equation (2.52) yields

$$\mathscr{S}(\mathscr{Z}) = \mathscr{G}(\mathscr{Z})\mathscr{D}(\mathscr{Z})\,. \tag{2.54}$$

As a consequence of assumption (2.51) the relaxation equations for any variable Q could be solved rigorously through the use of the Laplace-transform formalism, even in the case when the λ_{ij} are comparable with τ_c^{-1}, that is even when the conditions (2.37) or (2.40) are not fulfilled. It is abundantly clear from the formal treatment given above that this conclusion is totally erroneous.

In fact, it is very seldom that this assumption is made in the crude way (2.51). It is usually introduced in a much more subtle and rigorous-looking way that will be examined later (paragraph 2·5).

2·2.3. *Calculation of the time evolution of all elements of the density matrix*. Spin-lattice relaxation theory was initially developed with the language of wave functions; its expression within the formalism of the density matrix has represented a definite progress. In particular, the compact expression of the density matrix evolution (due to relaxation and in the adiabatic approximation) under the form

$$\frac{d}{dt}\sigma_{\alpha\beta} = \sum_{\varepsilon,\gamma} R_{\alpha\beta,\varepsilon\gamma}\sigma_{\varepsilon\gamma}\,, \tag{2.55}$$

where $\sigma_{\alpha\beta} = \langle\alpha|\sigma|\beta\rangle$ is a matrix element of σ in the rotating frame, has been a landmark in relaxation theory.

This has been done by REDFIELD[12], and the matrix $R_{\alpha\beta,\varepsilon\gamma}$ is known as the Redfield relaxation matrix. It acts in a space whose number of dimensions n^2 is the square of that of the Hilbert space of the spin system. It is the now very popular Liouville space, on which more will be said later on (paragraph 2.5).

Expression (2.55) is compact and elegant, and very useful in discussing relaxation. It has often been misused, which is why it deserves some comments.

The physical variables that can be experimentally observed are theoretically described by linear combinations of matrix elements of σ. The number of these variables is limited, and is much less than that of the matrix elements of σ. It is only the variation of these physical variables that is of any interest. The detailed calculation of all n^2 matrix elements of σ as a function of time is an unnecessarily complicated task that is completely useless. It is only relatively recently that such a calculation has become possible, for still relatively small spin systems, thanks to the availability of large computers. In the majority of cases, it is much better, and much simpler, to compute the evolution of the expectation values of a limited set of operators.

2˙2.4. *Guesses about the form of the density matrix.* As seen in eq. (2.25), the relaxation part of the evolution equation of a physical variable involves the expectation values of double commutators whose calculation is sometimes a little complicated. It is tempting (and often done) to simplify the calculation as illustrated below.

Consider, for instance, the longitudinal relaxation of a spin I_i. According to eq. (2.25), the derivative $d\langle I_{iz}\rangle/dt$ involves, among others, the following term:

(2.56) $$\langle[[I_{iz}, V_\alpha], V_\beta^\dagger]\rangle(t) = \mathrm{Tr}\{[[I_{iz}, V_\alpha], V_\beta^\dagger]\sigma(t)\} =$$
$$= \mathrm{Tr}\{[I_{iz}, V_\alpha], [V_\beta^\dagger, \sigma(t)]\},$$

where the last line is a consequence of eq. (2.24).

It is always possible to write $\sigma(t)$ in the form

(2.57) $$\sigma(t) = \sum_j \xi_j(t) I_{jz} + P,$$

where P is orthogonal to the I_{jz}:

$$\mathrm{Tr}(I_{jz} P) = 0.$$

We then have

$$\langle I_{jz}(t)\rangle = \mathrm{Tr}\{\sigma(t) I_{jz}\} = \xi_j(t) \mathrm{Tr}(I_{jz}^2),$$

i.e. $\xi_j(t)$ is proportional to $\langle I_{jz}\rangle(t)$.

If now we *assume* that the evolution of $\langle I_{iz}\rangle$ depends only on the various $\langle I_{jz}\rangle$, *i.e.* that the term P on the right-hand side of eq. (2.57) is irrelevant to the Zeeman relaxation, the term (2.56) takes the form

$$\sum_j \xi_j(t) \mathrm{Tr}\{[I_{iz}, V_\alpha][V_\beta^\dagger, I_{jz}]\}.$$

This brings about an enormous simplification in relaxation calculations but is very dangerous, as discussed below.

If we had assumed that the evolution of $\langle I_{iz}\rangle$ depended only on $\langle I_{iz}\rangle$, we would have completely missed cross-relaxation. In fact, whereas the basis of relaxation through dipolar interactions in liquids was clearly formulated in the celebrated BPP paper [8] in 1948, it is only in 1955 that cross-relaxation was discovered by SOLOMON [11]. (This cross-relaxation, described by the Solomon equations, is nowadays referred to as NOE, for nuclear Overhauser effect. The phenomenon predicted by OVERHAUSER [15] was the modification of the nuclear steady-state magnetization upon saturation of the conduction electron spin resonance in metals, a case where the nuclear relaxation is due to the electron-nuclear scalar contact interaction; he did not consider other types of couplings, nor the dynamics of cross-relaxation.)

Even when all spin operators are included in eq. (2.57), the procedure misses important interference effects between dipolar and anisotropic chemical-shift interactions (ref. [16] and references therein), or between the dipolar interactions of different spin pairs [17]. These interference effects provided useful structural information.

As another example, consider the longitudinal relaxation of two identical spins $I > 1/2$ through their dipolar coupling in a liquid. Writing

$$(2.58) \qquad I_\alpha = I_{1\alpha} + I_{2\alpha} \qquad (\alpha = x, y, z)$$

use of eq. (2.25) yields

$$(2.59) \quad \frac{d}{dt} \langle I_z \rangle = -2J_1(\omega) \{\langle [T_1, T_1^\dagger] \rangle(t) - \langle [T_1, T_1^\dagger] \rangle_{eq}\} - $$
$$- 4J_2(2\omega) \{\langle [T_2, T_2^\dagger] \rangle(t) - \langle [T_2, T_2^\dagger] \rangle_{eq}\}.$$

The trivial calculation of the commutators yields the following operators:

$$[T_1, T_1^\dagger] \to I_z, I_z I_{1z} I_{2z}, I_z(I_{1+} I_{2-} + I_{1-} I_{2+}),$$

$$[T_2, T_2^\dagger] \to I_z, I_z I_{1z} I_{2z}.$$

Terms other than I_z yield no observable signals, but, since $d\langle I_z \rangle/dt$ depends on their expectation values, their relaxation must be investigated. It can be shown that the evolution of a product of n operators depends, among others, on products of $n' > n$ operators. Each spin has $2I + 1$ states, that is $2I$ independent traceless diagonal operators. Therefore, the full description of the longitudinal relaxation of the two spins I involves a system in excess of $(2I)^2$ coupled linear equations. This case has never been analysed, to my knowledge, if only because the relaxation of spins $I > 1/2$ is in most cases dominated by the quadrupole mechanism.

2'2.5. *Use of the memory function formalism.* The memory function method initially developed by MORI [18] and, somewhat independently, by LADO et al. [19] is described in ref. [5] and [20]. Its use for relaxation is touched upon in ref. [5] and thoroughly discussed in ref. [21]. It uses the formalism of Liouville spaces and goes schematically as follows.

We choose a set Q_i of «observables of interest» and we get for their evolution a system of equations of the form (neglecting precession terms)

$$\text{(2.60)} \qquad \frac{d}{dt}\langle Q_i\rangle(t) = -\sum_j \int_0^t K_{ij}(t-t')\langle Q_j\rangle(t')\,dt'.$$

The K_{ij} are called memory functions.

In fact, one should add on the right-hand side terms proportional to $\langle Q_j\rangle_{eq}$, that we assume here for simplicity to vanish, plus terms that can be equated to zero by a proper choice of initial conditions. What is seducing in the system (2.60) is that it is *rigorous* (under the assumptions made above).

Two approaches can be used to solve it.

In the first approach, we use the method of Laplace transforms. Let $\theta_i(\mathcal{Z})$ and $\varphi_{ij}(\mathcal{Z})$ be the Laplace transforms of $\langle Q_i\rangle(t)$ and $K_{ij}(t)$, respectively. The system (2.60) yields

$$\text{(2.61)} \qquad \mathcal{Z}\theta_i(\mathcal{Z}) - \langle Q_i\rangle(0) = -\sum_j \varphi_{ij}(\mathcal{Z})\theta_j(\mathcal{Z}),$$

whose solution is trivial if the $K_{ij}(t)$ are known, and have simple forms.

The flaw in this approach is the same as in paragraph 2·2.2: one has to use ensemble averages, which yields, in place of eq. (2.61),

$$\text{(2.62)} \qquad \mathcal{Z}\,\overline{\theta_i(\mathcal{Z})} - \overline{\langle Q_i\rangle}(0) = -\sum_j \overline{\varphi_{ij}(\mathcal{Z})\theta_j(\mathcal{Z})}$$

and, since in general one has

$$\text{(2.63)} \qquad \overline{\varphi_{ij}(\mathcal{Z})\theta_j(\mathcal{Z})} \neq \overline{\varphi_{ij}(\mathcal{Z})}\cdot\overline{\theta_j(\mathcal{Z})},$$

the system (2.62) has no simple solution.

The second approach consists in using in eq. (2.60) a time t much longer than the decay time of the memory functions, and much shorter than the decay time of the physical variables $\langle Q_j\rangle(t)$. We may then replace in eq. (2.60) $\langle Q_j\rangle(t')$ by $\langle Q_j\rangle(t)$ and extend the integration to infinity. This yields the system

$$\text{(2.64)} \qquad \frac{d}{dt}\langle Q_i\rangle(t) = -\sum_j \lambda_{ij}\langle Q_j\rangle(t),$$

where

$$\text{(2.65)} \qquad \lambda_{ij} = \int_0^\infty K_{ij}(\tau)\,d\tau$$

is the same for all members of the Gibbs ensemble.

Apart from the precession and thermal-equilibrium terms, that are ignored here for simplicity, the system (2.64) looks very similar to the system (2.27). There is, however, an enormous difference: whereas the operators Q_j on the right-hand side of eq. (2.27), entering the evolution of $\langle Q_i \rangle$, were *deduced* from the calculation of the double commutators in eq. (2.25), those on the right-hand side of eq. (2.64) have been *guessed*. The flaw is similar to that of paragraph 2˙2.4.

Why is this treatment faulty, since the system (2.60) is rigorous, whatever the arbitrary choice of the operators of interest? As seen in paragraph 2˙2.2, since we have to consider an ensemble average, it is necessary to choose the time t long enough, although still short compared with the evolution time of the $\langle Q_j \rangle$'s. Equation (2.60) then yields

$$(2.66) \qquad \frac{\mathrm{d}}{\mathrm{d}t} \langle Q_i \rangle(t) = - \sum_j \left\{ \int_0^t K_{ij}(\tau) \mathrm{d}\tau \right\} \times \langle Q_j \rangle(t).$$

The critical assumption is that the function $K_{ij}(\tau)$ decays fastly enough to allow the extension of the integral to infinity (eq. (2.65)). This will be correct if the choice of the operators Q_j is good. It will be wrong if the choice of the Q_j is bad. It is left to the reader to prove it on a specific example.

Although this method is elegant and has the appearance of rigour, it suffers from the fact that it requires *a priori* intelligent guesses about which variables should be included into the «set of interest».

3. – Spin temperature.

The purpose of this lecture is to describe the main features associated with the concept of spin temperature, as they evolved during the 60's, and eventually reached the mature state of a spin temperature theory. As will be seen below, the spin temperature concept, in its most useful form, applies essentially to solids, and spin temperature theory is now an essential tool in the practice of NMR in solids.

We limit ourselves throughout to the case when the only spin-spin interactions are the ever-present dipole-dipole interactions. A selection of general references to the subject of spin temperature is: ABRAGAM [1], JEENER [22]. HEBEL [23], GOLDMAN [24], SLICHTER [2], in addition to the «historical» references: POUND [25], PURCELL and POUND [26], REDFIELD [27], ABRAGAM and PROCTOR [28].

3˙1. *Spin temperature in zero field.* – We begin by considering the case when no external field is applied, so that the spin Hamiltonian reduces to the dipolar Hamiltonian $\mathscr{H}_\mathrm{D}$, which is time-independent in a solid. The number of spins being very large, of the order of the Avogadro number, the energy spectrum of $\mathscr{H}_\mathrm{D}$ is quasi-continuous.

Let us consider the evolution of the density matrix σ, as described by eq. (1.7). In a basis of eigenstates $|i\rangle$ of the Hamiltonian, of eigenvalues $\mathcal{H}_i$, we obtain for the evolution of the diagonal and off-diagonal matrix elements of σ, respectively,

(3.1) $$\langle i|\sigma(t)|i\rangle = \langle i|\sigma(0)|i\rangle,$$

(3.2) $$\langle i|\sigma(t)|j\rangle = \langle i|\sigma(0)|j\rangle \exp[-i(\mathcal{H}_i - \mathcal{H}_j)t].$$

The diagonal elements are constant in time, and the off-diagonal elements oscillate without decaying.

If we consider the expectation value of an operator Q whose only nonvanishing matrix elements are off-diagonal, we have according to eq. (3.2)

(3.3) $$\langle Q\rangle(t) = \mathrm{Tr}\{Q\sigma(t)\} = \sum_{i,j} \langle j|Q|i\rangle\langle i|\sigma(0)|j\rangle \exp[-i\Omega_{ij}t],$$

where $\Omega_{ij} = \mathcal{H}_i - \mathcal{H}_j$.

It is a sum of terms oscillating with a quasi-continuous spread of frequencies and, through destructive interference, it decays to zero.

As for an operator Q with only diagonal matrix elements, its expectation value is time-independent.

These facts led to the following approximations:

    i) After a time of the order of T_2, defined loosely as the time of decay of all expectation values of observables, it is assumed that the phases of the off-diagonal matrix elements of σ are randomly distributed, and that for all practical purposes they can be forgotten. This is known as the random-phase approximation.

    ii) Except in possible pathological cases, the expectation value of a diagonal operator Q is the same as if the diagonal matrix elements were those of the most probable configuration consistent with the actual value of the energy. This is known as the assumption that the energy is the only constant of the motion.

The most probable configuration is a Boltzmann distribution of probabilities in the various energy levels of the Hamiltonian. The corresponding density matrix is

(3.4) $$\sigma = \xi \exp[-\beta\mathcal{H}_\mathrm{D}].$$

The coefficient β has the dimension of an inverse temperature. More precisely, if $\mathcal{H}_\mathrm{D}$ is expressed in frequency units, so that the energy is

(3.5) $$E = \hbar\langle\mathcal{H}_\mathrm{D}\rangle,$$

β is of the form

(3.6) $$\beta = \hbar/k_\mathrm{B}T,$$

where k_B is Boltzmann's constant and T is the spin temperature. The usual practice is to call β the inverse spin temperature. Its value is adjusted so as to have

(3.7) $$\langle \mathcal{H}_D \rangle = \text{Tr}\{\mathcal{H}_D \sigma\}/\text{Tr}\{\sigma\} = E/\hbar.$$

As far as the coefficient ξ of eq. (3.4) is concerned, it is adjusted so as to have

(3.8) $$\text{Tr}\{\sigma\} = 1.$$

It is known nowadays that the off-diagonal elements of σ are not at random, and that eq. (3.3) describes the building-up of multiple-quantum coherences that can be detected by suitable excitations of the system. As far as the expectation value of an off-diagonal operator is concerned, its decay is not irreversible, as was long thought, and in some cases it can reappear in the form of an echo through the refocussing of the matrix elements of eq. (3.2) by proper spin manipulations. In the absence of a tricky refocussing process used on purpose, the random-phase approximation keeps a practical validity, and the form (3.4) for σ provides a pragmatically valid description of the properties of the system.

Most of NMR is performed in the high-temperature limit, such that for all eigenvalues $\mathcal{H}_i$ of $\mathcal{H}_D$ one has

$$\beta \mathcal{H}_i \ll 1.$$

In this limit, eq. (3.4) can be expanded to the first order in β:

(3.9) $$\sigma = \frac{1}{n}(1 - \beta \mathcal{H}_D),$$

where n is equal to the number of dimensions of the Hilbert space, so as to satisfy eq. (3.8). If we use reduced traces:

$$\text{Tr}\, A = \frac{1}{n} \sum_i \langle i|A|i \rangle,$$

we can drop the factor $1/n$ and use

(3.9') $$\sigma = 1 - \beta \mathcal{H}_D.$$

The high-temperature expression (3.9') of σ, or its extension to a Hamiltonian $\mathcal{H}$ with quasi-continuous spectrum as in the following, yields an enormous simplification in NMR calculations, which explains in part the great variety of NMR methods that were developed in connection with spin temperature theory.

We have so far disregarded the coupling of the spins with the lattice. Insofar as the time necessary for reaching internal equilibrium is of the order of T_2 (typically a few tens of microseconds), the spin temperature concept can have meaning only if the spin-lattice relaxation time T_1 is much longer than T_2. This is the case in solids, where T_1 ranges from a fraction of a millisecond to hours, months or years, depending on the system and on the lattice temperature.

3'2. *Spin temperature in low field.* – When an external field is applied, the spins experience a Zeeman interaction in addition to the dipolar interactions. Their Hamiltonian is

(3.10) $$\mathcal{H} = \omega_0 I_z + \mathcal{H}_D ,$$

where $\omega_0 = -\gamma H_0$ is the Larmor frequency. If $\Delta\omega_D$ is the typical frequency associated with the dipolar interactions, the low-field case is defined by the condition

(3.11) $$\omega_0 \sim \Delta\omega_D ,$$

i.e. the Zeeman coupling is low enough for the spectrum of $\mathcal{H}$ to retain its quasi-continuous character. The same arguments as before can be used, leading to the conclusion that after a time of the order of T_2 the system reaches a state of internal equilibrium characterized by a temperature. This is described by a density matrix

(3.12) $$\sigma = 1 - \beta\mathcal{H} .$$

The energy of the system is (in frequency units)

(3.13) $$\langle\mathcal{H}\rangle = \mathrm{Tr}\{\sigma\mathcal{H}\} = -\beta\,\mathrm{Tr}\{\mathcal{H}^2\} =$$
$$= -\beta\{\omega_0^2\,\mathrm{Tr}(I_z^2) + \mathrm{Tr}(\mathcal{H}_D^2)\} = -\beta\,\mathrm{Tr}\,I_z^2\{\omega_0^2 + D_L^2\} ,$$

where the local frequency D_L is defined through

(3.14) $$D_L^2 = \mathrm{Tr}(\mathcal{H}_D^2)/\mathrm{Tr}(I_z^2) .$$

One can also define a local field B_L as

(3.15) $$B_L = D_L/\gamma .$$

As a last remark, the energy spectrum has an upper, as well as a lower, bound and, if the initial energy is high enough, one may end up with a distribution of populations in the most probable configuration where the population of a level is higher the higher its energy. In order to fit the Boltzmann formula for populations

(3.16) $$p_i \propto \exp[-E_i/k_B T] ,$$

the case referred to above corresponds to a *negative* spin temperature. The only reason why negative temperatures are not found in usual thermodynamic systems is that their Hamiltonian contains a term of kinetic energy, whose spectrum extends to infinity.

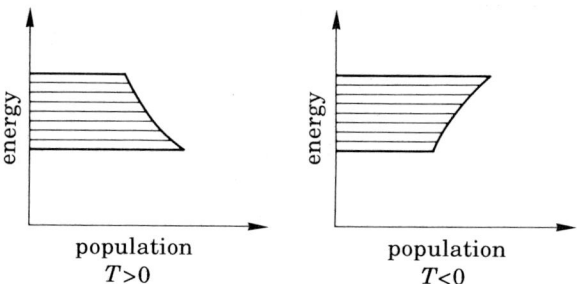

Fig. 1. – Boltzmann distribution of populations at positive and negative spin temperatures.

Examples of Boltzmann distributions at positive and negative temperature are shown in fig. 1.

3˙3. *Spin temperature in high field.* – We suppose now that the field is high, in the sense that

(3.17) $$B \gg B_L$$

which corresponds to

$$\omega_0 \gg \Delta\omega_D .$$

Let us split the dipolar Hamiltonian $\mathcal{H}_D$ into the part $\mathcal{H}_D'$ that commutes with I_z and the part $\mathcal{H}_D''$ that is off-diagonal with respect to I_z:

(3.18) $$\mathcal{H} = \omega_0 I_z + \mathcal{H}_D' + \mathcal{H}_D'' .$$

The nondiagonal part $\mathcal{H}_D''$ connects states of I_z differing by $\Delta I_z = \pm 1, \pm 2$, whose Zeeman energy differs by $\pm\omega_0$ and $\pm 2\omega_0$. Since the matrix elements of $\mathcal{H}_D''$, of the order of $\Delta\omega_D \approx D_L$, are much smaller than ω_0, it can be shown by a perturbation calculation that the eigenvalues and eigenstates of $\mathcal{H}$ depend very little on $\mathcal{H}_D''$, which can, therefore, for practical purposes, be discarded. We then use in place of (3.18) a Hamiltonian of the form

(3.19) $$\mathcal{H} = \omega_0 I_z + \mathcal{H}_D' ,$$

which is a sum of two commuting operators.

The expectation value of an operator Q varies according to

$$\frac{d}{dt} \langle Q \rangle = \langle -i[Q, \mathcal{H}] \rangle .$$

If we choose for Q either the Zeeman interaction $Z = \omega_0 I_z$ or the truncated dipolar interaction $\mathcal{H}_D'$, they both commute with $\mathcal{H}$ and their expectation value is time-independent. We then have *two* constants of the motion instead of one in zero or low field.

The most probable configuration consistent with the conservation of these quantities corresponds to a density matrix

(3.20) $$\sigma = \xi \exp[-\alpha\omega_0 I_z - \beta\mathcal{H}_D'].$$

The constants α and β are the inverse Zeeman and dipolar temperatures, respectively.

In the high-temperature limit, eq. (3.20) reduces to

(3.21) $$\sigma = 1 - \alpha\omega_0 I_z - \beta\mathcal{H}_D'.$$

The forms (3.20) or (3.21) for σ were derived from purely statistical arguments. Their plausibility can be checked by examining the various transitions that can take place within the spin system.

i) *Transitions affecting the Zeeman order.* Let us consider the levels for which the values of I_z^i of individual spins are $m+1$, m and $m-1$ (fig. 2), differing in Zeeman energy by ω_0 (in frequency units). The truncated dipolar Hamiltonian is a sum of two-particle operators of the form

(3.22) $$\mathcal{H}_D' \sim 2 I_z I_z' - \frac{1}{2}(I_+ I_-' + I_- I_+').$$

The last term on the right-hand side is able to induce flip-flop transitions whereby one spin flips from level m to level $m+1$, whereas another spin flips from level m to level $m-1$. Such a transition takes place at constant energy and at constant value of $\langle I_z \rangle = \sum_i \langle I_z^i \rangle$, and produces the following change of the Zeeman level populations:

$$\Delta p(m+1) = +1, \quad \Delta p(m) = -2, \quad \Delta p(m-1) = +1.$$

We see then that flip-flop transitions are able to redistribute at constant $\langle I_z \rangle$ the populations of the various Zeeman levels. The most probable distribution, reached eventually, is a Boltzmann one, characterized by the constant α in eq. (3.20).

ii) *Transition affecting the dipolar order.* Each Zeeman level is broadened by the dipolar interactions and there may be at a given time a distribution

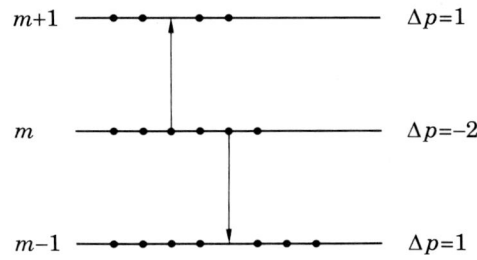

Fig. 2. – Flip-flop transition rearranging the populations of the Zeeman levels at constant energy and magnetization.

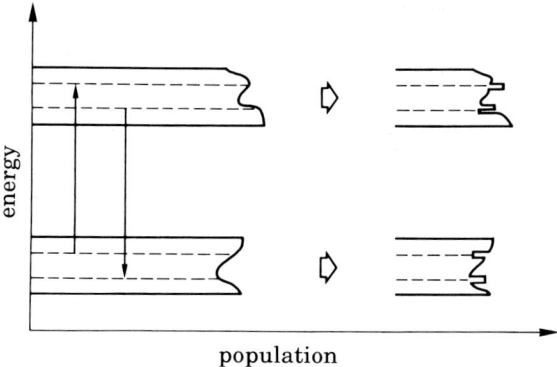

Fig. 3. – Flip-flop transition rearranging the populations of the dipolar levels at constant energy and magnetization.

of populations of the dipolar levels within each Zeeman manifold. Let us consider a flip-flop transition of two spins between levels m and $m + 1$, such as shown in fig. 3. Although it takes place at constant energy, it is able to change the populations within the dipolar levels of each Zeeman manifold, and end up with the most probable (Boltzmann) distribution of population among these levels, characterized by the constant β in eq. (3.20). However, there is no exchange between $\langle Z \rangle = \omega_0 \langle I_z \rangle$ and $\mathcal{H}_D'$, because such an exchange (under the effect of $\mathcal{H}_D''$) must conserve the total energy, and $\langle Z \rangle$ can change only in steps of ω_0 which is much more than can be accommodated by the dipolar energy in high field.

There is, therefore, no reason why α should be equal to β.

Since both the spectra of $\omega_0 I_z$ and $\mathcal{H}_D'$ are limited upwards as well as downwards, each of the Zeeman or dipolar temperatures may be either positive or negative.

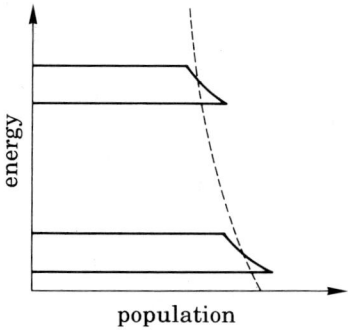

Fig. 4. – Boltzmann distribution of populations with different Zeeman and dipolar temperatures. In this example they are both positive.

An example of distribution with unequal temperatures is shown in fig. 4. According to eq. (3.21), we have

$$\langle I_z \rangle = \text{Tr}(I_z \sigma) = -\alpha \omega_0 \text{Tr}(I_z^2), \tag{3.23}$$

$$\langle \mathcal{H}_D' \rangle = -\beta \text{Tr}(\mathcal{H}_D'^2) = -\beta D^2 \text{Tr}(I_z^2), \tag{3.24}$$

where the local frequency D in high field is defined through

$$D^2 = \text{Tr}(\mathcal{H}_D'^2)/\text{Tr}(I_z^2). \tag{3.25}$$

We define likewise a local field in high field through

$$B_L' = D/\gamma. \tag{3.26}$$

It is seen from eqs. (3.23) and (3.24) that, as a result of the high-temperature approximation, the Zeeman energy $\omega_0 \langle I_z \rangle$ depends only on the Zeeman inverse temperature α, and the dipolar energy $\langle \mathcal{H}_D' \rangle$ depends only on the inverse dipolar temperature β.

The considerations developed above can be generalized to systems with unequally spaced levels: systems containing different spin species, or involving quadrupole as well as Zeeman interactions. There are as many independent temperatures at equilibrium as independent constants of the motion.

3'4. *R.f. irradiation in high field.* – We come back to a system containing one spin species only and subjected to Zeeman plus dipolar interactions.

If the system is irradiated with a r.f. field of frequency ω close to the Larmor frequency ω_0, the r.f. field will induce transitions that will modify both the Zeeman and the dipolar energies, and the density matrix will not remain of the form (3.21). However, the induced transitions are subjected to restrictions that result in the existence of a new constant of the motion. This is shown as follows.

Let us analyse what happens when the spin system absorbs one r.f. photon: Its energy increases by

$$\delta E = \hbar \omega. \tag{3.27}$$

The only elementary transition that the r.f. field can produce is the flip of an individual spin corresponding to $|\Delta m| = 1$. In the present case, it produces an increase of the Zeeman energy equal to

$$\delta \langle Z \rangle = \hbar \omega_0. \tag{3.28}$$

Since the total energy must be conserved, the dipolar energy is modified through flip-flop transitions by the amount

$$\delta \langle \mathcal{H}_D' \rangle = \delta E - \delta \langle Z \rangle = \hbar(\omega - \omega_0), \tag{3.29}$$

whence, according to eqs. (3.28) and (3.29),

$$(3.30) \qquad \delta\langle\mathcal{H}_D'\rangle + \frac{\omega_0 - \omega}{\omega_0}\delta\langle Z\rangle = \delta\{\langle\mathcal{H}_D'\rangle + (\omega_0 - \omega)\langle I_z\rangle\} = 0.$$

The expectation value of $(\omega_0 - \omega)I_z + \mathcal{H}_D'$ does not vary: it is a constant of the motion. In the steady state, the density matrix is, therefore, of the form

$$(3.31) \qquad \sigma = \exp[-\beta[(\omega_0 - \omega)I_z + \mathcal{H}_D']].$$

This result can be obtained by a different route, as follows. If we use a frame rotating at the frequency ω around the direction of the steady field B_0, the applied rotating field B_1 looks static. The equation of evolution of the density matrix, as viewed from this rotating frame, is

$$(3.32) \qquad \frac{d}{dt}\tilde{\sigma} = -i[\mathcal{H}_{\text{eff}}, \tilde{\sigma}],$$

i.e. it is governed by an effective time-independent Hamiltonian of the form

$$(3.33) \qquad \mathcal{H}_{\text{eff}} = (\omega_0 - \omega)I_z + \omega_1 I_x + \mathcal{H}_D' = \Delta I_z + \omega_1 I_x + \mathcal{H}_D'$$

with $\omega_1 = -\gamma B_1$.

This is a «low field» Hamiltonian, in an effective field of transverse component B_1 and longitudinal component

$$(3.34) \qquad b = -\Delta/\gamma = B_0 + \omega/\gamma.$$

By the same argument as in actual low field, the equilibrium form of the density matrix in this reference frame is

$$(3.35) \qquad \tilde{\sigma} = \xi\exp[-\beta(\Delta I_z + \omega_1 I_x + \mathcal{H}_D')].$$

When ω_1 is very small, the term $\omega_1 I_x$ can be neglected. Since both I_x and $\mathcal{H}_D'$ are invariant by rotation around Oz, it is immaterial to stay in the rotating frame, and we recover from eq. (3.35) the same form as (3.31). In the general case, eq. (3.35) is said to define a spin temperature in the rotating frame.

The effective energy in the rotating frame is

$$(3.36) \qquad \langle\mathcal{H}_{\text{eff}}\rangle = -\beta(\Delta^2 + \omega_1^2 + D^2)\operatorname{Tr}(I_z^2).$$

3'5. *Value and evolution of the spin temperature.* – Let $\mathcal{H}$ be the constant Hamiltonian of the system and σ_i its initial, nonequilibrium density matrix. The latter will evolve towards a final equilibrium form

$$(3.37) \qquad \sigma_f = 1 - \beta\mathcal{H}.$$

The evolution takes place at constant energy:

(3.38) $$\mathrm{Tr}(\sigma_i \mathscr{H}) = \mathrm{Tr}(\sigma_f \mathscr{H}) = -\beta \mathrm{Tr}(\mathscr{H}^2),$$

whence the value of the inverse temperature:

(3.39) $$\beta = -\frac{\mathrm{Tr}(\sigma_i \mathscr{H})}{\mathrm{Tr}\,\mathscr{H}^2}.$$

Among the various ways of starting from nonequilibrium it is possible, once equilibrium has been achieved with respect to a given Hamiltonian,

to produce a sudden change of the density matrix, by r.f. pulses;

to produce a sudden change of the Hamiltonian, for instance in the rotating frame by changing the amplitude, and/or phase, and/or frequency of the r.f. field.

We analyse this latter case below.

3˙5.1. *Sudden change of the Hamiltonian*. Starting from equilibrium with a Hamiltonian equal to $\mathscr{H}_i$, that is with a density matrix

(3.40) $$\sigma_i = 1 - \beta_i \mathscr{H}_i,$$

where β_i is the initial inverse spin temperature, the Hamiltonian is suddenly changed to the form $\mathscr{H}_f$, and one waits for the establishment of a new equilibrium, of density matrix

(3.41) $$\sigma_f = 1 - \beta_f \mathscr{H}_f.$$

The value of the new inverse temperature β_f is obtained from eq. (3.39), with $\mathscr{H} = \mathscr{H}_f$ and σ_i of the form (3.40).

This yields

(3.42) $$\frac{\beta_f}{\beta_i} = \frac{\mathrm{Tr}\{\mathscr{H}_i \mathscr{H}_f\}}{\mathrm{Tr}\{\mathscr{H}_f^2\}}.$$

3˙5.2. *Adiabatic variation of the Hamiltonian*. A completely different evolution results from an adiabatic variation of the Hamiltonian. Such a variation is defined as being so slow that the system is at all times very close to equilibrium. The evolution then takes place at constant entropy. The simplest way of obtaining the form of the entropy at high temperature is through the thermodynamic expression

(3.43) $$dE = T\,dS,$$

i.e. to within a constant

(3.43') $$dS = \beta\,dE,$$

whence, by integration,

$$(3.44) \quad S - S_0 = \int_0^E \beta \, dE' = |\beta E'|_0^E - \int_0^\beta E \, d\beta' .$$

In the high-temperature limit, we have

$$E = -\beta \operatorname{Tr}(\mathcal{H}^2),$$

whence

$$(3.45) \quad S = \mathrm{const} - \frac{1}{2} \beta^2 \operatorname{Tr}(\mathcal{H}^2).$$

During an adiabatic variation of the Hamiltonian, the inverse spin temperature varies according to

$$(3.46) \quad \beta \propto \frac{1}{\sqrt{\operatorname{Tr}(\mathcal{H}^2)}} .$$

In the rotating frame, for instance, we obtain

$$(3.47) \quad \beta \propto \frac{1}{\sqrt{\Delta^2 + \omega_1^2 + D^2}} .$$

The slow variation can be that of Δ, ω_1 or D.

We consider in more details the so-called adiabatic demagnetization in the rotating frame (in short ADRF)[29] which consists in the following sequence.

　i) Start the irradiation of the system at a distance from resonance:

$$(3.48) \quad \Delta \gg \omega_1, D .$$

　ii) Decrease Δ slowly to zero, by a sweep of either the radiofrequency or the external field.

　iii) Decrease ω_1 slowly to zero.

We have initially

$$(3.49) \quad \mathcal{H}_i \simeq \Delta I_z , \qquad \sigma_i = 1 - \beta_i \Delta I_z ,$$

and finally

$$(3.49') \quad \mathcal{H}_f = \mathcal{H}_D' , \qquad \sigma_f = 1 - \beta_f \mathcal{H}_D' .$$

The initial inverse temperature β_i is related to the lattice inverse temperature β_L through

$$(3.50) \quad \langle I_z \rangle_i = -\beta_i \Delta \operatorname{Tr}(I_z^2) = -\beta_L \omega_0 \operatorname{Tr}(I_z^2),$$

whence

(3.51) $$\beta_i = \beta_L(\omega_0/\Delta).$$

According to condition (3.48) and relations (3.49), (3.49') and (3.50), eq. (3.47) then yields, for the final dipolar inverse temperature,

(3.52) $$\beta_f = \beta_L(\omega_0/D) \gg \beta_L.$$

The effect of ADRF is to turn Zeeman order into secular dipolar order. The condition for the demagnetization to be adiabatic is

(3.53) $$\frac{d\Delta}{dt} \ll \omega_1^2.$$

The increase of entropy during a quasi-adiabatic ADRF is analysed in ref. [24], Chap. 5.

The production of dipolar order from Zeeman order can be made much more quickly, through the use of pulses [30], but not so efficiently as by ADRF.

3˙6. *What gives rise to a r.f. signal?* – NMR studies proceed through the observation of NMR signals. The question of the origin and shape of NMR signals is, therefore, central to NMR and it goes much beyond the subject of spin temperature.

The present subsection analyses some features of transient NMR signals, that is those following excitation but observed in the absence of excitation. They qualify to the name of free-induction decay signals, even when they differ from what is usually called FID. The motivation for this analysis in the present lecture is that these signals are at the heart of some experimental tests of spin temperature theory to be described in the next subsection.

What one measures in the r.f. coil of a NMR probe is a transverse precessing magnetization, proportional to a spin component either along Ox or along Oy in the frame rotating at the Larmor frequency:

$$\langle I_x \rangle(t) = \mathrm{Tr}\{\tilde{\sigma}(t) I_x\},$$

or

$$\langle I_y \rangle(t) = \mathrm{Tr}\{\tilde{\sigma}(t) I_y\},$$

where $\tilde{\sigma}(t)$ is the density matrix in the rotating frame. The effective Hamiltonian in that frame being limited to the secular spin-spin interactions, that we suppose to be purely dipolar, we have

(3.54) $$\tilde{\sigma}(t) = \exp[-i\mathscr{H}_D' t]\sigma(0)\exp[i\mathscr{H}_D' t],$$

where $\sigma(0)$ is the initial density matrix.

The most obvious and most common case is that when $\sigma(0)$ contains a term in

I_x or I_y. Such a case happens, for instance, when one starts from thermal equilibrium:

$$\sigma_{eq} = 1 - \beta_L \omega_0 I_z$$

and applies a pulse, say of 90°, around Oy. Then

$$\sigma(0) = 1 - \beta_L \omega_0 I_x ,$$

and the FID signal is

(3.55) $$\mathcal{S} = \langle I_x \rangle = \text{Tr}\{\tilde{\sigma}(t) I_x\} = -\beta_L \omega_0 \, \text{Tr}(I_x^2) G(t),$$

where the FID shape $G(t)$ is equal to

(3.56) $$G(t) = \text{Tr}\{\exp[-i\mathcal{H}_D' t] I_x \exp[i\mathcal{H}_D' t] I_x\}/\text{Tr}\{I_x^2\}$$

with $G(0) = 1$. This shape should more properly bear the name of Zeeman FID shape.

If the pulse is around $-Ox$, the signal is along Oy:

(3.57) $$\mathcal{S} = \langle I_y \rangle = -\beta_L \omega_0 \, \text{Tr}(I_y^2) G(t).$$

The Zeeman FID signal is, therefore, in quadrature with the pulsed r.f. field. Its amplitude yields a measurement of $\langle I_z \rangle$ prior to the pulse. All this is well known.

Let us now look at the derivative of $G(t)$. We obtain from eq. (3.56)

(3.58) $$\frac{dG}{dt} = \text{Tr}\{\exp[-i\mathcal{H}_D' t](-i[\mathcal{H}_D', I_x])\exp[i\mathcal{H}_D' t] I_x\}/\text{Tr}(I_x^2).$$

By inspection of this formula, we conclude that, if the initial density matrix is equal to

(3.59) $$\sigma(0) = \lambda i[I_x, \mathcal{H}_D'] + P,$$

where P yields no signal, one will observe a FID signal along Ox:

(3.60) $$\mathcal{S} = \langle I_x \rangle = \lambda \, \text{Tr}(I_x^2) \frac{dG}{dt} .$$

Likewise, if $\sigma(0)$ is of the form

(3.61) $$\sigma(0) = \lambda i[I_y, \mathcal{H}_D'] + P,$$

one will observe the same kind of signal, but along Oy.

Now, according to sect. 1, $\mathcal{H}_D'$ is a linear combination of tensorial operators $T_2^0(I_i, I_j)$, i.e. it is a tensorial operators T_0.

According to the properties of 2nd-rank tensorial operators (eq. (1.46)), we have

(3.62) $$\begin{cases} i[I_x, \mathcal{H}_D'] = i\sqrt{3/2}\,(T_1 + T_{-1}), \\ i[I_y, \mathcal{H}_D'] = \sqrt{3/2}\,(T_1 - T_{-1}). \end{cases}$$

From the observation of a signal proportional to dG/dt, and from its phase, it is then possible to determine the initial values of $\langle T_1 \rangle$ and $\langle T_{-1} \rangle$.

As an application of this result, we can measure the secular dipolar energy $\langle \mathcal{H}_D' \rangle = \langle T_0 \rangle$ by determining the part of σ that is proportional to $\mathcal{H}_D'$ [31]. The way to do this is to transform $\mathcal{H}_D' = T_0$ partly into T_1 and T_{-1}. For instance, a pulse of angle θ around Oy transforms, according to eq. (1.54a), T_0 into

$$\tilde{T}_0 = -\sqrt{\frac{3}{2}}\,cs\,(T_1 - T_{-1}) + \ldots.$$

The maximum signal is obtained for $\theta = \pi/4$. It is along Oy, that is *in phase* with the pulsed r.f. field. The density matrix must contain no interfering terms.

3˙7. Verification of spin temperature theory.

3˙7.1. General.

High temperature. All consequences of spin temperature theory in the high-temperature domain have been thoroughly tested, as described, *e.g.*, in ref. [24], in zero and low field, in high field with or without r.f. irradiation, as regards the form of σ and its incidence on signal shapes, the effect of sudden and adiabatic variations, the effect of spin-lattice relaxation, the dynamics of establishment of thermal equilibrium in the rotating frame, adiabatic cross-over of resonance frequencies, etc.

Low temperature. Nonlinear effects in spin temperature have been studied, when the inverse temperatures α and β are so high that it is no longer permissible to expand the density matrix

(3.63) $$\sigma = \xi \exp[-\alpha \omega_0 I_z - \beta \mathcal{H}_D']$$

to the first order in α and β. In that case the Zeeman and dipolar energies, $\omega_0 \langle I_z \rangle$ and $\langle \mathcal{H}_D' \rangle$, are nonlinear functions of both α and β.

When $\alpha = 0$ and β is large enough, one observes a symmetry breaking in the form of nuclear magnetic ordering in a variety of forms: ferromagnetism, antiferromagnetism, helimagnetism, in agreement with the predictions of approximate theories.

The general reference for the low-temperature domain is [32].

All these studies, both at high and low spin temperature, provide detailed verifications of the general validity of the concept of spin temperature. But ...

Limitations. One of the basic assumptions of the temperature concept, namely the random-phase approximation, is not valid: under the effect of a static Hamiltonian, the evolution of the off-diagonal elements of the density matrix is deterministic, and their phases are *not* random. In usual thermodynamic systems there is no way to prove it by recovering lost signals, and it has become a matter of philosophical profound thought to wonder that time-reversal-invariant elementary laws give rise to irreversibility in macroscopic systems. Spin systems are particular in that it is possible to mimic time reversal by proper spin manipulations and to refocus the transverse elements of the density matrix. This was most dramatically shown by the «magic sandwich» experiment[33]. Much more sophisticated manipulations are described, *e.g.*, in ref.[34] and in other lectures of this school. Besides enlarging the possibilities of NMR, as a tool for the study of condensed matter, these experiments contribute to the progress of philosophy by proving that macroscopic systems are *not* irreversible, after all.

3˙7.2. *The case of effective Hamiltonians.* We come back to high temperatures and describe in this paragraph illustrations of the validity of the concept of spin temperature in the particular case when the evolution of the spin system is produced by an effective Hamiltonian.

We begin by recalling the stroboscopic analysis of systems subjected to a periodic Hamiltonian[20, 35] of period τ_c. The evolution from 0 to τ_c is described by a unitary operator:

$$(3.64) \qquad U(0; \tau_c) = T \exp\left[-i \int_0^{\tau_c} \mathscr{H}(t) \, dt\right],$$

where T is the Dyson chronological operator. The evolution operator from $n\tau_c$ to $(n+1)\tau_c$ is likewise

$$(3.65) \qquad U(n\tau_c; (n+1)\tau_c) = T \exp\left[-i \int_{n\tau_c}^{(n+1)\tau_c} \mathscr{H}(t) \, dt\right].$$

Since $\mathscr{H}(t)$ is periodic with period τ_c, the operations (3.64) and (3.65) are equal, so that

$$(3.66) \qquad U(0; n\tau_c) = U(0; \tau_c)^n \, .$$

The operator $U(0; \tau_c)$, being unitary, can always be written under the form

(3.67) $$U(0; \tau_c) = \exp[-i\overline{\mathscr{H}}\tau_c],$$

where $\overline{\mathscr{H}}$ is a Hermitian operator. We obtain from eq. (3.66)

(3.68) $$U(0; n\tau_c) = \exp[-i\overline{\mathscr{H}}n\tau_c].$$

If the system is observed only at the stroboscopic times $t = n\tau_c$, its state is found to be the same as if it had evolved under the effect of a Hamiltonian equal to $\overline{\mathscr{H}}$, which is for that reason called the effective Hamiltonian (in the next section, it will be called the stroboscopic Hamiltonian).

The cases of interest are those when the period τ_c is so short that $U(0; \tau_c)$ differs little from unity. In that case the effective Hamiltonian is, to the lowest order, equal to the average of the time-dependent Hamiltonian over a period:

(3.69) $$\overline{\mathscr{H}} = \frac{1}{\tau_c}\int_0^{\tau_c}\mathscr{H}(t)\,dt.$$

The approximation can be pushed to higher orders, and $\overline{\mathscr{H}}$ can be calculated in the form of an expansion in powers of τ_c, known as the Magnus expansion.

The major motivation for using periodic Hamiltonians (with a modulation produced either by multipulses or by mechanical sample rotation) has been to study the evolution of a spin system under the effect of a tailored Hamiltonian. It can, however, also be used in connection with spin temperature theory. Since the evolution at the stroboscopic times $n\tau_c$ is the same as that due to the effective Hamiltonian $\overline{\mathscr{H}}$ (to which a time $\overline{T}_2$ can be associated), it can be expected that, for

(3.70) $$n\tau_c \gg \overline{T}_2,$$

the system will reach a steady state corresponding to a density matrix

(3.71) $$\sigma = 1 - \beta\overline{\mathscr{H}}.$$

This will be true only at the stroboscopic times $n\tau_c$. In the case of multipulses, for instance, the state of the system changes wildly during a cycle and, if observed stroboscopically at different phases of the cycle, say $t = n\tau_c + \theta$, the «steady-state» density matrix will be very different from eq. (3.71).

We present two illustrations of the validity of eq. (3.71).

In the first case [36], following an ADRF on the ^{19}F spins of a CaF_2 sample, the system was subjected to pulse of successive angles θ and $-\theta$ around Oy in the rotating frame, at regular time intervals τ (fig. 5). The two forms of the

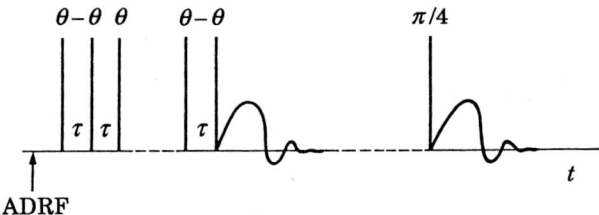

Fig. 5. – Effective Hamiltonian and spin temperature. 1st example. Pulse sequence and observation periods (after ref.[36]).

Hamiltonian over the period $\tau_c = 2\tau$ being $\mathscr{H}_1 = \mathscr{H}_D'$ and

(3.72) $$\mathscr{H}_2 = \exp[-i\theta I_y]\mathscr{H}_D' \exp[i\theta I_y],$$

the average Hamiltonian $\overline{\mathscr{H}}$ is, according to eq. (1.54a),

(3.73) $$\overline{\mathscr{H}} = \frac{1}{2}(\mathscr{H}_1 + \mathscr{H}_2) = \frac{1+3c^2}{4}T_0 - \sqrt{\frac{3}{8}}\,cs(T_1 - T_{-1}) +$$
$$+ \sqrt{\frac{3}{32}}\,s^2(T_2 + T_{-2}).$$

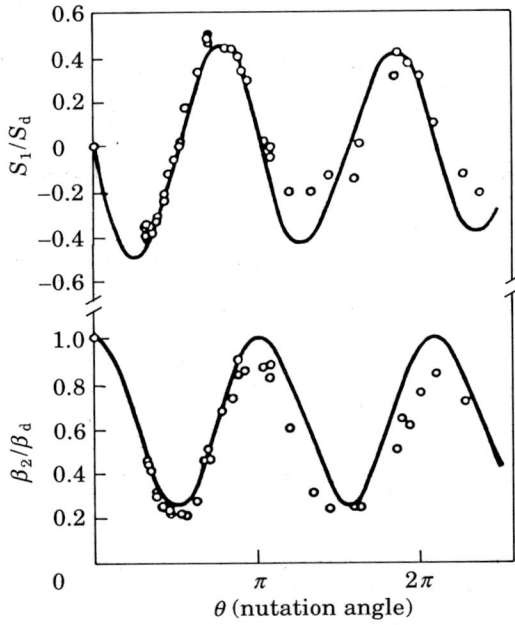

Fig. 6. – Dipolar-signal amplitudes after the pulse train of fig. 5 (top) and dipolar inverse temperature several times T_2 later (bottom) as a function of the pulse angle θ. The solid curves are theoretical (after ref.[36]).

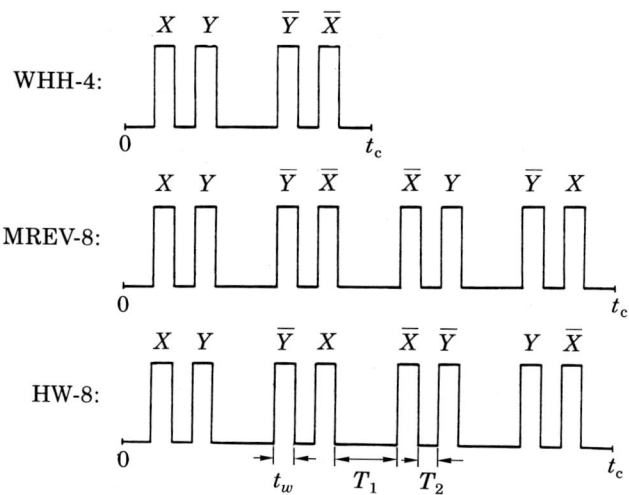

Fig. 7. – Pulse cycles for three sequences used for averaging out homonuclear dipolar interactions.

The term in $T_1 - T_{-1}$ of the density matrix (3.71) is determined from the FID signal observed at the end of the pulse sequence. As far as the term in T_0 is concerned, it is determined as follows. One waits a time longer than T_2 after the end of the pulse sequence, so as to allow the density matrix to reach the new

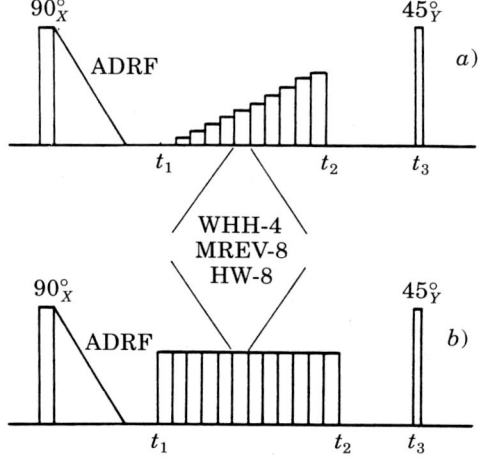

Fig. 8. – Effective Hamiltonian and spin temperature. 2nd example. Basic experimental procedure: a) adiabatic variation of the dipolar Hamiltonian, b) sudden change of this Hamiltonian.

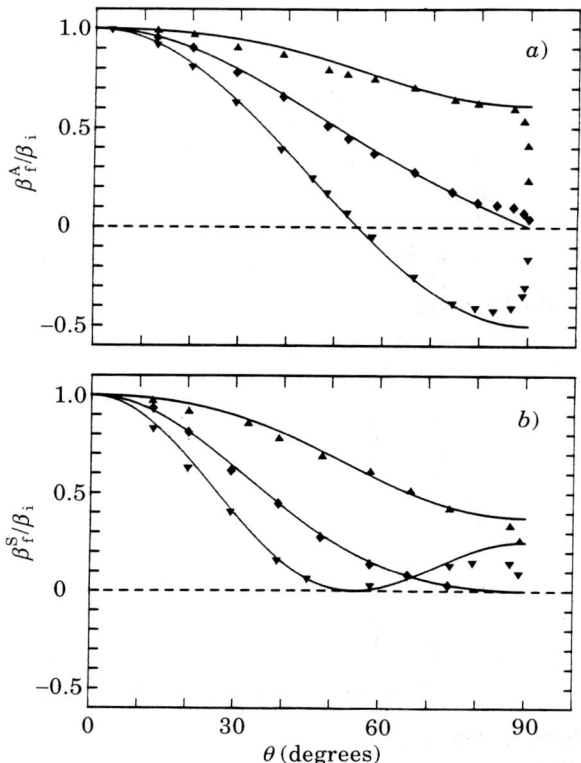

Fig. 9. – Dipolar inverse temperature several times T_2 after the end of the pulse sequence of fig. 8, corresponding to the term in T_0 of the density matrix after an adiabatic (a)) or sudden (b)) variation of the Hamiltonian as a function of the angle θ. The solid curves are theoretical: ◆ WHH-4; ▲ MREV-8, ▼ HW-8 (after ref. [38]).

equilibrium form

(3.74) $$\sigma_f = 1 - \beta_f \mathcal{H}_D',$$

where, according to eq. (3.42), the inverse temperature β_f is related to β through

(3.75) $$\frac{\beta_f}{\beta} = \frac{\mathrm{Tr}\{\mathcal{H}_D' \overline{\mathcal{H}}\}}{\mathrm{Tr}\{\mathcal{H}_D'^2\}} = \frac{1 + 3c^2}{4}$$

and β_f is measured from the signal following a $\pi/4$ pulse. (Both signals are schematically shown in fig. 5.) The calibrated results are shown in fig. 6 and compared with the theoretical expectation. The agreement is very good. The departure from this simple theory at large nutation angles θ is due mostly to the finite duration of the pulses.

The second case concerns a series of experiments on the protons of adaman-

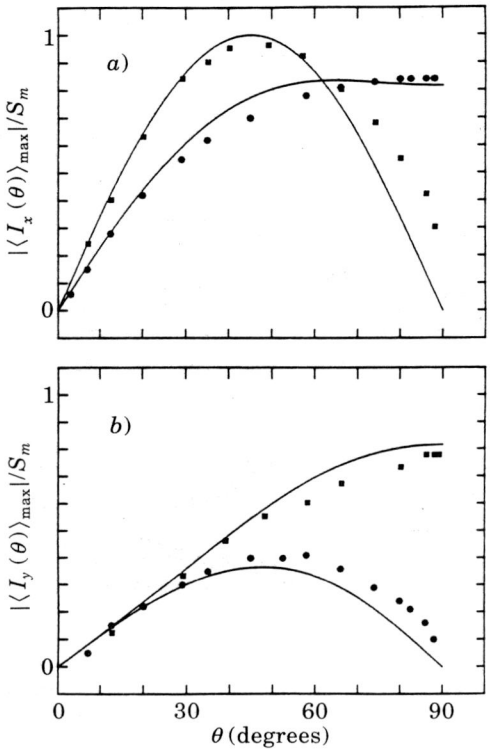

Fig. 10. – Dipolar signals following the adiabatic pulse sequences of fig. 8a), corresponding to different combinations of the terms in T_1 and T_{-1} of the density matrix as a function of the final angle θ. The solid curves are theoretical. The signal vanishes as expected from theory, for the sequences ■ MREV-8 along Ox and ● HW-8 along Oy (after ref.[38]).

tane [37, 38] in which these spins were subjected to multipulse sequences of the types WHH-4, MREV-8 and HW-8, whose basic sequences are shown in fig. 7. The cycles end in the middle of the large windows. For suitable pulse angles ($\pi/2$ for δ pulses), their result is that the truncated dipolar interactions $\mathcal{H}_D'$ yield a vanishing lowest-order contribution to $\overline{\mathcal{H}}$. These sequences are used for high resolution in solids, as discussed at length in ref.[20] and [35]. In the general case, the effective Hamiltonian is a sum of 2nd-rank tensorial components that depend on the pulse angle θ.

Experimentally, following an ADRF, one of these pulse sequences was applied, either with pulse angle starting from 0 and growing slowly up to a value θ (fig. 8a)), or directly with pulse angle equal to θ (fig. 8b)). The former case corresponds to an adiabatic evolution of the effective Hamiltonian, and the latter case to a sudden variation.

The terms in T_0 in the final density matrix were measured by waiting a time

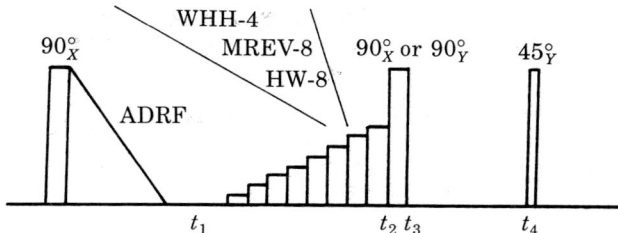

Fig. 11. – Modified procedure for observing the terms in $T_2 + T_{-2}$ of the density matrix after an adiabatic pulse sequence. The signals are observed after the pulse at time t_4 (after ref. [38]).

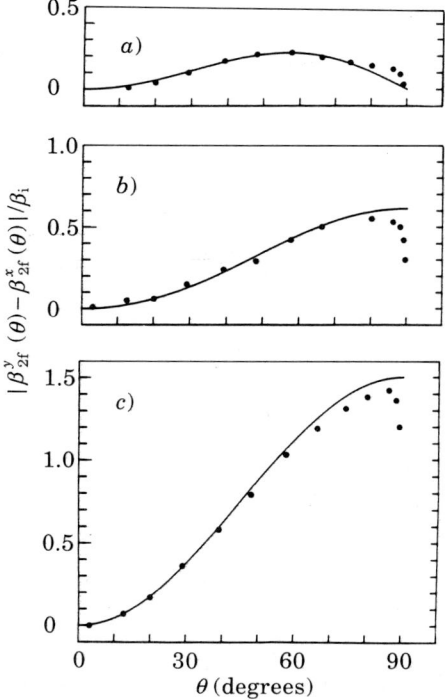

Fig. 12. – Combination of dipolar inverse temperatures at time t_4 of fig. 11, yielding the term in $T_2 + T_{-2}$ of the density matrix as a function of the final angle θ, of adiabatic pulse sequences. The solid curves are theoretical (after ref. [38]): a) WHH-4, b) MREV-8, c) HW-8.

longer than T_2 and observing the signal after a $\pi/4$ pulse. The results are shown in fig. 9 for the three pulse sequences, together with theoretical predictions. The results depart from theory only for θ very close to $\pi/2$, that is when the effective Hamiltonian $\overline{\mathscr{H}}$ nearly vanishes to the lowest order, and its form is dominated by higher-order terms in the Magnus expansion.

The terms in T_1 and T_{-1} were measured for the adiabatic pulse sequences by observing the components of the signals along Ox and Oy following the end of the sequences.

The results are shown in fig. 10, together with the theoretical curves.

As far as the terms in $T_2 + T_{-2}$ are concerned, in the density matrix (3.71) at the end of an adiabatic sequence, their measurement requires a special procedure. The pulse sequence is followed by a $\pi/2$ pulse either along Ox or along Oy. One then waits a time longer than T_2, in order that the density matrix will reach the form (3.74), and measures β_f from the signal following a $\pi/4$ pulse. The sequence is shown in fig. 11.

A detailed calculation not given here[38] shows that in the difference of the density matrices resulting from $\pi/2$ pulses around Ox and Oy, following an adiabatic sequence, the term in T_0 is proportional to that of $T_2 + T_{-2}$ at the end of the adiabatic sequence. The latter can, therefore, be determined from the difference of the signals measured at the end of the experimental sequences depicted in fig. 11. The results are shown in fig. 12, together with the theoretical curves.

The overall agreement between theory and experiment provides a further proof of the applicability of the spin temperature concept to the case of effective Hamiltonians, proves the feasibility of adiabatic evolution of such a Hamiltonian and the detailed form of $\overline{\mathcal{H}}$ as a function of final pulse angle θ, in eq. (3.71), for all three sequences used.

4. – The case of a periodic Hamiltonian.

4˙1. *Introduction.* – We have referred in the preceding section to the use of periodic Hamiltonians for the purpose of high resolution in solids, as described in detail in ref.[20] and [35], and in a more elementary fashion in ref.[32].

There are essentially two different methods of production and use of periodic Hamiltonians. The first one consists in the application of multipulse sequences and the observation of the transverse magnetization at stroboscopic times $n\tau_c$, in appropriate windows between pulses. The density matrix at these times is of the form

(4.1) $$\sigma(n\tau_c) = \exp[-i\overline{\mathcal{H}}_S n\tau_c]\sigma(0)\exp[i\overline{\mathcal{H}}_S n\tau_c],$$

where $\overline{\mathcal{H}}_S$ is the stroboscopic Hamiltonian, which can be calculated in the form of a series in increasing powers of the cycle time τ_c through the Magnus expansion. (In fact, the real expansion parameter is $\Delta\omega\tau_c$, where $\Delta\omega$ is the typical frequency of the instantaneous Hamiltonian.)

The pulse sequences are chosen in such a way that in principle the homonuclear dipolar interactions yield a vanishing contribution to the stroboscopic Hamiltonian, and the effective Zeeman interactions in the rotating frame are only scaled down, by a known factor. These methods are mostly used for high resolution of systems with abundant nuclear spins, typically protons in organic solids, in the form of either single crystals or powders. In the latter case, the spectrum of each chemically distinct nuclear species has an average frequency, width and shape characteristic of its average and anisotropic chemical shifts.

The second method is magic-angle spinning, or MAS [39, 40], where the sample is subjected to fast mechanical spinning around an axis making an angle θ with the direction of the external magnetic field B_0. This results in a periodic modulation of the orbital part of the various interactions. When the angle θ is given the «magic» value

$$(4.2) \qquad \theta = \cos^{-1}(1/\sqrt{3}),$$

the average value of 2nd-order spherical harmonics Y_2^0 vanishes, and their periodic parts involve the frequencies $\pm \Omega$ and $\pm 2\Omega$, where Ω is the spinning frequency.

In contradistinction with the case of multipulse excitation, the FID signal is observed continuously, that is with a time resolution much shorter than the period

$$(4.3) \qquad \tau_c = 2\pi/\Omega$$

of the Hamiltonian modulation.

If the frequency Ω is very fast compared with the frequency width $\Delta\omega$ of the Hamiltonian in the rotating frame, the effect of these periodic terms is negligible on the evolution of the system. These are essentially the dipolar and anisotropic chemical-shift interactions. The spectrum then consists of a series of sharp lines, at frequencies determined by the average chemical shifts of the various kinds of nuclei. At lower spinning frequencies, one observes, in addition to these peaks, series of sidebands whose frequency separation is equal to Ω.

Let us consider, for instance, the case of nuclei with negligible dipolar interactions. This is approximately valid for ^{13}C nuclei in organic solids, decoupled from the protons, thanks to their low isotopic abundance ($\sim 1\%$) and moderate gyromagnetic ratio (~ 10 MHz per tesla). In a substance containing several different chemical types of carbon, the ^{13}C spectrum after proton decoupling is a superposition of individual powder spectra, whose overlap may completely inhibit the extraction of information. Under fast magic-angle spinning, the observation of sharp lines brings back some information: number of different chemical sites, through their average chemical shift, and their relative concentration. However, all information on the anisotropy of the nuclear shielding tensors is lost. This last information can, however, be retrieved from the relative intensities of the sidebands at lower spinning frequencies, *provided* one is able to calculate these relative intensities for a powder, as a function of spinning frequency and of the principal values of the nuclear shielding tensor. This is precisely the case when dipolar interactions are negligible, so that each nuclear spin can be considered as isolated. The effective Hamiltonian of a given spin species, in the rotating frame, is of the form

(4.4) $$\mathscr{H}(t) = \partial(t) I_z .$$

The Hamiltonians at various times commute with each other, and the evolution operator at time t is

(4.5) $$U(t) = \exp\left[-iI_z \int_0^t \partial(t') \, \mathrm{d}t'\right].$$

Detailed calculations have been made by HERZFELD and BERGER [41], and their results presented in the form of charts of practical usefulness. The success of MAS for ^{13}C in organic solids has motivated an extension of its use to many other nuclear species, with two consequences. Firstly, for nuclei $I > 1/2$ with quadrupole interactions, the only observable transition, $+1/2 \leftrightarrow -1/2$, is broadened by 2nd-order quadrupole effects which involve spherical harmonics of order 0, 2 and 4. It is only those of order 2 that are averaged to zero by the MAS. This case is analysed in other lectures of this school, and it will not be considered further. The second consequence is that, when the magnetic spins under study do not have a small isotopic abundance, their dipolar interactions may no longer be negligible. They are likely to contribute to the intensity of the spinning sidebands. It is useful to calculate these contributions, firstly in order to derive from experiment the correct principal values of the nuclear shielding tensors, and, secondly, if possible to derive information on the relative orientation of interspin vectors and shielding-tensor principal axes.

The main problem associated with such an analysis is the fact that the secular dipolar homonuclear Hamiltonians at different phases of the sample spinning do not commute with one another, which makes it much more difficult to calculate the evolution operator.

Besides its obvious practical usefulness for the exploitation of MAS spectra, the general problem of the evolution of a spin system subjected to a periodic Hamiltonian has a fundamental character. This problem was addressed by BUISHVILI and MENABDE [42], who solved it to the lowest order. A slightly different formulation was given by MEHRING (ref. [20], p. 310), leading to the same result.

The extension of the theory to higher orders of approximation [43] within the formalism of Mehring is the subject of the present section.

4'2. Fast modulation.

We use a frame rotating at about the centre of gravity of the average Larmor frequencies of the various homonuclear spins under study, and we assume in the present section that the effective Hamiltonian in this frame has a frequency spread $\Delta\omega$ much smaller than the spinning frequency Ω. The successive approximations for the evolution of the system will correspond to increasing powers in $\Delta\omega/\Omega$.

4'2.1. The basic equations.

The effective Hamiltonian in the rotating frame is

$$(4.6) \qquad \mathcal{H} = \mathcal{H}_0 + \sum_{m \neq 0} \mathcal{H}_m \exp[im\Omega t],$$

where $\mathcal{H}_0$ and $\mathcal{H}_m$ are small compared with Ω. The effect of the oscillating terms on the evolution of the system will be small, and this evolution will be essentially determined by $\mathcal{H}_0$. This is the first-order approximation, which presided to the design of MAS.

To go further, we write the density matrix $\sigma(t)$ of the system under the form

$$(4.7) \qquad \sigma(t) = \sigma_0(t) + \sum_{m \neq 0} \sigma_m(t) \exp[im\Omega t],$$

where σ_0 and σ_m evolve slowly on the time scale Ω^{-1}, and furthermore one has, for each $m \neq 0$,

$$(4.8) \qquad \sigma_m \ll \sigma_0.$$

This has a meaning only when one compares their contributions to the expectation value of a physical variable. Equation (4.8) simply expresses the fact, physically evident, that the fast oscillation of small interactions produces small oscillations on the value of physical variables.

By inserting the forms (4.6) and (4.7) of $\mathcal{H}$ and σ into the Liouville-von Neumann equation (2.2), and separating the different orders m, we obtain, for $m = 0$,

$$(4.9) \qquad \frac{d\sigma_0}{dt} = \dot{\sigma}_0 = -i[\mathcal{H}_0, \sigma_0] - i \sum_{m \neq 0} [\mathcal{H}_{-m}, \sigma_m]$$

and, for $m \neq 0$,

$$(4.10) \qquad \dot{\sigma}_m + im\Omega\sigma_m = -i[\mathcal{H}_m, \sigma_0] - i \sum_{p \neq 0} [\mathcal{H}_{m-p}, \sigma_p].$$

The separation of $\sigma(t)$ into the various terms σ_m is not as obvious as it may seem. Consider, for instance, the modulus of a particular matrix element

$\langle i|\sigma(t)|j\rangle$, and its Fourier transform

(4.11) $$\langle i|\Gamma(\omega)|\rangle = \int_0^\infty \langle i|\sigma(t)|j\rangle \exp[i\omega t]\,dt\,.$$

According to eq. (4.7), this Fourier transform consists of narrow peaks centred at the frequencies $p\Omega$. However, since the σ_m are time-dependent, the various peaks have a finite width, and their wings «spill» over the neighbouring peaks. As a consequence, a given σ_m also yields some oscillation at the frequencies $m \pm 1$, $m \pm 2$, etc., so that the definition of the various σ_m suffers from some arbitrariness. This arbitrariness disappears if they are *defined* by eqs. (4.9) and (4.10).

4˙2.2. The next-order approximation. Since the evolution of σ_m is slow on the time scale Ω^{-1}, $\dot\sigma_m$ is much smaller than $im\Omega\sigma_m$, on the left-hand side of eq. (4.10) and, in a first step, it may be ignored. One then obtains, to the lowest order of approximation,

(4.12) $$\sigma_m = -\frac{1}{m\Omega}[\mathscr{H}_m, \sigma_0],$$

where we have used eq. (4.8). Inserting the result (4.12) into eq. (4.9) yields

(4.13) $$\dot\sigma_0 = -i[\mathscr{H}_0, \sigma_0] + i\sum_{m\neq 0}\frac{1}{m\Omega}[\mathscr{H}_{-m},[\mathscr{H}_m, \sigma_0]].$$

The last term on the right-hand side can be modified as follows. First of all, by changing m into $-m$, we have

(4.14) $$\sum_m \frac{1}{m\Omega}[\mathscr{H}_{-m},[\mathscr{H}_m, \sigma_0]] = -\sum_m \frac{1}{m\Omega}[\mathscr{H}_m,[\mathscr{H}_{-m}, \sigma_0]].$$

Secondly, according to the general and well-known relation

(4.15) $$[A,[B,C]] = [[A,B],C] + [B,[A,C]],$$

we have

(4.16) $$[\mathscr{H}_m,[\mathscr{H}_{-m}, \sigma_0]] = [[\mathscr{H}_m, \mathscr{H}_{-m}], \sigma_0] + [\mathscr{H}_{-m},[\mathscr{H}_m, \sigma_0]].$$

We then obtain from eq. (4.14)

(4.17) $$\sum_{m\neq 0}\frac{1}{m\Omega}[\mathscr{H}_{-m},[\mathscr{H}_m, \sigma_0]] = -\sum_{m\neq 0}\frac{1}{2m\Omega}[[\mathscr{H}_m, \mathscr{H}_{-m}], \sigma_0].$$

The evolution of σ_0 is then, according to eq. (4.13),

(4.18) $$\dot\sigma_0 = -i[\overline{\mathscr{H}}, \sigma_0],$$

where the effective Hamiltonian $\overline{\mathscr{H}}$ is equal to

$$(4.19) \qquad \overline{\mathscr{H}} = \mathscr{H}_0 + \sum_{m \neq 0} \frac{1}{2m\Omega} [\mathscr{H}_m, \mathscr{H}_{-m}].$$

This is the result of Buishvili and Menabde [42], recovered by MEHRING [20]. The point that has been strongly emphasized by these authors is that this effective Hamiltonian is *not* the same as the stroboscopic Hamiltonian $\overline{\mathscr{H}}_S$ to the same order of approximation. It is easy to verify that the Magnus expansion yields

$$(4.20) \qquad \overline{\mathscr{H}}_S = \overline{\mathscr{H}} + \sum_{m \neq 0} \frac{1}{m\Omega} [\mathscr{H}_0, \mathscr{H}_m].$$

These expressions coincide only when $\mathscr{H}_0 = 0$.

4˙2.3. *The general expansion.* When expanding the approximation to higher orders, it is no longer permissible to neglect the term $\dot{\sigma}_m$ on the left-hand side of eq. (4.10). This may seem too bad, because, otherwise, the expansion of σ_m could have been obtained by a straightforward iteration procedure. The way out is to express σ_m in the form of a sum:

$$(4.21) \qquad \sigma_m = \lambda_m + \mu_m,$$

where λ_m is chosen so as to allow its expansion through such an iteration. As will become clear later on, the calculation of the remaining part, μ_m, can be made step by step, in a systematic fashion.

We then define λ_m through

$$(4.22) \qquad im\Omega\lambda_m = -i[\mathscr{H}_m, \sigma_0] - i \sum_{p \neq 0} [\mathscr{H}_{m-p}, \lambda_p],$$

whence, by iteration,

$$(4.23) \qquad \lambda_m = -\frac{1}{m\Omega}[\mathscr{H}_m, \sigma_0] + \sum_{p \neq 0} \frac{1}{mp\Omega^2}[\mathscr{H}_{m-p}, [\mathscr{H}_p, \sigma_0]] + \ldots.$$

This is written symbolically as

$$(4.24) \qquad \lambda_m = \sum_{k \neq 0} \lambda_m^{(k)},$$

where $\lambda_m^{(k)} \propto \Omega^{-k}$, and is a function of the various $\mathscr{H}_p$ and of σ_0.

According to eqs. (4.10) and (4.21) we have

$$(4.25) \qquad \mu_m = \frac{i}{m\Omega}(\dot{\lambda}_m + \dot{\mu}_m) - \frac{1}{m\Omega}[\mathscr{H}_{m-p}, \mu_p].$$

By analogy with eq. (4.24), we can write, for any quantity Q expanded in powers of Ω^{-1},

$$Q = \sum_k Q^{(k)}, \tag{4.26}$$

where $Q^{(k)} \propto \Omega^{-k}$, and eq. (4.25) yields

$$\mu_m^{(k)} = \frac{i}{m\Omega}(\dot{\lambda}_m^{(k-1)} + \dot{\mu}_m^{(k-1)}) - \frac{1}{m\Omega}\sum_{p\neq 0}[\mathscr{H}_{m-p},\mu_p^{(k-1)}]. \tag{4.27}$$

An example of operator Q is $\dot{\sigma}_0$ for which we have, according to eqs. (4.9) and (4.21),

$$\dot{\sigma}_0^{(k)} = -i[\mathscr{H}_0,\sigma_0]\delta_{0,k} - i\sum_{m\neq 0}[\mathscr{H}_{-m},(\lambda_m^{(k)}+\mu_m^{(k)})]. \tag{4.28}$$

As for $\dot{\lambda}_m^{(k)}$, since the only dependence of λ_m on time is through σ_0, it is a sum of several terms, as explained below. Let us write

$$\lambda_m^{(p)} = f_m^{(p)}(\sigma_0), \tag{4.29}$$

which is proportional to Ω^{-p}. If, in its time derivative, we select the term $\dot{\sigma}_0^{(q)}$, proportional to Ω^{-q}, we obtain a term proportional to $\Omega^{-(p+q)}$, which contributes to $\dot{\lambda}_m^{(p+q)}$. This contribution is written symbolically $\dot{\lambda}_m^{(p;q)}$. Since for $\lambda_m^{(k)}$ we have $k \geq 1$, and, for $\dot{\sigma}_0^{(k)}$, $k \geq 0$, we obtain

$$\dot{\lambda}_m^{(k)} = \sum_{q=1}^{k} \dot{\lambda}_m^{(q;k-q)}. \tag{4.30}$$

To the lowest order, we have, according to eqs. (4.23) and (4.28),

$$\dot{\lambda}_m^{(1)} = \frac{1}{m\Omega}[\mathscr{H}_m,[\mathscr{H}_0,\sigma_0]]. \tag{4.31}$$

The lowest order for μ_m is $k = 2$, and we obtain from eq. (4.27)

$$\mu_m^{(2)} = -\frac{1}{m^2\Omega^2}[\mathscr{H}_m,[\mathscr{H}_0,\sigma_0]] \tag{4.32}$$

from which it is possible to calculate step by step $\mu_m^{(3)}$, $\mu_m^{(4)}$, etc.

In the rest of this lecture, we limit the explicit calculations to $k = 2$. According to eqs. (4.21), (4.23) and (4.32), we have, to this order,

$$\sigma_m = -\frac{1}{m\Omega}[\mathscr{H}_m,\sigma_0] + \frac{1}{m^2\Omega^2}[[\mathscr{H}_0,\mathscr{H}_m],\sigma_0] + \tag{4.33}$$

$$+ \sum_p{}' \frac{1}{mp\Omega^2}[\mathscr{H}_{m-p},[\mathscr{H}_p,\sigma_0]],$$

where $m \neq 0$ everywhere, and the primed sum corresponds in addition to $p \neq 0$ and $p \neq m$. We have used eq. (4.15) for deriving (4.33).

The evolution of the average density matrix σ_0. By inserting the form (4.33) for σ_m into eq. (4.9), we obtain

$$(4.34) \quad \dot{\sigma}_0 = -i[\mathcal{H}_0, \sigma_0] - i\sum_m \frac{1}{2m\Omega} [[\mathcal{H}_m, \mathcal{H}_{-m}], \sigma_0] -$$

$$-i\sum_m \frac{1}{m^2\Omega^2} [\mathcal{H}_{-m}, [[\mathcal{H}_0, \mathcal{H}_m], \sigma_0]] -$$

$$-i\sum_{m,p}' \frac{1}{3mp\Omega^2} [[[\mathcal{H}_{-m}, \mathcal{H}_{m-p}], \mathcal{H}_p], \sigma_0].$$

The form of the last term on the right-hand side is derived from the term

$$-i\sum_{m,p}' \frac{1}{mp\Omega^2} [\mathcal{H}_{-m}, [\mathcal{H}_{m-p}, [\mathcal{H}_p, \sigma_0]]]$$

by a procedure similar to that used for the derivation of eq. (4.17): systematic use of eq. (4.15) and change of the names of the indices. It is, however, much more complicated in the present case.

One should note that, when $\mathcal{H}_0 = 0$, eq. (4.34) takes the form

$$(4.35) \quad \dot{\sigma}_0 = -i[\overline{\mathcal{H}}, \sigma_0]$$

with

$$(4.36) \quad \overline{\mathcal{H}} = \frac{1}{2}\sum_m \frac{1}{m\Omega} [\mathcal{H}_m, \mathcal{H}_{-m}] + \frac{1}{3}\sum_{m,p}' \frac{1}{mp\Omega^2} [[\mathcal{H}_{-m}, \mathcal{H}_{m-p}], \mathcal{H}_p],$$

so that we have

$$(4.37) \quad \sigma_0(t) = \exp[-i\overline{\mathcal{H}}t]\sigma_0(0)\exp[i\overline{\mathcal{H}}t].$$

The evolution of σ_0 is determined by a unitary transformation and there only remains to determine its initial form $\sigma_0(0)$, which is *not* equal to $\sigma(0)$, as will be seen shortly.

On the other hand, when $\mathcal{H}_0$ does not vanish, the evolution of σ_0 is not through a unitary transformation, and is not evident *a priori*. It can be found through the use of an auxiliary operator, as shown next.

The operator ρ. We have seen above that the stroboscopic density matrix

$$\sigma(n\tau_c) = \sigma_0(n\tau_c) + \sum_m \sigma_m(n\tau_c)$$

evolves under the influence of an effective Hamiltonian $\overline{\mathcal{H}}_S$ that can be calculated through the Magnus expansion. We now define for any time t the operator

$$\rho(t) = \sigma_0(t) + \sum_m \sigma_m(t), \tag{4.38}$$

which coincides with σ for $t = n\tau_c$. It will be shown later that its evolution is governed by $\overline{\mathscr{H}}_S$. In any case, it is initially equal to $\sigma(0)$ and is, therefore, adapted at predicting the signal amplitudes.

According to eq. (4.33), its form up to terms in Ω^{-2} is

$$\rho = \sigma_0 - \sum_m \frac{1}{m\Omega} [\mathscr{H}_m, \sigma_0] + \sum_m \frac{1}{m^2\Omega^2} [[\mathscr{H}_0, \mathscr{H}_m], \sigma_0] + \tag{4.39}$$

$$+ \sum_{p,m}{}' \frac{1}{mp\Omega^2} [\mathscr{H}_{m-p}, [\mathscr{H}_p, \sigma_0]].$$

Being an expansion in powers of Ω^{-1}, this expression can be inverted so as to yield the expression of σ_0 as a function of ρ. One obtains the following result:

$$\sigma_0 = \rho + \sum_m \frac{1}{m\Omega} [\mathscr{H}_m, \rho] - \tag{4.40}$$

$$- \sum_m \frac{1}{m^2\Omega^2} \{[\mathscr{H}_m, [\mathscr{H}_{-m}, \rho]] + [[\mathscr{H}_0, \mathscr{H}_m], \rho]\} +$$

$$+ \sum_{m,p}{}' \frac{1}{m(m-p)\Omega^2} [\mathscr{H}_{-m}, [\mathscr{H}_p, \rho]],$$

whence also, when inserted into eq. (4.33),

$$\sigma_m = -\frac{1}{m\Omega} [\mathscr{H}_m, \rho] + \frac{1}{m^2\Omega^2} \{[\mathscr{H}_m, [\mathscr{H}_{-m}, \rho]] + [[\mathscr{H}_0, \mathscr{H}_m], \rho]\} - \tag{4.41}$$

$$- \sum_p{}' \frac{1}{m(m-p)\Omega^2} [\mathscr{H}_{-m}, [\mathscr{H}_p, \rho]].$$

MAS sideband intensities. If, starting from thermal equilibrium, we apply a $\pi/2$ pulse, we have

$$\rho(0) = \sigma(0) \propto I_x.$$

The integrated amplitude of the sideband of order m, in the Fourier transform of the FID signal $\langle I_+ \rangle(t)$, is equal to

$$\mathscr{I}_m = \text{Tr}\{I_+ \sigma_m(0)\}. \tag{4.42}$$

In order that the lowest-order approximation for σ_m (eq. (4.41)) yield a contribution to the sideband, $\mathscr{H}_m$ must be proportional to I_z, *i.e.* it must correspond to an anisotropic chemical shift:

$$\mathcal{H}_m = \delta_m I_z \, .$$

This contribution is proportional to

$$\mathcal{I}_m \propto -\frac{\delta_m}{m\Omega} \mathrm{Tr}\{I_+[I_z, I_x]\} = \frac{\delta_m}{m\Omega} \mathrm{Tr}(I_y^2) \, .$$

This may be observed in a single crystal, but not in a powder where the average of δ_m vanishes [41]. For a powder, the lowest-order form of the sideband signal is

(4.43) $$\mathcal{I}_m \propto \frac{1}{m^2 \Omega^2} \mathrm{Tr}\{[I_+, \mathcal{H}_m][\mathcal{H}_{-m}, I_x]\} \, .$$

Contributions to this signal arise not only from the anisotropic chemical shift, but also from the dipolar interaction terms:

$$\mathcal{H}_m = \sum_{i,j} F_{ij} \left[2 I_z^i I_z^j - \frac{1}{2}(I_+^i I_-^j + I_-^i I_+^j) \right]$$

with no interference between them because one of them is linear and the other one bilinear in spin operators. The form of F_{ij} depends on the order m.

The important result is that the dipolar interactions do contribute to the sideband intensities, and that this contribution can be calculated. This contribution exists even in the absence of chemical-shift anisotropy.

In a MAS experiment, the modulated parts $\mathcal{H}_m$ of the Hamiltonian correspond to $-2 \leq m < 2$ (except for the second-order shift of the central Larmor frequency for quadrupolar nuclei), and the expressions above are only valid for these values of m. In order to calculate the intensities of the sidebands further removed from the central line, it is necessary to expand σ_m to orders higher than Ω^{-2}.

The calculation above is too restrictive in so far as the most interesting practical cases correspond to modulated chemical shifts that are not small compared with Ω. This will be analysed in a later section.

Evolution of the operator ρ. We want to show that the evolution of the operator ρ is governed at all times by the stroboscopic operator $\overline{\mathcal{H}}_S$, i.e. that we have

(4.44) $$\rho(t) = \exp[-i\overline{\mathcal{H}}_S t]\rho(0)\exp[i\overline{\mathcal{H}}_S t] \, .$$

We know this to be true at the stroboscopic times $t = n\tau_c$. The plausibility of eq. (4.44) can be established by the following argument. Let us write $\rho(t)$ under the form

(4.45) $$\rho(t) = \exp[-i\overline{\mathcal{H}}_S t]\rho(0)\exp[i\overline{\mathcal{H}}_S t] + \varepsilon(t)$$

with $\varepsilon(n\tau_c) = 0$.

If the operator $\varepsilon(t)$ does not vanish identically, it is a periodic function of time, and it can be written

(4.46) $$\varepsilon(t) = \varepsilon_0(t) + \sum_{m \neq 0} \varepsilon_m(t) \exp[im\Omega t].$$

For $m \neq 0$, this is in contradiction with the fact that σ_0 and the σ_m evolve slowly and do not exhibit fast oscillations. We must, therefore, have $\varepsilon_m \equiv 0$. As for $\varepsilon_0(t)$, which is constrained not to be periodic, to evolve slowly and to vanish at the stroboscopic times, the only possibility for it is to vanish too: $\varepsilon_0 \equiv 0$.

The validity of eq. (4.44) can be established directly within the approximation used in this lecture. The calculation is cumbersome and we limit ourselves in giving its principle. It goes through the following steps.

1) The operator ρ is a function of σ_0, given by eq. (4.39):

$$\rho = f(\sigma_0),$$

whence

$$\dot{\rho} = f(\dot{\sigma}_0).$$

2) The time derivative $\dot{\sigma}_0$ is a function of σ_0, given by eq. (4.34), whence

$$\dot{\rho} = g(\sigma_0).$$

3) The operator σ_0 is a function of ρ, given by eq. (4.40), whence

$$\dot{\rho} = h(\rho).$$

This last expression contains simple, double and triple commutators. Through a systematic use of eq. (4.15) and changes in the names of the subscripts, one finally obtains

(4.47) $$\dot{\rho} = -i[\overline{\mathcal{H}}_S, \rho],$$

which is equivalent to eq. (4.44).

The form of $\overline{\mathcal{H}}_S$ is

(4.48) $$\overline{\mathcal{H}}_S = \mathcal{H}_0 + \sum_m \frac{1}{m\Omega} \left\{ [\mathcal{H}_0, \mathcal{H}_m] + \frac{1}{2}[\mathcal{H}_m, \mathcal{H}_{-m}] \right\} +$$

$$+ \sum_m \frac{1}{m^2\Omega^2} \left\{ \frac{1}{2}[\mathcal{H}_m, [\mathcal{H}_{-m}, \mathcal{H}_m]] + [[\mathcal{H}_0, \mathcal{H}_m], (\mathcal{H}_0 - \mathcal{H}_{-m})] \right\} +$$

$$+ \sum_{m,p}' \frac{1}{mp\Omega^2} \left\{ \frac{1}{2} [\mathcal{H}_p, [\mathcal{H}_{-m}, \mathcal{H}_m]] + [\mathcal{H}_{m-p}, [\mathcal{H}_p, \mathcal{H}_0]] + \right.$$

$$\left. + \frac{1}{3} [\mathcal{H}_{-m}, [\mathcal{H}_{m-p}, \mathcal{H}_p]] \right\}.$$

4'3. *Large anisotropic chemical shifts.* – In MAS experiments, it often happens that the periodic terms of $\mathcal{H}$ arising from the anisotropic chemical shifts are not small compared with the spinning frequency Ω. This shows up by a rich pattern of sidebands. The treatment given above for analysing the effect of the dipolar interactions on the sideband intensities must then obviously be modified. As stated above, we still assume the dipolar interactions to be small compared with Ω.

Let I^α be different kinds of spins in the sample. The Hamiltonian in the rotating frame is of the form

(4.49) $\quad \mathcal{H} = \sum_\alpha \delta_0^\alpha I_z^\alpha + \sum_{\alpha, m} \delta_m^\alpha I_z^\alpha \exp[im\Omega t] + \sum_m \mathcal{H}'_{Dm} \exp[im\Omega t].$

In MAS, there is no static dipolar term.

The observed signal is

(4.50) $\quad \langle I_+ \rangle(t) = \text{Tr}\left\{ \sum_\alpha I_+^\alpha \sigma(t) \right\}.$

The gist of the method is to use the fact that a trace is invariant under a unitary transformation, *i.e.* that, if $V(t)$ is a unitary operator, eq. (4.50) may be replaced by

(4.51) $\quad \langle I_+ \rangle(t) = \text{Tr}\left\{ \sum_\alpha (V(t) I_+^\alpha V^\dagger(t))(V(t) \sigma(t) V^\dagger(t)) \right\} = \text{Tr}\left\{ \sum_\alpha \tilde{I}_+^\alpha(t) \tilde{\sigma}(t) \right\}.$

This is the exact analogue of the use of an interaction representation: if the unitary operator $V(t)$ is defined through

(4.52) $\quad \dfrac{d}{dt} V(t) = iA(t) V(t),$

where $V(0) = 1$, and $A(t)$ is a Hermitian operator, we have

(4.53) $\quad \dfrac{d}{dt} \tilde{\sigma}(t) = \dfrac{d}{dt} (V(t) \sigma(t) V^\dagger(t)) =$

$$= -i[(\tilde{\mathcal{H}}(t) - A(t)), \tilde{\sigma}(t)] = -i[\mathcal{H}'(t), \tilde{\sigma}(t)].$$

The evolution of $\tilde{\sigma}(t)$ in this representation is governed by the effective Hamiltonian $\mathcal{H}'(t) = \tilde{\mathcal{H}}(t) - A(t)$. What we want is to be able to describe $\tilde{\sigma}(t)$ by an expansion analogous to that used for $\sigma(t)$ in the case of fast modulation. The condition is that $\mathcal{H}'(t)$ be small compared with Ω, and that its (unavoidably present) oscillating terms contain no other fundamental frequency than Ω. Another condition for the problem to be tractable is that the forms of both $\widetilde{\mathcal{H}}(t)$ and $\widetilde{I}_+(t)$ be as simple as possible. These conditions dictate the form of the operator $A(t)$.

The second condition implies that $A(t)$ consists only of longitudinal terms. In order to fulfil the first conditions, it is obvious that $A(t)$ must include all oscillating Zeeman terms of the Hamiltonian (4.49). This, however, is not enough, because the effective static frequencies δ_0^α may differ from one another by large amounts, and $\mathcal{H}'(t)$ which would contain the static part of $\mathcal{H}$ would not be small. Neither would it be possible to include this part into $A(t)$, so as to remove it from $\mathcal{H}'(t)$, because then both $\mathcal{H}'(t)$ and $\widetilde{I}_+(t)$ would contain terms oscillating at frequencies which are *not* multiples of Ω; nor would it be possible to expand $\tilde{\sigma}(t)$ under a form analogous to eq. (4.7), and the usability of the whole treatment of subsect. 4·2 would be spoiled. The way out is through a compromise, that is the choice

$$(4.54) \qquad A(t) = \sum_\alpha I_z^\alpha \left\{ p_\alpha \Omega + \sum_m \delta_m^\alpha \exp[im\Omega t] \right\},$$

where the p_α are integers. The effective Hamiltonian $\mathcal{H}'(t)$ then contains the effective static Zeeman frequencies

$$(4.55) \qquad \delta_0'^\alpha = \delta_0^\alpha - p_\alpha \Omega,$$

and the p_α are chosen in principle so as to minimize these shifts. A similar representation has been used for studying the MAS spectra of pairs of spins [44, 45].

The effective Hamiltonian $\mathcal{H}'(t)$ governing the evolution of $\tilde{\sigma}(t)$ is then

$$(4.56) \quad \mathcal{H}' = \sum_\alpha \delta_0'^\alpha I_z^\alpha + V(t) \left[\sum_m \mathcal{H}'_{Dm} \exp[im\Omega t] \right] V^\dagger(t) =$$

$$= \mathcal{H}_0' + \sum_q \mathcal{H}_q' \exp[iq\Omega t].$$

The oscillating terms are purely dipolar, that is small and, if the static term $\mathcal{H}_0'$ is also small, the treatment given above can be used. The density matrix $\tilde{\sigma}(t)$ is of the form

$$(4.57) \qquad \tilde{\sigma} = \tilde{\sigma}_0 + \sum_m \tilde{\sigma}_m \exp[im\Omega t].$$

As for $\widetilde{I}_+^\alpha(t)$, it is equal to

$$(4.58) \qquad \widetilde{I}_+^\alpha(t) = I_+^\alpha \exp\left[i\left(p_\alpha \Omega t + \sum_m \delta_m^\alpha \int_0^t \exp[im\Omega t'] \, dt' \right) \right].$$

As discussed at length by HERZFELD and BERGER [41], the exponential is a sum of terms oscillating at multiples of Ω, weighted by Bessel functions of $\delta_m^\alpha/m\Omega$, that is,

$$(4.59) \qquad \tilde{I}_+^\alpha(t) = I_+^\alpha \sum_p \xi_p^\alpha \exp[ip\Omega t].$$

In the absence of dipolar interactions, the Hamiltonian $\mathscr{H}'$ would be time-independent, and $\tilde{\sigma}(t)$ would solely consists of $\tilde{\sigma}_0 \propto I_x$. The intensity of the sideband of order p for the spins α would then be equal to ξ_p^α (to within a trivial constant). When dipolar interactions are taken into account, this intensity is

$$(4.60) \qquad \mathscr{S}_p^\alpha = \sum_q \xi_q^\alpha \operatorname{Tr}\{I_+^\alpha \tilde{\sigma}_{p-q}(0)\},$$

i.e. it is modified. This may be at the origin of difficulties sometimes encountered in determining the principal values of the nuclear shielding tensors from the charts of Herzfeld and Berger.

The dipolar interactions create other problems that are considered next.

4'3.1. Spin-spin relaxation. – The width of the various sidebands is determined by the effective Hamiltonian $\mathscr{H}'$ governing the evolution of $\tilde{\sigma}(t)$, and more precisely by the dipolar terms of $\mathscr{H}'$. The latter are sandwiched between the operators $V(t)$ and $V^\dagger(t)$, which has the consequence that some of the dipolar terms, that were oscillating in the rotating frame, become time-independent in the new representation. This is only possible for dipolar flip-flop terms that do not commute with the operator $A(t)$ which determines the evolution of $V(t)$. According to eqs. (4.53) and (4.54), it is elementary to show that one has, for such flip-flop terms,

$$V(t)\mathscr{H}'_{Dm}V^\dagger(t) = \mathscr{H}'_{Dm}\sum_l \eta_m^l \exp[il\Omega t],$$

which yields a time-independent part $\eta_m^{-m}\mathscr{H}'_{Dm}$. This has been thoroughly studied, both experimentally and theoretically, in the case of pairs of homonuclear spins with different chemical shifts [46, 47] and for pairs of like spins observed stroboscopically [48]. Since the part of the formerly oscillating dipolar interactions that become time-independent in the interaction representation consists of flip-flop terms, these terms will actually produce flip-flops only between those spins whose static Zeeman frequencies $\delta_0'^\alpha$ and $\delta_0'^\beta$ are equal or nearly equal. For unlike spins, the condition is, according to eq. (4.55),

$$(4.61) \qquad \delta_0^\alpha - \delta_0^\beta = (p_\beta - p_\alpha)\Omega,$$

that is, it corresponds to spins whose Larmor frequencies differ by a multiple of the spinning frequency. Under such «rotational resonance», one observes indeed drastic effects of broadening of the lines and of increased rate of polarization transfer between the spins [46, 47]. For like spins, for which the condition

(4.62) $$\delta_0^\alpha = \delta_0^\beta$$

is always satisfied, whatever Ω, the condition for flip-flop terms of the form

$$I_+^\alpha I_-^\beta + I_-^\alpha I_+^\beta$$

to become time-independent in the interaction representation is that these flip-flop terms be affected by the passage to this representation, *i.e.* that they do not commute with the operator $A(t)$ (eq. (4.54)). This means that there must be at least some m for which

(4.63) $$\delta_m^\alpha - \delta_m^\beta \neq 0,$$

which corresponds to the case of like spins with different orientations of their nuclear-shielding-tensor principal axes [48].

4˙3.2. *Application of the theory. Generalities.* The condition for the preceding expansion to be valid is that the dipolar interactions be small compared with the spinning frequency. For practical applications, the largest number that can be considered as small is about $1/3$. The above condition is then

(4.64) $$\frac{\Delta\omega}{\Omega} \leqslant \frac{1}{3}.$$

In powders, the contribution of $\tilde{\sigma}_m(0)$ ($m \neq 0$) to the sideband pattern intensity is of order $(\Delta\omega/\Omega)^2$, relative to that of $\tilde{\sigma}_0(0)$, that is at most of the order of $1/10$. In the observed pattern, which is the superposition of those originating from $\tilde{\sigma}_0$ and the various $\tilde{\sigma}_m$, the relative intensities of different sideband pairs will depart from those of independent spins by a few per cent at most. Although quite visible, and definite witness of the effect of dipolar interactions, the sideband intensities will have to be known with high accuracy if these departures are to be used for quantitative structural information.

As a last remark, the contribution of the dipolar interactions to the sideband intensities is *a priori* different for crystals of different orientations. Therefore, the sideband intensity pattern in a powder sample is *not* the simple addition of the dipolar contribution to that of the sole chemical-shift anisotropy. An estimation of the atomic positions from the sideband intensity requires extensive simulation computations: one must use models for the positions of the various nuclei and the values and orientations of the nuclear shielding tensors, compute the sideband intensities from the theory developed in this lecture for each crystal orientation, and then perform the powder average. The practical aspects of the corresponding computation programs have yet to be explored.

REFERENCES

[1] A. ABRAGAM: *The Principles of Nuclear Magnetism* (Clarendon Press, Oxford, 1961).
[2] C. P. SLICHTER: *Principles of Magnetic Resonance*, 3rd edition (Springer-Verlag, Berlin, 1990).
[3] U. HAEBERLEN: unpublished lecture at the symposium on *High Resolution* NMR *in Solids in honor of* J. S. WAUGH (MIT, Cambridge, Mass., 1989); W. KUHN: Ph. D. Thesis (Heidelberg, 1983).
[4] T. P. DAS and E. L. HAHN: *Nuclear quadrupole resonance spectroscopy*, in *Solid State Physics*, edited by F. SEITZ and D. TURNBULL, Suppl. 1 (Academic Press, New York, N.Y., 1958), p. 1.
[5] A. ABRAGAM and M. GOLDMAN: *Nuclear Magnetism: Order and Disorder* (Oxford University Press, Oxford, 1982).
[6] K. W. ZILM and J. M. MILLAR: *Solid state and solution* NMR *of nonclassical transition metal polyhydrides*, in *Advances in Magnetic and Optical Resonance*, edited by W. S. WARREN, Vol. 15 (Academic Press, New York, N.Y., 1990), p. 163.
[7] M. GOLDMAN: *Quantum Description of High-Resolution* NMR *in Liquids* (Oxford University Press, Oxford, 1988).
[8] N. BLOEMBERGEN, E. M. PURCELL and R. V. POUND: *Phys. Rev.*, **73**, 679 (1948).
[9] R. K. WANGSNESS and F. BLOCH: *Phys. Rev.*, **89**, 278 (1953).
[10] F. BLOCH: *Phys. Rev.*, **102**, 104 (1956).
[11] I. SOLOMON: *Phys. Rev.*, **99**, 559 (1955).
[12] A. G. REDFIELD: *IBM J. Res. Develop.*, **1**, 19 (1957); *Theory of relaxation processes*, in *Advances in Magnetic Resonance*, edited by J. S. WAUGH, Vol. **1** (Academic Press, New York, N.Y., 1965), p. 1.
[13] J. F. JACQUINOT and M. GOLDMAN: *Phys. Rev. B*, **8**, 1944 (1973).
[14] D. AILION and C. P. SLICHTER: *Phys. Rev.*, **137A**, 235 (1965).
[15] A. W. OVERHAUSER: *Phys. Rev.*, **92**, 411 (1953).
[16] M. GOLDMAN: *J. Magn. Reson.*, **60**, 437 (1984).
[17] C. DALVIT and G. BODENHAUSEN: *Measurement of dipole-dipole cross correlation by triple-quantum filtered two-dimensional exchange spectroscopy* and references therein, in *Advances in Magnetic Resonance*, edited by W. S. WARREN, Vol. 14 (Academic Press, New York, N.Y., 1990), p. 1.
[18] H. MORI: *Prog. Theor. Phys.*, **33**, 423 (1965).
[19] F. LADO, J. D. MEMORY and G. W. PARKER: *Phys. Rev. B*, **4**, 1406 (1971).
[20] M. MEHRING: *High Resolution* NMR *in Solids*, 2nd edition (Springer-Verlag, Berlin, 1983).
[21] D. KIVELSON and K. OGAN: *Spin relaxation theory in terms of Mori's formalism*, in *Advances in Magnetic Resonance*, edited by J. S. WAUGH, Vol. 7 (Academic Press, New York, N.Y., 1974), p. 72.
[22] J. JEENER: *Thermodynamics of spin systems in solids*, in *Advances in Magnetic Resonance*, edited by J. S. WAUGH, Vol. 3 (Academic Press, New York, N.Y., 1968), p. 206.
[23] L. C. HEBEL jr.: *Spin temperature and nuclear relaxation in solids*, in *Solid State Physics*, edited by F. SEITZ and D. TURNBULL, Vol. 15 (Academic Press, New York, N.Y., 1963), p. 409.
[24] M. GOLDMAN: *Spin Temperature and Nuclear Magnetic Resonance in Solids* (Oxford University Press, Oxford, 1970).

[25] R. V. Pound: *Phys. Rev.*, **81**, 156 (1951).
[26] E. M. Purcell and R. V. Pound: *Phys. Rev.*, **81**, 279 (1951).
[27] A. G. Redfield: *Phys. Rev.*, **98**, 1787 (1955).
[28] A. Abragam and W. G. Proctor: *Phys. Rev.*, **109**, 1441 (1958).
[29] S. R. Hartmann and E. L. Hahn: *Phys. Rev.*, **128**, 2042 (1962).
[30] J. Jeener and P. Broekaert: *Phys. Rev.*, **157**, 232 (1967).
[31] J. Jeener, H. Eisendraht and R. van Steenwinkel: *Phys. Rev.*, **133A**, 478 (1964).
[32] A. Abragam and M. Goldman: *Nuclear Magnetism: Order and Disorder* (Oxford University Press, Oxford, 1982).
[33] W. K. Rhim, A. Pines and J. S. Waugh: *Phys. Rev. B*, **3**, 684 (1971).
[34] A. Pines: *Lectures on pulsed* NMR, in *Physics of* NMR *Spectroscopy in Biology and Medicine*, edited by B. Maraviglia (North-Holland, Amsterdam, 1988), p. 43.
[35] U. Haeberlen: *High-resolution* NMR *in solids*, in *Advances in Magnetic Resonance*, edited by J. S. Waugh, Suppl. 1 (Academic Press, New York, N.Y., 1976), p. 1.
[36] A. Pines and J. S. Waugh: *Phys. Lett. A*, **47**, 337 (1974).
[37] L. Quiroga, J. Virlet and M. Goldman: *J. Magn. Reson.*, **51**, 540 (1983).
[38] L. Quiroga: Thesis, Note CEA-N-2340 (1983).
[39] E. R. Andrew, A. Bradbury and R. G. Eades: *Nature (London)*, **182**, 1659 (1958).
[40] I. J. Lowe: *Phys. Rev. Lett.*, **2**, 285 (1959).
[41] J. Herzfeld and A. E. Berger: *J. Chem. Phys.*, **73**, 6021 (1980).
[42] L. L. Buishvili and M. G. Menabde: *Ž. Èksp. Teor. Fiz.*, **77**, 2435 (1979) (English translation: *Sov. Phys. JETP*, **50**, 1176 (1979)).
[43] M. Goldman: *J. Magn. Reson.* (to be published).
[44] Z. H. Gan and D. M. Grant: *Mol. Phys.*, **67**, 1419 (1989).
[45] A. Kubo and C. A. McDowell: *J. Chem. Phys.*, **92**, 7156 (1990).
[46] D. P. Raleigh, G. S. Harbison, T. G. Neiss, J. E. Roberts and R. G. Griffin: *Chem. Phys. Lett.*, **138**, 285 (1987).
[47] M. H. Levitt, D. P. Raleigh, F. Creuzet and R. G. Griffin: *J. Chem. Phys.*, **92**, 6347 (1990).
[48] M. Matti Maricq and J. S. Waugh: *J. Chem. Phys.*, **70**, 3300 (1979).

Double Resonance.

CH. P. SLICHTER

Department of Physics and Department of Chemistry
University of Illinois at Urbana-Champaign
1110 West Green Street, Urbana, IL 61801-3080

1. – What is double resonance, and why do it?

As the name implies, double resonance involves simultaneously bringing two distinct transitions into resonance at the same time. For example, one might simultaneously excite the resonances of two different nuclear species present in a sample or simultaneously excite two different transitions of a multilevel system. There are many reasons for so doing. They include polarizing nuclear spins, enhancing sensitivity, simplifying spectra, unraveling complex spectra and generating coherent radiation (*e.g.*, masers and lasers).

Over the years, double resonance has developed into a rich and varied field. There are so many schemes that it would take a very thick book to describe them all. It is possible and useful, therefore, to divide them into a small number of categories each of which is based on a common set of principles. We shall do so in this lecture[1].

The first category, which we call *Pound-Overhauser double resonance*, involves a set of energy levels whose populations are normally held in thermal equilibrium by spin-lattice relaxation processes. Application of a strong radio-frequency field to one transition will disturb the populations from thermal equilibrium. The spin-lattice relaxation processes will then repopulate all the levels. The combined effect may then produce unusual population differences between energy levels which may produce useful effects. For example, an upper energy level may achieve a larger population than a lower one, the condition necessary for production of induced emission of energy. Examples of this class of double resonance are the Overhauser effect (including what are today called the nuclear Overhauser effect, NOE, and its two-dimensional form, NOESY), the solid effect, electron-nuclear double resonance (ENDOR), and masers and lasers.

The second class of double resonance we call *cross-relaxation double reso-*

nance. The basic idea is that, if two spin systems can exchange energy (*i.e.* cross-relax), one can detect the absorption of energy by one spin system through its effect on the spin temperature (energy level populations) of the second system. Various schemes involving cycling of the static magnetic field as well as the so-called Hartmann-Hahn method of cross-polarization are examples.

The third class of double resonance we label *spin coherence double resonance*. It depends on there existing spin-spin couplings which cause the orientation of one nuclear species to affect the precession frequency of another. Then, if we change the orientation of the first nucleus, we change the precession frequency of a nucleus to which it is coupled. Examples of this category are spin tickling experiments, decoupling methods, spin-echo double resonance (SEDOR), coherence transfer and two-dimensional Fourier transfer NMR (2D-FT NMR).

2. – Discovery of the Pound-Overhauser family.

Historically, the Pound-Overhauser family represents the first double-resonance experiments. POUND [2] studied ^{23}Na in NaNO$_3$. The three magnetic transitions of the ^{23}Na nucleus (whose spin, I, is $3/2$) acted on by a strong static magnetic field are split by an axially symmetric electric-field gradient. POUND studied how the intensity of one transition was affected by prior saturation of a different transition to show that the mechanism of spin-lattice relaxation arose from time-dependent electric-field gradients.

OVERHAUSER studied the mechanism of spin-lattice relaxation of conduction electrons in metals for his Ph.D. thesis [3]. At that time, the conduction electron spin resonance had not yet been observed. He concluded that the resonance should be observable, and in the process discovered that, if the electron spin resonance were saturated, the electron-nuclear spin-spin coupling should lead to polarization of the nuclear spins, achieving a polarization as large as one would have at thermal equilibrium if the nuclear gyromagnetic ratio, γ_n, were replaced by the electron spin gyromagnetic ratio, γ_e.

Shortly thereafter, the conduction electron spin resonance was discovered in Li and Na by GRISWOLD, KIP and KITTEL [4]. Within a few months, CARVER and SLICHTER [5] verified Overhauser's proposal by observing the effect on the size of the ^{7}Li NMR signal of saturating the conduction electron spin resonance (ESR) in Li metal. In order to obtain enough power to saturate the lithium, they used a low static magnetic field (30.0 G) which put the ESR at 84 MHz and the ^{7}Li NMR at 50 kHz.

Verification of Overhauser's proposal came as a great surprise to most of the resonance community at that time, and stimulated numerous investigations of other circumstances in which «Overhauser effects» could be observed.

3. – A model system.

To discuss the various families of double resonance, it is convenient to utilize a simple model system of two spins, I and S, both of spin $1/2$ which are coupled to a static magnetic field H_0 in the z direction and to each other by a scalar coupling $A\mathbf{I}\cdot\mathbf{S}$. The Hamiltonian, $\mathcal{H}$, is then

$$(3.1) \qquad \mathcal{H} = -\gamma_S \hbar H_0 S_z - \gamma_I \hbar H_0 I_z + A\mathbf{I}\cdot\mathbf{S} .$$

We shall assume that H_0 is strong enough so that we need keep only the portion $AI_z S_z$ of the spin-spin coupling, giving then

$$(3.2) \qquad \mathcal{H} = -\gamma_S \hbar H_0 S_z - \gamma_I \hbar H_0 I_z + AI_z S_z .$$

(Alternatively, one would take (3.2) as the model Hamiltonian, but for some purposes it is important to consider forbidden transitions which can arise from exact solutions of (3.1) but are absent from (3.2).)

Since both I_z and S_z commute with the Hamiltonian of eq. (3.2), we can take their eigenvalues m_I and m_S as good quantum numbers, obtaining for the energy eigenvalues

$$(3.3) \qquad E = -\gamma_S \hbar H_0 m_S - \gamma_I \hbar H_0 m_I + Am_I m_S$$

with corresponding eigenfunctions

$$(3.4) \qquad \psi_{m_S m_I} = |m_S m_I\rangle ,$$

which we label by the abbreviations $|++\rangle$, $|+-\rangle$, $|-+\rangle$, $|--\rangle$ to signify whether $m_S = +1/2$ or $-1/2$ and $m_I = +1/2$ or $-1/2$.

In the following pages we designate the four wave functions as

$$(3.5) \quad \psi_1 = |++\rangle, \quad \psi_2 = |-+\rangle, \quad \psi_3 = |+-\rangle, \quad \psi_4 = |--\rangle .$$

In discussing problems involving a nucleus and an electron, we will utilize «I» to stand for the nucleus, and «S» to stand for the electron. However, for a nucleus γ can be either positive or negative, whereas the gyromagnetic ratio of an electron is always negative. Therefore, we will replace γ_S by $-\gamma_e$ for electron spins, with γ_e a positive quantity.

Then, for electrons

$$(3.6) \qquad E = +\gamma_e \hbar H_0 m_S - \gamma_I \hbar H_0 m_I + Am_I m_S .$$

4. – The electron-nuclear case: the Overhauser effect.

Although OVERHAUSER treated the case of conduction electron spins interacting with the nuclei of a metal, we can understand the essence of his ideas using the energy levels of the model system given by eq. (3.6). Application of an alternating field H_{0x} perpendicular to H_0 can then induce transitions among these energy levels. The alternating field couples to both the nuclear and electron spin magnetic moments, giving a perturbing Hamiltonian

(4.1) $$\mathcal{H}_{\text{pert}} = (-\gamma_I \hbar I_x + \gamma_e \hbar S_x) H_{0x} \cos \omega t ,$$

where ω is the frequency of the applied alternating magnetic field, and where we have taken the x direction along the direction of the alternating field.

The nonvanishing matrix elements of $\mathcal{H}_{\text{pert}}$ between the wave functions of eq. (3.4) require that either

(4.2a) $$\Delta m_I = 0 , \quad \Delta m_S = \pm 1 ,$$

or

(4.2b) $$\Delta m_I = \pm 1 , \quad \Delta m_S = 0 .$$

These transitions may be labeled electron spin resonance (eq. (4.2a)) or nuclear magnetic resonance (eq. (4.2b)), respectively.

They occur at frequencies ω_e and ω_n respectively given by

(4.3a) $$\omega_e = \left| \gamma_e H_0 + \frac{A}{\hbar} m_I \right| ,$$

(4.3b) $$\omega_n = \left| \gamma_n H_0 - \frac{A}{\hbar} m_S \right| .$$

The corresponding absorption lines are doublets. For example, assuming $|\gamma_e \hbar H_0| \gg |A|$, the electron spin resonance is split into two lines spaced by $A/\hbar$ in frequency about the value $\gamma_e H_0$ of a free-electron spin.

The electron spin resonance corresponds to transitions between states 1 and 2 or 3 and 4. The nuclear resonance represents transitions between states 1 and 3 or 2 and 4. The transitions are shown in an energy level diagram of fig. 1.

In thermal equilibrium, the various states 1, 2, 3 and 4 are occupied according to a Maxwell-Boltzmann law. Let us denote the probability of occupation of state i by p_i. We then expect that there will in general be transitions induced

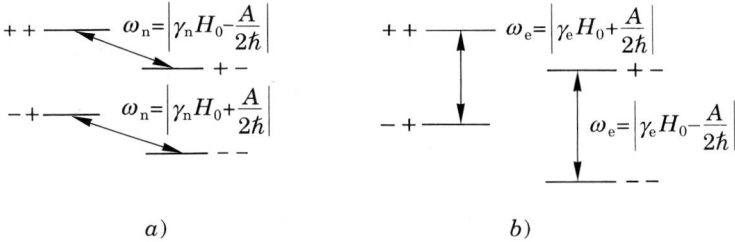

Fig. 1. – The energy level diagram and allowed transitions of a system consisting of an electron of spin $S = 1/2$ coupled to a nucleus of spin $I = 1/2$ acted on by an external static magnetic field H_0: a) nuclear-resonance transitions, b) electron spin resonance transitions. The figure assumes γ_n is negative.

between state i and other states j by the spin-lattice relaxation processes. To study the Overhauser effect, we consider that we are applying an alternating magnetic field which is tuned to the electron spin resonance.

In a metal, an electron spin couples to many nuclei since the electron wave function extends throughout the crystal. One does not then see the electron spin resonance split by the nuclear-spin orientations as in eq. (4.3a). Then, when one induces electron spin transitions, one is doing something akin to simultaneously exciting both transitions of eq. (4.3a).

We could do this for the model system by employing two oscillators, one tuned to the frequency

$$(4.4a) \qquad \omega = \gamma_e H_0 + \frac{a}{2},$$

the other to

$$(4.4b) \qquad \omega = \gamma_e H_0 - \frac{a}{2},$$

where we defined a as

$$(4.5) \qquad a \equiv \frac{A}{\hbar}.$$

Denoting the rate of electron spin flips induced by the applied alternating field as W_e, we then get a set of rate equations which describe the competition between spin-lattice relaxation and the saturating tendency of the applied alternating fields. Let us consider the situation shown in fig. 2, in which the spin-lattice relaxation processes are shown with dashed arrows, and the transitions induced by the applied alternating field are given by solid lines:

$$(4.6a) \qquad \frac{dp_1}{dt} = p_2 W_{21} - p_1 W_{12} + (p_2 - p_1) W_e,$$

$$(4.6b) \qquad \frac{dp_2}{dt} = p_1 W_{12} - p_2 W_{21} + p_3 W_{32} - p_2 W_{23} + (p_1 - p_2) W_e,$$

$$(4.6c) \qquad \frac{dp_3}{dt} = p_2 W_{23} - p_3 W_{32} + p_4 W_{43} - p_3 W_{34} + (p_4 - p_3) W_e,$$

$$(4.6d) \qquad \frac{dp_4}{dt} = p_3 W_{34} - p_4 W_{43} + (p_3 - p_4) W_e.$$

To solve such equations in general we make use of the concept of detailed balance to say that the thermal-equilibrium populations at temperature T, $p_i(T)$, and the W_{ij}'s are related by

$$(4.7) \qquad p_i(T) W_{ij} = p_j(T) W_{ji}$$

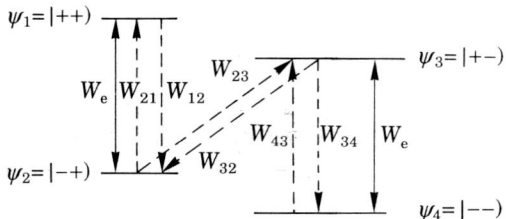

Fig. 2. – The Overhauser effect. The thermally induced transitions W_{ij} shown attempt to maintain thermal equilibrium. An applied alternating field induces electron spin transitions at a rate W_e between the states 1 and 2 and between 3 and 4.

with

(4.8) $$\frac{p_i(T)}{p_j(T)} = \exp\left[\frac{E_j - E_i}{k_B T}\right].$$

We define a quantity B_{ji} as

(4.9) $$p_i(T) = p_j(T) B_{ji}.$$

Now, the normalization of the probabilities gives one the condition

(4.10) $$\sum_i p_i = 1.$$

The thermal-average expectation value of I_z, $\langle I_z \rangle$, is given by

(4.11) $$\langle I_z \rangle = \sum_i \langle i | I_z | i \rangle p_i = \frac{1}{2}(p_1 + p_2 - p_3 - p_4).$$

If then we saturate the electron spin transitions 1, 2 and 3, 4, we produce the condition

$$p_1 = p_2, \qquad p_3 = p_4.$$

But, adding eq. (4.6a) and (4.6b), we get that in steady state for *any* value of W_e

(4.12) $$p_3 W_{32} - p_2 W_{23} = 0.$$

We, therefore, have

(4.13) $$p_3 = p_2 B_{23},$$

(4.14) $$2p_2 + 2p_3 = 1,$$

(4.15) $$\langle I_z \rangle = \frac{1}{2}(2p_2 - 2p_3),$$

so that

(4.16) $$\langle I_z \rangle = \frac{1}{2} \frac{1 - B_{23}}{1 + B_{23}}.$$

But

(4.17) $$\begin{cases} E_2 = E_{-+} = \left(\gamma_e H_0 \frac{1}{2} - \gamma_I H_0 \frac{1}{2} - \frac{a}{4} \right) \hbar, \\ E_3 = E_{+-} = \left(\gamma_e H_0 \left(-\frac{1}{2} \right) + \gamma_I \frac{H_0}{2} - \frac{a}{4} \right) \hbar, \end{cases}$$

giving

(4.18) $$E_2 - E_3 = (\gamma_e H_0 - \gamma_I H_0) \hbar.$$

It is useful to consider the high-temperature approximation. Then, expanding the Boltzmann factors, and keeping the terms linear in $1/T$, we get

(4.19) $$\langle I_z \rangle = \frac{1}{2} \frac{(\gamma_e - \gamma_I) \hbar H_0}{2 k_B T}.$$

On the other hand, if $W_e = 0$, the system is in thermal equilibrium. Then all the p_i's are at their thermal-equilibrium values. In the high-temperature limit

(4.20a) $$\langle I_z \rangle = \left[\frac{1}{2} [(p_1 - p_3) + (p_2 - p_4)] \right],$$

(4.20b) $$\langle I_z \rangle = \frac{1}{2} \frac{E_3 - E_1 + E_4 - E_2}{4 k_B T},$$

(4.20c) $$\langle I_z \rangle = \frac{1}{2} \frac{\gamma_I \hbar H_0}{2 k_B T}.$$

Therefore, contrasting eq. (4.19) with eq. (4.20c) we see that saturation of the electron spin resonance has produced a spin polarization which is equivalent to having the nuclear gyromagnetic ratio γ_I replaced by the much larger number $\gamma_e - \gamma_I$.

There is a simple physical picture which enables one to see how this result comes about. Let us neglect the nuclear Zeeman energy and the electron-nuclear coupling A compared to the electron spin Zeeman energy $\gamma_e \hbar H_0$. Let us further define a quantity ε

(4.21) $$\varepsilon = \gamma_e \hbar H_0 / 2 k_B T.$$

Then we can denote the thermal-equilibrium populations of states 1, 2, 3 and 4 by the following diagram:

diagram 4.1

$$\frac{1}{4}(1-\varepsilon) \underline{\qquad} 1 \qquad \frac{1}{4}(1-\varepsilon) \underline{\qquad} 3$$

$$\frac{1}{4}(1+\varepsilon) \underline{\qquad} 2 \qquad \frac{1}{4}(1+\varepsilon) \underline{\qquad} 4 \, .$$

If we then saturate transitions 1, 2 and 3, 4, we make $p_1 = p_2$, $p_3 = p_4$ but we maintain the ratio of p_2 to p_3 at the thermal-equilibrium value as we have explained above (eq. (4.7) and eq. (4.8)).

It is then easy to see that we can satisfy these conditions by the populations given below:

diagram 4.2

$$\frac{1}{4}(1+\varepsilon) \underline{\qquad} 1 \qquad \frac{1}{4}(1-\varepsilon) \underline{\qquad} 3$$

$$\frac{1}{4}(1+\varepsilon) \underline{\qquad} 2 \qquad \frac{1}{4}(1-\varepsilon) \underline{\qquad} 4 \, .$$

Comparing diagram 4.2 with 4.1, we see that diagram 4.2 is what one would get if one turned diagram 4.1 on its side.

In these diagrams, the upper levels (1 and 3) are for electron spin up, the lower levels (2 and 4) are for the electron spin down.

On the other hand, the left-hand levels (1 and 2) are for nuclear spin up, the right-hand levels (3 and 4) are for nuclear spin down. Thus diagram 4.1 corresponds to polarized electron spins (negligible polarized nuclear spins). Diagram 4.2 corresponds to unpolarized electron spins, but polarized nuclear spins. The nuclei are polarized with the electron Boltzmann factor.

In our model, the Overhauser polarization results from the fact that, though we equalize populations 1 and 2 and equalize populations 3 and 4, the levels 2 and 3 maintain thermal equilibrium. Diagram 4.2 is thereby produced, with its strong left-right population asymmetry. Basically, the saturating field pumps the systems from level 2 to level 1, depleting level 2. Spin-lattice relaxation then pumps the system from level 3 to level 2, and we then pump the system from level 4 to level 3. All told there is a flow from right to left in the diagram.

However, in our equations we did not include all possible thermal-relaxation processes. For example, if we had set $W_{23} = 0$, but had included a W_{14}, the diagram under saturation would become

diagram 4.3

$$\frac{1}{4}(1-\varepsilon) \underline{\qquad} 1 \qquad \frac{1}{4}(1+\varepsilon) \underline{\qquad} 3$$

$$\frac{1}{4}(1-\varepsilon) \underline{\qquad} 2 \qquad \frac{1}{4}(1+\varepsilon) \underline{\qquad} 4,$$

which reverses the sense of nuclear polarization with respect to diagram 4.2.

Clearly if *both* W_{14} and W_{23} are present, they will compete with one another, reducing the net nuclear polarization from its size from one mechanism alone. One cannot be sure, it turns out, of having one of these two W_{ij}'s totally dominant compared to the other.

The so-called solid effect [6, 7] is a polarization scheme based on the fact that one can almost always be sure that the spin-lattice relaxation time of electron spins is much the fastest relaxation time in the problem, so that W_{12}, W_{21}, W_{34} and W_{43} are by far the largest W_{ij}'s.

This fact led JEFFRIES and ABRAGAM independently to suggest that saturation of an otherwise forbidden transition (such as the 2-3 transition) would then lead to a reliable polarization. If one saturates the 2-3 transition, then it is easy to see that diagram 4.4 will result.

Diagram 4.4

$$\frac{1}{4}(1-2\varepsilon) \underline{\qquad} 1 \qquad \frac{1}{4} \underline{\qquad} 3$$

$$\frac{1}{4} \underline{\qquad} 2 \qquad \frac{1}{4}(1+2\varepsilon) \underline{\qquad} 4.$$

This diagram has a strong population difference of the left-hand (nuclear spin up) state compared to the right-hand (nuclear spin down) states.

Bloembergen's [8] three-level maser, which is also the basis for all lasers, is

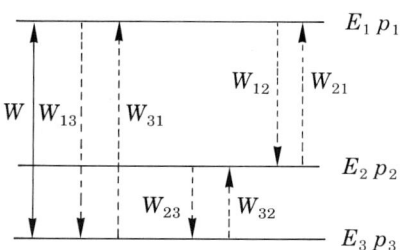

Fig. 3. – Bloembergen's three-level maser. The thermal relaxation is shown by the dashed arrows, the induced transition by the two-headed solid arrow. If the populations p_1 and p_3 are equalized by saturation, p_2 reaches a value which causes an inverted population either of p_1 relative to p_2 or of p_2 relative to p_3, depending on the relative strength of thermal rates between states 1 and 2 *vs.* 2 and 3.

another system whose operation belongs to the Pound-Overhauser category. Figure 3 shows the three-level system, with a saturating signal W applied between levels 1 and 3. Suppose W is sufficiently large that $p_1 = p_3$. If the W_{12} and W_{21} relaxation rates are very high compared to the W_{23} and W_{32} rates, p_3 and p_1 will be in thermal equilibrium, which will cause p_2 to exceed p_1. The situation $p_2 > p_1$ is a population inversion and can thus produce a net stimulated emission if one excites the p_2-p_3 transition with an alternating field.

5. – Cross-relaxation double resonance.

The basic idea of cross-relaxation double resonance arises from the intuitive feeling that, if one pumps energy into one system of spins, the energy should be able to leak into a second system of spins if the two spin systems are intermingled. A moments reflection makes one realize that the key question is how strong is the thermal contact between the two spin systems.

To answer this question it is useful to consider fig. 4 which shows the energy levels of two spin-1/2 nuclei. If the spins are originally in the states marked with a «x» ($m_S = -1/2$, $m_I = +1/2$), they may simultaneously flip to the state designated with the «o» ($m_S = +1/2$, $m_I = -1/2$). Such a process could be induced by the dipolar coupling through the term $I^- S^+$. The inverse process can arise from the $I^+ S^-$ term in the nuclear dipolar coupling.

If the nuclei have identical gyromagnetic ratios, the energy level spacings for the two nuclear species are identical. However, it is unlikely that two different nuclear species have identical γ's. Therefore, ordinarily a process such as $x \to o$ will not conserve energy. We may still conserve energy if the energy levels have sufficient width that there is an overlap in energy.

One way of achieving overlap is to work at low static magnetic field H_0 where the line widths (which often are independent of H_0) are comparable to H_0. Or one may cycle H_0 between large and small values to allow the spin systems to exchange energy (in low H_0), then return to a large H_0 to inspect what happened.

A simple and clever scheme, however, was discovered by HARTMANN and

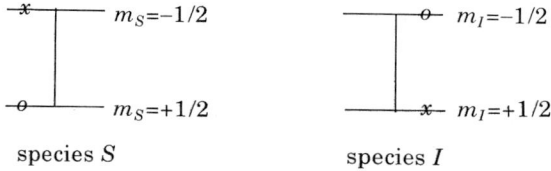

Fig. 4. – A mutual spin flip from the x to the o state exchanges energy between the I spins and the S spins. Shown vertically are the allowed energy levels. For this example the energy spacing is assumed to be the same for the two species, and both nuclei have spin 1/2.

HAHN [9]. Actually, a special case of what was later called the Hartmann-Hahn condition was earlier discovered experimentally, and explained theoretically, by BLOEMBERGEN and SOROKIN [10]. Their experiment and its relation to the Hartmann-Hahn concept are explained in ref. [1]. A discussion of the Hartmann-Hahn scheme in terms of spin temperature is given by LURIE and SLICHTER [11].

For the purposes of our lecture, it is convenient to think of the two spin systems, I and S, acted on simultaneously by two alternating fields. One alternating (rotating magnetic) field of amplitude (in the rotating frame) $H_1)_I$ is at frequency Ω_I close to the I-spin resonance frequency $\omega_I = \gamma_I H_0$. The other rotating magnetic field of amplitude (in the rotating frame) $H_1)_S$ is at frequency Ω_S close to the resonance frequency $\omega_S = \gamma_S H_0$ of the S spins.

We assume that ω_I and ω_S are sufficiently different that effectively the I spins experience only $H_1)_I$ and the S spins experience only $H_1)_S$.

In the rotating frame Ω_I the I spins then experience an effective magnetic field $H_{\text{eff}})_I$

$$(5.1a) \qquad H_{\text{eff}})_I = \left(H_0 - \frac{\Omega_I}{\gamma_I}\right)\mathbf{k} + H_1)_I \mathbf{i}$$

and the S spins similarly have

$$(5.1b) \qquad H_{\text{eff}})_S = \left(H_0 - \frac{\Omega_S}{\gamma_S}\right)\mathbf{k} + H_1)_S \mathbf{i}.$$

Suppose we adjust Ω_I and Ω_S so that both the I and S spins are exactly at resonance. Then the I-spin magnetization will precess about $H_1)_I$ at angular frequency

$$(5.2a) \qquad \omega_1)_I = \gamma_I H_1)_I ,$$

and the S-spin magnetization will precess about $H_1)_S$ at angular frequency

$$(5.2b) \qquad \omega_1)_S = \gamma_S H_1)_S .$$

But the I and S spins are coupled to each other by their nuclear magnetic dipole interaction

$$(5.3) \qquad \mathscr{H}_{IS} = \frac{\gamma_I \gamma_S \hbar^2}{r_{IS}^3}(1 - 3\cos^2\theta_{IS}) I_z S_z ,$$

where r_{IS} is the internuclear distance, and θ_{IS} is the angle between H_0 and the internuclear vector r_{IS}.

Classically, the precession of the I-spin magnetization M_I, about $H_{\text{eff}})_I$ causes I_z to oscillate in time at $\omega_1)_I$. But eq. (5.3) shows that this produces a magnetic field of the S spins perpendicular to $H_{\text{eff}})_S$, their direction of quantization.

Thus it can induce transitions of the S spins among their states of quantization under $H_{\text{eff}})_S$. The S-spin energy level spacing, ΔE, in the rotating frame is

(5.4) $$\Delta E = \gamma_S \hbar H_1)_S = \hbar \omega_1)_S .$$

When $\omega_1)_I = \omega_1)_S$, the rotating I-spin magnetization produces an alternating field on the S spins exactly able to flip the S spins relative to $H_1)_S$.

We can view this quantum-mechanically be examining fig. 4, and treating those as the energy levels of the two spin species in their respective rotating frames.

When

(5.5) $$\gamma_I H_1)_I = \gamma_S H_1)_S ,$$

the $x \to 0$ or $0 \to x$ transition conserves energy. Recalling that the spins are quantized along their respective H_1's, we recognize that $\mathscr{H}_{IS}$ of eq. (5.3) can induce the transition.

Thinking of the spin temperatures θ_I and θ_S in the respective rotating frames, we see that, starting from an arbitrary initial condition, the spins will undergo mutual flips until their spin temperatures become the same [11]. These ideas can be put on a firm quantum-mechanical basis using a formal transformation of the Hamiltonian to the doubly rotating frame [1].

The original idea of Hartmann and Hahn was, then, to utilize this special condition to facilitate the transfer of energy between two different spin systems.

To appreciate the idea, suppose we have two spin systems. We consider them in the double rotating frames in which each spin species has a time-independent Hamiltonian as long as $H_1)_I$ or $H_1)_S$ is kept constant. The energy of an isolated system with a time-independent Hamiltonian is conserved. The existence of the term $\mathscr{H}_{IS}$ (eq. (5.3)) couples the two spin systems together, so it is only their *total* energy which is conserved.

Suppose we have created magnetization $\langle M_I \rangle$ lying along $H_1)_I$ and $\langle M_S \rangle$ lying along $H_1)_S$ where $\langle M_I \rangle$ and $\langle M_S \rangle$ are the thermal-average expectation values of the magnetization components along the respective effective fields. Then the total energy, E_{total}, in the rotating frame is

(5.6) $$E_{\text{total}} = - \langle M_I \rangle H_1)_I - \langle M_S \rangle H_1)_S .$$

If we express the magnetization in terms of a spin temperature θ, the magnetization obeys Curie's law

(5.7) $$\begin{cases} \langle M_I \rangle = (C_I H_1)_I / \theta , \\ \langle M_S \rangle = (C_S H_1)_S / \theta , \end{cases}$$

where C_I and C_S are the respective Curie constants

(5.8) $$C_I = \frac{N_I \gamma_I^2 \hbar^2 I(I+1)}{3k_B},$$

etc., where N_I is the number of I nuclei per unit volume. If we start with the system initially magnetized at a temperature θ_i in the rotating frame, and the S spin initially unmagnetized ($\langle M_S \rangle = 0$), then the initial energy is

(5.9) $$E_{\text{total}} = -\langle M_I \rangle H_1)_I = -\frac{C_I(H_1)_I^2}{\theta_i}.$$

We expect that eventually the two systems will come to a final common spin temperature, θ_f. In the process, the total energy does not change. Therefore,

(5.10) $$E_{\text{total}} = -\frac{C_I(H_1)_I^2 + C_S(H_1)_S^2}{\theta_f}.$$

As a result

(5.11a) $$\frac{\theta_i}{\theta_f} \equiv \frac{C_I(H_1)_I^2 + C_S(H_1)_S^2}{C_I(H_1)_I^2},$$

(5.11b) $$\frac{\theta_i}{\theta_f} \equiv 1 + \varepsilon,$$

where

(5.12) $$\varepsilon = \frac{C_S(H_1)_S^2}{C_I(H_1)_I^2}.$$

If the H_1's satisfy the Hartmann-Hahn condition,

(5.13) $$\varepsilon = \frac{N_S S(S+1)}{N_I I(I+1)}.$$

We have, thus, that

(5.14) $$\frac{\langle M_I \rangle_f}{\langle M_I \rangle_i} = \frac{\theta_i}{\theta_f} = \frac{1}{1+\varepsilon},$$

so that the mixing has diminished $\langle M_I \rangle$.

If $N_S \ll N_I$, the charge in $\langle M_I \rangle$ will be small. However, if we turn off $H_1)_S$ momentarily, while keeping $H_1)_I$ on, $\langle M_I \rangle_S$ will rapidly decay to zero, but, as REDFIELD has shown (see ref.[1], Chap. 6), $\langle M_I \rangle$ will remain locked. We then switch $(H_1)_S$ back on again, once again creating the starting condition of zero $\langle M_S \rangle$, nonzero $\langle M_I \rangle$. Once again the spins mix, reducing $\langle M_I \rangle$ by another factor

$1/(1 + \varepsilon)$. If we do the same thing N times we will produce $\langle M_I \rangle_N$ given by

$$(5.15a) \qquad \frac{\langle M_I \rangle_N}{\langle M_I \rangle_i} = \left(\frac{1}{1+\varepsilon}\right)^N,$$

which for small ε is

$$(5.15b) \qquad \simeq \exp[-N\varepsilon].$$

This process can be repeated as long as $\langle M_I \rangle$ can be kept spin-locked. That is a time determined by spin-lattice relaxation processes, typically T_1. Thus the maximum N can be made of order

$$(5.16) \qquad N_{\max} \simeq \frac{T_1}{T_{\mathrm{mix}}},$$

where T_{mix} is the time for the two spins to mix to arrive at a common spin temperature. One expects this to be of the order of the inverse of their dipolar-coupling frequency, a typical T_2.

6. – Spin coherence double resonance—S-flip-only echoes.

For the Pound-Overhauser schemes to work, one needs certain relaxation times to be sufficiently strong. That is, the detailed spin-lattice relaxation mechanisms play a key role in determining what is possible. For cross-relaxation double resonance to work, there must be cross-relaxation mechanisms of sufficient strength, but the spin-lattice processes must be sufficiently long that they do not compete. We turn now to the third family of double-resonance schemes, spin coherence double resonance, which require for their success typically that neither spin-lattice nor spin-spin relaxation times be too rapid. Spin coherence double resonance requires coherent spin precession of one species of frequencies which can be changed by flipping the spins of a second species. The second species must be able to maintain its precession frequency without interruption by flips induced either by the lattice (T_1 processes) or resulting from the mutual flip with a third spin (a T_2 process).

To illustrate the basic principles underlying the various schemes, we utilize the Hamiltonian of eq. (3.2) with the energy levels of eq. (3.3). It is then easy to see that, in analogy with the discussion of sect. 4, an alternating magnetic field can cause either the I spins or the S spins to flip, with selection rules

$(6.1a) \qquad I\text{-spin NMR:} \quad \Delta m_I = \pm 1, \quad \Delta m_S = 0;$

$(6.1b) \qquad S\text{-spin NMR:} \quad \Delta m_I = 0; \quad \Delta m_S = \pm 1.$

The corresponding resonance frequencies, ω, occur at

$(6.2a) \qquad I\text{-NMR:} \quad \omega_I = \omega_{0I} - am_S,$

$(6.2b) \qquad S\text{-NMR:} \quad \omega_S = \omega_{0S} - am_I,$

where

(6.3a) $$\omega_{0I} = \gamma_I H_0,$$

(6.3b) $$\omega_{0S} = \gamma_S H_0.$$

Equations (6.2) show that the I-spin and the S-spin resonances consist of $2S + 1$ or $2I + 1$ lines, respectively. For our example this means each resonance is a doublet.

If the S spins were polarized in the $m_S = 1/2$ state, the I-spin resonance would consist of a single line at $\omega_I = \omega_{0I} - a/2$. If we applied a π pulse to the S spins, flipping them to the $m_S = -1/2$ state, the I resonance would shift to $\omega_I = \omega_{02} + a/2$. Thus by observing the change in absorption pattern, we could use the I spins to detect flipping of the S spins.

However, in practice this scheme will not work since the states $m_S = +1/2$ and $-1/2$ are almost equally populated. Thus for every down S spin which we flip up there is an up S spin which we flip down. Therefore, a π pulse applied to the S spins does not change the intensity of either line of the I-spin doublet.

The S-spin π pulse, however, does produce an effect on the I spin, because it changes the precession frequency of any given I spin. As we see below, we can detect this change by pulse methods.

It is convenient to discuss the time average expectation value of the I-spin magnetization. We define the expectation value of the transverse components of the magnetization of the I spins as $\langle M_x(t) \rangle$ and $\langle M_y(t) \rangle$. They can be expressed in terms of the density matrix $\rho(t)$:

(6.4) $$\begin{cases} \langle M_x(t) \rangle = \text{Tr}[M_{Ix}\rho(t)], \\ \langle M_y(t) \rangle = \text{Tr}[M_{Iy}\rho(t)]; \end{cases}$$

their complex sum $\langle M_{Ix}(t) \rangle + i\langle M_{Iy}(t) \rangle$ is convenient to use. It is given by

(6.5) $\langle M_{Ix}(t) \rangle + i\langle M_{Iy}(t) \rangle = \text{Tr}[(M_{Ix} + iM_{Iy})\rho(t)] = \text{Tr}[M_I^+ \rho(t)] = \langle M_I^+(t) \rangle.$

These expressions, with appropriate choice of the x and y axes and of the definition of $\rho(t)$, apply to the laboratory reference frame, or to the I-spin rotating frame, or to the I-spin and S-spin doubly rotating frames.

The real part of $\langle M_I^+(t) \rangle$ is $\langle M_{Ix}(t) \rangle$, while the imaginary part is $\langle M_{Iy}(t) \rangle$.

Let us then think of having a system which is initially in thermal equilibrium. We apply a $\pi/2$ pulse to the I spins, with the rotating magnetic field oriented along the x-axis in the rotating frame. This puts $\langle M_I \rangle$ initially along the y-axis.

Half the I spins have $m_S = +1/2$, the other half have $m_S = -1/2$. Therefore, half the I-spin magnetization rotates in the left-handed sense at $\omega_{0I} - a/2$

in the laboratory frame (or $-a/2$ in the rotating frame in the left-handed sense), half at $\omega_{0I} + a/2$ in the left-handed sense in the laboratory (or $a/2$ in the left-handed sense in the rotating frame).

Remembering that $A(t) = |A| \exp[i\phi] \exp[i\omega t]$ corresponds to a complex vector $|A| \exp[i\phi]$ rotating in the right-handed sense at angular frequency ω, we write

(6.6) $$\langle M_I^+(t) \rangle = i\frac{C_0}{2}[\exp[iat/2] + \exp[-iat/2]],$$

where the first term comes from $m_S = +1/2$, the second from $m_S = -1/2$, and where C_0 is the initial I-spin magnetization.

If then at $t = \tau$ we apply a π pulse to the S spins, we cause the I spins to change precession rates. Therefore, for later times

(6.7) $\langle M_I^+(t) \rangle =$

$$= i\frac{C_0}{2}[\exp[iat/2]\exp[-ia(t-\tau)/2] + \exp[-ia\tau/2]\exp[+ia(t-\tau)/2]].$$

From eq. (6.7) we see that, at $t = 2\tau$,

(6.8) $$\langle M_I^+(t) \rangle = i\frac{C_0}{2}[1+1] = iC_0,$$

so that it returns to its value just after the $\pi/2$ pulse on the I spins.

We call such an echo an «S-flip-only echo». An example is shown in fig. 5 showing the ^{31}P NMR signal in $(C_2H_5O)_2PHO$ when the ^{1}H signal of the attached proton is flipped.

From eq. (6.7) we see that at $t = 2\tau$ we are back at the same situation as we were at $t = 0^+$ (the time just after the I-spin $\pi/2$ pulse). Clearly, if we apply a second S-spin π pulse at time $t = 3\tau$, we would again produce an echo, this time at $t = 4\tau$.

In this manner, a succession of π pulses applied to the S spins at τ, 3τ, 5τ, etc. will produce a sequence of I-spin echoes at 2τ, 4τ, 6τ, etc. As a result, the two components of the I-spin magnetization will stay within an angle $a\tau/2$ of the y-axis in the rotating frame, and $\langle M_{Iy} \rangle$ will vary between C_0 and $C_0 \cos(a\tau/2)$.

Clearly, if $a\tau/2$ is small enough, $\langle M_I \rangle$ will behave as though a were zero, i.e. as though the I and S spins were not coupled. Thus we have described a system for decoupling the I and S spins.

The first demonstration of decoupling (using, however, steady rather than pulsed r.f. decoupling fields) was performed by ROYDEN[12], and was closely followed by a related study by BLOOM and SHOOLERY[13].

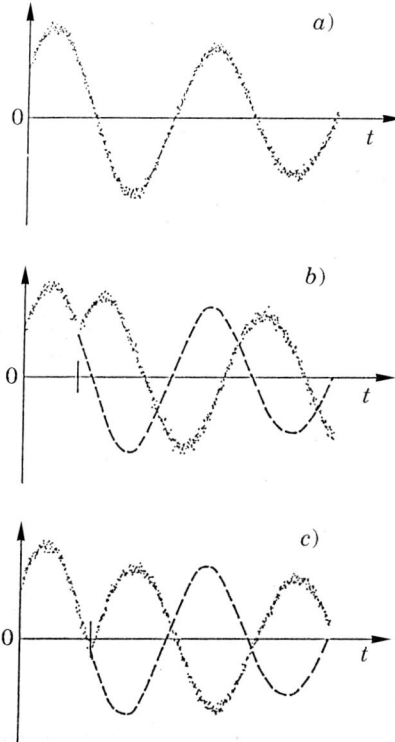

Fig. 5. – ^{31}P NMR signal *vs.* time for diethyl phosphite [$(C_2H_5O)_2$PHO]. *a*) The precession of the ^{31}P nuclei due to their coupling to the ^{1}H nuclei; *b*) and *c*) the precession of the ^{31}P nuclei with the ^{1}H nuclei inverted at the indicated times. The dashed lines in *b*) and *c*) show what the precession would have been had there been no ^{1}H inversion. (Data taken by Charles PENNINGTON with assistance from Jean-Philippe ANSERMET, Dale DURAND and David ZAX.) (Reprinted with permission from ref.[1], p. 301, © Springer-Verlag, Heidelberg.)

7. – Spin coherence double resonance—coherence transfer.

One of the most important present-day NMR techniques is the use of so-called 2-dimensional NMR to unravel complex spectra. The concept in its modern form was discovered by JEENER[14], and has subsequently been greatly extended by ERNST[15], FREEMAN[16] and many others. One of the most useful forms is Jeener's original invention, now called COSY, which involves transfer of spin-coherent magnetization from one spin to another. We turn now to an explanation of how that works.

The method for transferring coherence from one spin to another was discovered by FEHER[17] in 1956 when he was studying the ^{31}P hyperfine structure of P donors in Si. One has 4 energy levels corresponding to the 4 states of fig. 1. To describe how his method works, consider diagram 7.1, in which m_S gives the electron spin orientation, m_I the nuclear-spin orientation, and in which we have given the thermal-equilibrium populations of the levels, neglecting all but the electron spin energy in the Boltzmann factors, and utilizing eq. (4.21)

(4.21) $$\varepsilon = \gamma_e \hbar H_0 / k_B T.$$

The two electron spin transitions for $m_I = +1/2$ and $-1/2$ which occur at

(7.1a) $$\omega_e = \gamma_e H_0 + \frac{a}{2} \qquad \text{for } m_I = +1/2,$$

(7.1b) $$\omega_e = \gamma_e H_0 - \frac{a}{2} \qquad \text{for } m_I = -1/2$$

are resolved.

Diagram 7.1

	$m_I = +\frac{1}{2}$	$m_I = -\frac{1}{2}$
$m_S = +\frac{1}{2}$	$\frac{1}{4}(1-\varepsilon)$ ——— 1	$\frac{1}{4}(1-\varepsilon)$ ——— 3
$m_S = -\frac{1}{2}$	$\frac{1}{4}(1+\varepsilon)$ ——— 2	$\frac{1}{4}(1+\varepsilon)$ ——— 4.

There are also two nuclear-resonance lines at

(7.2a) $$\omega_n = |\gamma_I H_0 - a/2| \qquad \text{for } m_S = +1/2,$$

(7.2b) $$\omega_n = |\gamma_I H_0 + a/2| \qquad \text{for } m_S = -1/2.$$

For this system, FEHER discovered that the spin-lattice relaxation times were very long (of the order of 1 min), so that he could sweep across various transitions under conditions of adiabatic passage in which the electron (or nuclear) magnetization remains at all time along the effective magnetic field in the rotating frame. His method of polarization transfer then consists of first doing an adiabatic passage of the ESR through the $m_I = +1/2$ line (connecting states 1 and 2), thereby inverting the electron spin population of that line. The resulting populations are shown in diagram 7.2.

Diagram 7.2

$\frac{1}{4}(1+\varepsilon)$ ——— 1	$\frac{1}{4}(1-\varepsilon)$ ——— 3
$\frac{1}{4}(1-\varepsilon)$ ——— 2	$\frac{1}{4}(1+\varepsilon)$ ——— 4.

Next he does an adiabatic passage across the NMR transition (2, 4) corresponding to $m_S = -1/2$, interchanging the populations of states 2 and 4. The final population produced is shown in diagram 7.3.

Diagram 7.3

		$m_I = \frac{1}{2}$		$m_I = -\frac{1}{2}$	
$m_S = \frac{1}{2}$		$\frac{1}{4}(1 + \varepsilon)$ ____ 1		$\frac{1}{4}(1 + \varepsilon)$ ____ 3	
$m_S = \frac{1}{2}$		$\frac{1}{4}(1 + \varepsilon)$ ____ 2		$\frac{1}{4}(1 - \varepsilon)$ ____ 4 .	

We see that diagram 7.3 now puts the population difference horizontal as opposed to the starting condition of diagram 7.1, where the population difference was vertical.

The populations of diagram 7.3 are identical to those of diagram 4.2, which describes the conventional Overhauser effect. However, we note that the method of Feher does not make use of spin-lattice relaxation to achieve the polarization.

It is possible to do a pulse version of the Feher experiment. The inversion of the $m_I = +1/2$ ESR line (*i.e.* producing diagram 7.2) can be produced by applying 2 $\pi/2$ pulses to the electron spin system using an H_1 which is much larger than a/γ_e, so that the pulse flips all electrons regardless of their m_I.

Remember that an electron has a negative γ_S (*i.e.* $\gamma_e = -\gamma_S$). Thus the electron spin undergoes right-hand rotations about a magnetic field. One applies first an $X_S(\pi/2)$ pulse (*i.e.* a $\pi/2$ pulse with H_1 along the x-axis in the S-spin rotating frame). This puts the electron magnetization along the $-y_S$ axis. It immediately breaks into two counterrotating components, one (for electrons whose nucleus is in the state $m_I = +1/2$) rotating at $a/2$ clockwise, the other at $a/2$ counterclockwise (electrons whose nucleus is $m_I = -1/2$). When $a\tau/2 = \pi/2$, the first ($m_I = = 1/2$) component lies along the $+x_S$ axis, the other along the $-x_S$ axis.

Now we apply a $y_S(\pi/2)$ pulse. This rotates the $m_I = 1/2$ component from $+x_S$ to $-z_S$, and the $m_I = -1/2$ component from $-x_S$ to $+z_S$. We thereby produce the condition of diagram 7.2.

A vector diagram, drawn for two nuclei with positive γ's, is shown in fig. 6.

If one then applied an $x_I(\pi/2)$ followed when an $a\tau/2 = \pi/2$ by a $-y_I(\pi/2)$, we would invert the $m_S - 1/2$ NMR line, producing the situation of diagram 7.3.

In the COSY pulse sequence, one is working with a single nuclear species (*e.g.*, protons) which have J-couplings between protons which have chemical-shift differences. One does not need two oscillators (one for the I spins, the other for the S spins). Rather, the $\pi/2$ pulse from a single oscillator rotates both the I and the S spins by $\pi/2$. A second simplification experimentally

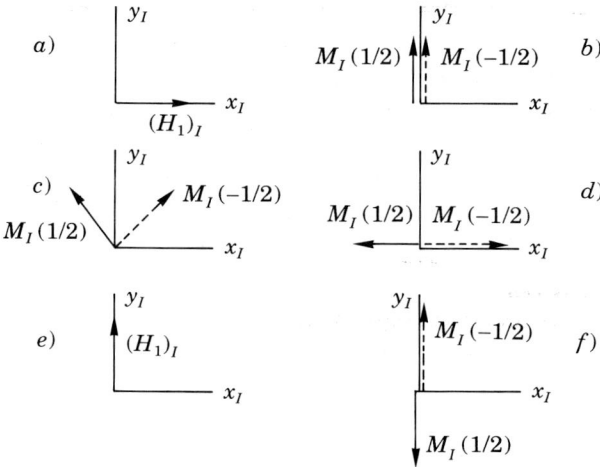

Fig. 6. – The effect of $(H_1)_I$ pulses tuned to resonance ($\omega_I = \omega_{0I}$) on the I-spin magnetization $M_I(m_S)$ for the two components $m_S = +1/2$ and $m_S = -1/2$. a) At $t = 0$, $(H_1)_I$ is applied along the x-axis in the rotating frame giving an $X_I(\pi/2)$ pulse; b) at $t = 0^+$, both $M_I(1/2)$ and $M_I(-1/2)$ lie along the y-axis; c) the two components $M_I(m_S)$ precess in opposite directions at angular frequencies $\pm a/2$; d) at a time $t = \pi/a$, $M_I(1/2)$ and $M_I(-1/2)$ point along the $-x$- and $+x$-axes, respectively; e) an $(H_1)_I$ lying along the y-axis gives a $Y_I(\pi/2)$ pulse; f) the I-magnetization vectors after the $Y_I(\pi/2)$, with $M_I(-1/2)$ pointing parallel to the z-direction, but $M_I(1/2)$ pointing antiparallel.

comes from the fact that there are a range of chemical shifts so that one is not ever able to get in a frame which is at rest with the precession frequencies of all spins. Therefore, adjusting phases so that the first pulse is an $X(\pi/2)$ and the second pulse a $Y(\pi/2)$ in the rotating frame has no utility since there are no unique x and y axes for all spins. Therefore, one simply applies two $\pi/2$ pulses of the same phase. One varies the time between them, defining a time t_1, and collects data after the end of the second pulse, obtaining the signal vs. the time t_2 after the second pulse. One then picks a new value for the pulse spacing t_1, and again collects data after the end of the second pulse. In this manner, one can obtain the complex (i.e. $\langle M_x \rangle$, $i\langle M_y \rangle$) signal $S(t_1, t_2)$. The two-dimensional Fourier transform gives a signal $G(\omega_1, \omega_2)$ which is the COSY signal.

8. – Spin echo double resonance (SEDOR).

In NMR studies of solids and of surfaces, one of the most useful double-resonance techniques has turned out to be so-called spin echo double resonance, frequently abbreviated as SEDOR. Invented by KAPLAN and HAHN [18], and further developed by EMSHWILLER, KAPLAN and HAHN[19], the technique for many years was left unused. In 1968 WALSTEDT and WERNICK[20] employed it to study the Knight shifts of ^{103}Rh atoms close to ^{59}Co atoms in dilute alloys of Co in Rh. In this section, we will review how SEDOR works, and describe several modern applications of the method.

The goal of SEDOR is to measure or detect the spin-spin coupling between two species, I and S, for spectra in which that coupling does not produce resolved splittings of the I-spin resonance. In an ordinary $\pi/2$-τ-π spin echo of the I spins, the S spins are not excited. One may then have either of two circumstances. If the orientation of the S spins is static over the period of the I-spin pulses and echo, the S spins contribute a static local field on the I spins, whose defocusing effect, following the I-spin $\pi/2$ pulse, is refocused by the I-spin π pulse. The echo amplitude thus is unaffected by the I-S spin coupling. If the S spins are reorienting rapidly compared to τ, they interfere with the phase coherence of the I-spin precession, and thus shorten the T_2 of the I spins. In the latter case, it is difficult to draw conclusions about the I-S coupling, although some techniques are available[21].

Spin echo double resonance is useful when the S-spin orientation is static in the absence of pulses applied to the S spins. The basic idea of the classic SEDOR is to flip the S spins at the time of the I-spin π pulse so that the I-spin refocusing usually produced by that pulse is prevented from occurring perfectly. One, therefore, has a pulse sequence which may be designated as

(8.1) $$X_I(\pi/2) \ldots \tau \ldots X_I(\pi) \ldots \text{echo},$$

$$\ldots \tau \ldots X_S(\pi),$$

where on the upper line we show the I-spin pulse sequence and on the lower line the S-spin pulse sequence, and where the fact that the two π pulses are coincident in time is indicated by writing $X_S(\pi)$ directly below $X_I(\pi)$. We observe the I-spin echo. By comparing the echo with and without the $X_S(\pi)$ pulse for various pulse spacings τ, one can deduce the I-S coupling.

This approach has two drawbacks. The first is that the I-spin T_2 processes cause the echo to vary with τ even in the absence of I-S coupling. The second is that for short τ's the echo gets close in time to the $X_I(\pi)$ pulse, causing problems with ring-down signals (*i.e.* residual voltages in the sample coil produced by the pulse).

Our group has employed an improved version of the SEDOR pulses. We hold the I-spin pulse spacing τ fixed, and vary the time T_S after the I-spin $\pi/2$ at which the S-spin π pulse is applied.

We will analyze the effect of this «improved» SEDOR using the model system of eq. (3.2).

Utilizing eq. (6.2a) we have then for the I spins two resonance frequencies ω_I given the two values of m_S

(8.2a) $$\omega_I = \omega_{0I} - \frac{a}{2}, \quad m_S = \frac{1}{2},$$

(8.2b) $$\omega_I = \omega_{0I} + \frac{a}{2}, \quad m_S = -\frac{1}{2},$$

where $\omega_{0I} = \gamma_I H_0$.

Consider, then, what happens if a $\pi/2$ pulse is applied to the I spins, with H_1 along the I-spin x-axis in the I-spin rotating frame, i.e. an $X_I(\pi/2)$ pulse. This will rotate the I-spin magnetization from its initial orientation along the static field H_0 (which we call the z_I-direction) into the positive y_I-direction in I-spin rotating frame. The I spin is coupled to an S spin. The S spins are equally likely to have $m_S = +1/2$. Then the I-spin magnetization will precess in a right-handed sense at angular velocity $a/2$ from the I-axis. Let us define θ as the angle from the $+y_I$ axes to the I-spin magnetization vector. θ will develop in time. At time T_S^-, just before the $X_S(\pi)$ pulse,

(8.3) $$\theta(T_S^-) = \frac{a}{2} T_S.$$

Following the $X_S(\pi)$ pulse, m_S changes to $-1/2$. Then the precession rate changes to $a'/2$, where

$$a' = -a.$$

At time τ^- just before the $X_I(\pi)$ pulse

(8.4) $$\theta(\tau^-) = \frac{a}{2} T_S + \frac{a'}{2}(\tau - \tau_S).$$

The $X_I(\pi)$ pulse rotates the I-spin magnetization about the x_I-axis, making $\theta(\tau^+)$, the orientation just after that pulse, become

(8.5) $$\theta(\tau^+) = \pi - \theta(\tau^-) = \pi - \left[\frac{a}{2} T_S + \frac{a'}{2}(\tau - T_S)\right].$$

In the interval from τ to 2τ, the I-spin magnetization continues to rotate at $a'/2$, so that at time 2τ we have

(8.6) $$\theta(2\tau) = \theta(\tau^+) + \frac{a'\tau}{2} = \pi - \left[\frac{a}{2} T_S + \frac{a'}{2}(\tau - T_S) - \frac{a'\tau}{2}\right] = \pi - (aT_S).$$

If M_{0I} is the magnitude of the *total* M_I thermal-equilibrium magnetization (1/2 of which goes with $m_S = +1/2$, the other half with $m_S = -1/2$), the components associated with a given initial m_S are then

(8.7)
$$\begin{cases} \langle M_{Iy}(2\tau) \rangle = \dfrac{M_{0I}}{2} \cos(\theta(2\tau)), \\ \langle M_{Ix}(2\tau) \rangle = -\dfrac{M_{0I}}{2} \sin(\theta(2\tau)). \end{cases}$$

We see from eq. (8.2) that, for the I spins with $m_S = -1/2$, we can immediately calculate $\theta(2\tau)$ simply by replacing a with $-a$.

Thus for the I spins with initial $m_S = -1/2$

(8.8) $$\theta(2\tau) = \pi + aT_S.$$

Comparing this result with eq. (8.6), we see that at $t = 2\tau$ the two components of I-spin magnetization form echoes which make equal but opposite angles with the y-axis, hence their x-components cancel, and their y-components add.

We, therefore, get that

(8.9a) $$\langle M_{Ix}(2\tau) \rangle = 0,$$

(8.9b) $$\langle M_{Iy}(2\tau) \rangle = \frac{M_0}{2}[\cos(\pi - aT_S) + \cos(\pi + aT_S)].$$

It is useful to think of this as arising from the sum of two complex numbers

(8.10) $$\langle M_I^+(2\tau) \rangle = -i\frac{M_0}{2}(\exp[-iaT_S] + \exp[+iaT_S]),$$

where the 1st term comes from I spins coupled to S spins initially with $m_S = +1/2$, the second term from those whose S spins were initially in $m_S = -1/2$.

Note that, if $T_S = 0$, we get the usual I-spin echo, formed along the negative y_I-axis (where the subscript I refers to the I-spin rotating frame).

We can immediately see the meaning of formula (8.10). If the S spin is in the state $m_S = -1/2$ for the entire time 2τ, the echo forms along the negative y_I-axis. If, however, the S spin spends a time T_S instead with $m_S = +1/2$, then during that time interval its precession rate differs by $a/2 - a'/2 = a$, so an extra angle aT_S is developed relative to focusing on the negative y_I-axis.

Note that we can consider that we have formed two echoes whose magnetization vectors make angles of $\pm aT_S$ with respect to the negative y_I-axis.

If the coefficient a varies from site to site (for example, if it arose from dipole-dipole coupling in a powder sample), we can think of a distribution function $f(a)$ such that the number of I spins dN_I with a between a and $a + da$ is given by

(8.11) $$dN_I = N_I f(a)\,da,$$

where N_I is the total number of I spins.

It is convenient to think of a as being both positive and negative in the formula so that $f(a)$ gives in fact the I-spin line shape. With this definition, $f(a)$ is an even function (since the I-spin spectrum has lines at $a/2$ and $-a/2$), and then

(8.12) $$\int_{-\infty}^{+\infty} f(a)\,da = 1.$$

Then

(8.13a) $$\langle M_{Iy}(2\tau)\rangle = -M_0 \int_{-\infty}^{+\infty} \cos(aT_S) f(a)\,da,$$

(8.13b) $$\langle M_{Iy}(2\tau)\rangle = -M_0 \int_{-\infty}^{+\infty} \exp[iaT_S] f(a)\,da,$$

where eq. (8.13b) follows since $f(a)$ is an even function of a.

Equation (8.13b) tells us that the SEDOR pattern is the Fourier transform of line shape function $f(a)$ which describes the spin-spin splitting. As we show below for some simple examples, this means that one can use it to determine molecular structures (*i.e.* to distinguish a ^{13}C in a CH group from a CH$_2$ or CH$_3$ group).

Suppose now that each line of the I-spin spectrum (corresponding to a given m_S) has a nonzero width. Then $f(a)$ is not a δ-function. One immediate consequence of (8.13b) is that, if τ is very long, and T_S is then also very long, the $\exp[iaT_S]$ factor oscillates rapidly, and the integral goes to zero. Physically, the spread in $f(a)$ corresponds to the statement that the I-spin vectors which refocus at $\theta = \pm aT_S$ away from negative y_I have a spread in angle. The longer T_S, the larger the spread. When the spread gets of the order π, the magnetization component $\langle M_{Iy}(2\tau)\rangle$ approaches zero.

A practical consequence is that at a time $T_S \sim \pi/\Delta a$, where Δa describes the individual line breadth, the SEDOR signal of the I spin is destroyed.

9. – SEDOR signals when there are several S spins.

An important application of SEDOR is to determine the structure of molecules. One may then, for example, observe the effect on a ^{13}C resonance (the I spins) of coupling to ^{1}H nuclei in a CH, a CH$_2$, or a CH$_3$ bond.

The Hamiltonian to describe this situation is

(9.1) $$\mathcal{H} = -\gamma_I \hbar H_0 I_z - \gamma_S \hbar H_0 \sum_k S_{zk} + \hbar I_z \sum_k a_k S_{zk},$$

where k denotes the different S spins coupled to the I spin.

As an example, suppose a_k is independent of k (i.e. all the S spins have identical coupling to the I spin). Then, defining

(9.2) $$F_z = \sum_k S_{zk},$$

we have

(9.3) $$\mathcal{H} = -\gamma_I \hbar H_0 I_z - \gamma_S \hbar H_0 F_z + \hbar a I_z F_z.$$

Both F_z and the S_{zk}'s commute with $\mathcal{H}$. The eigenvalues of F_z, m_F, are related to the eigenvalues m_{Sk} by

(9.4) $$m_F = \sum_k m_{Sk},$$

so for two spin-1/2 S spins

(9.5) $$m_F = 1, 0, -1$$

with their being two eigenfunctions of $m_F = 0$ composed as linear combinations of the two states $\alpha(1)\beta(2)$ and $\beta(1)\alpha(2)$, where α (β) designate $m_S = +1/2$ ($-1/2$).

For three S spins

(9.6) $$m_F = 3/2, \quad 1/2, \quad -1/2, \quad -3/2,$$

there is one state $(\alpha(1)\alpha(2)\alpha(3))$ for $m_F = 3/2$, but 3 states (linear combinations of $\alpha(1)\alpha(2)\beta(3)$, $\alpha(1)\beta(2)\alpha(3)$ and $\beta(1)\alpha(2)\alpha(3)$) for $m_F = 1/2$.

It is straightforward to show that the SEDOR signal for the two-S-spin case is

(9.7a) $$\langle M_{Iy}(2\tau) \rangle = -\frac{M_0}{4}(\exp[i2aT_S] + 2 + \exp[-i2aT_S]),$$

(9.7b) $$\langle M_{Iy}(2\tau) \rangle = -\frac{M_0}{4}(\exp[iaT_S] + \exp[-iaT_S])^2,$$

whereas the SEDOR signal for the three-S-spin case is

(9.8a) $$\langle M_{Iy}(2\tau) \rangle =$$
$$= -\frac{M_0}{8}[(\exp[i3aT_S] + 3\exp[iaT_S] + 3\exp[-iaT_S] + \exp[-i3aT_S])],$$

(9.8b) $$\langle M_{Iy}(2\tau) \rangle = -\frac{M_0}{8}(\exp[iaT_S] + \exp[-iaT_S])^3.$$

The important point of these formulae is that the build-up of the SEDOR destruction occurs most rapidly from the terms $\exp[iaT_S]$, $\exp[i2aT_S]$, $\exp[i3aT_S]$ for 1, 2, 3 S spins, so that the short-time behavior is characteristic of the number of S spins coupled to the I spin.

At long times, one cannot neglect the linebreadth. All three cases will lead to eventual destruction. This can be simulated by multiplying each exponential $\exp[iaT_S]$ (or $\exp[-iaT_S]$) in eq. (8.10), eq. (9.7b) or eq. (9.8b) by a Gaussian

$$\exp[-(2\tau/\tau_0)^2],$$

where τ_0 is determined by the spread Δa.

If a arises form a conventional dipole-dipole coupling, the line shape $f(a)$ is the famous Pake doublet powder pattern[22] when I and S are a pair of spin-1/2 nuclei.

10. – Examples of SEDOR studies.

Figure 7 shows an example of the application of SEDOR to the study of dilute magnetic alloys[23-25] by BOYCE *et al.* They studied dilute alloys of Co in Cu. Shown is the effect on the ^{59}Co signal produced of applying π pulses near the ^{63}Cu resonance frequency $\gamma_{63}H_0$. A direct observation of the ^{63}Cu NMR reveals a broad NMR line, located approximately at the position found for pure

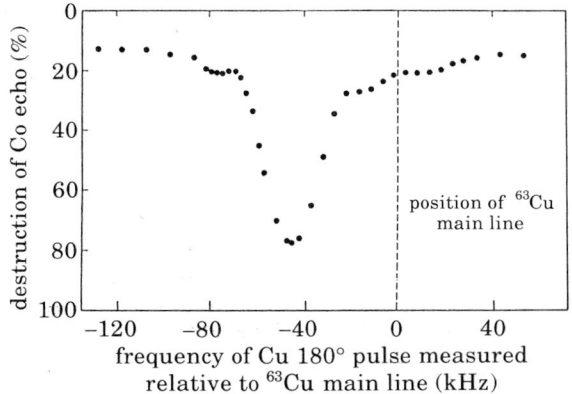

Fig. 7. – Boyce's observation of SEDOR in a Cu powder containing 0.54 at.% Co near the ^{63}Cu resonance frequency of pure Cu at 9907 G and 1.5 K, showing the satellite due to the $1/2 \leftrightarrow -1/2$ transition of the first neighbor. The parameters used are $\tau = 250$ µs, $H_1)_S \simeq$ $\simeq 5$ G for a π pulse of the Cu $1/2 \leftrightarrow -1/2$ transition. 100 echoes were averaged for each point and the dots are larger than the scatter and drift. Note that the much more abundant ^{63}Cu nuclei a long distance from the Co, the so-called Cu main line, do not show up in the Co SEDOR. (Reprinted with permission from ref.[24], © 1972 American Physical Society.)

Cu metal, but substantially broader owing to the magnetic susceptibility of the Co atoms which leads to an induced polarization of the conduction electrons via the RKKY mechanism (see ref. [1]). This broad ^{63}Cu NMR line is called the Cu «main line». Cu nuclei in the near vicinity of a Co atom have, however, a different shift. In this figure we see the effect of the Cu atoms which are the first neighbors of the Co atoms. Since there are 0.54 at.% Co, and 12 first neighbors, the Cu nuclei in that shell are only 6% of the total Cu in the sample.

Figure 8, from Boyce's thesis [23], shows the same line on a much broader frequency scale. Since ^{63}Cu is a spin-3/2 nucleus, the m_I 3/2-to-1/2 and the m_I − 1/2-to- − 3/2 transitions have a first-order quadrupole splitting, whereas the m_I 1/2-to- − 1/2 transition has only a second-order quadrupole effect. We see here the singularities at ± 105 kHz and ± 380 kHz from the first-order quadrupole effect, and the narrow line at 0 kHz arising from the + 1/2-to- − − 1/2 transition.

We were unable to detect the second or higher shells probably because the Cu nuclei in those shells could undergo mutual spin flips with more distant Cu shells, thus limiting the time that the ^{63}Cu nuclei could maintain their spin orientation.

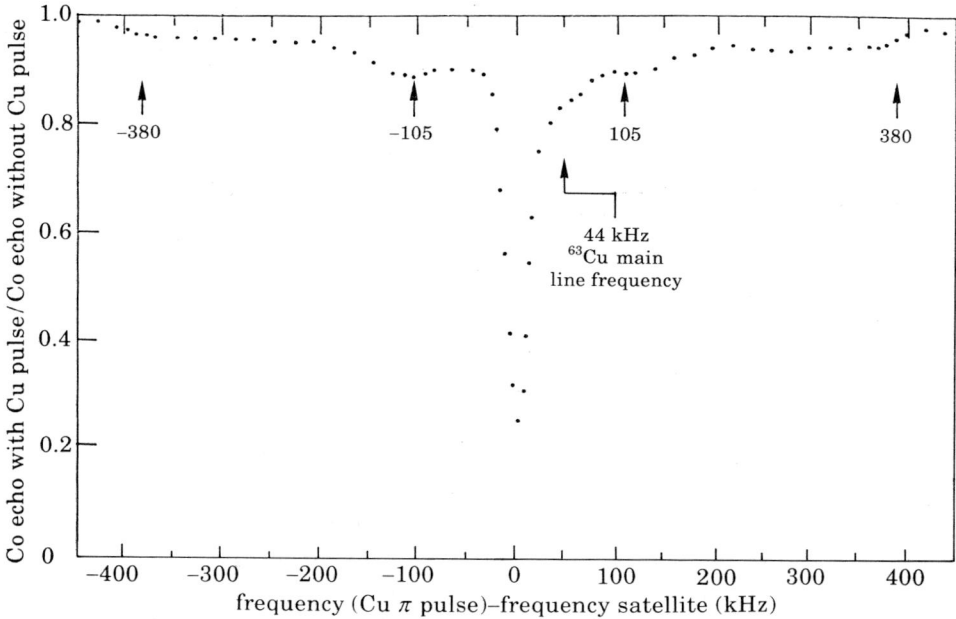

Fig. 8. – SEDOR spectrum of Cu-0.54% Co at 9907 G and 1.5 K, showing the powder pattern due to the satellite transitions of the first neighbors. The parameters used are $\tau = 350$ μs and $H_1)_S$ is about 5 G and is a π pulse for the 3/2 ↔ 1/2 transition. The horizontal scale is the frequency of the Cu pulse measured relative to the frequency of the large satellite of fig. 7. 100 echoes were averaged for each point and a few points were taken at each frequency.

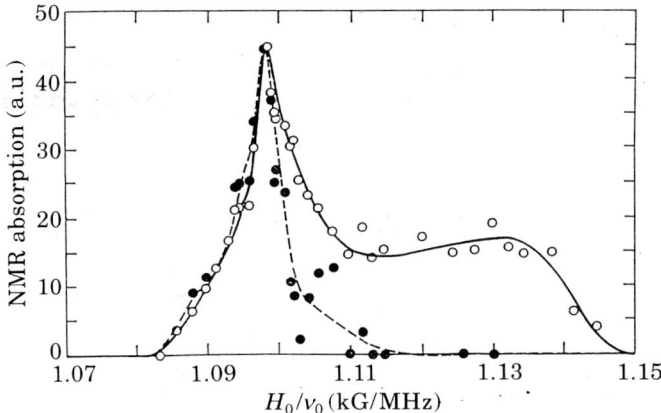

Fig. 9. – Measurement of Makowka et al. of the spin-echo (○) and SEDOR (●) line shapes of ^{195}Pt for a sample of small particles of Pt metal supported on Al_2O_3 whose surface is coated with ^{13}CO molecules. The metal particles have diameters of a few tens of Å. The ^{195}Pt spin echo gives the total line shape of all the ^{195}Pt in the particle. The SEDOR data, involving ^{195}Pt-^{13}CO double resonance, and an add-subtract method, give the NMR line shape of the ^{195}Pt nuclei in close proximity to the CO molecules, i.e. the surface layer of Pt atoms. (Reprinted with permission from ref.[26], © 1987 American Chemical Society.)

We now turn to some examples of applications of SEDOR to the study of metal surfaces. Figure 9 shows data of Makowka et al. [26] showing how by SEDOR they were able to resolve the resonance of the surface layer of Pt atoms on small metal clusters of Pt. The sample consisted of small metal clusters, perhaps 20 Å in diameter, attached to the surface of a fine powder of Al_2O_3. These samples are characteristic of real catalysts. The cluster size of the sample studied in fig. 9 was shown by hydrogen chemisorption to have 26% of all the Pt atoms in the surface. The ^{195}Pt NMR line shape as determined by spin echoes extends from H_0/ν_0 of about 1.084 to about 1.145, about 4.4 kG at the 8.4 T magnetic field used, a truly broad line! By comparing the line shape of samples with various size clusters, we realized that the absorption at $H_0/\nu_0 = 1.145$ arose from the center of the cluster since this is the position of the NMR line in bulk Pt metal. We also found that the peak near $H_0/\nu_0 = 1.10$ was relatively more prominent the smaller the cluster size. Since decreasing the cluster diameter increases the fraction of cluster atoms in the surface layer, we concluded that the peak near $H_0/\nu_0 = 1.10$ arose from atoms at or near the surface since it became relatively more prominent for samples with smaller-diameter clusters.

In order to isolate the contribution of the surface ^{195}Pt atoms, MAKOWKA absorbed CO enriched in ^{13}C on the Pt surface. These molecules bond to the surface at the C end. Thus the ^{13}C nuclear moment exerts a magnetic field on the ^{195}Pt surface atoms. The magnetic field exerted at the second and deeper layers

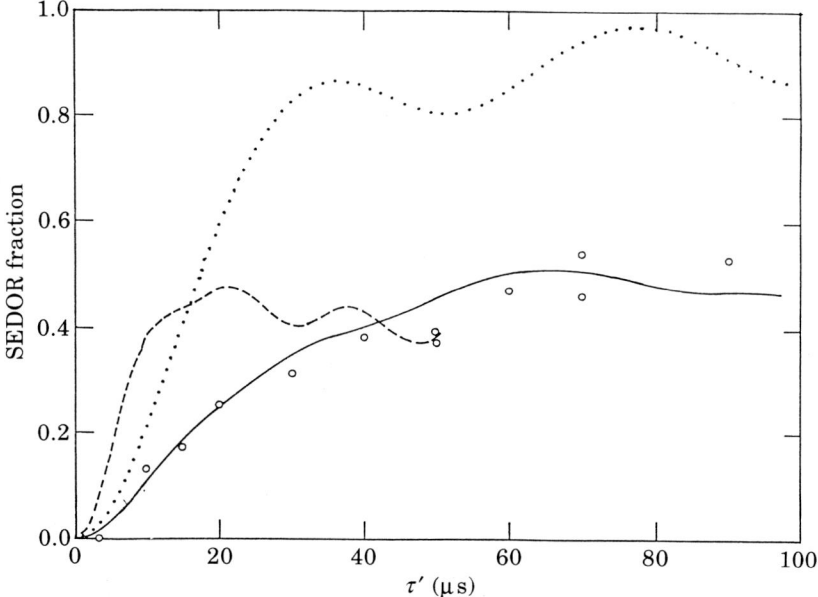

Fig. 10. – Bonding of hydrogen atoms to carbon in ethylene: The fractional destruction of the ^{13}C spin echo produced by flipping ^{1}H nuclei in a SEDOR experiment vs. the time, τ', of application of the ^{1}H pulse. Data at 77 K. Solid line: CCH$_3$ species with rapid CH$_3$ rotation about the C-C axis; dashed line: CCH$_3$ without CH$_3$ rotation; dotted line: CHCH$_3$ with rotating CH$_3$. (Reprinted with permission from ref.[27], © 1985 American Chemical Society.)

is much smaller owing to the $1/r^3$ form of the nuclear dipole magnetic coupling. MAKOWKA produced ^{195}Pt echoes in which on alternate echoes he applied π pulses to the ^{13}C nuclei. These pulses changed the echo height of the surface ^{195}Pt nuclei, but did not affect the echoes of deeper layers. By subtracting the echoes with ^{13}C pulses from the echoes without ^{13}C pulses, he canceled the signal from ^{195}Pt nuclei beyond the first layer, and was left with just the signal of the first layer.

Figure 10 shows the use of SEDOR to determine the structure of ethylene, C$_2$H$_4$, adsorbed at room temperature on Pt clusters. The data by PO-KANG WANG et al. [27] eliminate various competing structures, and show that at 77 K the CH$_3$ group is rotating rapidly.

The plots show the so-called SEDOR fraction vs. the time of the ^{1}H pulse, where the SEDOR fraction (SF) is defined as

$$\text{SF} = 1 - \frac{\text{echo height with SEDOR}}{\text{echo height without SEDOR}}.$$

Thus, if the echo completely destroys the SEDOR signal, SF goes to 1, where-

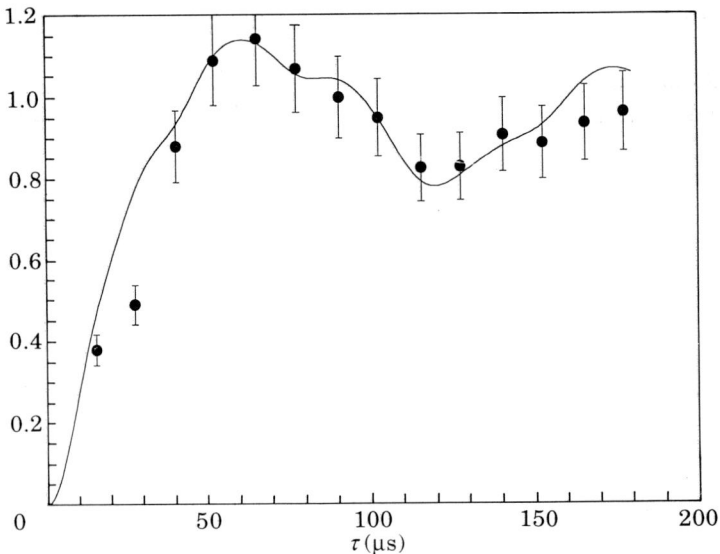

Fig. 11. – ^{13}C-^{1}H SEDOR signal for ^{13}CH$_3$I illustrating comparison of theoretical with an experimental curve. —— scaled CH$_3$ fit, ● CH$_3$I (SF). The vertical axis is scaled by 95% to obtain the fit, representing inhomogeneity in the ^{1}H flipping field. The time axis is scaled by 90% of its value assuming normal CH single-bond length and CH$_3$ tetrahedral-bond angles.

as, if the SEDOR flip produces *no* effect, SF = 0. The latter is the situation for short time.

Figure 11, obtained by SAKAIE and KLUG in my laboratory, shows the ^{13}C echo destruction in a ^{13}C-^{1}H SEDOR experiment at 77 K for the molecule ^{13}CH$_3$I to test the SEDOR theoretical expression against experiment for a molecule of known structure. If the alternating fields are not quite uniform, one does not get perfect π pulses throughout the sample. In this experiment, in which a single coil was used to excite both the ^{13}C and ^{1}H resonances, it was necessary to assume that the maximum SEDOR signal observed was 95% of the theoretical value for perfectly homogeneous alternating fields. One also needs to know the structure perfectly to get the time scale. To fit the data, SAKAIE and KLUG assumed that the HCH bond angle was that of a perfect tetrahedron, but that the CH bond length was 97% of the normal CH single-bond length.

Lastly, fig. 12 shows some theoretical SEDOR patterns calculated by SAKAIE and KLUG for 4 simple hydrocarbons adsorbed on Pt. Plotted is the SEDOR fraction for destruction of the ^{13}C signal by protons. Note first of all that one can tell immediately by the fact that the SEDOR fraction goes to a value near to 0.5 rather quickly for CCH and CCH$_2$ that only half the C are bonded to H atoms. For HCCH, the SEDOR fraction shoots up to near 1. The fact that it

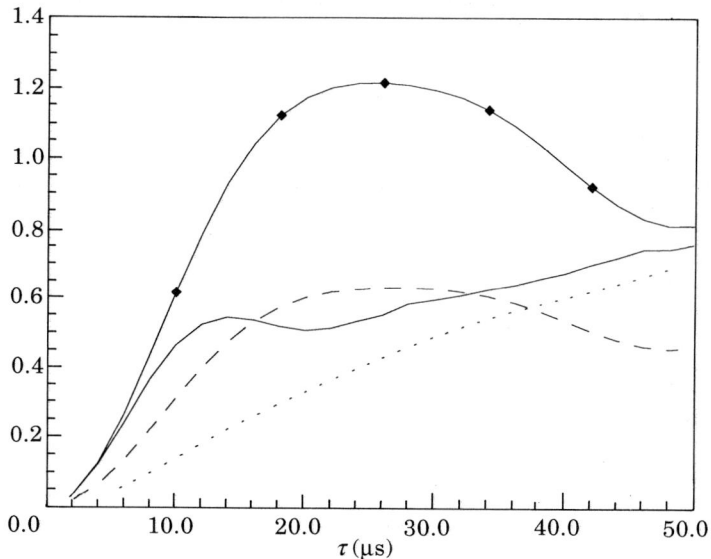

Fig. 12. – Theoretical SEDOR patterns for destruction of the ^{13}C echo by ^{1}H flips for 4 different molecules adsorbed on a metal surface: -♦- HCCH, —— CCH$_2$, --- CCH, ----- CCH$_3$. Both C nuclei are assumed to be ^{13}C (τ is the time at which the π pulse is applied to the hydrogen).

exceeds 1 shows that the ^{13}C echo is formed on the positive y_I-axis for τ about 25 μs.

Note that the first peak of the CCH$_2$ spectrum occurs at about 13 μs, half the time to reach the peak in the HCCH spectrum, representing the effect of flipping 2 hydrogen spins bonded to a single carbon in CCH$_2$, compared to 1 hydrogen bonded to a carbon in HCCH. The much slower rise of SF for CCH$_3$ results from the assumption that the CH$_3$ group is spinning about the C-C bond axis which reduces the ^{13}C-^{1}H dipolar coupling. Were the orientation of the CH$_3$ group fixed, the first peak would occur at even shorter τ than for CCH$_2$.

The slow rise for CCH$_2$ occurring beyond 20 μs comes from SEDOR destruction of the ^{13}C atom which is not directly bonded to the hydrogens. The time scale for that destruction is quite long owing to the great distance between the ^{13}C and ^{1}H atoms for that case.

SEDOR enables one to study the spin-spin couplings between unlike nuclei. The spin-spin couplings between like nuclei can be obtained from the envelope of the spin echo (the use of an echo eliminates magnetic-field broadening such as arises from chemical-shift anisotropy). Thus a combination of these two methods gives one reliable method of determining bond lengths and molecular structures from molecules if the molecule is not too complicated. For complex molecules in the solid, one can often resolve the resonances of various sites in a molecule by use of magic-angle spinning. This is especially true for nuclei such as ^{13}C. If one does coherence transfer from protons to ^{13}C, followed by proton decoupling, one can resolve the isotropic chemical shifts. GULLION and SCHAEFER [28] have invented a technique which combines SEDOR with magic-angle spinning to measure dipolar couplings such as ^{13}C-^{15}N nuclei and thus to obtain accurate bond distances. They call the technique rotational-echo double resonance (REDOR). In a system containing protons, they utilize Hartmann-Hahn cross-polarization to polarize natural-abundance ^{13}C nuclei of a spinning sample, holding the protons then decoupled by a ^{1}H decoupling field. The ^{13}C is then released to precess freely when its coupling field is turned off. π pulses are applied to the ^{15}N spins synchronously with the magic-angle spinning, two pulses per cycle, so that the ^{13}C precesses for different time intervals with the ^{15}N in one orientation as opposed to the other. Varying the relative time intervals enables them to deduce the strength of the ^{13}C-^{15}N coupling.

This approach to measuring distances (in a solid) has substantial advantages over the method employed for liquids of using nuclear Overhauser polarization information since the latter involves an unknown correlation time.

REFERENCES

[1] The treatment in this lecture is based on a much more extensive review in the third edition of the author's textbook: C. P. SLICHTER: *Principles of Magnetic Resonance*, 3rd edition (Springer-Verlag Publishers, Heidelberg, 1990).
[2] R. V. POUND: *Phys. Rev.*, **79**, 685 (1950).
[3] A. W. OVERHAUSER: *Phys. Rev.*, **91**, 476 (1953); **92**, 411 (1953).
[4] T. W. GRISWOLD, A. F. KIP and C. KITTEL: *Phys. Rev.*, **88**, 951 (1952).
[5] T. R. CARVER and C. P. SLICHTER: *Phys. Rev.*, **92**, 212 (1953); **102**, 975 (1956).
[6] C. D. JEFFRIES: *Phys. Rev.*, **106**, 164 (1957); **117**, 1056 (1960).
[7] A. ABRAGAM, J. COMBRISSON and I. SOLOMON: *C. R. Acad. Sci.*, **247**, 2237 (1958).
[8] N. BLOEMBERGEN: *Phys. Rev.*, **104**, 324 (1956).
[9] S. R. HARTMANN and E. L. HAHN: *Phys. Rev.*, **128**, 2042 (1962).
[10] N. BLOEMBERGEN and P. SOROKIN: *Phys. Rev.*, **100**, 865 (1958).
[11] F. M. LURIE and C. P. SLICHTER: *Phys. Rev.*, **133**, A1108 (1964).
[12] V. ROYDEN: *Phys. Rev.*, **97**, 543 (1955).
[13] A. L. BLOOM and J. N. SHOOLERY: *Phys. Rev.*, **97**, 1261 (1955).
[14] J. JEENER: *Lectures*, AMPERE *International Summer School, Basko Potze, Yugoslavia* (1971), unpublished.
[15] W. P. AUE, E. BARTHOLDI and R. R. ERNST: *J. Chem. Phys.*, **64**, 2229 (1976).
[16] R. FREEMAN and G. A. MORRIS: *Bull. Magn. Reson.*, **1**, 5 (1979).
[17] G. FEHER: *Phys. Rev.*, **103**, 500 (1956); G. FEHER and E. A. GERE: *Phys. Rev.*, **103**, 501 (1956).
[18] D. E. KAPLAN and E. L. HAHN: *J. Phys. Radium*, **19**, 821 (1958).
[19] M. EMSHWILLER, E. L. HAHN and D. KAPLAN: *Phys. Rev.*, **118**, 414 (1960).
[20] R. E. WALSTEDT and J. H. WERNICK: *Phys. Rev. Lett.*, **20**, 856 (1968).

[21] J. PH. ANSERMET, C. P. SLICHTER and J. H. SINFELT: *J. Chem. Phys*, **88**, 9 (1988).
[22] G. E. PAKE: *J. Chem. Phys.*, **16**, 327 (1948).
[23] J. B. BOYCE: Thesis, University of Illinois (1972).
[24] D. V. LANG, J. B. BOYCE, D. C. LO and C. P. SLICHTER: *Phys. Rev. Lett.*, **29**, 776 (1972).
[25] D. V. LANG, D. C. LO, J. B. BOYCE and C. P. SLICHTER: *Phys. Rev. B*, **9**, 3077 (1974).
[26] C. D. MAKOWKA, C. P. SLICHTER and J. H. SINFELT: *Phys. Rev. Lett.*, **49**, 379 (1982); *Phys. Rev. B*, **31**, 5663 (1985).
[27] P.-K. WANG, C. P. SLICHTER and J. H. SINFELT: *J. Phys. Chem. Lett.*, **89**, 3606 (1989).
[28] T. GULLION and J. SCHAEFER: *J. Magn. Reson.*, **81**, 196 (1989).

Explanatory note.

At the Enrico Fermi School, Prof. SLICHTER distributed both the manuscript of his lecture and also reprints of five short articles. These articles were selected because they described three aspects of double resonance and their first experimental verification: the Overhauser effect, spin decoupling and coherence transfer. They have value today as teaching aids because of the simplicity of their description of both the central phenomenon and its observation. They are reproduced here with the kind permission of the *Physical Review*, the journal in which they appeared.

Polarization of nuclear spins in metals (*)(**).

T. R. CARVER(†) and C. P. SLICHTER

Department of Physics, University of Illinois - Urbana, Ill.

When a metal such as lithium is placed in a static magnetic field, two sets of Zeeman levels are produced associated with the nuclear spins [1, 2] and the conduction electron spins [3, 4], respectively. The corresponding resonance transitions have been observed. With magnetic resonances in general it is found that, when the amplitude of the alternating magnetic field is increased, the population difference between the Zeeman levels decreases, a phenomenon known as saturation. For example, we have found previously [4] that, for the conduction

---

(*) Reprinted with permission from *The Physical Review*, **92**, 212 (1953), © 1953 American Physical Society.
(**) This work was supported in part by the U. S. Office of Naval Research.
(†) National Science Foundation Predoctoral Fellow.

electrons in lithium, an alternating field of some 5 G produces saturation. On the basis of a detailed calculation of the nucleus-electron interaction, OVERHAUSER [5] has predicted that for metals the saturation of the conduction electrons should simultaneously *increase* the population difference between the nuclear Zeeman levels by a factor of several thousand, and has proposed this as a method of polarizing nuclear spins. Since the strength of the nuclear-resonance absorption is proportional to the population difference between adjacent nuclear Zeeman levels, the nuclear resonance forms a convenient method of measuring the degree of nuclear alignment. We have verified Overhauser's theory by observing the enhancement of the nuclear resonance in metallic lithium produced by electron saturation.

The experiment was performed in a static magnetic field of 30.3 G provided by a small end-corrected solenoid. The sample containing 5 cm^3 of lithium dispersed in oil was placed in the tank coil of a 50 W oscillator operating at 84 MHz/s, the Larmor frequency for the electrons in the magnetic field. Measurements indicated an alternating field of about 4 G. The nuclear resonance was detected using a 50 kHz/s crystal-controlled oscillator and a twin-T bridge, the 50 kHz/s signal being converted to 600 kHz/s and detected in a communica-

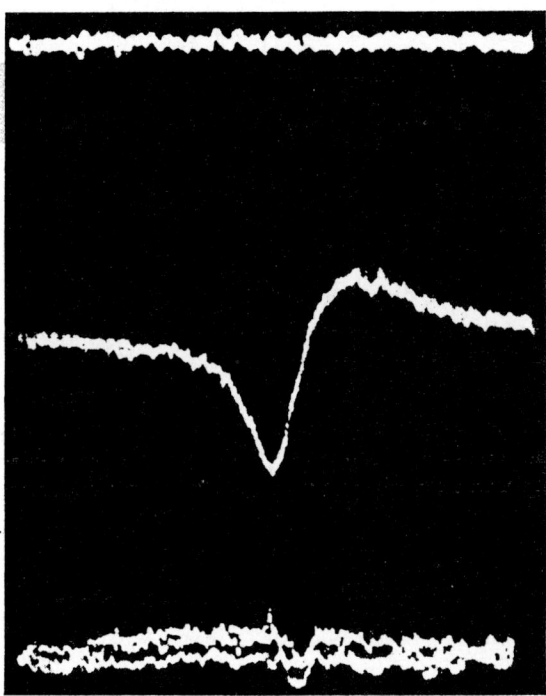

Fig. 1. – Oscilloscope pictures of 50 kHz/s nuclear-resonance absorption *vs.* static magnetic field. Field excursion 0.2 G. Top line: ^{7}Li resonance (lost in noise). Middle line: ^{7}Li resonance enhanced by electron saturation. Bottom line: proton resonance in glycerin sample.

tions receiver. The signal was observed on an oscilloscope or with a 30 c.p.s. lock-in amplifier. The r.f. tank coil of 1 turn, the 270-turn nuclear-resonance coil and the solenoid were oriented mutually perpendicular, and the array was placed in a copper box to shield the detection apparatus from r.f. radiation. The 84 MHz/s oscillator could be switched on or off without disturbing the bridge balance.

The accompanying oscilloscope photographs (fig. 1) summarize the results. The top line shows the appearance of the ordinary lithium nuclear resonance, which is so weak at these frequencies as to be completely lost in noise. The second line was photographed after the electron saturating oscillator was turned on. The ^{7}Li resonance now appears strongly. For comparison, the proton line in glycerin (also at 50 kHz/s) is shown in line three. In all three cases the amplifier gain settings and degree of bridge balance were identical. The very weak glycerin resonance, however, is produced by a sample of eight times the number of nuclei as the lithium. The enhancement of the lithium resonance, therefore, confirms Overhauser's theory strikingly.

On the basis of comparison with the glycerin resonance, we estimate the nuclear-population difference to be increased nearly 100-fold in our experiment. This figure is somewhat smaller than the maximum amount predicted, showing that either complete saturation was not achieved, or that other nuclear-relaxation processes short-circuit the alignment partially. One process we believe to be important is the relaxation produced by self-diffusion of the nuclei [2]. We have also observed a small enhancement in sodium under conditions of incomplete saturation. With both metals the saturating fields produced intense heating of the samples.

* * *

We are indebted to Dr. A. W. OVERHAUSER for interesting discussions and a most stimulating association during his stay at Illinois. Mr. R. SCHUMACHER assisted in construction of the equipment. The results of Mr. D. HOLCOMB and Dr. R. NORBERG on nuclear-relaxation times in lithium were helpful in searching for the nuclear resonance. We wish to thank Dr. R. M. BROWN for several useful suggestions concerning low-frequency nuclear resonance.

REFERENCES

[1] H. S. GUTOWSKY and B. R. MCGARVEY: *J. Chem. Phys.*, **20**, 1472 (1952).
[2] D. F. HOLCOMB and R. E. NORBERG: to be published.
[3] T. W. GRISWOLD, A. F. KIP and C. KITTEL: *Phys. Rev.*, **88**, 951 (1952).
[4] T. R. CARVER, D. F. HOLCOMB and C. P. SLICHTER: to be published.
[5] A. W. OVERHAUSER: *Phys. Rev.*, **89**, 689 (1953); **91**, 476 (1953); *Phys. Rev.* (to be published).

Measurement of the spin and gyromagnetic ratio of ^{13}C by the collapse of spin-spin splitting (*).

V. ROYDEN

Varian Associates - Palo Alto, Cal.

A precise and interesting measurement of the ratio of the resonance frequency of ^{13}C to that of ^{1}H has been obtained in the course of some experiments on indirect spin-spin interactions between nuclei in molecules [1-3]. The multiplet structure in the nuclear resonance of a given nucleus caused by spin-spin interaction with another nucleus of like or different atomic species can be reduced to a single line by irradiating the second nucleus with an r.f. magnetic field of its own resonance frequency [4, 5].

The transmitter section of a standard nuclear induction probe was modified so that it tuned to both 30 MHz/s and 7.5 MHz/s simultaneously. The proton spectrum of CH$_3$I enriched with 51 percent ^{13}C was observed at 30 MHz/s under « slow passage » conditions with the high-resolution spectrometer.

The oscilloscope trace showed three peaks, of which the central one was caused by those protons attached to ^{12}C (fig. 1). The two outer proton peaks were separated by 11 mG from the central peak as a result of the spin-spin interaction between ^{13}C and its companion ^{1}H nuclei. The number and relative amplitudes of the peaks give a spin of 1/2 for the ^{13}C nucleus, verifying the earlier result obtained by JENKINS from hyperfine-structure measurements [6].

An auxiliary transmitter was link-coupled into the transmitter section of the probe and its frequency varied slightly around 7.544 MHz/s. At the ^{13}C resonance frequency, the two side peaks of the 30 MHz/s proton spectrum coalesced with the central peak, which doubled in amplitude and became quite sharp (fig. 2). The frequency of the auxiliary transmitter at which this occurred and the frequency of the 30 MHz/s transmitter were then counted with a frequency counter.

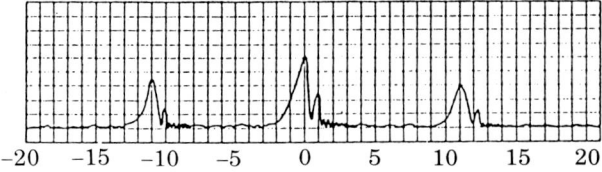

Fig. 1. – Proton spectrum at 30 MHz/s of CH$_3$I enriched with 51 percent ^{13}C. The two outside lines are split by the spin-spin interaction between ^{13}C and ^{1}H.

(*) Reprinted with permission from *The Physical Review*, **96**, 543 (1954), © 1954 American Physical Society.

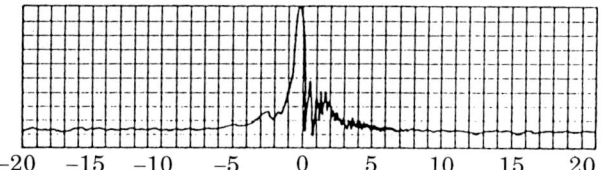

Fig. 2. – Proton spectrum at 30 MHz/s of CH_3I enriched with 51 percent ^{13}C with strong $7\frac{1}{2}$ MHz/s r.f. coupled into the transmitter. The side peaks are almost completely collapsed.

A comparison of the resonant frequency of ^{13}C in CH_3I to that of ^1H in the same molecule gave

$$\nu(^{13}C)/\nu(^1H) = 0.251\,443\,1 \pm 0.000\,000\,5.$$

This result is in excellent agreement with that previously obtained by Poss[7]. The proton resonances in CH_3I and in mineral oil occur, within experimental error, at the same frequency; hence a direct comparison of the above result with that of Poss is justified.

The ^{13}C line is a quadruplet, being split into $2I + 1$ equally spaced lines, where I refers to the total spin of $3/2$ for the three protons in CH_3I. Since the side peaks of the proton resonance split out symmetrically as a function of frequency as the auxiliary oscillator is tuned above and below the coalition frequency, that frequency was assumed to be at the center of the ^{13}C quadruplet.

It is interesting to note that although the 7.5 MHz/s r.f. field had a peak-to-peak amplitude of 0.75 G, a shift in frequency of 30 to 40 Hz was sufficient to broaden the central peak markedly and to decrease its amplitude proportionately. This is more sensitive by a factor of ten than one is led to expect on the basis of r.f. line broadening. The crystal-controlled auxiliary oscillator could be varied about 300 Hz above and below the ^{13}C frequency, a variation sufficient for the irradiated doublet to be in part reconstituted. Nine measurements of the ^{13}C and ^1H frequencies in CH_3I were made, the maximum discrepancy between any two ^{13}C measurements being 20 Hz.

The magnetic field was swept with a 50 mG recurrent sawtooth sweep, and it was suggested that the measurement of the frequency ratio might be in question by the amount by which the position of the proton resonance in the magnetic field was undetermined. The uncertainty was minimized by keeping the sweep frequency low enough to ensure «slow passage» through the resonance.

The incorporation of both the nucleus to be measured and the reference nucleus into a single sample, and the measurement of both resonance frequencies with a single coil give assurance that any field difference at the two nuclei must be a chemical shift inherent in the molecule since all external experimental factors are the same. A merit of this technique is that it was possible to determine the ^{13}C frequency by observing a sensitive change in the strong narrow ^{1}H resonance with a high-resolution spectrometer. No sweep field calibration is involved, the measurement being reduced to the simultaneous determination of two frequencies. Both ^{1}H and ^{19}F give strong sharp lines and make excellent companions for more refractory nuclei.

* * *

The author wishes to thank C. REILLY of Shell Development Research Laboratories, Emeryville, California for supplying the sample of ^{13}CH$_3$I used in this experiment.

REFERENCES

[1] N. F. RAMSEY and E. M. PURCELL: *Phys. Rev.*, **85**, 143 (1952).
[2] E. L. HAHN and D. E. MAXWELL: *Phys. Rev.*, **84**, 1246 (1951).
[3] H. S. GUTOWSKY, D. W. MCCALL and C. P. SLICHTER: *Phys. Rev.*, **84**, 589 (1951).
[4] J. SHOOLERY and M. PACKARD: to be published.
[5] F. BLOCH, J. T. ARNOLD and W. A. ANDERSON have reported by private communication the application of this technique to the protons in ethyl alcohol.
[6] F. A. JENKINS: *Phys. Rev.*, **72**, 169 (1947).
[7] H. L. POSS: *Phys. Rev.*, **75**, 600 (1949).

Effects of perturbing radiofrequency fields on nuclear-spin coupling (*).

A. L. BLOOM and J. N. SHOOLERY

Varian Associates - Palo Alto, Cal.

1. – Introduction.

When a molecule in a liquid contains several nuclei that have different Larmor frequencies in a given magnetic field, it is often found that the magnetic-resonance line for a given nucleus is split into a multiplet whose separation is independent of the applied field [1]. The origin of this line splitting has been discussed by RAMSEY [2] and by RAMSEY and PURCELL [3], who show that a coupling between nuclei transmitted via the surrounding electron cloud produces an $I_1 \cdot I_2$ type of interaction. Recently, experiments have been performed in this

---

(*) Reprinted with permission from *The Physical Review*, **97**, 1261 (1955), © 1955 American Physical Society.

laboratory and at Stanford University in which the spin-spin interaction has been disturbed by the application of a strong r.f. in the vicinity of resonance of one of the nuclei [4-6]. If the r.f. is strong enough, the spin-spin coupling can be completely destroyed, but for intermediate r.f. amplitudes the multiplet structure of the spectrum from the undisturbed nucleus often increases in complexity. The purpose of this paper is to discuss the more complex multiplet structure and its behavior as a function of the disturbing r.f.

2. – Theory.

We shall consider the simplest possible case, that of a system of two nuclei with different gyromagnetic ratios γ_1 and γ_2, respectively. A steady magnetic field H_0 is applied in the z direction which, in the absence of any spin-spin interaction, would provide Larmor angular frequencies $\gamma_1 H_0$ and $\gamma_2 H_0$ for the respective nuclei. The γ's are here defined to be observed gyromagnetic ratios and thus include the effects of any chemical shifts (*i.e.* induced local fields) that may be present. With the spin-spin interaction, the Hamiltonian for the system is

$$(1) \qquad \mathscr{H} = -\hbar[\gamma_1(\mathbf{I}_1 \cdot \mathbf{H}) + \gamma_2(\mathbf{I}_2 \cdot \mathbf{H}) + J(\mathbf{I}_1 \cdot \mathbf{I}_2)].$$

We are concerned with the usual experimental conditions of nuclear magnetic resonance where H is predominantly in the z direction, the Larmor frequencies and their difference are of the order of MHz/s, and $J/2\pi$ is a kHz/s or less. Under these conditions only the expectation values for spin in the z direction are effective in the spin-spin interaction, and the allowed transitions between energy levels of eq. (1) occur at $\gamma_1 H_0 + m_2 J$, $\gamma_2 H_0 + m_1 J$ ($m = I, I-1, ..., -I$); thus each of the original Larmor frequencies is split into a field-independent multiplet. We now propose to do the following experiment: A strong r.f. magnetic field H_2, rotating in the (x, y)-plane with angular frequency ω_2 in the vicinity of $\gamma_2 H_0$ is impressed on nucleus 2; simultaneously the transitions in the vicinity of $\gamma_1 H_0$ are investigated by producing resonance with a weak r.f. field H_1 whose angular frequency is ω_1. The field H_1 is to be weak enough so that it does not appreciably affect the line width of the observed transitions.

The problem is most easily solved by transforming to a coordinate system rotating with H_2. In the rotating frame of reference, terms in the Hamiltonian including H_2, H_0 and $\mathbf{I}_1 \cdot \mathbf{I}_2$ are time-independent and the effect of H_1 can be considered as a perturbation inducing transitions between otherwise well-defined energy levels. Since the effect of H_2 and the spin-spin coupling term is at best only a small perturbation on the precession of the spin of nucleus 1, the resonance radiation will be predominantly polarized in the (x, y)-plane and spectrum frequencies calculated in the rotating frame can be transferred to the laboratory frame merely by adding to them the factor $\omega_2/2\pi$.

The dynamics of magnetic-resonance problems in a rotating coordinate system has been discussed by RABI, RAMSEY and SCHWINGER [7] and by others. For a nucleus with gyromagnetic ratio γ_i, the H_0 field in the z direction must be replaced by the «effective» z component $(H_0 - \omega/\gamma_i)$. Thus the Hamiltonian for our problem, including only stationary terms, can be described in the rotating frame as

$$(2) \quad \mathcal{H}' = -\hbar[\mathbf{I}_1 \cdot \mathbf{k}(\gamma_1 H_0 - \omega_2) + \mathbf{I}_2 \cdot \mathbf{k}(\gamma_2 H_0 - \omega_2) +$$
$$+ \mathbf{I}_1 \cdot \mathbf{i}(\gamma_1 H_2) + \mathbf{I}_2 \cdot \mathbf{i}(\gamma_2 H_2) + J\mathbf{I}_1 \cdot \mathbf{I}_2].$$

For simplicity, and because it is obviously valid in the experimental case considered here, we shall assume the first term large compared to the others involving $\mathbf{I}_1$. This is equivalent to saying that the nucleus-1 spin is quantized only in the z direction. Spin 2, on the other hand, is quantized in the direction of its effective field, at an angle Θ from the z-axis. To determine Θ we consider $J\mathbf{I}_1 \cdot \mathbf{I}_2$ in terms of an equivalent magnetic field acting at the site of the nucleus. Refer to fig. 1. We have

$$(3) \quad a(m_1) = [(\gamma_2 H_0 + Jm_1 - \omega_2)^2 + \gamma_2^2 H_2^2]^{1/2},$$

$$(4) \quad \cos\Theta(m_1) = (\gamma_2 H_0 + Jm_1 - \omega_2)/a(m_1),$$

$$(5) \quad \sin\Theta(m_1) = \gamma_2 H_2/a(m_1);$$

then the energy levels in the rotating frame are

$$(6) \quad W(m_1, m_2) = \hbar[m_1(\gamma_1 H_0 - \omega_2) + m_2(\gamma_2 H_0 - \omega_2 + m_1 J) \times$$
$$\times \cos\Theta(m_1) + m_2 \gamma_2 H_2 \sin\Theta(m_1)],$$

where m_1 is the z component of spin of nucleus 1 and m_2 is now the nucleus-2 spin component in the direction of the effective field.

Restricting ourselves for the time being to $I_1 = I_2 = 1/2$, let us call the

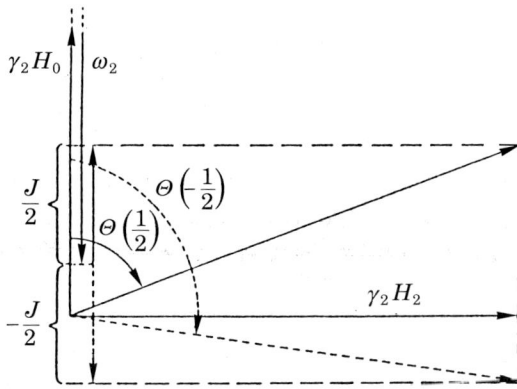

Fig. 1. – Vector diagram of forces acting on I_2 in the rotating coordinate system, showing the effect of different orientations of I_1. The special case shown is $I_1 = I_2 = 1/2$.

$W(m_1 = +1/2)$ initial states and the $W(m_1 = -1/2)$ final ones. Of the four possible transitions only two are allowed by selection rules in the unperturbed case where $H_2 = 0$. In our experiment, however, all four transitions may be observable because Θ changes during the transition, as indicated in fig. 1; thus the final states of m_2 may not be orthogonal to the initial ones. The relative transition probability for a single nucleus is given by the square of the matrix element connecting initial and final states of nucleus 2, and for $I_2 = 1/2$ it is

$$P = \cos^2(\xi/2), \qquad (7)$$

where ξ is the angle between initial and final states. However, the observed line intensity depends not only on P but also on the population distribution among the various energy levels, and the presence of H_2 changes the distribution from the usual Boltzmann distribution among the energy levels. The type of spin-spin interaction considered here generally does not enter into the thermal-relaxation process to any significant extent, and the population distribution can then be calculated as if the two species of nuclei were completely separate and noninteracting. Thus the total spin population for each value of m_1 is governed by the Boltzmann distribution, but at each value of m_1 the population difference between adjacent levels of m_2 is proportional, not to the Boltzmann distribution value M_0, but to M_z, where M_z is the «slow passage» population difference [8] and is a function of H_2 and Θ. An example of a calculation making use of this change in distribution is given at the end of sect. 4.

Recently, OVERHAUSER [9] has shown that the hyperfine coupling between electronic and nuclear spins in metals provides a means for greatly enhancing the net nuclear polarization in double-resonance experiments, and BLOCH [4] and KORRINGA [10] have predicted a similar effect in nonconducting paramagnetic substances. In order that the «Overhauser effect» take place it is necessary that the primary means of thermal relaxation available to the nucleus (here equivalent to our nucleus 1) be through the spin-spin coupling. However, in the experimental case considered here the thermal-relaxation process for each species of nucleus is almost entirely through direct interaction with the lattice, as the J-coupling plays an insignificant role; thus it follows that an enhancement of polarization analogous to that predicted by OVERHAUSER will *not* take place.

In the general case, including spins greater than $1/2$, the spin-spin multiplet in the absence of H_2 contains $2I_2 + 1$ lines. When H_2 is applied, the selection rule ($\Delta m_2 = 0$ in the laboratory frame) no longer holds strictly, as explained above, and in addition transitions starting from different m_1 levels may no longer be superimposed, since eq. (6) is not a linear function of m_1. Thus there may be as many as $(2I_1)(2I_2 + 1)^2$ lines in the multiplet when H_2 is present. In addition, it appears possible that the selection rule ($\Delta m_1 = \pm 1$)

is partially broken down by the effect of H_2, making possible the observation of multiple-quantum transitions.

3. – Calculations.

We present some calculated results for the simple case of two nuclei each of spin 1/2, showing how the multiplet varies as a function of experimental conditions. For simplicity we define the four possible transitions by subscripts a, b, c, d and the energies of transition in the laboratory frame as follows:

$$(8a) \qquad \Delta W_a = W\left(\frac{1}{2}, \frac{1}{2}\right) - W\left(-\frac{1}{2}, \frac{1}{2}\right) + \hbar\omega_2 ,$$

$$(8b) \qquad \Delta W_b = W\left(\frac{1}{2}, -\frac{1}{2}\right) - W\left(-\frac{1}{2}, -\frac{1}{2}\right) + \hbar\omega_2 ,$$

$$(8c) \qquad \Delta W_c = W\left(\frac{1}{2}, \frac{1}{2}\right) - W\left(-\frac{1}{2}, -\frac{1}{2}\right) + \hbar\omega_2 ,$$

$$(8d) \qquad \Delta W_d = W\left(\frac{1}{2}, -\frac{1}{2}\right) - W\left(-\frac{1}{2}, \frac{1}{2}\right) + \hbar\omega_2 .$$

Case 1. $\omega_2 = \gamma_2 H_0$ (center of nucleus-2 doublet). ω_1 and amplitude of H_2 variable.

In this case $\Theta(-1/2) = -\Theta(1/2)$. The transition angular frequencies become

$$(9) \qquad \begin{cases} \Delta W_a/\hbar = \Delta W_b/\hbar = \gamma_1 H_0 , \\ \Delta W_c/\hbar = \gamma_1 H_0 + \frac{1}{2}J' , \\ \Delta W_d/\hbar = \gamma_1 H_0 - \frac{1}{2}J' , \end{cases}$$

where

$$(10) \qquad J' = (J^2 + 4\gamma_2^2 H_2^2)^{1/2} .$$

The relative transition probabilities are

$$(11) \qquad \begin{cases} P_a = P_b = 4\gamma_2^2 H_2^2/(J')^2 , \\ P_c = P_d = J^2/(J')^2 . \end{cases}$$

When $H_2 = 0$, $J' = J$ and transitions c and d form the doublet [11]. As the r.f. level H_2 is increased, the original doublet lines are spread apart and weakened; simultaneously a new line appears at the center frequency $\gamma_1 H_0$ and grows at the expense of the doublet. When $\gamma_2 H_2 \gg J$ there is only the single line due to transitions a and b, twice as intense as each of the original doublet lines. This result for strong H_2 agrees with the one obtained by using the concept of an «averaging out» of the spin-spin interaction due to rapid transitions in the laboratory frame.

Case 2. $\omega_2 = \gamma_2 H_0$, ω_1 fixed at $\gamma_1 H_0$, a sweep field ΔH is used to investigate the spectrum.

This case is generally more amenable to study in the laboratory than case 1. By substituting $H_0 + \Delta H$ for H_0 in eq. (6) and solving for transitions at $\Delta W = \hbar \omega_1$ in eq. (8), we obtain lines at

$$\Delta H = 0, \tag{12a}$$

$$(\Delta H)^2 = \frac{J^2(\gamma_1^2 - \gamma_2^2) + 4\gamma_1^2 \gamma_2^2 H_2^2}{4\gamma_1^2(\gamma_1^2 - \gamma_2^2)}. \tag{12b}$$

The intensity of the line at $\Delta H = 0$ is given by P_a in eq. (11); for the other lines it must be determined with the aid of eq. (7). If $\gamma_1 > \gamma_2$, the results are similar to those of case 1. If $\gamma_1 < \gamma_2$, the line structure collapses with increasing H_2, giving only a single line for $4\gamma_2 H_2 \geqslant J$.

Case 3. $\gamma_2 H_2 \gg \frac{1}{2} J$, ω_1 and ω_2 variable.

By $\gamma_2 H_2 \gg \frac{1}{2} J$, we imply that to good approximation $\Theta(1/2) = \Theta(-1/2) = \Theta_0$, where Θ_0 is the effective field angle on nucleus 2 in the absence of any spin-spin interaction. In this case there are no more than two lines, located at

$$\Delta W/\hbar = \gamma_1 H_0 \pm \frac{1}{2} J \cos \Theta_0. \tag{13}$$

4. – Experimental results.

To check the theory, an experiment was performed on the spin-spin interaction between ^{19}F and ^{31}P, with the former taken to be nucleus 1. Both nuclei have spin 1/2 and their gyromagnetic ratios differ by more than a factor of two, thus the approximations used in the theory are valid to a high order of accuracy. The molecule studied was Na_2PO_3F in aqueous solution; since the Na atoms are ionically bonded or ionized in solution, they do not contribute to the spin interaction. The field-independent splitting of the fluorine resonance in this molecule has been reported previously by GUTOWSKY, MCCALL and SLICHTER [1].

The apparatus used was a Varian Associates V-4300 High Resolution Spectrometer and V-4012 Electromagnet operating at about 7500 G. The transmitter coil input circuit was modified so it was resonant at both the fluorine Larmor frequency of 30.00 MHz/s and the phosphorous frequency of 12.91 MHz/s. An auxiliary oscillator and power amplifier provided r.f. power for the H_2 field at the latter frequency. Both oscillators were crystal controlled; the one providing

H_2 was tunable over a range of several hundred cycles. The receiver coil and amplifiers were tuned only to the 30 MHz/s fluorine signal. The V-4300 Spectrometer utilizes a slow linear magnetic-field sweep and the signal intensity is displayed directly on the oscilloscope or recording meter.

Before the splitting could be studied as a function of H_2, the interaction factor J had to be determined. An audiofrequency sine wave modulation of the magnetic field was superimposed on the sawtooth sweep with the result that satellites were produced on each side of each doublet line. The satellites occur because a field modulation is in every way equivalent to a frequency modulation of the r.f. insofar as the equations of motion of the nuclear ensemble are concerned [8]. The satellites thus have positions and intensities determined by the «sidebands» of the equivalent frequency modulation. The splitting $J/2\pi$ was thus determined as twice the value of the audiofrequency which superimposed the first satellites from each line in the center of the doublet. This measurement gave $J/2\pi = (860 \pm 4)$ Hz, which when converted to equivalent magnetic-field difference gives a splitting of 0.214 G. The value obtained by GUTOWSKY *et al.* [1] by measurement of the field difference was 0.195 G. We have not ex-

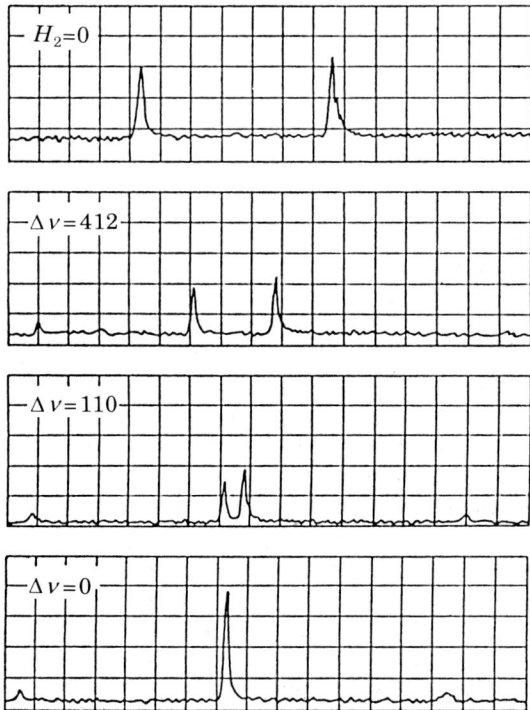

Fig. 2. – Effect on the fluorine spectrum of varying the phosphorous radiofrequency, with H_2 fixed at 0.464 G. The top trace shows the unperturbed spectrum ($H_2 = 0$) for comparison. We define $\Delta\nu = (\gamma_2 H_0 - \omega_2)/2\pi$.

plored the reasons for the discrepancy between our results and those of Gutowsky, but it should be pointed out that the accuracy of our method depends only on the calibration of the audio-oscillator and on the linewidth. In our measurements the audiofrequency was counted directly on a Hewlett-Packard 524-A frequency counter, and with sufficiently slow sweep the two satellites were clearly resolved if the frequency was varied by 5 Hz on either side of its center value.

When the perturbing r.f. field was turned on at maximum amplitude, the doublet separation was found to be a function of ω_2, as expected. This is illustrated in fig. 2. The angular frequency ω_2 was taken to be in the center of the phosphorous doublet when the two lines coalesced into one, and the deviations $\Delta\nu$ in fig. 2 were measured from this value. The frequencies were determined by counting the crystal fundamental frequency with the 524-A counter. Owing to the existence of chemical shifts which are not known to sufficient accuracy, one cannot set ω_2 for $\Delta\nu = 0$ in advance of actually seeing the spectrum.

Figure 3 shows the relative separation as a function of ω_2, compared with the theoretical separation calculated with the aid of eq. (13). In this calculation $\gamma_2 H_2$ was taken to be 5000. However, it should be pointed out that H_2 was not strong enough to justify entirely the approximation $\theta = \theta_0$; this is evident in fig. 2 from the fact that all four lines are visible instead of only the center two. A more exact calculation shows that the asymmetry of the intensities in some of the traces is a real effect and the direction of asymmetry can be reversed by changing the sign of $\Delta\nu$; this has been verified experimentally.

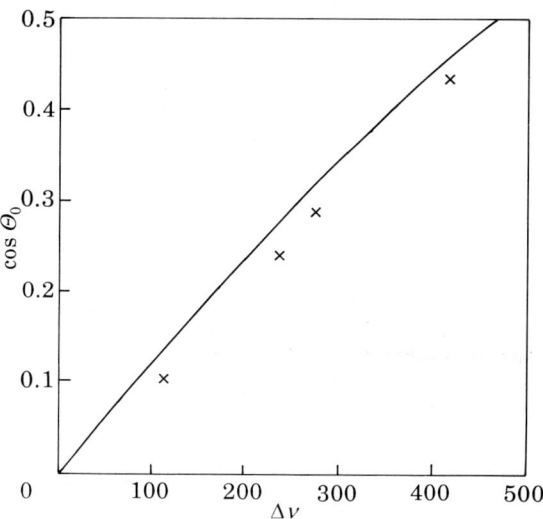

Fig. 3. – Measured and calculated splitting as a function of the perturbing-field frequency deviation, $\Delta\nu$, from the «center» frequency. In the calculation, we assume $\gamma_2 H_2 = 5000$.

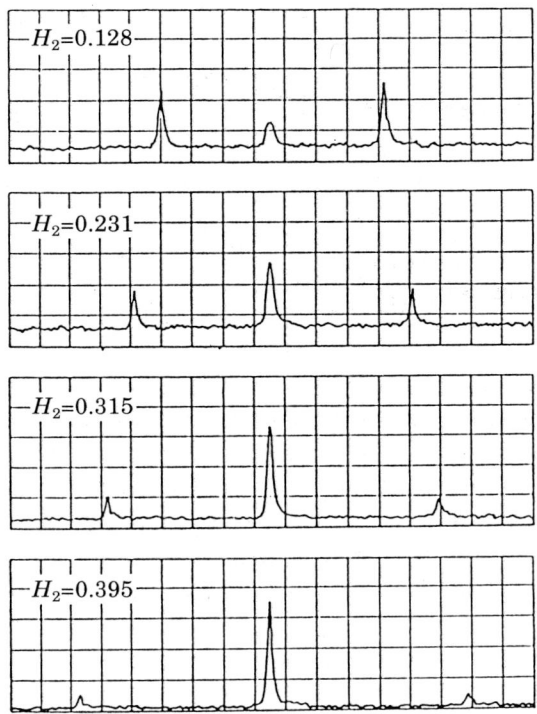

Fig. 4. – Effect on the fluorine spectrum of varying the phosphorous r.f. amplitude H_2, with $\omega_2 = \gamma_2 H_0$. The top and bottom traces of fig. 2 are also part of this series.

With $\Delta \nu$ set at zero, the amplitude of H_2 was varied and typical results were as shown in fig. 4. At each setting of the H_2 attenuator, the alternating field $2H_2 \cos(\omega_2 t)$ was investigated by means of a small search coil placed in the probe and connected to an electronic voltmeter. The search coil measurements were subject to large systematic errors owing to uncertainties in the effective area of the coil, and in addition were not very reproducible because it was difficult to place the coil in exactly the same spot each time, but the results indicate the order of magnitude of the field. Measurements were taken on the spectrometer traces of the relative separation of the outside lines (with the separation at $H_2 = 0$ taken as unity), and of the ratio of the center line peak amplitude to the average amplitude of the outside lines.

From the relative line splitting, we calculated H_2 with the aid of eq. (12b) and compared it with the search coil measurements. Assuming the calculated H_2 to be correct, we calculated the relative intensity of the center line (eq. (11)) and the outside lines, obtained by first calculating $\theta(1/2)$ and $\theta(-1/2)$. To these intensities a correction had to be made for changes in the population distribution. From the argument given in sect. 2 it is easy to show that for the line at $\Delta H > 0$ the ratio of the true to the uncorrected line intensity is given by

$$(14) \qquad 1 - (\gamma_2/2\gamma_1 M_0)\left[M_z\left(m_1 = \frac{1}{2}\right) - M_z\left(m_1 = -\frac{1}{2}\right)\right],$$

with an analogous expression for the other outside line (the center line requires no correction). An exact expression for M_z requires a knowledge of T_1 and T_2, which we did not have; however, under the circumstance that $\gamma_2^2 H_2^2 T_1 T_2 \gg 1$ we could use, to excellent approximation,

$$(15) \qquad M_z = M_0 \cos^2 \Theta.$$

The correction was relatively small, amounting at most to about 15 percent.

The results of the measurements and calculations are summarized in table I. The line intensities are summarized as the ratios of the center to the average outside line amplitudes; uncertainties are based on the estimated r.m.s. noise level. In all calculations we used $\gamma_1 = 2.52 \cdot 10^4$, $\gamma_2 = 1.083 \cdot 10^4$ and $J = 5400$.

TABLE I. – *Relative line splitting and intensity as a function of the perturbing-field amplitude.*

Relative splitting measured	H_2 (G) from search coil	H_2 (G) from splitting	Intensity ratio, measured	Intensity ratio from splitting
1.00	0	0	0	0
1.15	0.13	0.128	0.40 ± 0.03	0.60
1.26	0.18	0.172	0.85 ± 0.05	1.21
1.43	0.26	0.231	1.76 ± 0.08	2.37
1.72	0.34	0.315	4.45 ± 0.33	4.83
1.90	0.39	0.364	6.30 ± 0.63	6.38
2.02	0.42	0.395	8.5 ± 1.1	7.69
2.09	0.49	0.414	10.9 ± 1.7	8.50
2.24	0.48	0.451	14.8 ± 3.0	10.2
2.29	0.49	0.464	12.0 ± 1.2	10.7
2.30	0.51	0.466	18.5 ± 4.6	10.9

5. – Discussion.

A comparison of the experimental and calculated results in fig. 3 and table I seems to indicate that the concept of a Hamiltonian which is made stationary by transforming to a rotating coordinate system provides an adequate working model for prediction of the spectrum in magnetic double-resonance experiments. Since the lattice motions have about the same appearance in the rotating frame as in the laboratory frame, one would also predict that there should be no gross changes in linewidth due to the presence of the perturbing r.f.; this is also borne out by the experiment. Although it is probable that there are small changes in natural linewidth owing to changes in the matrix elements which affect the relaxation probability [12], such changes would not be observable in our experiments because other work has shown that the observed linewidth is due almost entirely to instrumental effects, *i.e.* H_0 field inhomogeneity and rate of sweep.

The intensity measurements were based entirely on peak height, rather than on area, because it is much more difficult to measure the area accurately. It is, therefore, evident that any instrumental effects which preferentially broaden either the center or the outside lines will introduce systematic discrepancies between the observed and the calculated intensity ratios. Thus, if H_2 were not set exactly in the center of the phosphorous doublet, then the center line would in reality be a closely spaced doublet and an amplitude measurement would not give the full intensity. Such an effect, which is most important when H_2 is small, is noticeable in the traces of fig. 4 and almost certainly accounts for the systematic discrepancies in table I for values of H_2 less than 0.3 G. There is apparently a systematic discrepancy at large values of H_2 whose origin is not completely understood; however, it can be shown that an inhomogeneity in the H_2 field of the order of one percent would broaden the outside lines enough to account for the observed discrepancy. From the known geometry of the probe, one would expect inhomogeneities of this order of magnitude to be present.

The data of this experiment, and other data taken in H_2 fields as high as 0.7 G, show that in the limit of very large H_2 the intensity of the center line becomes exactly twice the height of the original doublet lines, within experimental error. Had an enhancement of the type predicted by OVERHAUSER [9] been present, the center line intensity would have been increased still further by a factor $1 + |\gamma_2/\gamma_1|$ or about 1.4. The absence of even a partial enhancement agrees with the statements made in sect. 2 regarding the absence of the Overhauser effect in connection with the indirect nuclear spin-spin coupling.

* * *

We wish to thank Dr. M. E. PACKARD for advice and encouragement, and V. ROYDEN for assistance in performing the experiment. The Na_2PO_3F was kindly lent by Ozark-Mahoning Company, Tulsa, Oklahoma. We are indebted to Prof. F. BLOCH for several stimulating discussions on this subject.

REFERENCES

[1] H. S. Gutowsky, D. W. McCall and C. P. Slichter: *J. Chem Phys.*, **21**, 279 (1953). References to earlier experimental work are given here.
[2] N. F. Ramsey: *Phys. Rev.*, **91**, 303 (1953).
[3] N. F. Ramsey and E. M. Purcell: *Phys. Rev.*, **85**, 143 (1952).
[4] F. Bloch: *Phys. Rev.*, **93**, 944 (1954).
[5] V. Royden: *Phys. Rev.*, **96**, 543 (1954).
[6] For other experiments involving double resonance see J. Brossel and F. Bitter: *Phys. Rev.*, **86**, 308 (1952); T. R. Carver and C. P. Slichter: *Phys. Rev.*, **92**, 212 (1953).
[7] I. I. Rabi, N. F. Ramsey and J. Schwinger: *Rev. Mod. Phys.*, **26**, 167 (1954).
[8] F. Bloch: *Phys. Rev.*, **70**, 460 (1946).
[9] A. W. Overhauser: *Phys. Rev.*, **92**, 411 (1953).
[10] J. Korringa: *Phys. Rev.*, **94**, 1388 (1954).
[11] Note that in the limit $H_2 = 0$ the effective field reverses direction during a transition, for $\omega_2 = \gamma_2 H_0$. Thus it requires a double transition (m_1 and m_2 both changing) in the rotating frame to agree with the selection rule ($\Delta m_2 = 0$) in the laboratory frame.
[12] J. P. Lloyd and G. E. Pake: *Phys. Rev.*, **94**, 579 (1954).

Method of polarizing nuclei in paramagnetic substances (*).

G. Feher

Bell Telephone Laboratories - Murray Hill, N.J.

Overhauser [1] has shown that a saturation of the electron spin resonance leads to a large enhancement of the nuclear polarization. A necessary condition for this enhancement is that the nuclei relax via the electrons whose resonance is being saturated. A scheme which is applicable to substances which exhibit resolved hyperfine lines was proposed by Bardeen, Slichter and Pines [2]. It requires that the predominant relaxation process for the nuclei results from a modulation of the $a(\mathbf{I} \cdot \mathbf{S})$ hyperfine interaction.

The scheme proposed in this paper, applicable to substances which show a resolved hyperfine structure, places no requirements on the detailed relaxation mechanism of either the electron or the nucleus. It requires, however, that one sweep through a certain fraction of the external magnetic field in a time short compared to either relaxation time. The method is illustrated in fig. 1, which shows the energy levels of a system with $I = 1/2$, $J = 1/2$ vs. applied magnetic

---

(*) Reprinted with permission from *The Physical Review*, **103**, 500 (1956), © 1956 American Physical Society.

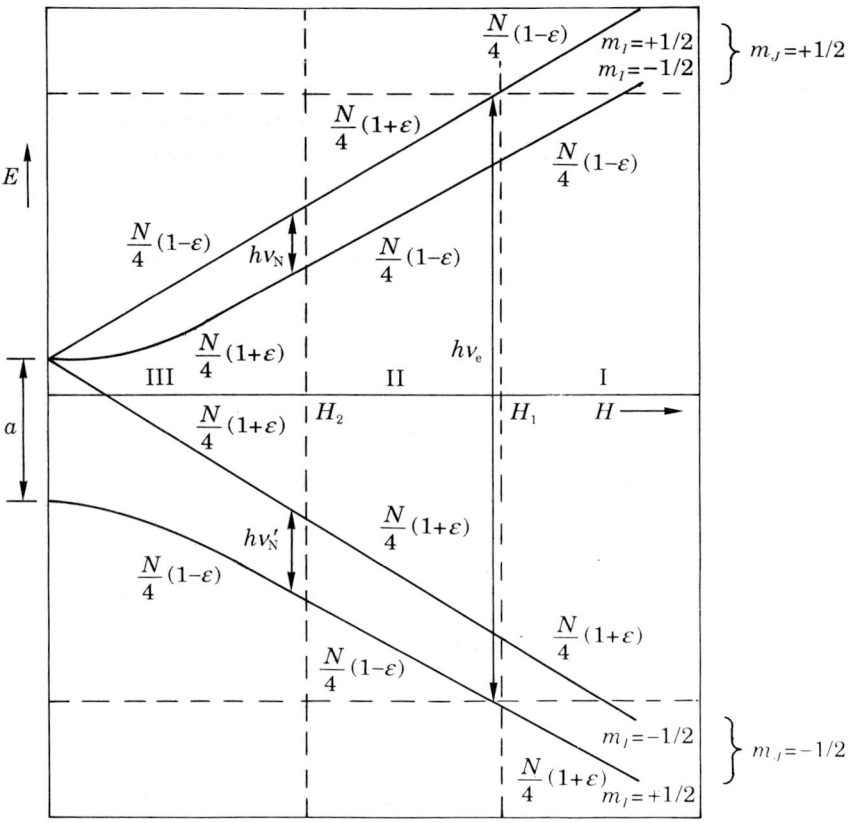

Fig. 1. – Energy levels and their populations for a system with $I = 1/2$, $J = 1/2$.

field as given by the Breit-Rabi[3] formula. This system is placed in a microwave magnetic field of frequency ν_e and radiofrequency field ν_N, both being perpendicular to the external magnetic field H. For $H > H_1$ (see fig. 1, region I), the population of both upper levels is given by $(1/4)N(1-\varepsilon)$, where N is the total number of electrons and $2\varepsilon \simeq g_e\mu_0 H/kT$ is the electronic Boltzmann factor. (We are neglecting the nuclear Boltzmann factor $g_n\mu_0 H/kT$ which is approximately 10^3 times smaller.) If we sweep through H_1, we will induce electronic transitions between the $m_I = +1/2$ levels. If we do this under adiabatic fast-passage conditions[4], the net magnetization of the electrons responsible for this transition will be turned through 180°. This reversal of the magnetization results in a reversal of the Boltzmann factor as indicated in region II of fig. 1. At this stage, each set of levels corresponding to the same m_J exhibits a nuclear polarization and we could perform a nuclear-resonance experiment in which the signal would be proportional to the electronic rather than the nuclear Boltzmann factor. However, the total population of both $m_I = +1/2$ levels equals that of the $m_I = -1/2$ levels, so that the sample as a whole does not exhibit a net polarization as yet. In order to obtain a net polarization we have to turn over the population of *only* one set of levels. This may be accomplished either by having a fixed radiofrequency and sweeping the magnetic field through H_2 in an adiabatic fast passage or keeping a fixed magnetic field and sweeping the radiofrequency. This is made possible by the fact that the spacing of the upper set of levels is different from the lower set ($\nu'_N > \nu_N$) as may be seen from the Breit-Rabi formula [3]. In order for only one transition to occur, the difference in level spacings has to be at least equal to the nuclear line width. The degree of nuclear polarization η obtained in region III will be given by

$$\eta = (N_{-1/2} - N_{+1/2})/(N_{-1/2} + N_{+1/2}) \simeq \varepsilon \simeq g_e \mu_0 H/2kT,$$

or more precisely by

(1) $$\eta = \mathrm{tgh}(g_e \mu_0 H/2kT).$$

For $T = 1$ K, $H = 10^4$ Oe and $g_e = 2$, we get a polarization of $\eta \simeq 0.7$. This polarization will, of course, decay with a characteristic time comparable to the nuclear-relaxation time but may be re-established by successive magnetic-field sweeps.

It is worth noting that from the difference in the level spacings given by [3]

(2) $$h(\nu' - \nu) = a(1 + x^2)^{1/2} - ax + 2g_I \mu_0 H,$$

where

$$x = (g_I + g_J)\mu_0 H/a,$$

and a is the hyperfine interaction constant, one may obtain the absolute value of the nuclear magnetic moment without having to know the electron wave function at the nucleus. Similar methods have been proposed and applied in molecular-beam experiments [5].

The transitions $\Delta m_I = \pm 1$, $\Delta m_J = 0$ will also affect the electron resonance line. This provides us with a sensitive method of studying nuclear-resonance phenomena by observing the behavior of the electron spin resonance line.

For the sake of simplicity, the case $I = 1/2$, $J = 1/2$ was treated. One can easily extend the arguments for larger values of I and J, in which case a nuclear polarization of any 2 adjacent levels may be realized.

* * *

I am indebted to Prof. J. BARDEEN, Prof. D. PINES and Prof. C. P. SLICHTER for sending us a preprint of their work [2] and discussing it with us. I would also like to achnowledge helpful discussions with Dr. P. W. ANDERSON.

REFERENCES

[1] A. OVERHAUSER: *Phys. Rev.*, **92**, 411 (1953).
[2] J. BARDEEN, D. PINES and C. P. SLICHTER: private communication; to be published.
[3] G. BREIT and I. I. RABI: *Phys. Rev.*, **38**, 2072 (1931).
[4] F. BLOCH: *Phys. Rev.*, **70**, 460 (1946).
[5] P. KUSCH, S. MILLMAN and I. I. RABI: *Phys. Rev.*, **57**, 765 (1940).

Polarization of phosphorous nuclei in silicon (*).

G. FEHER and E. A. GERE

Bell Telephone Laboratories - Murray Hill, N.J.

In the preceding letter a scheme for polarizing nuclei was described. This letter deals with the experimental verification of the scheme.

The experiments were performed on a phosphorous-doped silicon crystal having a room temperature resistivity of 0.3 $\Omega \cdot$ cm ($\simeq 3 \cdot 10^{16}$ centers/cm^3). FLETCHER et al. [1] were the first to observe a resolved hyperfine structure in a similar sample arising from the interaction of the donor electron with the magnetic moment of the phosphorus nucleus. Subsequent studies [2] showed a very long electron spin relaxation time which made this kind of sample ideal for the testing of the polarization method. The external field at which the transitions were observed was about 3000 Oe and the temperature of the sample was 1.25 K. The electron spin resonance line was observed with a balanced-bridge superheterodyne detection scheme [3] which was sensitive to the real part of the susceptibility, χ'. The rectified output from the i.f. amplifier was fed directly into the recorder. This system avoids the necessity of modulating the magnetic field and thus eliminates unnecessary complications which may easily cause a misinterpretation of experimental results [4]. The microwave cavity was made out of Pyrex with a thin silver coating on the inside. This permitted the r.f. field necessary for the Δm_I transitions to penetrate. It was supplied by a coil wound around the cavity. Both the r.f. and microwave magnetic fields at the sample were of the order of a tenth of an oersted. This insured adiabatic fast-passage conditions for all of the transitions induced.

In order to prove that a polarization of the nuclei has taken place, we have to show that the population of the levels corresponds to the value predicted by theory. Since the amplitude of the electron spin resonance line is proportional to the population difference between two levels, it was used as a probe to investigate the occupancy of the levels.

(*) Reprinted with permission from *The Physical Review*, **103**, 501 (1956), © 1956 American Physical Society.

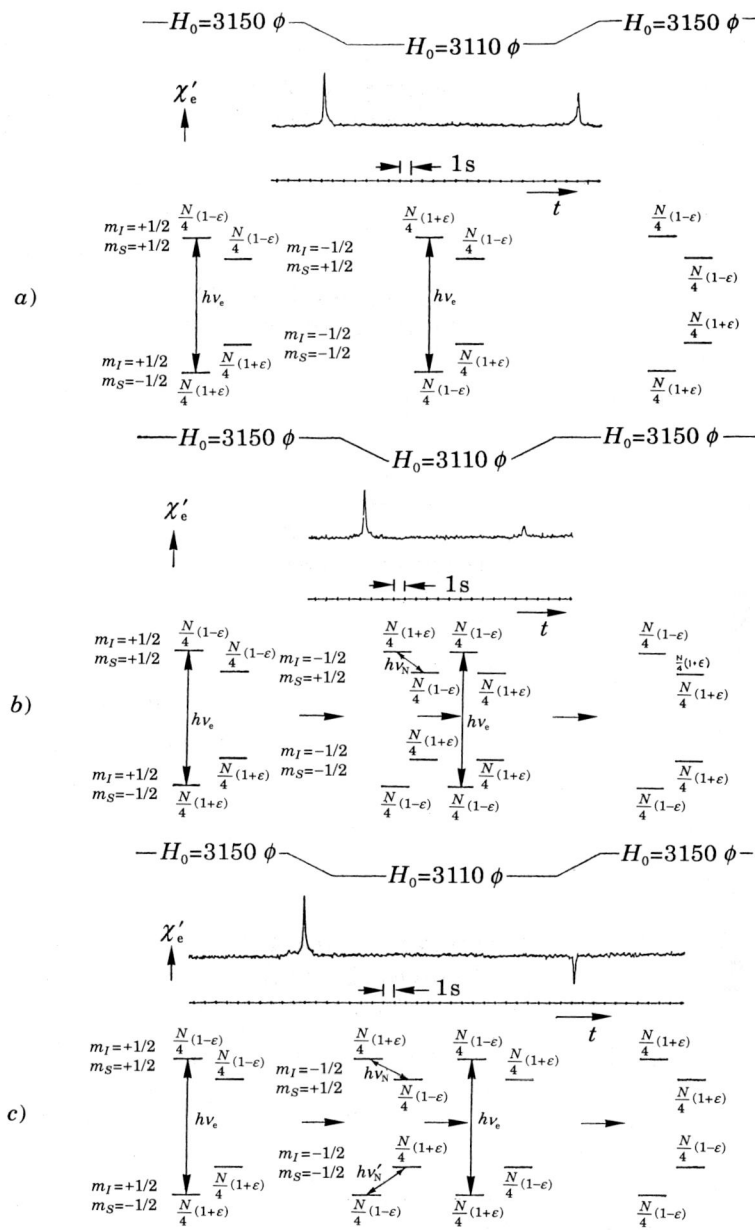

Fig. 1. – Electron spin resonance lines in phosphorus-doped silicon under adiabatic fast-passage conditions. The electron line is being used as a probe to investigate the population of the levels after different transitions have been induced. a) No radiofrequency field applied. For infinite relaxation times, the amplitude of the line after the second passage should be the same as after the first. b) After first passage through the line, the radiofrequency field is swept from 52 to 54 MHz/s to induce the $h\nu_N$ transition, after which nuclear polarization is established. c) After first passage through the line, the radiofrequency field is swept from 50 to 65 MHz/s to cover both the $h\nu_N$ and $h\nu'_N$ transitions.

Figure 1 shows the experimental results. The predicted populations are shown below each recorder tracing. The transitions which are induced at each stage are indicated by arrows. The time variation of the external magnetic field is shown above the tracing.

In fig. 1a) no r.f. was applied. From the theory of adiabatic fast passage [5], we would expect the electron spin resonance line to have equal amplitude and sign when the time between two successive passages through the line is short in comparison to the relaxation time. In our case this condition was not completely fulfilled since the relaxation time of the sample was of the order of a minute whereas the time between two sweeps was approximately 20 s. This explains, partially at least, the experimentally observed reduction in amplitude.

Figure 1b) shows the effect of inducing nuclear transition between the $m_s = +1/2$ states. This was accomplished by sweeping the radiofrequency generator from 52 to 54 MHz/s. A similar result was obtained by sweeping the radiofrequency generator from 64 to 66 MHz/s, which induced transitions between the $m_s = -1/2$ states. Since this nuclear adiabatic fast passage reverses the population of the levels with the same m_s, the population of the levels with different m_s has been equalized. We should, therefore, not expect a signal when sweeping back through the line. Experimentally we find a small residual amplitude which again can be explained by the finite relaxation time. It is this case that exhibits the nuclear polarization given by $\eta \simeq \varepsilon \simeq g_e \mu_0 H/2kT$. This may be readily seen from the figure by merely adding the occupancy of the $m_I = +1/2$ levels and comparing it with that of the $m_I = -1/2$ levels.

Figure 1c) shows what happens if we induce both nuclear transitions by sweeping the r.f. through the (52 ÷ 65) MHz/s range. We see from the level diagram that the population of the levels is just opposite to the ones in fig. 1a). We should observe, therefore, a reversal of the electron spin line, which indeed is verified experimentally.

It should be noted that it is the magnetic field in which the nuclei come to thermal equilibrium that enters into the polarization formula. This may be several times larger than the magnetic field in which the spin transitions occur, as long as the field is changed to the resonance field in a time short compared to the relaxation time.

REFERENCES

[1] R. C. Fletcher, W. A. Yager, G. L. Pearson, A. N. Holden, W. T. Read and F. R. Merritt: *Phys. Rev.*, **94**, 1392 (1954).
[2] G. Feher, R. C. Fletcher and E. A. Gere: *Phys. Rev.*, **100**, 1784 (1955); G. Feher and R. C. Fletcher: *Bull. Am. Phys. Soc.*, Ser II, **1**, 125 (1956).
[3] G. Feher: *Rev. Sci. Instrum.* (to be published).
[4] A. Honig: *Phys. Rev.*, **96**, 234 (1954); A. Honig and J. Combrisson: *Phys. Rev.*, **102**, 917 (1956).
[5] F. Bloch: *Phys. Rev.*, **70**, 460 (1946).

Lectures on Pulsed NMR (2nd Edition).

L. EMSLEY and A. PINES

Materials Sciences Division, Lawrence Berkeley Laboratory, and
Department of Chemistry, University of California - Berkeley, CA 94720

Introduction.

These lectures are an expanded version of those given at the Fermi School in 1986 and published in 1988[1]. The combined lectures of the two authors at the present school encompass, first of all, the previous topics of multiple-quantum NMR, spin decoupling, interactions of spins with quantized fields, cross-polarization, coherent averaging and zero-field NMR. Secondly, in addition to expanding the discussion on some of the above topics, the present lectures treat several other topics (including some recent developments), among them new sample spinning techniques, group theory and molecular dynamics, unitary bounds on spin dynamics and geometric phases. Within the scope of this school it is impossible to provide a comprehensive treatment of all the topics above, and we apologize, therefore, in advance that we have omitted important concepts and contributions from many workers in the field. Our hope is that we can provide a glimpse into some of the fundamental physics of pulsed NMR, with examples, mainly from our own laboratory, of the applications of modern techniques to condensed-matter problems.

1. – Multiple-quantum NMR.

1'1. *Dipolar couplings and molecular structure.* – In high magnetic field, the dipolar coupling d_{jk} between two spins j and k on a molecule depends on the distance r_{jk} between the spins and on the angle θ_{jk} of the j-k vector relative to the field [2-5]:

$$(1.1) \qquad d_{jk} \sim \frac{1}{r_{jk}^3}(3\cos^2\theta_{jk} - 1).$$

Measurements of time-averaged dipolar couplings from the NMR spectrum, therefore, can provide information about molecular structure and dynamics [6-

8]. Figure 1.1 illustrates the sensitivity of NMR spectra to structure by showing an experimental spectrum of benzene oriented in a liquid crystal, as well as simulated spectra for different possible structures of symmetric six-carbon frameworks. This figure, provided by Z. LUZ of the Weizmann Institute of Science, demonstrates clearly how precise the determination of the authentic hexagonal structure is in this simple pedagogical case.

1`2. *Onset of spectral complexity.* – The case of benzene, with six proton spins, exemplifies a resolved tractable spectrum of dipolar couplings [9, 10]. The situation rapidly becomes more complex, however, as the number of spins is increased. This effect is illustrated in fig. 1.2, also courtesy of Z. LUZ. Beyond ten or so spins, depending on symmetry, the exponential increase in the number of NMR transitions renders the spectrum intractably complex. The

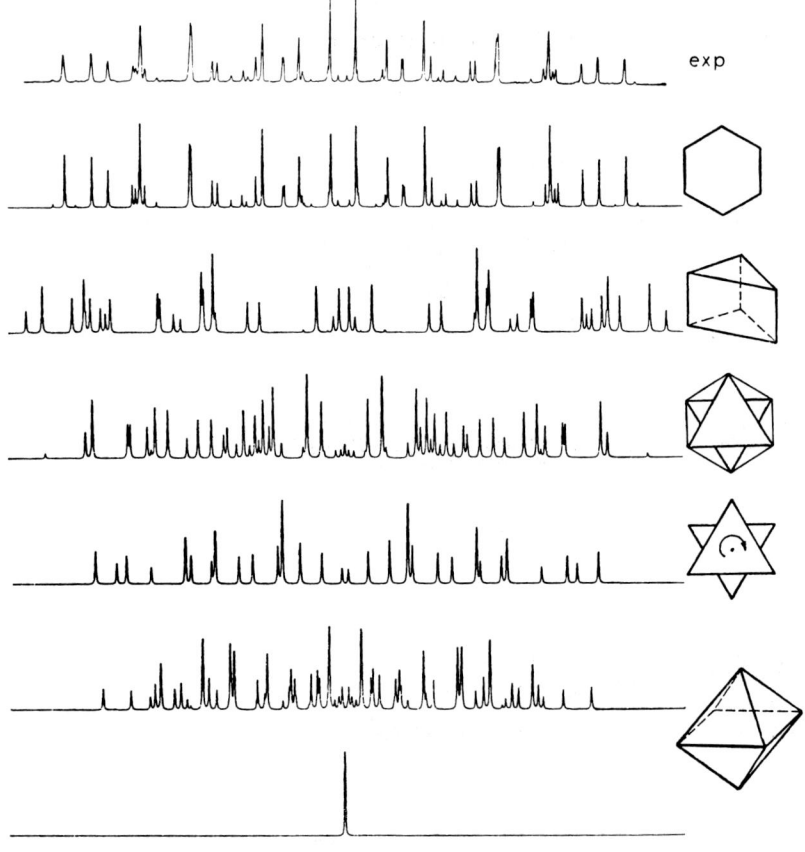

Fig. 1.1. – Experimental (exp) one-quantum proton NMR spectrum of benzene oriented in a liquid crystal, compared with simulated spectra of symmetric six-carbon structures. The isotropic spectrum, consisting of one line (bottom), is compatible with all these structures. (Courtesy of Z. LUZ.)

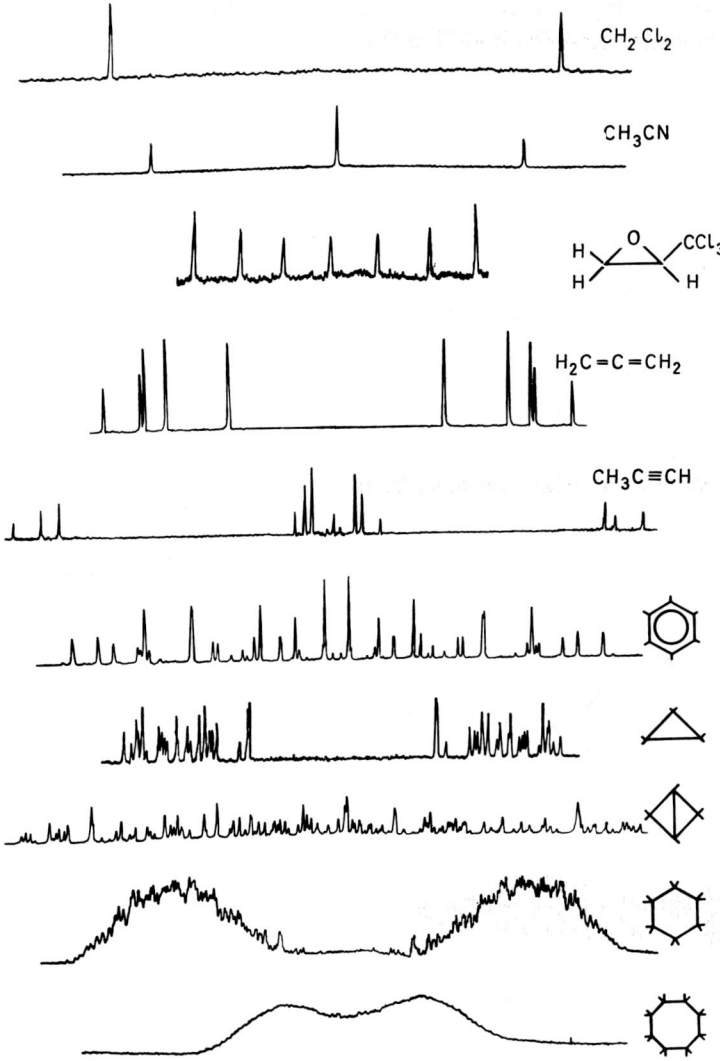

Fig. 1.2. – Proton NMR spectra of oriented molecules with various numbers of spins. The spectral complexity increases exponentially with the size of the spin system. (Courtesy of Z. Luz.)

case of eight spins is illustrated in fig. 1.3 and 1.4, which show the spectrum of n-hexane-d_6 (methyl groups deuterated) in isotropic solution (fig. 1.3), where the dipolar couplings are averaged to zero (and where, at this resolution, the chemical shifts are essentially the same), compared to the spectrum of the same molecule oriented in a liquid crystal (fig. 1.4).

A rough estimate of the complexity can be made by looking at fig. 1.5, which shows the energy level diagram for a group of N coupled spins $I = 1/2$ in high

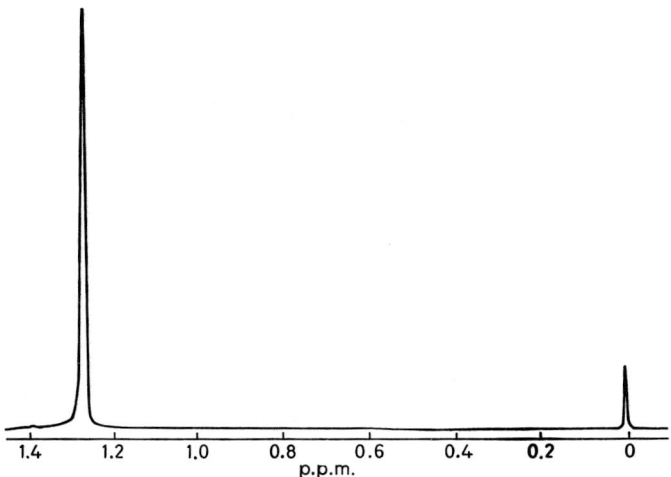

Fig. 1.3. – Proton NMR spectrum of n-hexane-d_6 in isotropic solution. All protons have roughly the same chemical shift, and give one line at this level of resolution. The line at the right is from a TMS standard. (Adapted from *Chem. Phys.*, **108**, 179 (1986), with permission.)

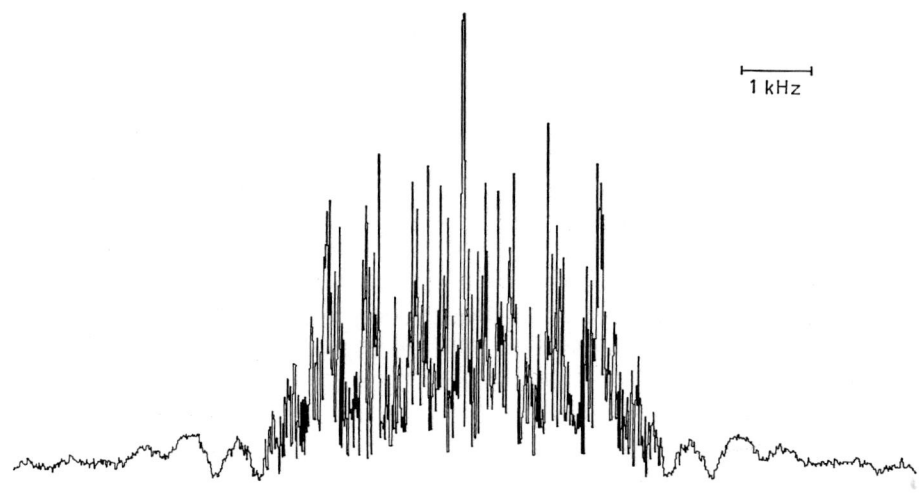

Fig. 1.4. – Proton NMR spin echo spectrum of oriented n-hexane-d_6 (deuterated methyls) with deuterium spin decoupling, to be compared with the isotropic spectrum in fig. 1.3. Even for this system of eight protons, there are about three thousand transitions, and the spectrum is intractable. (Adapted from *Chem. Phys.*, **108**, 179 (1986), with permission.)

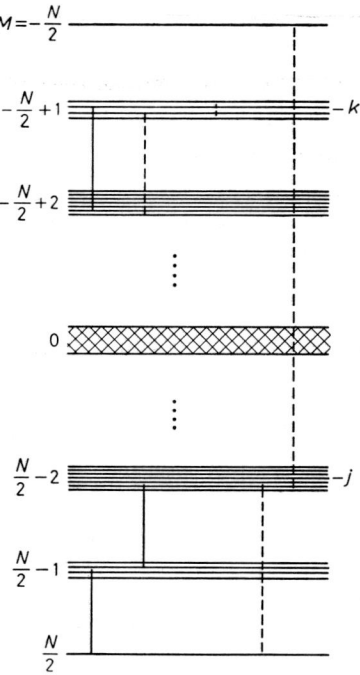

Fig. 1.5. – Energy level diagram of a system of N coupled spins $I = 1/2$ in high magnetic field. The groups of energy levels are characterized by the magnetic Zeeman quantum number M. The splittings in each group are due to chemical shifts and couplings between the spins. The solid vertical lines indicate allowed ($\Delta M = \pm 1$) one-quantum transitions and the dashed vertical lines depict some «forbidden» multiple-quantum ($\Delta M = \pm n$, $n = 0, 1, ..., N$) transitions. (Adapted from *J. Chem. Phys.*, **73**, 2084 (1980), with permission.)

field. The conventional one-quantum NMR transitions, shown as solid vertical lines, are subject to the selection rule

$$(1.2) \qquad |\Delta M| = 1 \, .$$

The number of energy levels with magnetic quantum number M is $\binom{N}{N/2 + M}$ and, if one is restricted to transitions between neighboring M's, then

$$(1.3) \qquad Z_1 = \sum_M \binom{N}{N/2 + M}\binom{N}{N/2 + M + 1} = \binom{2N}{N + 1}$$

is an upper bound to the number of one-quantum transitions. For $N = 4$ this number is 56, for $N = 8$ it is 11 440, and for $N = 12$ it is 2 496 144 [11].

1˙3. *Simplification by multiple-quantum transitions.* – Having encountered the problem of spectral complexity, we may also discern in fig. 1.5 a possible solution. The dashed vertical lines indicate *multiple-quantum* transitions [12-18] in which several spins flip together subject to the general rule

(1.4) $$|\Delta M| = n.$$

As n, the number of quanta, increases, the number of transitions decreases and the spectra should, therefore, become simpler. Generalization of eq. (1.3) to n-quantum transitions yields

(1.5) $$Z_n = \binom{2N}{N+n},$$

and for zero-quantum transitions

(1.6) $$Z_0 = \binom{2N}{N} - 2^N.$$

Examining these expressions, we can see that the dependence of the number of transitions on N for high n ($n = N, N-1, N-2$) is roughly

(1.7) $$\begin{cases} Z_N \sim 1, \\ Z_{N-1} \sim N, \\ Z_{N-2} \sim N^2. \end{cases}$$

Thus the number of $(N-2)$-quantum transitions is quadratic in the number of spins, and, since the maximum number of couplings is

(1.8) $$Z_d = \frac{N(N-1)}{2} \sim N^2,$$

the $(N-2)$- and $(N-1)$-quantum spectra should contain sufficient information to determine the d_{jk}.

1˙4. *Analogy to chemical isotopic labeling.* – There are other ways to simplify spectra. One conceptually simple but synthetically demanding method is to isotopically label spins at selected positions in the molecule. For example, deuteration of all but two positions on a molecule dramatically reduces the number of lines in the proton spectrum and thus facilitates the direct determination of the d_{jk}'s. We may then ask about the relationship of this simplification to the simplification realized in multiple-quantum spectra. Note that the number of isotopically substituted species with single labels is at most N and that the number of isotopically doubly labeled species is at most $N(N-1)/2$, the same as the number of $(N-1)$- and $(N-2)$-quantum transitions, respectively. In effect, when we flip $N-1$ out of N spins, the remaining spin (or, equivalently, the one labeled spin) can be in any one of N positions. When we flip $N-2$, the remaining two can be in $N(N-1)/2$ configurations. Hence the $(N-p)$-quantum spectrum is a superposition of spectra from all possible arrangements of p isotopic labels.

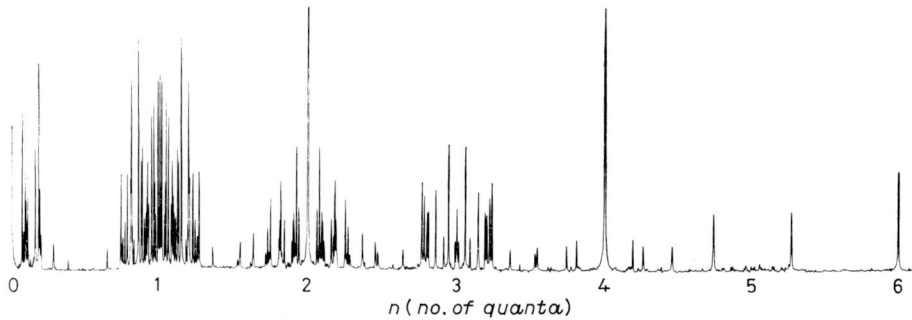

Fig. 1.6. – Multiple-quantum NMR spectra of oriented benzene, showing the progressive simplifications as the number of quanta increases from 1 to 6. The spectra were averaged over a range of preparation and mixing times. We can see one six-quantum transition, two five-quantum transitions and seven four-quantum transitions, as anticipated in the text. (Courtesy of G. DROBNY.)

As an example, consider oriented benzene, whose one-quantum spectrum was shown in fig. 1.1. The 6-quantum spectrum should contain one line, since there is only one way to absorb six quanta and thereby flip all six spins. The 5-quantum spectrum arises from, roughly speaking, flips of five spins in the field of a sixth. Since this last spin may be up or down, and since all positions on the molecule are equivalent, the 5-quantum spectrum should contain just one doublet. In general there will be a doublet for each inequivalent position or each different singly labeled species. Next we have the 4-quantum spectrum, which involves the flip of four spins in the field of the remaining two. These two can be ortho, meta, or para (the three doubly labeled isomers), so we expect three triplets (seven lines). Figure 1.6 clearly shows these predictions realized in the multiple-quantum NMR spectra of oriented benzene.

1`5. *Obtaining multiple-quantum spectra.* – Since multiple-quantum transitions are not directly observable with an NMR coil, they must be detected indirectly using coherence transfer [19, 20] by two-dimensional spectroscopic methods [5, 16]. The general scheme of such an experiment is

(1.9)   preparation(τ)-evolution(t_1)-mixing(τ')-detection(t_2).

During the time period t_2, direct detection of allowed one-quantum frequencies provides information about the multiple-quantum frequencies in the earlier time period t_1. Two-dimensional Fourier transform then yields a spectrum which contains the directly detected one-quantum spectrum parallel to the frequency axis ω_2 and the multiple-quantum spectrum parallel to the frequency axis ω_1. An example is shown in fig. 1.7. The multiple-quantum spectrum of fig. 1.6 is the projection of such a two-dimensional spectrum, in absolute-value mode, onto the ω_1 axis. Figure 1.7b) clearly shows the seven four-quantum lines of benzene and their connections to the one-quantum transitions. Note that, to

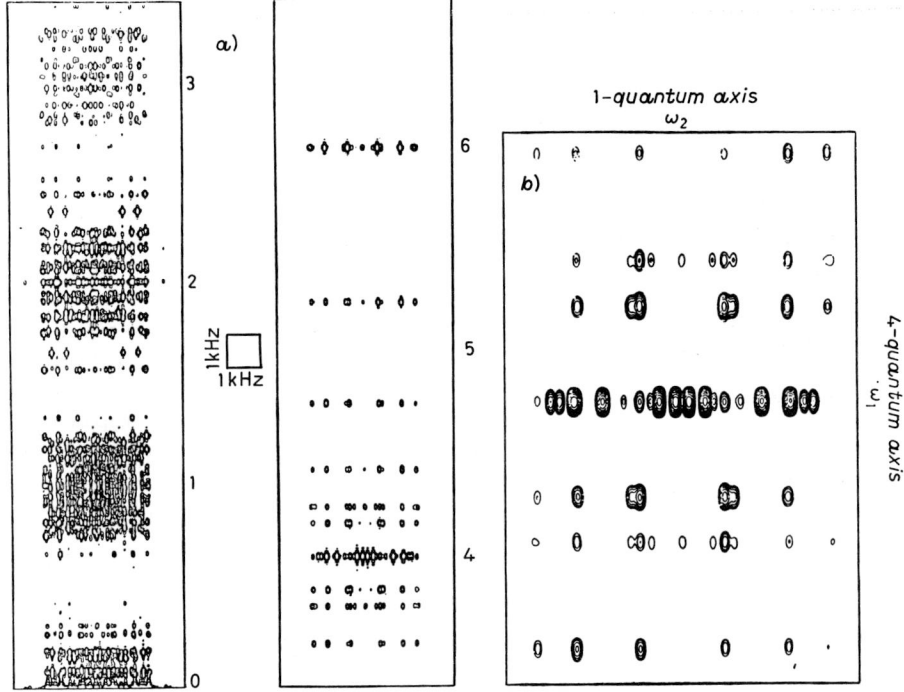

Fig. 1.7. – Contour plot of the two-dimensional multiple-quantum spectrum of oriented benzene and expansion of the four-quantum region. These plots show how coherence is transferred between the multiple-quantum evolution period (vertical axis) and the one-quantum detection period (horizontal axis). The preparation and mixing times were nine milliseconds. (Courtesy of G. DROBNY.)

obtain uniform multiple-quantum intensities, multiple-quantum spectra are typically averaged over a range of values of the preparation time $\tau(=\tau')$ [21].

1˙6. *Theory of multiple-quantum* NMR: *preliminaries.* – Since we will be interested only in the multiple-quantum spectrum, that is, the projection onto the ω_1 axis, we need detect just the integrated ω_2 signal, which is given by the value of the magnetization at the first point in t_2. Furthermore, since the system usually begins in equilibrium at high temperature and in high field, where its density operator is proportional to I_z, we shall find it convenient to monitor I_z at $t_2 = 0$ following the mixing. This formulation, outlined in fig. 1.8, frames the problem in its most symmetric form with no loss of generality. U is the preparation propagator, $V(t_1)$ the evolution propagator, and U' the mixing propagator. The value of $\langle I_z \rangle$ is detected at $t_2 = 0$ as a function of t_1. For simplicity of notation, we suppress the subscript in t_1 and write t, since only this one dimension will be relevant to the remaining discussion.

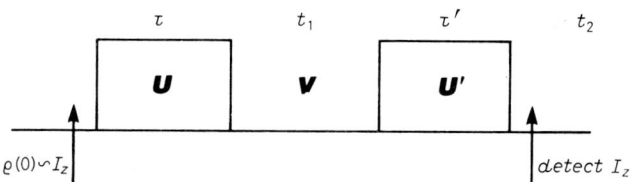

Fig. 1.8. – Timing diagram for basic multiple-quantum sequence. The multiple-quantum coherences are prepared from an initial high-temperature equilibrium state $\rho(0) \sim I_z$ by the propagator $U(\tau)$; they evolve under the propagator $V(t_1)$, and after mixing by $U'(\tau')$ they are detected indirectly as z magnetization. This scheme is analogous to detecting x and y components in t_2 by adding or omitting pulses from U'. By using different U', quadrature phases can be detected in t_1.

In general [22, 23], the initial quantum statistical state of a system is characterized by a density operator $\rho(0)$, which we depict as a *ket* $|\rho(0)\rangle$ (and its dual *bra* $\langle\rho(0)|$) in Liouville space. Under the influence of the propagator $\boldsymbol{U}(t)$ the system develops into a state $|\rho(t)\rangle$, i.e.

(1.10) $$|\rho(t)\rangle = \boldsymbol{U}(t)|\rho(0)\rangle.$$

Diagrammatically, this transformation looks as follows:

(1.11) $$|\rho(0)\rangle \xrightarrow{\boldsymbol{U}(t)} |\rho(t)\rangle.$$

$\boldsymbol{U}(t)$ is a superoperator [24], or superpropagator, which we shall continue to term propagator for short. (This usage should be distinguished from the propagator $U(t)$ that acts on quantum-mechanical kets in Hilbert space; context will normally render this distinction evident, but, for clarity, superoperators will be denoted in bold type.) $\boldsymbol{U}(t)$ is derived from the Hamiltonian superoperator or Liouvillian $\boldsymbol{H}$ by

(1.12) $$\boldsymbol{U}(t) = \boldsymbol{T}\exp\left[(-i/\hbar)\int_0^t \boldsymbol{H}(t')\,dt'\right],$$

where $\boldsymbol{T}$ is the Dyson time-ordering operator [25].

We recall that the density operator

(1.13) $$\rho = \overline{|\psi\rangle\langle\psi|}$$

is the ensemble average of the projectors of the states ψ for the constituent members of the ensemble, and furthermore that the ensemble average expectation value of any observable Q (what we measure or observe) is given by

(1.14) $$\overline{\langle Q \rangle} = \text{Tr}\{\rho Q\} = \langle \rho | Q \rangle,$$

where

(1.15) $$\langle A|B\rangle = \mathrm{Tr}\{A^\dagger B\}$$

is the scalar product in Liouville space.

1˙7. *Multiple-quantum signal.* – We shall be concerned with the detection of the normalized signal $\langle I_z \rangle$ as a function of the evolution time t which yields the multiple-quantum free-induction decay $f(t)$ which will contain τ and τ' as parameters. In the superoperator notation, the state of the system $|\rho(t)\rangle$ at the end of the mixing in the scheme of fig. 1.8 is given by

(1.16) $$|\rho(t)\rangle = U'(\tau')V(t)U(\tau)|\rho(0)\rangle.$$

If we now assume a high-temperature, high-field equilibrium initial condition [26],

(1.17) $$|\rho(0)\rangle = |I_z\rangle,$$

and detect $\langle I_z \rangle$, then eq. (1.16) yields the multiple-quantum free-induction decay

(1.18) $$f(t) = \frac{\langle I_z|U'(\tau')V(t)U(\tau)|I_z\rangle}{\langle I_z|I_z\rangle}.$$

Henceforth, we assume the signal to be normalized ($\langle I_z|I_z\rangle = 1$), so that

(1.19) $$f(t) = \langle I_z|U'(\tau')V(t)U(\tau)|I_z\rangle.$$

So now you see the general rule with superoperators: begin on the right with a ket for the original state, propagate from right to left, and end on the left with a bra for the detected state. From eq. (1.19), given the Hamiltonians for the three periods (τ, t, τ'), $f(t)$ can be evaluated, and from the Fourier transform of this signal the multiple-quantum spectrum can be obtained. Let us examine some examples.

1˙8. *Special case: one-quantum* FID *point by point.* – First we consider a «normal» one-quantum free-induction decay (FID) within the framework of fig. 1.8 and eq. (1.19). In such a normal FID, the transverse magnetization, say $\langle I_x \rangle$, is detected directly. However, we assume that the experiment is performed in a silly way, indirectly and point by point, by making $U(\tau)$ a $(\pi/2)_y$ pulse, $U'(\tau')$ a $(\pi/2)_{\bar{y}}$ pulse, and then detecting $\langle I_z \rangle$. This procedure is equivalent to detecting $\langle I_x \rangle$ vs. t. The scheme of fig. 1.8 then becomes

(1.20) $$P\text{-}t\text{-}P^\dagger\text{-detect}\,\langle I_z\rangle,$$

where P and $P^\dagger$ are the $\pi/2$ pulse superoperators. We now insert these expressions into eq. (1.19) and expand in a basis set of kets in Liouville space. Several basis sets may be considered, such as fictitious spin operators [2, 27-29], product operators [30-33], spherical tensors or multipole operators [34], and eigenoperators [16]. The non-Hermitian eigenoperators of $V(i)$, $|j\rangle\langle k|$, are in fact particularly well suited to these problems. For $j \neq k$ these correspond to coherences and for $j = k$ they correspond to populations. We represent the eigenoperators of $V(t)$ as kets written $|j - k\rangle$, so that

(1.21) $$V(t)|j - k\rangle = \exp[-i\omega_{jk}t]|j - k\rangle,$$

where ω_{jk} is the frequency of the transition between eigenstates j and k of the (time-dependent) evolution Hamiltonian. Hence, we have

(1.22) $$f(t) = \langle I_z | \boldsymbol{P}^\dagger V(t) \boldsymbol{P} | I_z \rangle = \sum_{j,k} \langle I_z | \boldsymbol{P}^\dagger | j - k\rangle \langle j - k | V(t) | j - k\rangle \langle j - k | \boldsymbol{P} | I_z \rangle.$$

We now use eq. (1.21) and evaluate the other matrix elements in eq. (1.22), for example,

(1.23) $$\langle j - k | \boldsymbol{P} | I_z \rangle = \text{Tr}\{|k\rangle\langle j| PI_z P^\dagger\} = \langle j | PI_z P^\dagger | k\rangle = (PI_z P^\dagger)_{jk} = I_{xjk},$$

where I_{xjk} is the (j, k)-th matrix element of the operator I_x. In this way we obtain the well-known result

(1.24) $$f(t) = \sum_{jk} |I_{xjk}|^2 \exp[-i\omega_{jk}t]$$

with its Fourier transform

(1.25) $$y(\omega) = \sum_{jk} |I_{xjk}|^2 \delta(\omega - \omega_{jk}),$$

giving the spectrum of $\langle I_x | I_x(t)\rangle$ the time autocorrelation function of I_x [35]. This spectrum yields the one-quantum frequencies, since $I_{xjk} \neq 0$ only for one-quantum transitions; consequently, we term I_x a one-quantum operator. An important point to note is that all lines in the spectrum have the same phase because of the appearance of squares of the absolute magnitudes of $(PI_z P^\dagger)_{jk} = I_{xjk}$.

1'9. *General case: multiple-quantum* FID. – We now expand eq. (1.19) in the $|j - k\rangle$ basis set for general U, U' to obtain

(1.26) $$f(t) = \langle I_z | U'(\tau') V(t) U(\tau) | I_z \rangle =$$
$$= \sum_{jk} \langle I_z | U'(\tau') | j - k\rangle \langle j - k | V(t) | j - k\rangle \langle j - k | U(\tau) | I_z \rangle =$$
$$= \sum_{jk} (I_z(-\tau'))_{kj} (I_z(\tau))_{jk} \exp[-i\omega_{jk}t],$$

where

(1.27) $$I_z(\tau) = U(\tau) I_z U^\dagger(\tau) \quad \text{and} \quad I_z(-\tau') = U'^\dagger(\tau') I_z U'(\tau')$$

are the effective preparation and mixing operators. These operators, however, are multiple-quantum operators, unlike I_x. As a consequence, the phases vary from line to line since $(I_z(-\tau'))_{kj} (I_z(\tau))_{jk}$ are, in general, *complex* numbers.

As an example, consider the simplest and most widely used multiple-quantum sequence, the three-pulse sequence

(1.28) $$\left(\frac{\pi}{2}\right)_y -\tau- \left(\frac{\pi}{2}\right)_{\bar{y}} -t- \left(\frac{\pi}{2}\right)_y \text{-detect} \langle I_x \rangle.$$

In the present framework, if the detection is to be matched to the preparation so that $\tau = \tau'$, then an additional $(\pi/2)_{\bar{y}}$ pulse must be inserted a time τ after the third $(\pi/2)_y$ to create

(1.29) $$U(\tau) = U'(\tau).$$

Here both U and U' derive from the Hamiltonian H_{xx}, transformed by the y pulses from the normal bilinear (*e.g.*, dipolar) Hamiltonian H_{zz}.

Inserting eq. (1.29) into eq. (1.26) yields

(1.30) $$f(t) = \sum_{jk} (I_z(\tau)_{jk})^2 \exp[-i\omega_{jk}t].$$

Note that we have $(...)^2$ not $|...|^2$ for the matrix element in eq. (1.30). For closely spaced multiple-quantum lines, the arbitrary phases will cause a cancellation of intensity, a crucial problem in complex molecules and solids that is illustrated schematically in fig. 1.9a).

1'10. *Time-reversal (conjugate) detection.* – One solution to the phase problem mentioned above follows from the realization that the integrated intensity of the multiple-quantum frequency spectrum is given by the first point, $f(t = 0)$, of the free-induction decay, *i.e.*

(1.31) $$f(0) = \langle I_z | U'(\tau') U(\tau) | I_z \rangle.$$

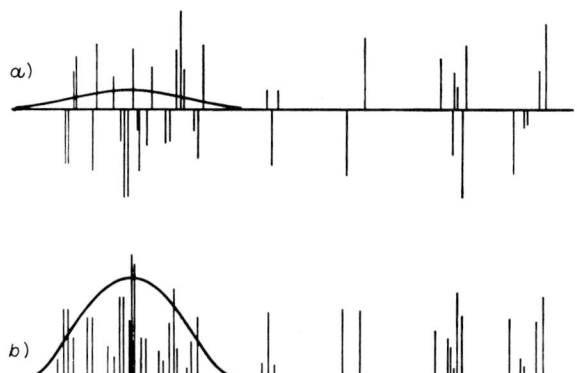

Fig. 1.9. – Effect of time-reversal detection. In *a*), the propagators U and U' are not conjugate (for example, $U = U'$ in the three-pulse multiple-quantum experiment $(x\text{-}\tau\text{-}\bar{x}\text{-}t_1\text{-}x\text{-}t_2)$), and the uncorrelated phases of nearby transitions can reduce the integrated intensity. In *b*), with time-reversal detection, $U = U'^\dagger$ and the lines are all in phase, thus restoring the full intensity.

Clearly, if we ensure that

(1.32) $$U'(\tau')U(\tau) = 1,$$

then we can recover the full intensity as in a spin echo [36], and all the lines will be necessarily in phase [37]. This phasing happened naturally in the one-quantum case of eq. (1.24), and for the multiple-quantum case the same effect can be achieved if *the Hamiltonian for the preparation period τ is the negative of that for the mixing period τ'*, namely

(1.33) $$U'(\tau') = U^\dagger(\tau).$$

Given eq. (1.32), we can evaluate eq. (1.26) as

(1.34) $$f(t) = \sum_{jk} |(I_z(\tau)_{jk})^2| \exp[-i\omega_{jk}t]$$

to yield «in-phase» lines with the maximum integrated intensity, as depicted schematically in fig. 1.9b). (We shall see later that a more general condition, allowing all lines within a multiple-quantum order to be in phase, is also possible.)

1˙11. *Effect of phase shifts.* – Consider the effect of a phase shift of φ, i.e. a rotation by φ around the z-axis. For example, if $U(\tau)$ derives from a secular (high-field) Hamiltonian plus r.f. pulses, then such a phase shift is induced by changing the phases of all r.f. pulses by φ. We define

(1.35) $$U_\varphi = R_z^\dagger(\varphi) U R_z(\varphi),$$

where $R_z(\varphi)$ is a z rotation superoperator

(1.36) $$R_z(\varphi) = \exp[-i\varphi I_z],$$

and we see that its effect on the eigenoperators is

(1.37) $$R_z(\varphi)|j-k\rangle = \exp[-i\varphi n_{jk}]|j-k\rangle,$$

where

(1.38) $$n_{jk} = M_k - M_j$$

is the number of quanta, or order, of the coherence or transition j-k. Thus *a phase shift of φ in any radiation that induces multiple-quantum transitions will be «seen» by an n-quantum transition as $n\varphi$* [29].

Suppose we now perform a multiple-quantum experiment with time-reversal detection as in subsect. 1˙10 but modify eq. (1.33) to obtain

(1.39) $$U(\tau) = U_\varphi'^\dagger(\tau) = R_z^\dagger(\varphi) U'^\dagger(\tau) R_z(\varphi),$$

so that the phase of the preparation sequence is shifted by φ. Inserting this propagator into eq. (1.26) and using eq. (1.37), we find that eq. (1.33) now becomes

(1.40) $$f_\varphi(0) = \sum_{jk} |I_z(\tau)_{jk}|^2 \exp[-i\omega_{jk}t]\exp[-i\varphi n_{jk}].$$

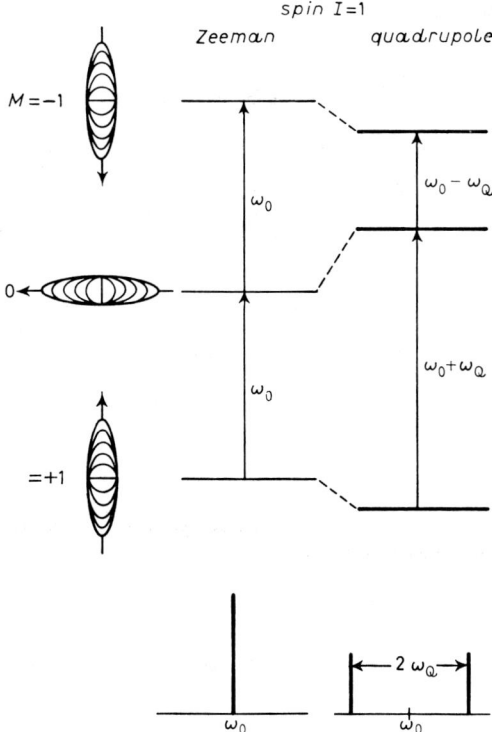

Fig. 1.10. – Deuterium energy levels in the laboratory frame showing the splitting of the one-quantum transitions by the electric-quadrupole interaction.

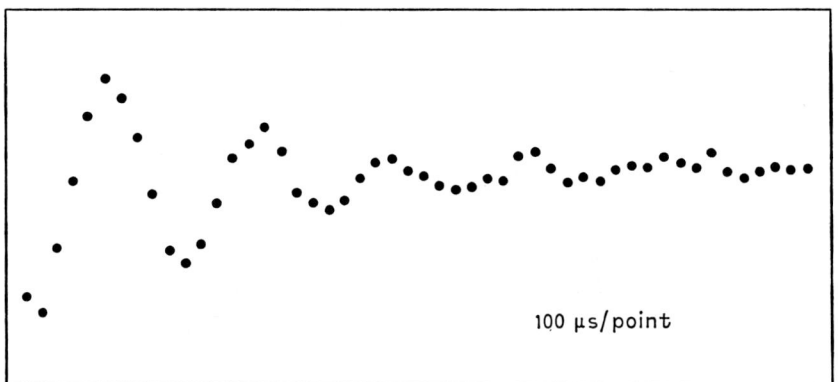

Fig. 1.11. – Double-quantum free-induction decay of deuterium in solid benzene-d_1 doped (10%) into benzene at -40 °C, with proton decoupling. In contrast, the normal one-quantum deuterium free-induction signal decays in tens of microseconds. (Courtesy of S. VEGA.)

Thus each complete n-quantum spectrum is phase shifted by $n\varphi$. For $t = 0$, we have

(1.41) $$f_\varphi(0) = \sum_{jk} |I_z(\tau)_{jk}|^2 \exp[-i\varphi n_{jk}],$$

so the integrated intensities of the multiple-quantum spectra for each τ may be determined by Fourier transformation with respect to φ. This determination may be made more efficiently by varying φ proportionately to τ to yield the full τ-dependence of the n-quantum intensities in one experiment[38].

1`12. *Time-proportional phase incrementation* (TPPI). – The phase behavior above provides a convenient way to separate multiple-quantum coherences from each other (as shown, for example, in the benzene spectrum of fig. 1.6)[39, 40]. In eq. (1.40), if we set

(1.42) $$\varphi = \Delta\omega t,$$

then we have

(1.43) $$f(t) = \sum_{jk} |I_z(\tau)_{jk}|^2 \exp[-it(\omega_{jk} + \Delta\omega n_{jk})],$$

so each n-quantum line is shifted by $n\Delta\omega$, thereby allowing a clear separation of orders in one experiment.

1`13. *Double-quantum NMR in solids.* – One area where double-quantum NMR has played a particularly useful role is in spin $I = 1$ systems, *e.g.* deuterium. The appropriate energy level diagram is depicted in fig. 1.10. Owing to electric quadrupolar coupling the nonspherical deuterium nucleus experiences an orientation-dependent splitting of its resonance lines leading to broad powder signals (~ 100 kHz) in solids. This inhomogeneous broadening obscures the small chemical shifts, a problem that can be overcome by detecting the double-quantum spectrum[41, 42] (transitions between $M = \pm 1$), which is unencumbered by electric-quadrupole broadening. This approach allowed the first measurement of the chemical-shielding anisotropy (~ -6 p.p.m.) of hydrogen in benzene[27]. An example of these effects is provided in fig. 1.11 and 1.12, which show the double-quantum free-induction decay and spectrum of solid benzene-d_1. The double-quantum coherence was prepared by irradiating the center of the spectrum with a weak pulse, for which the r.f. amplitude ω_1 is much less than the quadrupole coupling ω_Q, resonant only with the double-quantum transition at ω_0. The effective $\pi/2$ pulse is then characterized by the condition

(1.44) $$\frac{\omega_1^2}{\omega_Q}\tau = \frac{\pi}{2},$$

where τ is the pulse length. We shall expand on this idea later when we get to double-quantum decoupling in subsect. 3`3 and double-quantum excitation in subsect. 4`8.

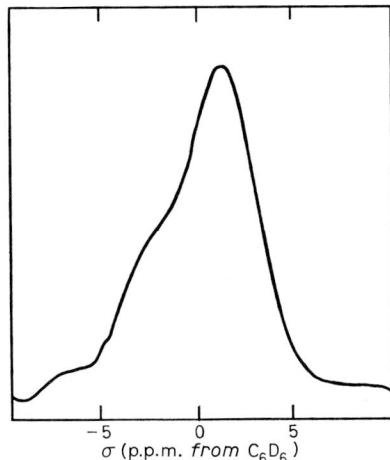

Fig. 1.12. – Fourier transform of the signal in fig. 1.11, showing the chemical-shift anisotropy of deuterium in benzene-d_1. The left edge of the spectrum corresponds to the benzene plane normal aligned with the magnetic field; the right peak corresponds to the perpendicular direction. (Courtesy of S. VEGA.)

The double-quantum transition of a single spin $I = 1$ is analogous to the full N-quantum spectrum of N coupled spins $I = 1/2$. We recall that the N-quantum spectrum arises from just one transition, unaffected by any spin-spin coupling.

1`14. *Double-quantum spin locking.* – Spin locking is a widely used technique for extending the lifetimes of coherences in the rotating frame, and is essential in various relaxation and cross-polarization experiments. It can be performed by applying a $(\pi/2)_y$ r.f. pulse to the z magnetization, and then shifting the r.f. phase by $\pi/2$ to x while irradiating continuously. The magnetization is thus aligned along the r.f. field and hence the term locking. The decay of the magnetization order now depends on spin-lattice processes in the rotating frame. An appealing question is whether one can lock *double-quantum* coherence in a similar fashion, even though the one-quantum spectrum may be extremely broad. In other words, can we create double-quantum coherence and then apply radiation to lock it? Suppose the double-quantum coherence is prepared with a weak pulse applied at the center of resonance. From the discussion in subsect. 1`11 we know a $\pi/4$ phase shift in the radiation is needed to deliver an effective phase shift of $\pi/2$ to the double-quantum transition. A pictorial representation of this behavior appears in fig. 1.13. The dipolar one-quantum coherences of fig. 1.13a) act like vectors (*i.e.* first-rank tensors), which are shown in the diagram as p_x and p_y orbitals that are created from a p_z orbital representing the equilibrium magnetization. In these circumstances, a $\pi/2$ phase shift takes p_x into p_y. The quadrupolar double-quantum coherences of fig. 1.13, however, act like second-rank tensors or $d_{x^2-y^2}$ and d_{xy} orbitals. A $\pi/4$ phase shift interchanges these orbitals, whereas a $\pi/2$ phase shift changes their

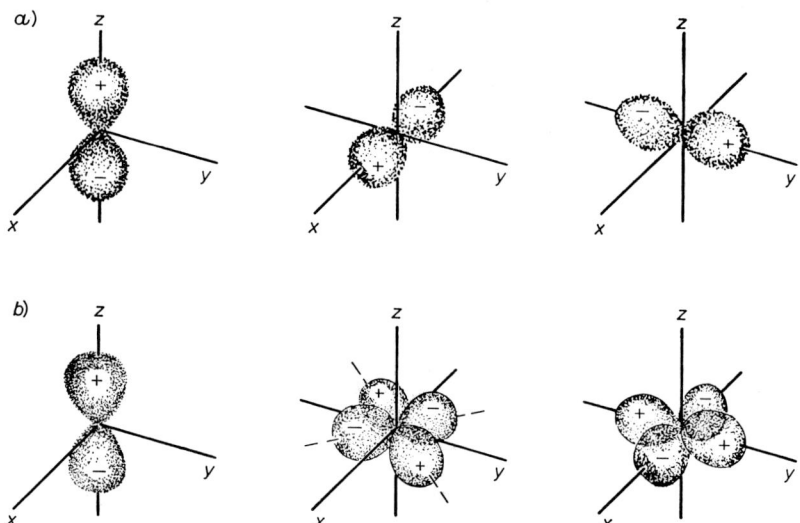

Fig. 1.13. – a) Symmetry of one-quantum dipole operators (analogous to p orbitals). They transform like vectors: a phase shift of $\pi/2$ interchanges p_x and p_y, and a phase shift of π changes their signs. b) Symmetry of double-quantum, quadrupole operators (analogous to p_z, d_{xy} and $d_{x^2-y^2}$ orbitals) involved in the fictitious spin $I = 1/2$ double-quantum transition ± 1 in fig. 1.10. The p_z orbital corresponds to z magnetization, and the two in-plane orbitals, corresponding to double-quantum coherence, transform like second-rank tensors: a phase shift of $\pi/4$ interchanges d_{xy} and $d_{x^2-y^2}$, and a phase shift of $\pi/2$ changes their signs. (Courtesy of J. MURDOCH.)

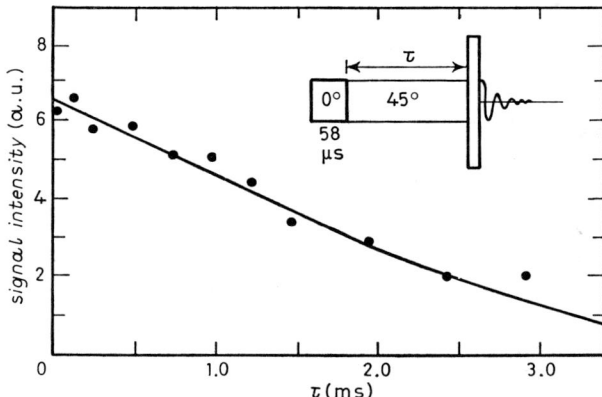

Fig. 1.14. – Double-quantum spin locking of deuterium in solid benzene-d_1. A «soft» $\pi/2$ pulse was applied at the ± 1 transition to create double-quantum coherence. The r.f. phase was then shifted by $\pi/4$ to spin lock the coherence for a prolonged period; the coherence was detected by a strong pulse which transformed it into observable one-quantum magnetization. (Courtesy of S. VEGA.)

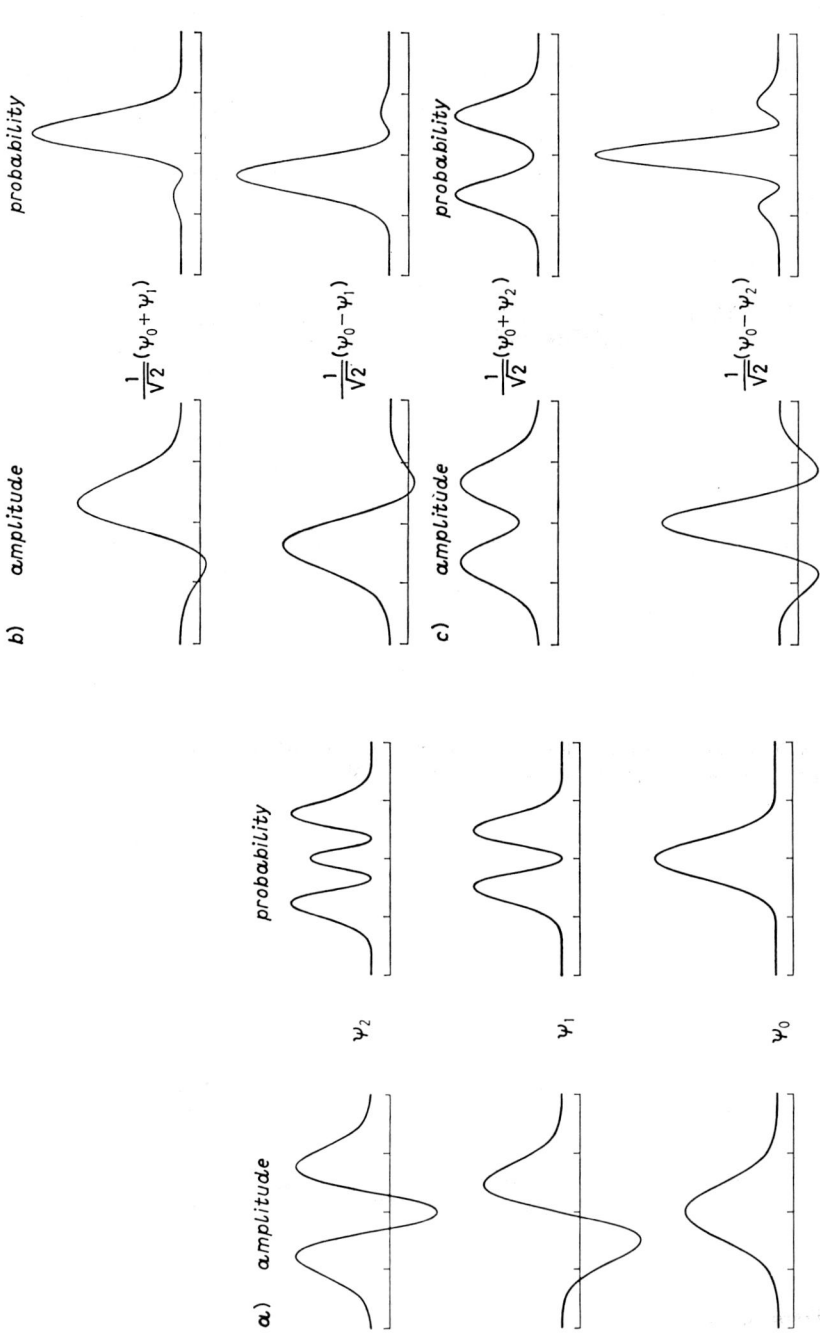

Fig. 1.15. – a) Three lowest wave functions $\psi_n(x)$ and probabilities $|\psi_n(x)|^2$ for a one-dimensional harmonic oscillator. b) Coherent superposition of ψ_0 and ψ_1 oscillating between $\psi_0 + \psi_1$ and $\psi_0 - \psi_1$. The probability density oscillates between left and right in analogy to the dipolar fictitious spin $I = 1/2$ discussed in the text. c) Coherent superposition of ψ_0 and ψ_2 displaying quadrupolelike oscillations. (Courtesy of E. L. HAHN.)

sign [43]. (Here recall our earlier discussion (subsect. 1˙10) of time reversal.)

The extension of a double-quantum free-induction decay time from a few hundred microseconds to a few milliseconds by a spin-locking experiment of this sort is shown in fig. 1.14. The technique has been used in double-quantum cross-polarization of broad deuterium and ^{14}N resonances in solids [27, 28]. A related technique involves the cross-polarization of the ^{14}N overtone [44].

Of course, these types of coherences do not occur exclusively in spin systems, but rather they are a general manifestation of the superposition of stationary states, which results in real or fictitious multipolar oscillations. For example, as shown in fig. 1.15, a coherent superposition between the ψ_0 and the ψ_1 states of a one-dimensional harmonic oscillator exhibits «dipolarlike» oscillations, and a coherent superposition between the ψ_0 and ψ_2 states exhibits «quadrupolarlike» oscillations as the superpositions evolve between the ± combinations [45].

1˙15. *Molecular structure by multiple-quantum NMR.* – In our previous discussions of multiple-quantum NMR, oriented benzene served as a convenient pedagogical prototype. Another example, perhaps closer to «real life», is given in fig. 1.16, which shows the normal one-quantum spectrum of a cyanobiphenyl liquid-crystal molecule containing 8 protons [18]. The result is to be contrasted with the six-quantum spectrum in fig. 1.17, where an analysis of the spectrum in terms of dipole couplings may be made to study the structure and dynamics of this flexible molecule. Using the structure

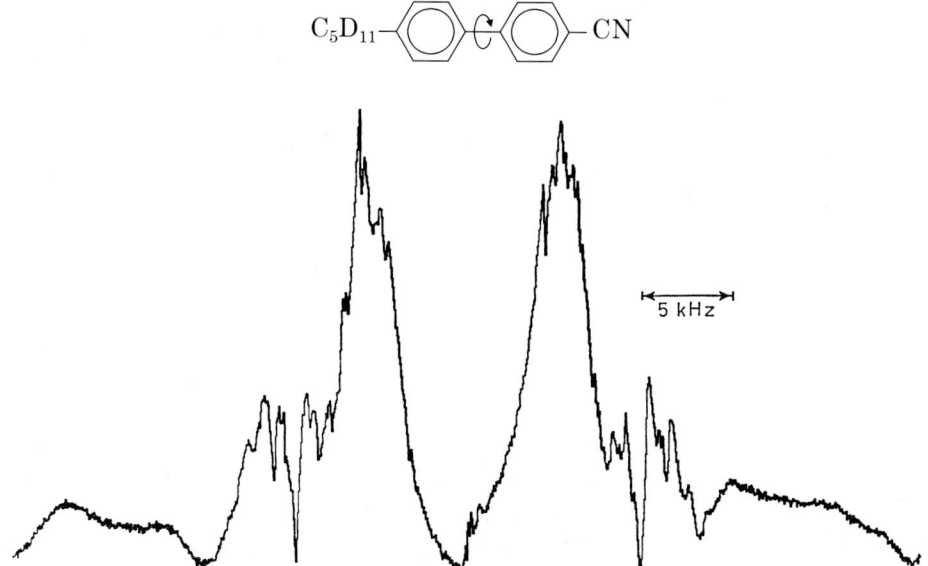

Fig. 1.16. – One-quantum NMR spectrum of the nematic liquid crystal *p*-pentyl-*p*-cyanobiphenyl-d_{11} (5CB) at 26.0 °C. The total frequency bandwidth is 50 kHz. (Adapted from *Mol. Phys.*, **53**, 333 (1984), with permission.)

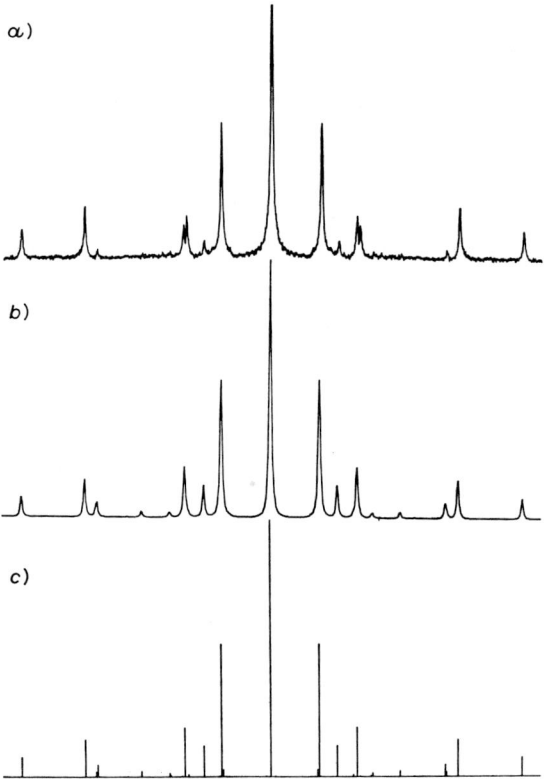

Fig. 1.17. – Six-quantum spectrum of 5CB, to be compared with fig. 1.16. This spectrum could be analyzed to yield the order tensor as well as the structure and dynamics of the biphenyl group. The lower plots are simulations (stick spectrum and slightly broadened spectrum) based on this structure assuming D_z symmetry for the spin Hamiltonian. The total frequency bandwidth is 44.2 kHz. Several spectra with different mixing times were used to obtain more uniform intensities for the lines. (Adapted from *Mol. Phys.*, **53**, 333 (1984), with permission.)

with jumps between the four equivalent conformations, the five-quantum spectrum can also be simulated, and this spectrum is compared with experiment in fig. 1.18.

A further application of multiple-quantum methods combines multiple-quantum filtering with two-dimensional correlation spectroscopy to investigate randomly deuterated molecules in liquid crystals. This method allows a separation of the different proton-containing isotopomers in situations where the normal proton spectra might be too complex (recall, for example, hexane-d_6 in fig. 1.4). An example of such simplified spectra, obtained for 80% randomly deuterated hexane, is shown in fig. 1.19 [46]. Spectrum 1.19a) was obtained with the

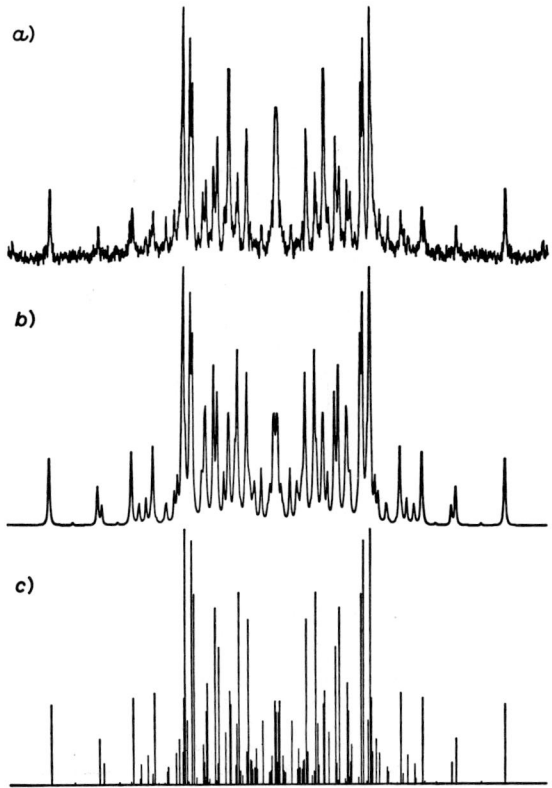

Fig. 1.18. – Five-quantum spectrum of 5CB showing the onset of complexity as we progress (or regress) from six quanta (fig. 1.17) towards one quantum (fig. 1.16). The simulation in the lower plots was produced from the structural and dynamic parameters derived from the six-quantum spectrum. The total frequency bandwidth is 62.5 kHz. Here, too, spectra from a range of preparation and mixing times were averaged together. (Adapted from *Mol. Phys.*, **53**, 333 (1984), with permission.)

pulse sequence

$$(1.45) \quad \left(\frac{\pi}{2}\right)_\varphi - t_1 - \left(\frac{\pi}{2}\right)_\varphi - \frac{\tau_1}{2} - \pi_x - \frac{\tau_1}{2} - \left(\frac{\pi}{2}\right)_x - \frac{\tau_2}{2} - \pi_x - \frac{\tau_2}{2} \text{-sample},$$

where φ is incremented in 90° steps while alternating the detector phase between 0° and 180° and increasing τ_1 after every fourth shot to reduce contributions from three-spin systems. Spectrum 1.19b) was obtained with the sequence

$$(1.46) \quad \left(\frac{\pi}{2}\right)_\varphi - \frac{\tau_1}{2} - \pi_\varphi - \frac{\tau_1}{2} - \left(\frac{\pi}{2}\right)_\varphi - t_1 - \left(\frac{\pi}{2}\right)_x - \frac{\tau_2}{2} - \pi_x - \frac{\tau_2}{2} \text{-sample},$$

where φ and the detector are again cycled. In this way, all 16 dipolar couplings of oriented hexane have been determined and assigned to positions on the molecules. Note that hexane is a formidable spin system, with 14 strongly cou-

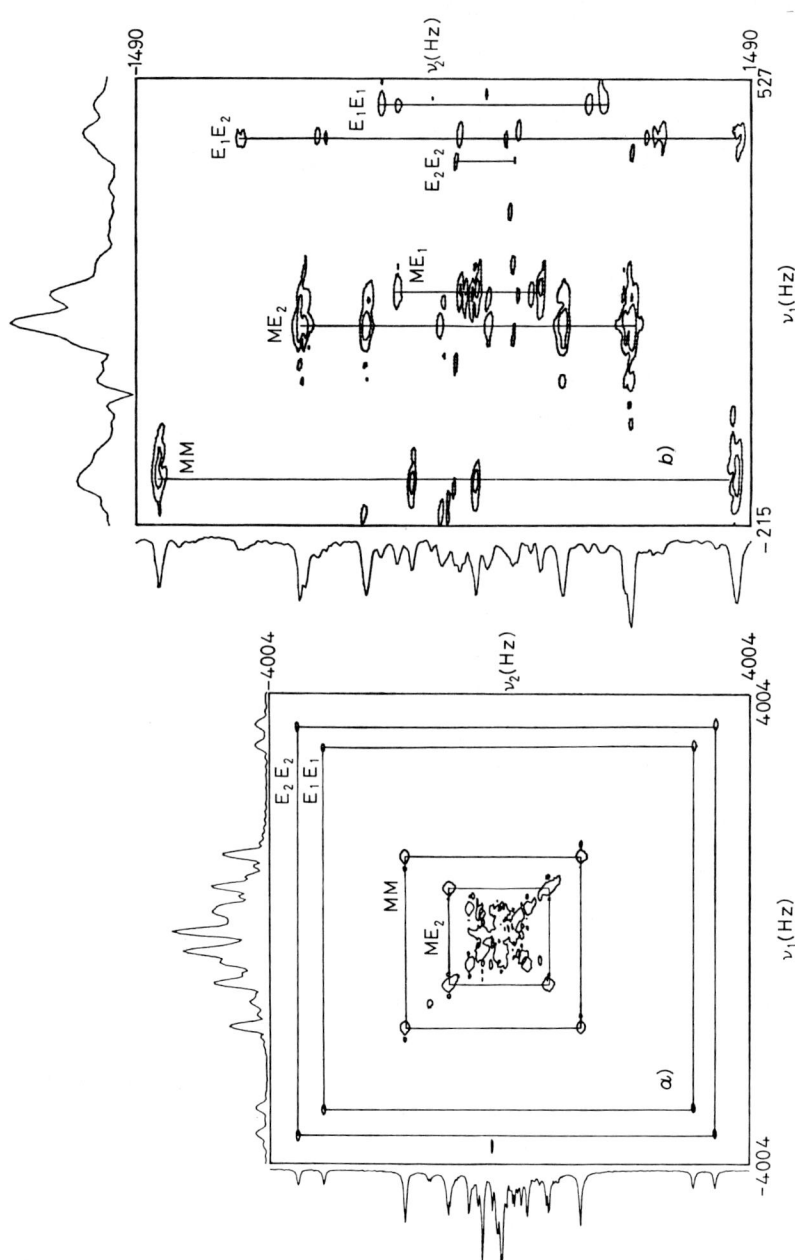

Fig. 1.19. – a) Double-quantum filtered two-dimensional correlation spectrum of 81% randomly deuterated n-hexane, 23 mole % in EK 11650, taken with the pulse sequence described in the text with deuterium decoupling. 128×1024 point data sets were collected at 360 MHz, with a spectral width of 16 667 Hz in both dimensions. The square patterns, which reveal dipole couplings in individual isotopomers, are shown for four of the sixteen proton pairs. b) Part of a double-quantum vs. single-quantum spectrum obtained using the pulse sequence described in the text. ν_1 is the double-quantum axis and ν_2 is the single-quantum axis. Vertical lines parallel to the ν_2 axis indicate the six possible double-quantum frequencies of molecules with two protons. M, E_1 and E_2 correspond to the methyl and the two inequivalent ethylene positions in hexane. All the dipole couplings in oriented hexane have been determined in this way. (Adapted from J. Am. Chem. Soc., **108**, 6813 (1986), with permission.)

pled protons, and that its conventional spectrum is considerably more complex than that shown in fig. 1.4. This technique can help determine conformational dynamics of flexible molecules of this kind [47, 48], and may be extended to molecules of biological interest.

1˙16. *Selective n-quantum excitation.* – High n-quantum transitions, though desirable, are generally of annoyingly low intensity. This problem can be partly overcome, however, by selective n-quantum excitation, in which a phase-shifted string of sequences with propagators U_φ is concatenated [16]. The total propagator U_T is given by

$$(1.47) \qquad U_T = U_{(n-1)\varphi} \ldots U_\varphi U_0 \,,$$

with each U_φ obtained from a basic U_0 by the transformation of eq. (1.35). If $\varphi = 2\pi/n$, then only n-quantum operators survive in the overall propagator, thereby making the propagator n-quantum selective. However, in order for this selectivity to be achieved, U must contain n-quantum operators. One approach is to construct a «sandwich»

$$(1.48) \qquad U_0 = Q^\dagger U Q \,,$$

where U describes a brief excitation and Q and $Q^\dagger$ describe a pair of lengthy sequences, related by time reversal, which induce multiple-quantum operators in U. An example of selective excitation of the 4-quantum spectrum of oriented benzene is shown in fig. 1.20.

Selective excitation can be viewed in the frequency domain as shown in fig. 1.21 and 1.22. Two-quantum excitation involves phase shifts of $2\pi/2 = \pi$, *i.e.* $0, \pi, 0, \pi, 0, \pi, \ldots$. This operator is symmetric in time and yields a symmetric

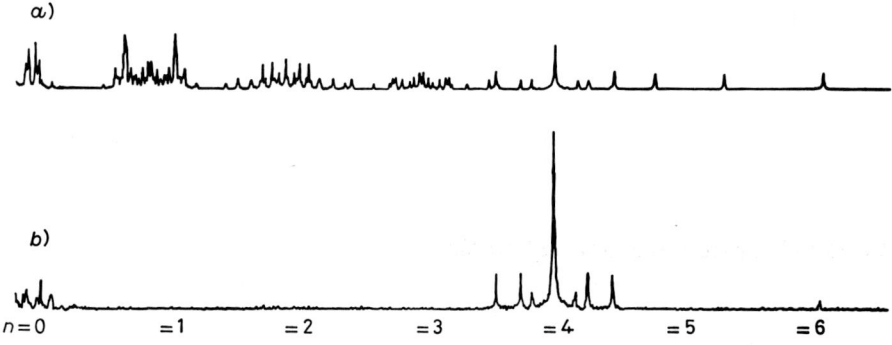

Fig. 1.20. – Multiple-quantum NMR spectra (ensemble averaged over a range of τ values) of oriented benzene *a)* with normal nonselective broad-band excitation and *b)* with four-quantum selective excitation. Selective n-quantum excitation enhances the n-quantum intensity, which, under nonselective excitation, may be low owing to the statistically low probability of absorbing many quanta. (Adapted from *J. Chem. Phys.*, **74**, 2808 (1981), with permission.)

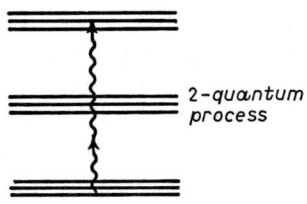

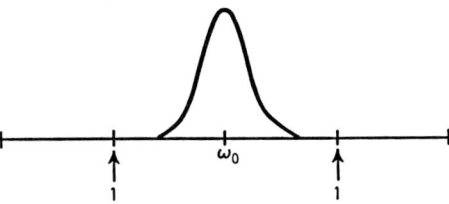

Fig. 1.21. – Symmetry of two-quantum selectivity by π phase shifting. The first resonant process involves one quantum each from the upper and the lower sidebands. (Adapted from *Adv. Magn. Reson.*, **11**, 111 (1983), with permission.)

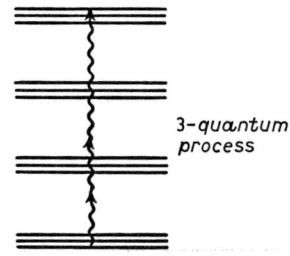

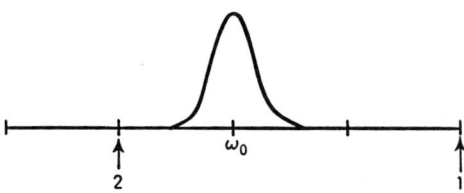

Fig. 1.22. – Symmetry of three-quantum selectivity by $2\pi/3$ phase shifts. The sidebands are now asymmetrically disposed about the resonance, and the first resonant process involves two quanta from the lower sideband and one from the upper sideband. (Adapted from *Adv. Magn. Reson.*, **11**, 111 (1983), with permission.)

nearest sideband structure about the resonance at ω_0 in fig. 1.21. The first resonant process is a 2-quantum transition involving one photon from each sideband. The phase shifts of $2\pi/3$ for 3-quantum excitation yield the sequence $0, 2\pi/3, 4\pi/3, 0, 2\pi/3, 4\pi/3, \ldots$ with the asymmetrical nearest sideband disposition shown in fig. 1.22. The first resonant process must now involve two photons from one sideband and one from the other. Bear in mind that this is a linear Fourier argument for the obviously nonlinear process of multiple-quantum excitation, so it is wrong in general. Nevertheless, it prescribes the correct symmetry and is useful if carefully applied.

1`17. *Multiple-quantum NMR in solids.* – Time-reversal detection as described in subsect. 1`10 is necessary in solids because the transition frequencies are distributed almost continuously. Consequently, there is an essentially complete cancellation of intensity if the phases of the line are random, as in fig. 1.9a). An alternative view of the same problem is depicted in fig. 1.23. Suppose we use the normal three-pulse sequence of eq. (1.28) to excite and observe multiple-quantum transitions in a solid. The integrated intensity of the multiple-quantum spectrum is given by the initial amplitude of the magnetization sampled in t_2 when $t_1 = 0$. But in a solid, after long τ the magnetization will have decayed to zero and hence little or no multiple-quantum intensity will be observed.

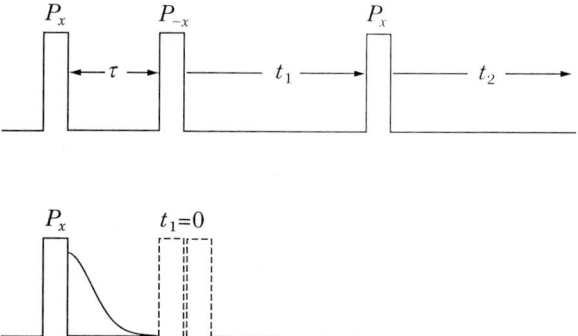

Fig. 1.23. – Pulse sequence for the three-pulse multiple-quantum scheme. If $t_1 = 0$ in a solid and τ is longer than the solid FID time, the signal is zero; hence the integrated multiple-quantum intensity will be zero as well. This illustration presents another view of the problem outlined in fig. 1.9.

Using the appropriate time-reversal detection, however, we can indeed obtain solid-state spectra with high n, in excess of 100 quanta [17]. An example of a suitable time-reversal excitation experiment is shown in fig. 1.24. The two pulses of this simple excitation sequence are replaced by a train of pulses whose average Hamiltonian (see sect. 2) is given by

(1.49) $$H_{\text{DQ}} = (H_{xx} - H_{yy})/3 \, ,$$

where DQ denotes a pure double-quantum operator. The nonsecular Hamiltonian H_{xx} was mentioned in subsect. 1.9.

If the phases of the x and $\bar{x}$ r.f. pulses in the lower sequence of fig. 1.24 are all shifted by $\pi/2$ to y and $\bar{y}$ pulses, then we simply exchange the x and y indices in eq. (1.49) and obtain

$$R_z^\dagger\left(\frac{\pi}{2}\right) H_{DQ} R_z\left(\frac{\pi}{2}\right) = -(H_{xx} - H_{yy})/3 \,. \tag{1.50}$$

Hence, from eq. (1.12), the unitary propagator U is shifted to its adjoint

$$U \to U^\dagger \,, \tag{1.51}$$

as required for detection by eqs. (1.33) or (1.39) for use with TPPI.

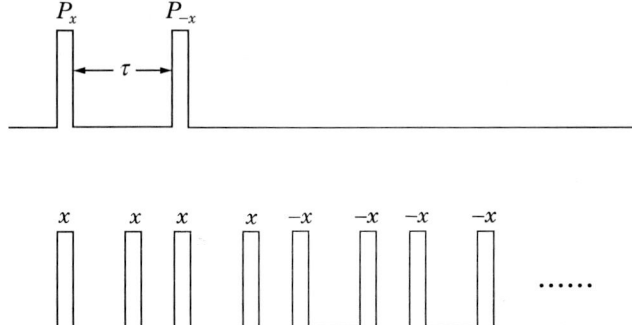

Fig. 1.24. – By replacing the pair of pulses with the lower pulse train and employing phase shifting, time-reversal detection can be used to recover the full multiple-quantum intensity in a solid.

Figure 1.25 shows multiple-quantum spectra obtained in solid hexamethylbenzene for different preparation and mixing times τ with the lower sequence of fig. 1.24 [18]. Solid hexamethylbenzene is an essentially infinite network of coupled spins, so we expect the multiple-quantum absorption to increase continuously with τ. If at any given τ we consider that a finite cluster of N spins has become involved, then, from eq. (1.5), the integrated intensity of the n-quantum transitions should go approximately as

$$Z_n = \binom{2N}{N+n} \sim \frac{4^N}{\sqrt{N\pi}} \exp[-n^2/N] \tag{1.52}$$

according to Stirling's approximation for large N and $n \ll N$. So we expect a roughly Gaussian distribution of intensities, a prediction borne out well in fig. 1.26, where even-ordered multiple-quantum spectra from a finite cluster of 21

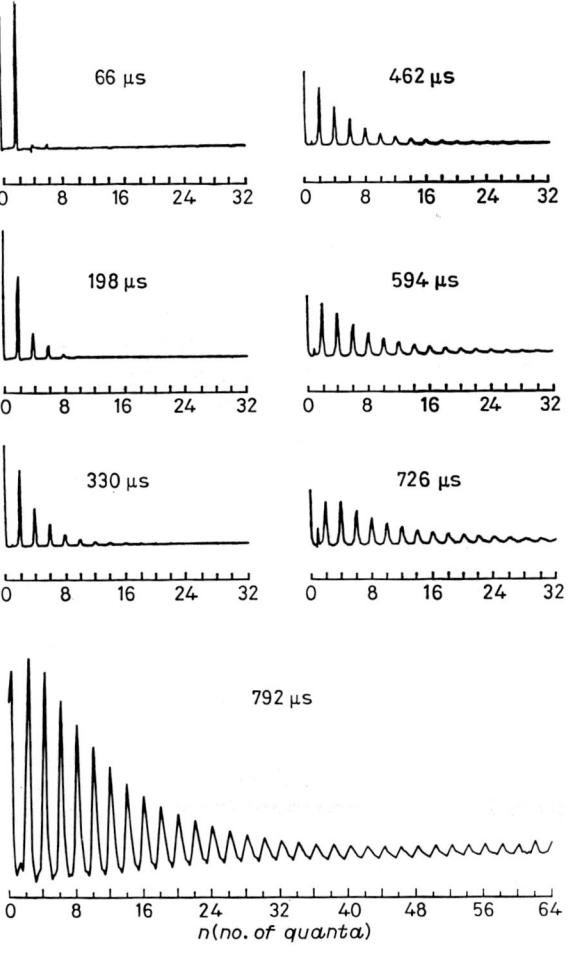

Fig. 1.25. – Multiple-quantum spectra of solid hexamethylbenzene obtained using the lower pulse sequence in fig. 1.24, with overall preparation and mixing times τ ranging from 66 to 792 µs. (The lowest trace, 792 µs, is expanded vertically.) As τ increases, more and more spins are correlated and more and more quanta are involved. In excess of one hundred quanta have been observed in this way. (Adapted from *J. Chem. Phys.*, **83**, 2015 (1985), with permission.)

spins are shown. The time dependence of the Gaussian width, also shown in fig. 1.26, can provide information on the distribution of atoms in materials, as depicted schematically in fig. 1.27. This method should be particularly useful for the study of clusters and, indeed, preliminary results have been realized in a number of cases, including inorganic solids, molecules in zeolite cavities [49], hydrogen in amorphous semiconductors [50] and atoms and molecules on sur-

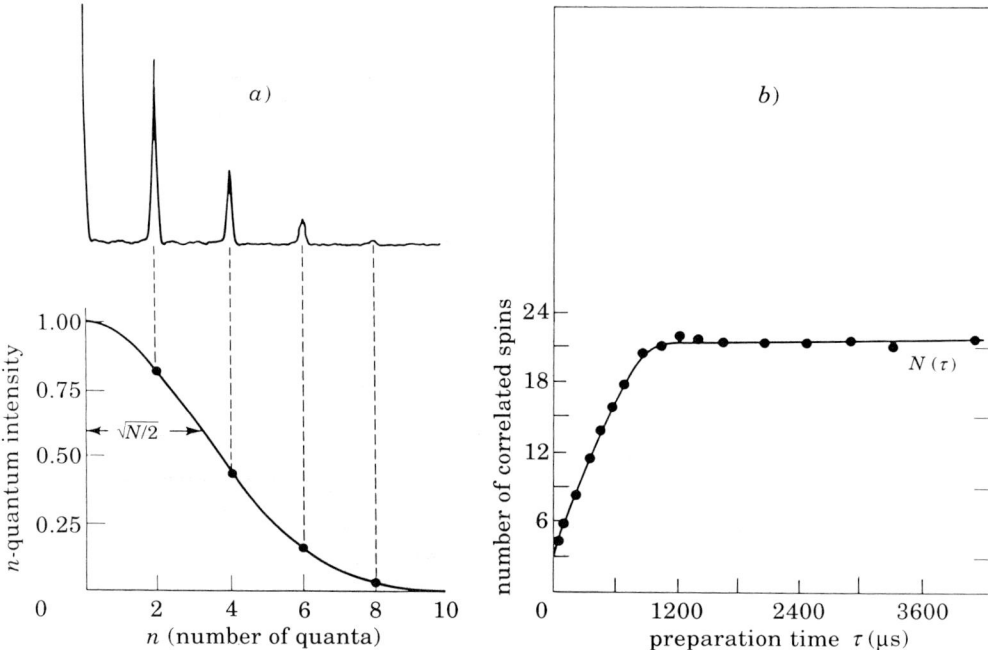

Fig. 1.26. – a) Multiple-quantum spectra of liquid crystal 5CB (containing 21 protons) showing a Gaussian distribution of intensities. b) Number of correlated spins $N(\tau)(\equiv$ maximum number of quanta) as a function of the preparation and mixing times τ in 5CB. $N(\tau)$ is extracted from an analysis of the Gaussian intensity distribution of a) as a function of τ. The plateau at 21 indicates that these molecules contain isolated clusters of 21 proton spins. (Adapted from J. Am. Chem. Soc., **108**, 7447 (1986), with permission.)

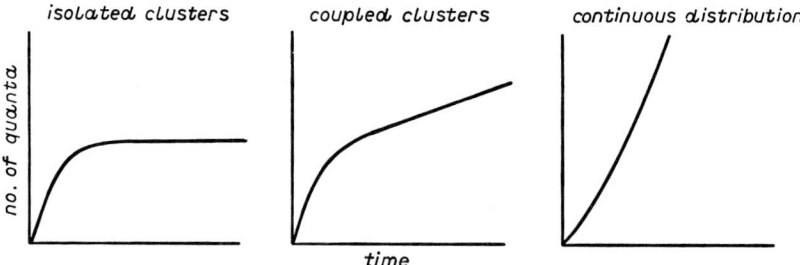

Fig. 1.27. – Schematic of $N(\tau)$, the number of correlated spins or maximum number of quanta, as a function of preparation time τ for various distributions of spins. This type of behavior has been observed in a variety of systems, including molecular crystals, molecules in zeolites, liquid crystals, absorbed species on surfaces and hydrogenated amorphous semiconductors.

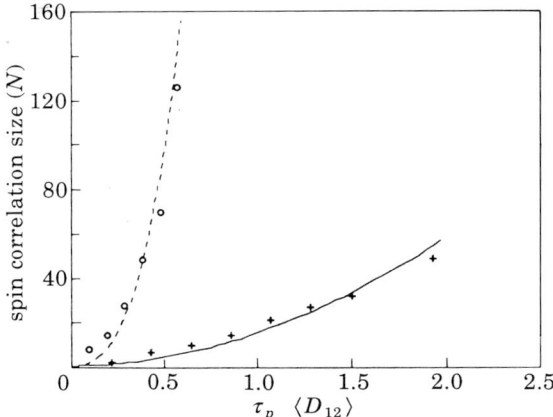

Fig. 1.28. – Fits of multiple-quantum dynamics models to data for CaH_2, representing a 3D system, and diamond powder, representing a surface or 2D system of hydrogen: ○ CaH_2, + diamond powder. The solid line refers to a 2D model while the dotted line represents a 3D model. (Adapted from *J. Phys. Chem.*, **96**, 8125 (1992), with permission.)

faces. For example, fig. 1.28 shows experimental results which allow a determination of the dimensionality of the system under consideration. In this case the growth of multiple-quantum coherence in the protons in a three-dimensional system (CaH_2) can easily be discriminated from the growth in the two-dimensional diamond powder system[51]. For extended coupled-spin systems, the more rigorous approach of Lacelle is to be preferred[51bis].

1˙18. *Selection rules in multiple-quantum dynamics.* – If the pure double-quantum Hamiltonian eq. (1.49) is used for preparation and detection, then for $\rho(0) = I_z$ only even-quantum transitions are excited. If we characterize the dynamical evolution of the system by the number of spins N and the number of quanta, or coherence order, n, then the dynamical selection rules

(1.53) $$\Delta N = \pm 1 \quad \text{and} \quad \Delta n = \pm 2$$

are imposed. With the initial condition for $\rho(0) = I_z$

(1.54) $$N = 1, \quad n = 0,$$

the states ($N = 4$, $n = 4$), ($N = 8$, $n = 8$), ... are not accessible, *i.e.* we have the surprising consequence that it should not be possible to observe 4-quantum coherence in a 4-spin system, etc.[52]. Multiple-quantum dynamics studies have recently been extended by means of Monte Carlo techniques[53, 54].

1˙19. *Total coherence transfer.* – Techniques related to multiple-quantum NMR are frequently useful for heteronuclear spectroscopy in isotropic liquids.

For example, to obtain narrow NMR spectra in the presence of inhomogeneous magnetic fields or susceptibility broadening. One technique, termed SHARP, uses echoes during the evolution of ^{13}C or ^{15}N coherence in the presence of coupled protons [55, 56]. This procedure should be useful in spatially selective NMR of heteronuclear systems with two r.f. coils, in particular where the relevant spatial region is in an inhomogeneous field or broadened by magnetic-susceptibility effects. An example, shown in fig. 1.29, demonstrates a simple version of spatial selectivity and high resolution with a surface coil on tubes con-

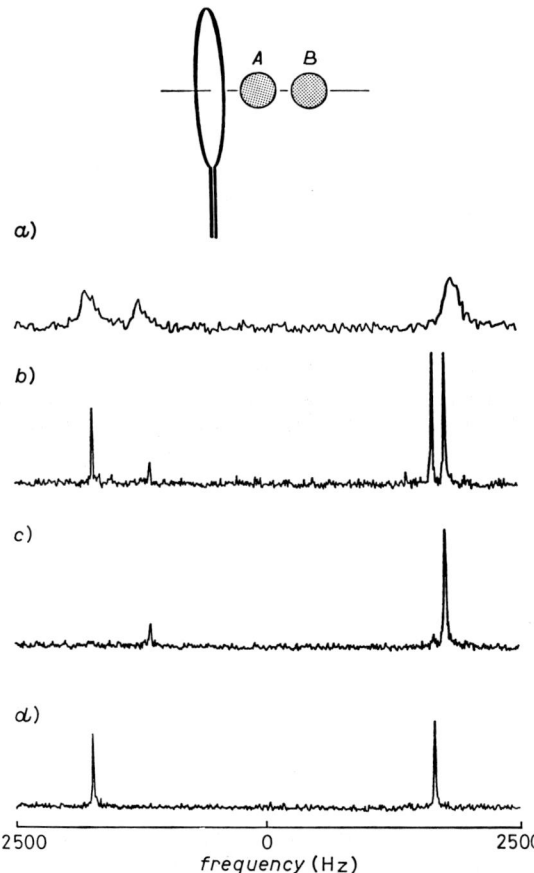

Fig. 1.29. – High-resolution surface coil experiment on two capillary tubes $A + B$ placed on axis at 2 and 4 mm from the coil. Tube A contained 3 μl of a ^{13}C-enriched alanine solution and B contained 2 μl ^{13}C-enriched ethanol. a) ^{13}C spectrum illustrating the inhomogeneous field. b) High-resolution SHARP spectrum, nonselective, implemented as described in the text. c) High-resolution SHARP spectrum, pulse times set for selectivity at A. d) SHARP spectrum, pulse times set for selectivity at B. (Adapted from *J. Am. Chem. Soc.*, **107**, 7193 (1985), with permission.)

taining ethanol and alanine. The pulse sequence used for this experiment was

$$(1.55) \quad \begin{cases} I(^1\text{H})\left(\dfrac{\pi}{2}\right)_x - \dfrac{\tau}{2} - \pi - \dfrac{\tau}{2} - x\left(\dfrac{4t_1}{5}\right) - \dfrac{\tau'}{2} - \pi - \dfrac{\tau'}{2} - \dfrac{t_1}{5}\text{-sample}\,(\pm), \\ S(^{13}\text{C}) \qquad \pi \quad \dfrac{\pi}{2}(\pm) \quad \dfrac{\pi}{2} \quad \pi \qquad \pi, \end{cases}$$

where $x(4t_1/5)$ implies continuous irradiation and the $\pm$ indicate that signals from experiments differing by 180° in phase are subtracted to further suppress nonsatellite peaks.

1˙20. *Bilinear rotation pulses.* – A further use of heteronuclear multiple-quantum coherence in heteronuclear spectroscopy in isotropic liquids is bilinear rotation decoupling (BIRD)[57], a method for the homonuclear decoupling of protons in weakly coupled liquid spin systems, which can be thought of as follows. Suppose a hard π pulse is applied to all protons and a π pulse is then applied to directly coupled ^{13}C-^{1}H pairs. This selective pulse restores to $+z$ all protons spins coupled directly to ^{13}C. The net effect is to invert all other spins, thus enabling them to be decoupled.

A pulse sequence that induces the bilinear rotation is

$$(1.56) \quad \begin{cases} I(^1\text{H})\left(\dfrac{\pi}{2}\right) - \dfrac{\tau}{2} - \pi - \dfrac{\tau}{2} - \left(\dfrac{\pi}{2}\right). \\ S(^{13}\text{C}) \qquad \pi \end{cases}$$

If the Hamiltonian for the coupled I-S system is $H_{IS} = JI_zS_z$, then eq. (1.56) produces a propagator of the form $U(\tau) = \exp[-i2\pi\tau JI_yS_z]$. For $\tau = 1/J$ this propagator is a π pulse only for the directly bonded (satellite) protons. Bilinear rotation sequences have also been used in a variety of experiments in liquids to discriminate between one-bond and long-range couplings[58].

2. – Coherent-averaging theory.

2˙1. *Introduction.* – A recurring theme throughout these lectures is the requirement for the preparation or evolution of a system under a specified Hamiltonian or under a Hamiltonian with a particular symmetry or transformation properties. For example, in decoupling we attempt to make the effective coupling Hamiltonian zero. In multiple-quantum spectroscopy of solids we required a time-reversed detection with $H \sim H_{xx} - H_{yy}$. It is clearly necessary on many occasions to implement a specific desired Hamiltonian, perhaps different from the natural unperturbed Hamiltonian of the system. In other words, whereas the system might naturally evolve under its propagator $U(t)$,

(2.1)

we apply a perturbation so that the system evolves under a different propagator $\overline{U}(t)$, due to a Hamiltonian $\overline{H}$, arriving perhaps at a different state at time t. The perturbation needed may be a sequence of pulses or a mechanical rotation. For incoherent perturbations $\overline{U}(t)$ may exist where $\overline{H}$ does not.

A theory that accounts well for the design of specific Hamiltonians (and, therefore, propagators) under coherent perturbation is coherent-averaging theory [59]. Suppose the Hamiltonian is time dependent (owing to, for example, an applied perturbation), as depicted in fig. 2.1. A question which arises is: can we find a time-independent Hamiltonian $\overline{H}$ which can induce evolution of the system through $\overline{U}(t)$ to the same state as $H(t)$ would induce at time t? The answer is yes, that $\overline{H}$ is the Magnus Hamiltonian [60], given by

(2.2) $$\overline{H} = \overline{H}^{(0)} + \overline{H}^{(1)} + \overline{H}^{(2)} + \dots,$$

where

(2.3) $$\overline{H}^{(0)} = \frac{1}{t} \int_0^t H(t') \, dt'$$

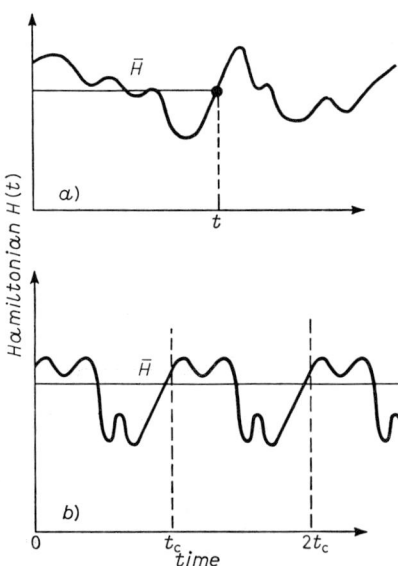

Fig. 2.1. – For a time-dependent Hamiltonian, the evolution can be often characterized by an effective time-independent average Hamiltonian $\overline{H}$. In general, if $H(t)$ is not periodic (a)), then $\overline{H}$ depends on t. If $H(t)$ is periodic (b)) then the same $\overline{H}$ is relevant for multiples of the period.

is the average Hamiltonian. The next two terms in this series are

$$(2.4) \qquad \overline{H}^{(1)} = \frac{-i}{2t} \int_0^t \int_0^{t'} [H(t'), H(t'')] \, dt'' \, dt' ,$$

and

$$(2.5) \qquad \overline{H}^{(2)} = -\frac{1}{6t} \int_0^t \int_0^{t'} \int_0^{t''} \{[H(t'), [H(t''), H(t''')]] + \\ + [H(t'''), [H(t''), H(t')]]\} \, dt''' \, dt'' \, dt' .$$

If eq. (2.2) converges rapidly, so that $\overline{H}^{(k)}$ are small for $k \neq 0$, then the average Hamiltonian $\overline{H}^{(0)}$ provides a good description of the system and the other $\overline{H}^{(k)}$ are correction terms. By applying a perturbation we have, therefore, taken a system which would have evolved under some Hamiltonian H and caused it to evolve effectively under $\overline{H}^{(0)}$ as in eq. (2.1). The trick is to implement the desired $\overline{H}^{(0)}$ by appropriate perturbations that can be realized experimentally, such as spin decoupling and time reversal. Note that, in general, $\overline{H}$ and $\overline{H}^{(0)}$ depend on t, but, if $H(t)$ is periodic, then the same $\overline{H}$ and $\overline{H}^{(0)}$ hold at all integer multiples of t_c. (If $H(t)$ is not periodic, the approach is still useful, since one can implement an $\overline{H}$ which provides a particular desired final state $|\bar{\rho}(t)\rangle$ at time t.)

Expressions (2.3) to (2.5) were used to calculate the time-reversal Hamiltonian of eq. (1.49) corresponding to the sequence of fig. 1.24, and they will be used repeatedly in the following sections. This approach has been useful in a variety of problems in different areas, for example truncation of couplings by high field [61-63], composite pulses [64-66], iterative schemes [67] and multiple-quantum operators [68, 69].

A *particularly useful simplification occurs when the time-dependent Hamiltonian H(t) commutes with itself at all times, i.e. [H(t), H(t')] = 0.* In this case all $\overline{H}^{(k)} = 0$ for all $k \neq 0$, as we can see from eqs. (2.4) and (2.5) for $k = 1, 2$, and the average Hamiltonian $\overline{H}^{(0)}$ is exact. This happens for Carr-Purcell trains [70] and for magic-angle spinning [71] involving chemical-shift interactions [72] (and related cases such as dipolar couplings of N spins in a one-dimensional array, of which a pair of spins is the simplest example), dipolar couplings averaged by uniaxial molecular motion, heteronuclear dipolar couplings, or first-order quadrupolar coupling of isolated spins. We shall see that in this way it is possible to narrow the spectrum by magic-angle spinning even for broad quadrupolar spectra in solids.

2˙2. *Multiple-pulse line narrowing*. – As an example of coherent averaging, we take the opportunity to review the WHH-4 multiple-pulse sequence [73] for removing the homonuclear dipolar interaction in solids. In this four-pulse se-

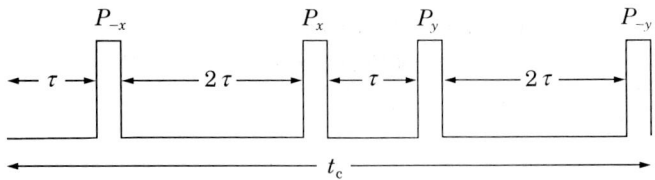

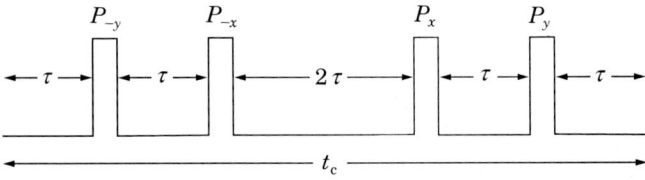

Fig. 2.2. – Pulse sequence for one cycle of the four-pulse WHH-4 multiple-pulse decoupling sequence. The signal is sampled in the center of the 2τ periods. Shifting the sequence by an amount τ to the left reveals the inherent symmetry, which ensures that all the odd-order correction terms in the Magnus Hamiltonian vanish.

quence, shown in fig. 2.2, the signal is observed stroboscopically during the sampling windows. The one-cycle propagator of the sequence can be written

(2.6) $$U(t_c) = U_z(\tau) P_{\bar{x}} U_z(2\tau) P_x U_z(\tau) P_y U_z(2\tau) P_{\bar{y}},$$

where

$$P_\alpha = \exp\left[-i\frac{\pi}{2} I_\alpha\right], \qquad P_{\bar{\alpha}} = \exp\left[-i\frac{\pi}{2} I_{\bar{\alpha}}\right]$$

and

$$U_\alpha(\tau) = \exp[-iH_{\alpha\alpha}\tau], \qquad \alpha = \pm x, \pm y, \pm z.$$

We now move into the so-called «toggling rotating frame» by applying a transformation corresponding to the operation produced by the r.f. pulses to obtain

(2.7) $$U(t_c) = U_z(\tau) U_y(2\tau) U_z(\tau) U_x(2\tau).$$

Note that by shifting the cycle from the first sampling window to the second (as illustrated in fig. 2.2), we can rewrite eq. (2.7) as

(2.8) $$U(t_c) = U_x(\tau) U_z(\tau) U_y(\tau) U_y(\tau) U_z(\tau) U_x(\tau).$$

This is exactly equivalent to the expression of eq. (2.6), but it reveals that the sequence is in fact symmetric about its center. We are now in a position to evaluate the different terms in the Magnus expansion, to find

(2.9) $$\overline{H}^{(0)} = \frac{1}{3}(H_{xx} + H_{yy} + H_{zz}),$$

which yields for the dipolar interaction

(2.10) $$\overline{H}_{\mathrm{D}}^{(0)} = 0,$$

and for the chemical-shift interaction

(2.11) $$\overline{H}_{\mathrm{CS}}^{(0)} = \sum_k \frac{1}{\sqrt{3}} (\omega_i^{(k)} + \Delta\omega) I_{111}^{(k)},$$

where $I_{111}^{(k)} = I_x^{(k)} + I_y^{(k)} + I_z^{(k)}$. Thus, to zeroth order, the homonuclear dipolar interaction is removed by the WHH-4 sequence, and the chemical-shift interaction is scaled by a factor $1/\sqrt{3}$.

Because of the reflection symmetry, $\overline{H}^{(1)}$ and all of the odd-order correction terms vanish [74, 75], leaving the second-order terms as the leading corrections,

(2.12) $$\overline{H}^{(2)} = \frac{\tau^2}{18} \{[(H_{zz} + 2H_{xx} + 2H_{yy}), [(H_{xx} + H_{yy}), H_{zz}]] -$$
$$- [(H_{yy} + 2H_{xx}), [H_{yy}, H_{xx}]]\},$$

which reduces to

(2.13) $$\overline{H}_{\mathrm{D}}^{(2)} = \frac{\tau^2}{18} [(H_{\mathrm{D}xx} - H_{\mathrm{D}zz}), [H_{\mathrm{D}yy}, H_{\mathrm{D}xx}]]$$

for the dipolar interaction, and in the case of the chemical-shift interaction

(2.14) $$\overline{H}_{\mathrm{CS}}^{(2)} = \frac{\tau^2}{18} \sum_k (\sigma_i^{(k)} + \Delta\omega)^3 (I_{yk} - 2I_{xk} + 4I_{zk}).$$

$\overline{H}_{\mathrm{D}}^{(2)}$ is responsible for the ultimate resolution available from the WHH-4 experiment on resonance (in the absence of pulse imperfections). Using coherent-averaging theory one can go on to calculate the effects of cross-terms between the chemical-shift and dipolar interactions, and the effect of pulse imperfections on the performance of the sequence in a relatively straightforward manner. In this way iterative schemes have been developed to compensate for such imperfections to higher and higher order, resulting in such advanced decoupling sequences as MREV-8 [76, 77], BLEW [78], BR-24 [79] and recent sequences from CORY [80].

2˙3. *Magic-angle spinning*. – The WHH-4 sequence and other multiple-pulse techniques present methods for removing dipolar interactions by averaging in spin space. In this subsection we shall briefly show how averaging of the spatial coordinates by sample spinning leads to an average Hamiltonian in which the dipolar interaction is also removed [81, 82]. In sect. 5 we shall return to the issue of averaging under various schemes of sample reorientation, but from a group-theoretical perspective, and the example presented here serves as a useful comparison.

The truncated Hamiltonian in the rotating frame representing the chemical shift has the form [83]

$$(2.15) \qquad H_0^{cs} = \gamma \hbar A_{00} T_{00} + \gamma \hbar A_{20} T_{20},$$

where A_{lm} and T_{lm} are irreducible spherical tensor components expressing the spatial and spin dependence of H, respectively. The heteronuclear dipolar and first-order quadrupolar interactions contain second-rank terms of the same form as those of eq. (2.15), and magic-angle spinning has an analogous effect on dipolar and first-order quadrupolar broadening as we derive here for the chemical shift. Note that the first term is isotropic, but that the second term depends on the orientation with respect to the external field.

Re-expressing $\mathbf{A}$ in a coordinate system fixed in the sample (rotor) which is rotating about an axis inclined by an angle Θ to the magnetic field, where the coordinates will be expressed by a_{lm}, we obtain

$$(2.16) \qquad H_0^{cs} = \gamma \hbar a_{00} \tau_{00} + \gamma \hbar \tau_{20} \sum_{m_1=-2}^{2} D_{m_1 0}^{(2)}(\Omega_{\text{sfc}}(t)) a_{2m}.$$

Because of the invariance of the Hamiltonian to rotations about the main field, this coordinate transform only depends on Θ and $\phi = \omega_r t$. The matrix $\mathbf{a}$ can be diagonalized by a rotation $\Omega_{\text{pas}} = (\alpha, \beta, \gamma)$ into its principal axis system, in which its components are written ρ_{lm},

$$(2.17) \qquad a_{lm_1} = \sum_{m_2=-2}^{2} D_{m_2 m_1}^{(l)}(\Omega_{\text{pas}}) \rho_{lm_2}.$$

In this picture the only nonvanishing components are ρ_{00}, ρ_{20} and $\rho_{2\pm2}$, which are conventionally expressed as the isotropic chemical shift $\rho_i = \rho_{00}$, an anisotropy $\delta = \sqrt{2/3}\,\rho_{20}$ and an asymmetry parameter $\eta = \sqrt{6}\,\rho_{2,\pm2}/\rho_{20}$. Applying eq. (2.17) to eq. (2.16) we finally have an expression for the time-dependent Hamiltonian in the rotating frame

$$(2.18) \qquad H_0^{cs} = \gamma \hbar a_{00} \tau_{00} + \gamma \hbar \tau_{20} \sum_{m_1=-2}^{2} D_{m_1 0}^{(2)}(\Omega_{\text{sfc}}(t)) \sum_{m_2=-2}^{2} D_{m_2 m_1}^{(2)}(\Omega_{\text{pas}}) \rho_{2m_2}$$

and expansion gives

$$(2.19) \qquad H_0^{cs} = \gamma \hbar \rho_i \tau_{00} + \gamma \hbar \sqrt{\frac{3}{2}} \delta \tau_{20} \left\{ \frac{1}{2}(3\cos^2\Theta - 1) \right\} \cdot$$
$$\cdot \left\{ \frac{1}{2}(3\cos^2\beta - 1) + \frac{\eta}{2} \sin^2\beta \cos 2\gamma \right\} + \gamma \hbar \sqrt{\frac{3}{2}} \delta \tau_{20} \xi_{\text{MAS}}(t),$$

where $\xi_{\text{MAS}}(t)$ is an orientation-dependent function of the rotor phase [83].

Equation (2.19) demonstrates the spinning-induced periodicity of the spin Hamiltonian. A single rotor period $\phi(0 \to 2\pi)$ is periodic in the sense discussed above, and we can determine the average of the time-dependent terms using the Magnus expansion. Performing the integration of eq. (2.3) yields $\bar{\xi} = 0$ and eq. (2.19) becomes

$$(2.20) \quad \overline{H}^{(0)} = \gamma\hbar\rho_i\tau_{00} + \gamma\hbar\sqrt{\frac{3}{2}}\delta\tau_{20}\left\{\frac{1}{2}(3\cos^2\Theta - 1)\right\} \cdot$$

$$\cdot \left\{\frac{1}{2}(3\cos^2\beta - 1) + \frac{\eta}{2}\sin^2\beta\cos 2\gamma\right\}.$$

If the angle of the spinning axis is chosen such that $3\cos^2\Theta - 1 = 0$, then we obtain a purely isotropic zeroth-order average Hamiltonian

$$(2.21) \quad \overline{H}^{(0)} = \gamma\hbar\rho_i\tau_{00}.$$

The conditions for the validity of this expression are that the series of eqs. (2.2)-(2.5) converges rapidly. In this case it converges as

$$(2.22) \quad \overline{H}^{(n+1)} \approx \frac{\Delta}{\omega_r}\overline{H}^{(n)},$$

where Δ is the «size» of the interaction being averaged. Therefore, eq. (2.20) will provide an accurate description when $\omega_r \gg \Delta$. Although the spinning speed currently reaches as high as 30 kHz [84], there are many examples where this is not sufficient and the criterion $\omega_r \gg \Delta$ is not met. However, a special situation arises if the form of the Hamiltonian is such that the commutator in eq. (2.4) vanishes identically. In this case eq. (2.4) and all higher terms (which depend on the same commutator) disappear. This will be true when the interaction is inhomogeneous, *e.g.* when the spin terms in the Hamiltonian remain the same and they are only scaled under rotation. In the inhomogeneous case eq. (2.19) reduces under spinning at the magic angle to

$$(2.23) \quad H_0 = \gamma\hbar\rho_i\tau_{00} + \gamma\hbar\sqrt{\frac{3}{2}}\tau_{20}\partial\xi_{\text{MAS}}(t),$$

and the spectrum breaks up into a narrow centerband at the isotropic chemical shift and a series of narrow sidebands evenly spaced at a separation equal to the rotor frequency. Thus, even at relatively low speeds, magic-angle spinning allows us to obtain high-resolution spectra. As examples of the effect of magic-angle spinning fig. 2.3 shows an early result of the deuterium spectrum of a mixture of solid hexamethylbenzene and ferrocene [85], and fig. 2.4 shows a recent ^{27}Al magic-angle spinning spectrum of the full quadrupolar manifold (central and satellite transitions) for α-Al$_2$O$_3$ which was used to determine the quadrupolar-coupling constants and asymmetry parameters for the two sites in this sample [86].

In order to average both homonuclear dipolar interactions and chemical-shift anisotropy it has proved necessary to combine the effects of multiple-pulse sequences and magic-angle spinning. The combined technique (CRAMPS) has provided resolved spectra for protons and fluorine [87, 88].

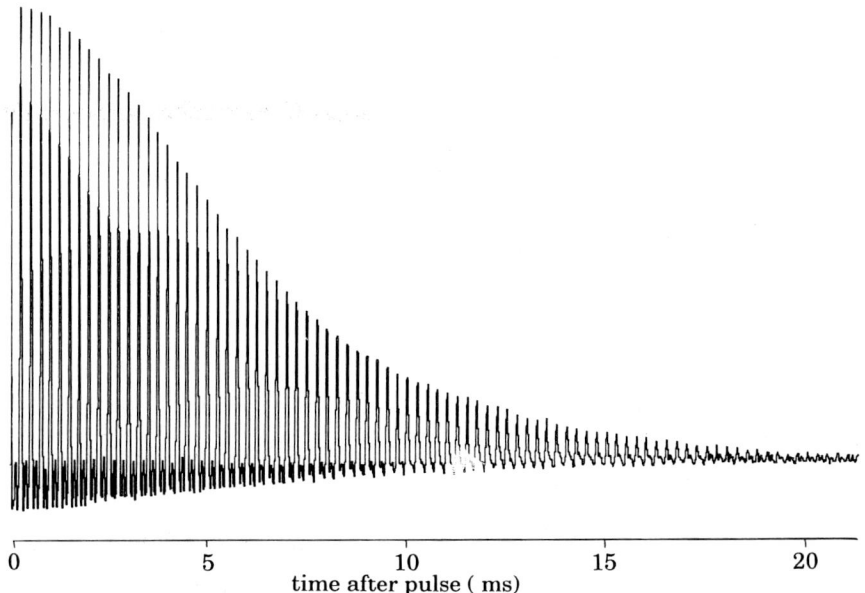

Fig. 2.3a. – Deuterium rotational echo train obtained from magic-angle spinning.

2˙4. *Double rotation*. – If interactions of different nonzero ranks are present, it is sometimes not possible to average all the anisotropic interactions to zero by spinning about a single axis [81, 89-91]. This occurs, for example, in the case of second-order quadrupolar broadening. In sect. 5 we shall demonstrate how to solve the problem of simultaneously averaging two interactions on the basis of symmetry. Here we consider averaging over more sophisticated trajectories than simply spinning around a single axis, for which the average of all interactions is zero [92, 93].

In the case of the central transition of a half-integer quadrupolar nucleus the Hamiltonian for a static sample is

$$(2.24) \qquad H_0^Q = \sum_{l=0,2,4} C_l A_{l0} T_{l0},$$

where the C_l are constants related to the quadrupolar coupling. The orientation dependence is expressed in terms of a sum of second- and fourth-rank tensors. We have seen above that after applying a rotation around a single axis the Hamiltonian is truncated along the rotation axis (if the spinning is fast enough), and the residual is then proportional to the Legendre polynomials $P_l(\cos\Theta)$. In this case $l = 2$ and 4, and spinning around a single axis will only partially narrow the line, since the roots of $P_2(\cos\Theta)$ and $P_4(\cos\Theta)$ do not coincide. However, applying another rotation relative to the first will further narrow the line. Solutions can be found which completely eliminate the second-order broadening with two rotations. In the double-rotation method (DOR) we spin around a first

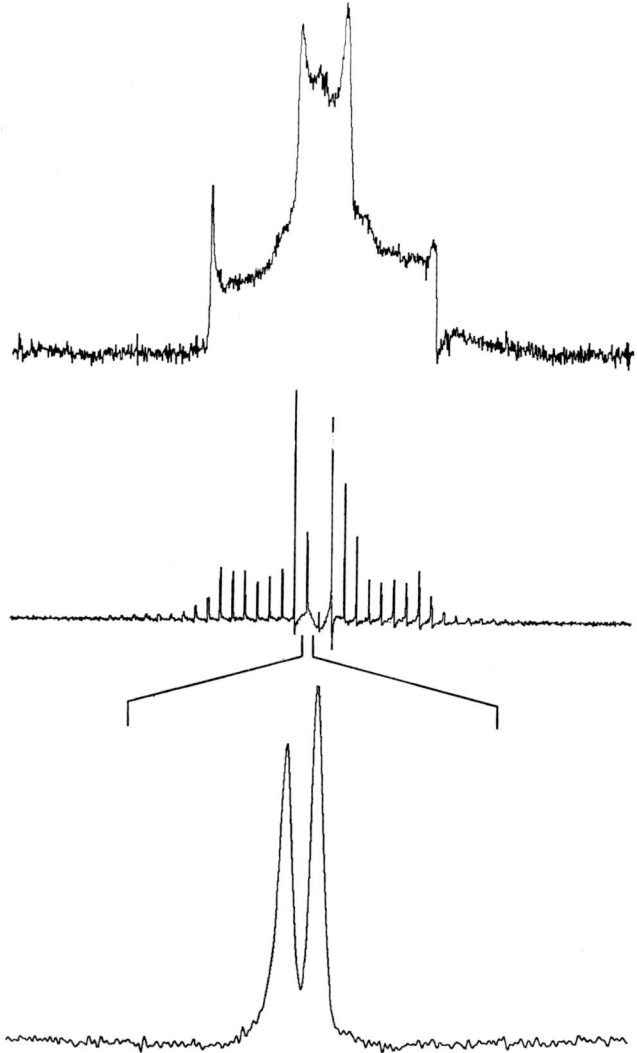

Fig. 2.3b. – Magic-angle spinning of deuterium in a solid mixture of deuterated hexamethylbenzene and deuterated ferrocene, demonstrating narrowing of the powder line even though the spinning frequency is considerably lower than the ~ 100 kHz quadrupolar broadened power linewidth. The two resonances at the bottom (expansion of the centerband in the middle trace) correspond to the isotropic chemical shifts of the two deuterated species. (Adapted from *Chem. Phys.*, **42**, 423 (1979), with permission.)

angle $\Theta_m^{(2)}$, the magic angle for P_2, and a second angle $\Theta_m^{(4)}$ which corresponds to one of the magic angles (30.56° or 70.12°) of the fourth-order Legendre polynomial. Rotation around two axes modifies eq. (2.24) to become

$$(2.25) \qquad \omega = \sum_{l=0,2,4} C_l \tau_{l0} \sum_{m_1, m_2 = -l}^{l} D_{m_1, 0}^{(l)}(\Omega_{r_1}(t)) D_{m_2, m_1}^{(l)}(\Omega_{r_2}(t)) a_{lm_2},$$

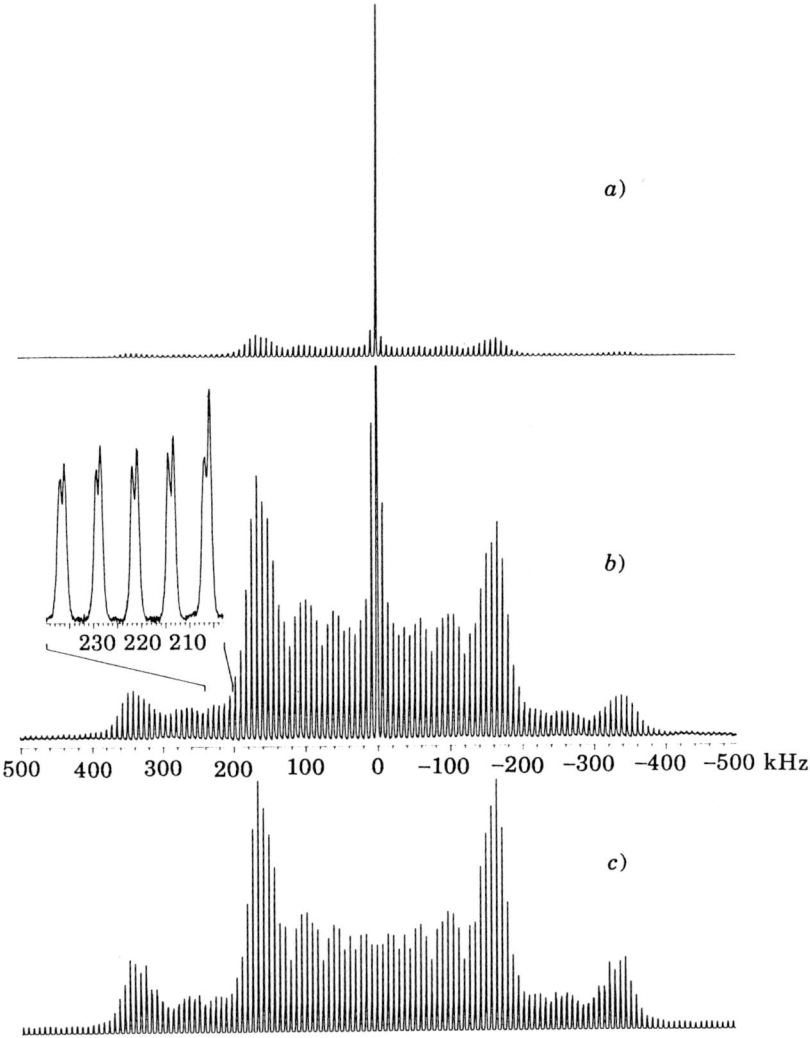

Fig. 2.4. – ^{27}Al MAS NMR spectra of the central and satellite transitions for α-Al$_2$O$_3$ spinning at 7.525 kHz. a) Experimental spectrum showing the relative intensities of the central and satellite transitions. b) Spectrum in a) with the vertical scale expanded by a factor of ten. The inset shows expansion of a region where the second-order quadrupole shift between the ($\pm 5/2$, $\pm 3/2$) and the ($\pm 3/2$, $\pm 1/2$) satellite transitions is clearly observed. c) Simulated spectrum for the satellite transitions obtained using a quadrupolar coupling $C_Q = 2.38$ MHz and $\eta = 0.00$. (Adapted from J. Magn. Reson., 85, 173 (1989), with permission.)

where $\Omega_{r_1}(t)$ and $\Omega_{r_2}(t)$ are sets of Euler angles which define the transforms to the laboratory frame from the first, outer rotor frame, and to the outer rotor frame from the second, inner rotor frame, and where

$$a_{lm} = \sum_{m'=-l}^{l} D_{m',m}^{(l)}(\Omega_{\text{pas}})\, q_{lm'}.$$

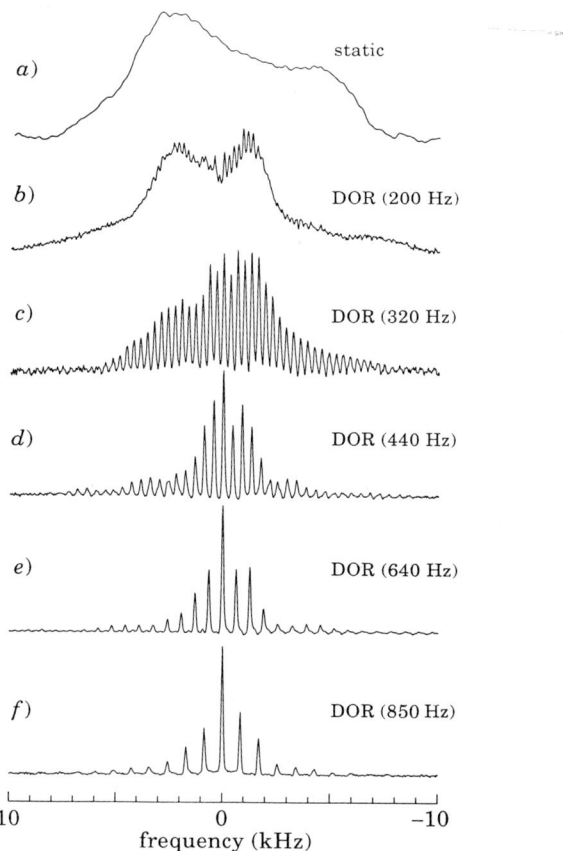

Fig. 2.5. – ^{23}Na DOR spectra of sodium oxalate spinning at various speeds, demonstrating the averaging of the second-order quadrupolar broadening and the effect of the time-dependent terms due to the slow outer rotor in eq. (2.25). There is only one type of ^{23}Na site in the crystal structure of this compound. (Adapted from *Solid State Nucl. Magn. Reson.*, **1**, 267 (1992), with permission.)

The $q_{lm'}$ are the elements of the quadrupolar-coupling tensor in its principal axis system.

After expansion of eq. (2.25) three terms result, similarly to the case of MAS:

$$(2.26) \quad \phi_{\text{DOR}}(t) = C_0 \, a_{00} \, \tau_{20} + \sum_{l=2,4} C_l \tau_{l0} \sum_{m=-l}^{l} d^{(l)}_{-m,0}(\Theta_{r_1}) \, d^{(l)}_{m,-m}(\Theta_{r_2}) \, a_{l,-m} + \xi_{\text{DOR}}(t),$$

where $\xi_{\text{DOR}}(t)$ is a time-dependent term which depends on the rotation axes and the rotation rates; it is dependent on both orientation and time. The first term in eq. (2.26) corresponds to the isotropic shift and is a scalar, independent of both orientation and time. The second term is dependent on orientation but not time. As in the MAS example, we can average over $\xi_{\text{DOR}}(t)$ *if at least one of the*

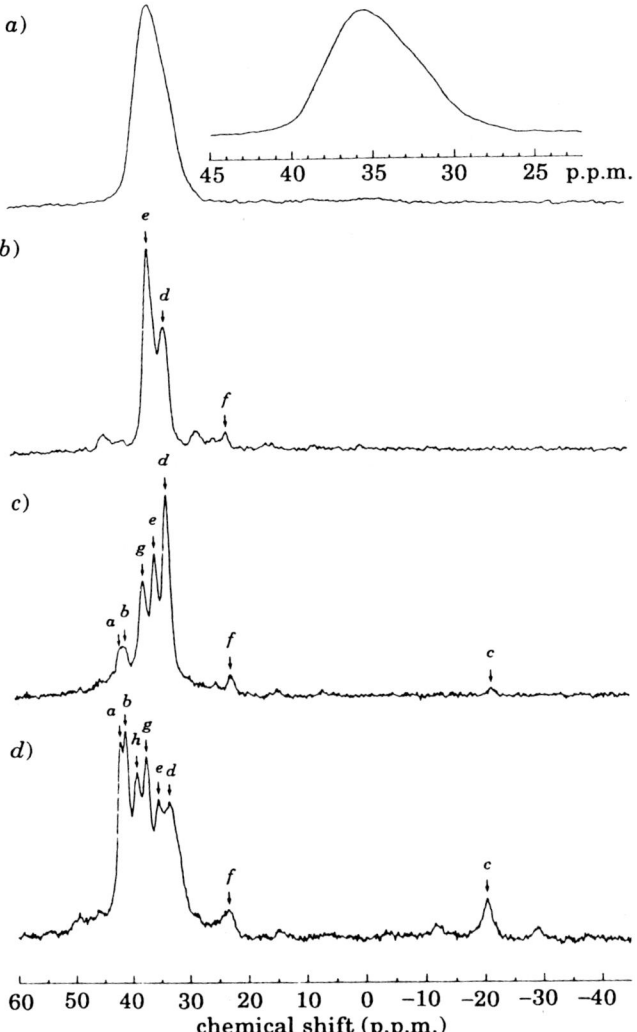

Fig. 2.6. – ^{27}Al NMR spectra of dehydrated and partially rehydrated VPI-5: *a)* MAS spectrum of dehydrated VPI-5, *b)* DOR spectrum of dehydrated VPI-5, *c)* DOR spectrum after two days of rehydration, *d)* DOR spectrum after 23 days of rehydration. (Adapted from *Nature (London)*, **346**, 550 (1990), with permission.)

double-rotation rates is large compared to the interaction strength, in which case the third term averages to zero. Since

$$(2.27) \qquad d_{00}^{(l)}(\Theta) = P_l(\cos\Theta),$$

we can eliminate the second term in eq. (2.26) if

$$(2.28) \qquad P_2(\cos\Theta_{r_1}) P_2(\cos\Theta_{r_2}) = 0$$

and

(2.29) $$P_4(\cos\Theta_{r_1})P_4(\cos\Theta_{r_2}) = 0.$$

The solutions are $\Theta_{r_1} = \Theta_m^{(2)}$ and $\Theta_{r_2} = \Theta_m^{(4)}$ or *vice versa*.

In the case where the spinning is not sufficiently fast to allow us to neglect $\xi_{\text{DOR}}(t)$, it leads to the presence of sidebands in an exactly analogous manner to the time-dependent term in the MAS example and this is demonstrated in the example of the ^{23}Na DOR spectra shown in fig. 2.5 [94]. As an example of the type of resolution that can be obtained for the important case of ^{27}Al by this method, fig. 2.6 shows a DOR spectrum of ^{27}Al in VPI-5 [95].

3. – Spin decoupling.

3`1. *Spin $I = 1/2$ pair.* – Maintaining a theme of multiple-quantum effects and coherent averaging, we shall show in this section how one may achieve double-quantum decoupling. However, to prepare ourselves we must first consider the decoupling of a spin $I = 1/2$ from an observed spin $S = 1/2$. The I spin is irradiated near resonance with an r.f. field of amplitude ω_1. In the rotating frame the Hamiltonian is

(3.1) $$H = H_0 + H_{IS},$$

where

(3.2) $$H_0 = -\Delta\omega I_z - \omega_1 I_x$$

describes the I fields in the rotating frame in frequency units and

(3.3) $$H_{IS} = dI_z S_z$$

is the I-S coupling. We now transform to a tilted frame with z along the effective I field. The tilt operator is given by

(3.4) $$T = \exp[-i\theta I_y],$$

where

(3.5) $$\cos\theta = \frac{\Delta\omega}{(\omega_1^2 + \Delta\omega^2)^{1/2}} = \frac{\Delta\omega}{\omega_e}.$$

In this frame

$$H = -\omega_e I_z + dS_z(I_z \cos\theta - I_x \cos\theta)$$

and

(3.6) $$\omega_e = \sqrt{\omega_1^2 + \Delta\omega^2}.$$

We now go into a frame defined by $-\omega_e I_z$ and take the average Hamiltonian of the I-S coupling term (as described in sect. 2), obtaining

(3.7) $$\overline{H}_{IS}^{(0)} = d \cos \theta S_z I_z ,$$

(3.8) $$\overline{H}_{IS}^{(1)} = d^2 \sin \theta \frac{1}{\omega_e} S_z^2 \left(\cos \theta I_x - \frac{1}{2} \sin \theta I_z \right),$$

and, at resonance ($\Delta \omega = 0$, $\theta = \pi/2$),

(3.9) $$\overline{H}_{IS}^{(2)} = \frac{d^3}{2\omega_1^2} S_z^3 I_x .$$

For a spin $S = 1/2$ the term $\overline{H}_{IS}^{(1)}$ in eq. (3.8) commutes with the spin vector S, so only $\overline{H}_{IS}^{(0)}$ and $\overline{H}_{IS}^{(2)}$ are relevant.

3˙2. *One-quantum offset and* r.f. *amplitude dependence.* – Away from resonance where $\Delta \omega \neq 0$, the effective coupling between I and S is dominated by $\overline{H}_{IS}^{(0)}$, and the decoupling efficiency δ_S (roughly the relative S linewidth) goes as

(3.10) $$\delta_S \sim \cos \theta = \frac{\Delta \omega}{(\omega_1^2 + \Delta \omega^2)^{1/2}} .$$

Thus for small $\Delta \omega$, more precisely $d \ll \Delta \omega \ll \omega_1$, the S linewidth should depend linearly on $\Delta \omega$.

At resonance ($\Delta \omega = 0$) the dominant term is $\overline{H}_{IS}^{(2)}$, and we expect

(3.11) $$\delta_S \sim \left(\frac{d}{\omega_1} \right)^2 ,$$

i.e. the decoupling efficiency should increase (the S linewidth should decrease) inverse quadratically with ω_1 ($\delta_S \sim 1/\omega_1^2$) for large ω_1.

We note here that the $\cos \theta$ factor is a scaling of an I_z term analogous to scaling of chemical shifts or resonance offsets due to strong irradiation off resonance. We have recognized that this scaling originates from the same source as the «mysterious» geometrical phase factor in certain closed-circuit adiabatic processes, which we shall discuss in sect. 9. As mentioned in subsect. 2˙1, more efficient and robust spin decoupling can be achieved by means of iterative schemes.

Iterative schemes rely on the realization that only $H^{(0)}$ of eq. (2.2) is important for heteronuclear decoupling [65, 96] and that $H^{(0)}$ can be made more and more insensitive to imperfections (such as offset) by constructing supercycles of composite 90° or 180° pulses. For example, the original MLEV-4 sequence consists of the cycle [97]

(3.12) $$RR\overline{R}R ,$$

where the basic element R of the supercycle is a composite 180° pulse such as

(3.13) $$R = (90°)_x\text{-}(180°)_y\text{-}(90°)_x$$

and $\overline{R}$ is the same as R with all the phases reversed. This sequence was subsequently iteratively expanded to yield the MLEV-16 decoupling sequence [98]

(3.14) $$R R \overline{R} \overline{R} \ \overline{R} R R \overline{R} \ \overline{R} \overline{R} R R \ R \overline{R} \overline{R} R .$$

Similar methods can be used to expand 90° composite pulses into supercycles, one of the most successful being WALTZ-16, which corresponds to the sequence [99, 100]

(3.15) $$\overline{3}4\overline{2}3\overline{1}\overline{2}4\overline{2}3 \ \overline{3}4\overline{2}3\overline{1}\overline{2}4\overline{2}3 \ 3\overline{4}2\overline{3}1\overline{2}4\overline{2}3 \ 3\overline{4}2\overline{3}1\overline{2}4\overline{2}3 ,$$

where the elements 1, 2, 3 and 4 correspond to 90°, 180°, 270° and 360° pulses, respectively. In fig. 3.1 we show the performance of WALTZ-16 with respect to offset from the transmitter frequency.

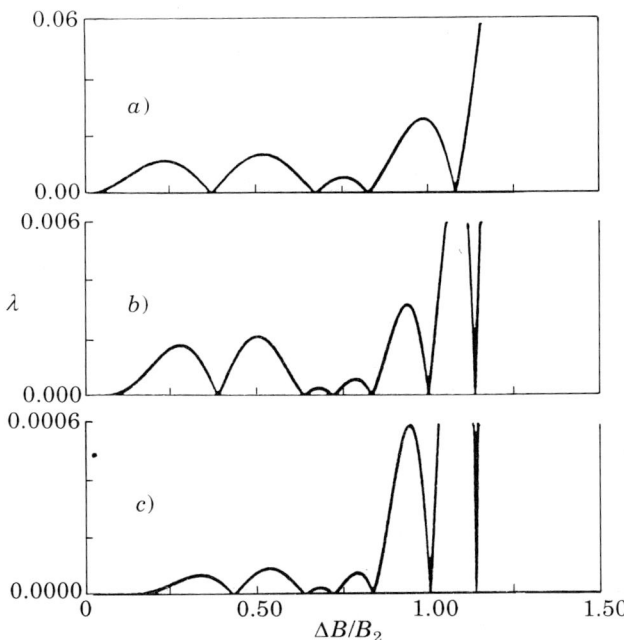

Fig. 3.1. – The predicted scaling factors delivered by the a) WALTZ-4, b) WALTZ-8 and c) WALTZ-16 sequences. Note the order-of-magnitude expansion of the vertical scale at each stage. The improvement is fastest near resonance and falls off towards the edge at each stage. (Adapted from *Prog. Nucl. Magn. Reson. Spectrosc.*, **19**, 47 (1987), with permission.)

3˙3. *Double-quantum decoupling.* – Imagine that we are required to decouple a spin $I = 1$, say a deuterium spin, from a spin $S = 1/2$, say a proton. Such would be the case, for example, in experiments on large biological molecules where isotopic substitution has been performed in order to simplify the spin system. If the deuterium quadrupole splitting is large (recall fig. 1.10), it may be impossible to cover the full I spectrum with a sufficiently large ω_1 field [101, 102]. However, if we irradiate at the unperturbed Zeeman resonance $\omega = \omega_0$, then the double-quantum transitions $M = 1 \leftrightarrow M = -1$ should make decoupling possible, since the $M = 0$ state exerts no z field at the S spin. A rough estimate of the condition can be made as follows. If the quadrupole coupling ω_Q is much larger than the I-S coupling (e.g., $\omega_Q \sim 100$ kHz for deuterium), then for normal one-quantum decoupling we would require

$$(3.16) \qquad \omega_1 \gg \omega_Q ,$$

a challenging technical demand with current NMR technology. This requirement is to be contrasted with the much less demanding condition

$$(3.17) \qquad \omega_1 \gg d$$

for the $S = 1/2$, $I = 1/2$ pair with dipolar couplings d, according to eq. (3.3).

If we excite the double-quantum transition, however, then according to second-order perturbation theory [103] (matrix element squared divided by energy difference) the transitions between $M = 1$ and $M = -1$ occur with an amplitude

$$(3.18) \qquad \omega_1^{DQ} \sim \frac{\omega_1^2}{\omega_Q} .$$

The effective double-quantum amplitude is to be compared with the coupling d; hence we now require

$$(3.19) \qquad \frac{\omega_1^2}{\omega_Q} \gg d ,$$

i.e.

$$(3.20) \qquad \omega_1 \gg \sqrt{d\omega_Q} ,$$

which is much more easily implemented than eq. (3.16) for the usual case where $d \ll \omega_Q$ [101, 102, 104-106].

3˙4. *Double-quantum offset and r.f. amplitude dependence.* – To appreciate the sensitivity of double-quantum decoupling to resonance offset $\Delta\omega$ and r.f. amplitude ω_1, we adopt eqs. (3.10) and (3.11), substituting ω_1^2/ω_Q for ω_1 and

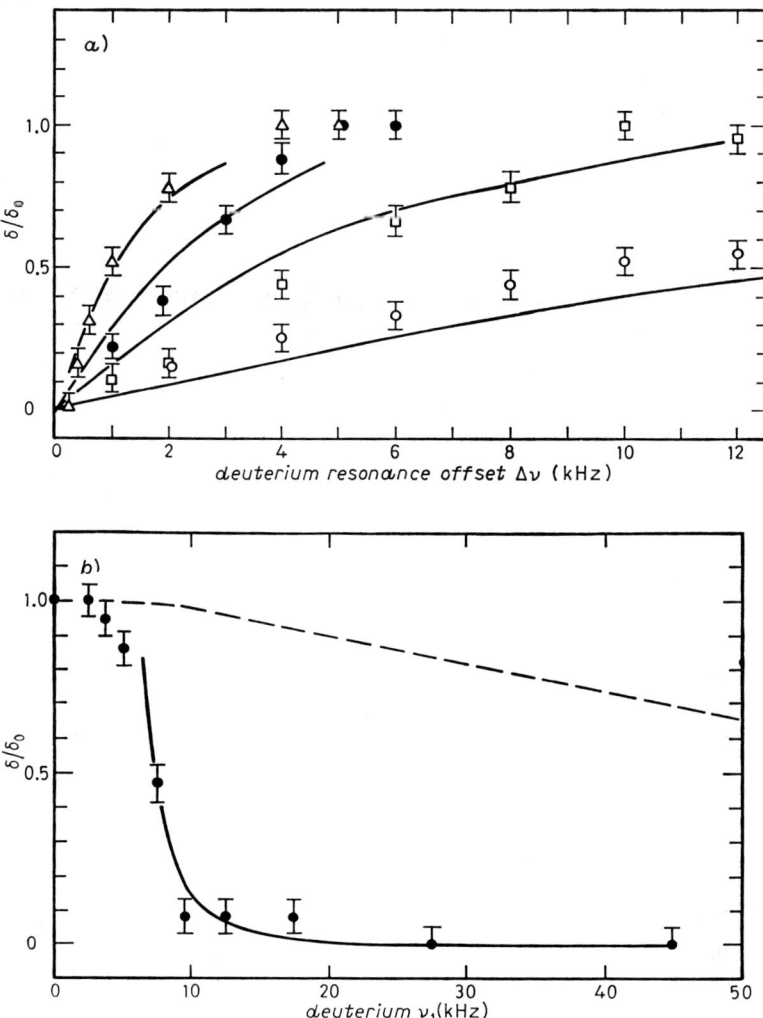

Fig. 3.2. – a) Proton linewidth in solid DMSO-d_6 (99%) at $-75\,°C$ under deuterium double-quantum decoupling as the deuterium spins are irradiated at various frequencies $\Delta\nu$ from resonance: $\triangle\ \nu_1 = 12.5\,\text{kHz}$, $\bullet\ \nu_1 = 17.0\,\text{kHz}$, $\square\ \nu_1 = 23.0\,\text{kHz}$, $\circ\ \nu_1 = 45.0\,\text{kHz}$. δ/δ_0 corresponds to δ_S in the text. Note the sensitivity of the decoupling to the I resonance condition, a characteristic of the double-quantum decoupling process. The solid lines are from theoretical calculations. b) Proton linewidth in solid DMSO-d_6 (99%) at $-75\,°C$ as the deuterium spins are irradiated at resonance with various values of ω_1 to induce double-quantum decoupling. The solid line shows the asymptotic $1/\omega_1^4$ behavior expected from theory. The dashed line shows the expected behavior if the double-quantum transition did not occur. (Adapted from *Phys. Rev. B*, **18**, 112 (1978), with permission.)

$2\Delta\omega$ for $\Delta\omega$. For double-quantum decoupling, the S linewidth is given by

$$\text{(3.21)} \qquad \hat{\delta}_S \sim \cos\theta = \frac{2(\Delta\omega)\omega_Q}{(\omega_1^4 + 4\Delta\omega^2\omega_Q^2)^{1/2}}$$

thus making the resonance condition for decoupling much more sensitive than in one-quantum decoupling, since $\Delta\omega$ is multiplied by ω_Q.

At resonance

$$\text{(3.22)} \qquad \hat{\delta}_S \sim \frac{d^2\omega_Q^2}{\omega_1^4},$$

so decoupling begins only when $\omega_1 \sim (d\omega_Q)^{1/2}$. The subsequent dependence on ω_1 is much more rapid, going as the inverse fourth power. All this behavior has been quantitatively verified, and examples are shown in fig. 3.2. Double-quan-

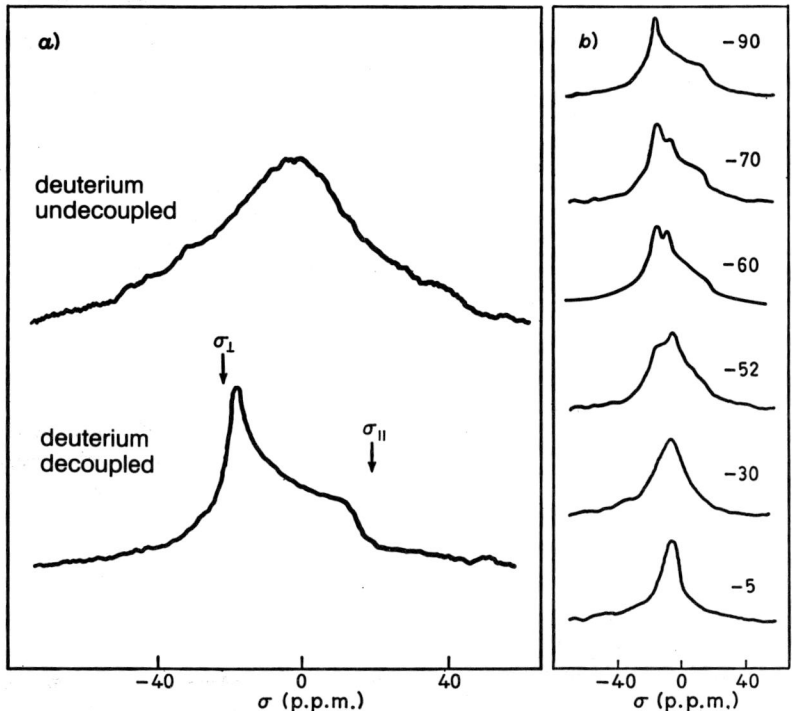

Fig. 3.3. – *a)* NMR spectra of residual protons in heavy ice (99% deuterium) at $-90\,°C$ with and without deuterium decoupling. The chemical-shift anisotropy of the protons in the hydrogen bonds is ~ 34 p.p.m. *b)* As the temperature is raised, the spectra exhibit motional narrowing, with lineshapes characteristic of tetrahedral jumps. This behavior is consistent with the protons hopping between the four hydrogen bonds around the oxygens. (Courtesy of D. E. WEMMER.)

tum decoupling has been used in our laboratory to obtain high-resolution spectra of diluted protons displaying, for example, proton chemical-shift anisotropy for the hydrogen bonds in solid heavy ice, and to observe the effects of the tetrahedral motion of the protons on the NMR lineshapes as ice is heated towards its melting point (see sect. 5). Typical proton spectra of 99% deuterated ice with deuterium double-quantum decoupling are shown in fig. 3.3. As the ice is heated, the effects of motional narrowing are observed [107]. A few degrees below the melting temperature the line is that of an isotropic sample. The intermediate spectra indicate that *the protons are jumping between the tetrahedrally disposed hydrogen bonds around the oxygens.*

3˙5. *Comment on the relationship between spin decoupling and multiple-quantum excitation.* – Consider a system of N I spins (*e.g.*, protons) coupled to a S spin (*e.g.*, ^{13}C). Several schemes involving composite pulses and iterative sequences have been devised to effect spin decoupling in such systems. These sequences, employing phase shifts and permutations, are reminiscent of some of the selective n-quantum schemes described in subsect. 1˙16. There must be some relationship between the two areas. Indeed, suppose we could engineer a pure $(N + 1)$-quantum selective scheme to excite the I spins. *Since N spins cannot absorb or emit more than N quanta, such an excitation would be tantamount to implementing the unit propagator, i.e.* decoupling. In this sense decoupling schemes employing π pulses and π phase shifts are two-quantum selective (good for decoupling one spin) and multiple-pulse schemes involving $\pi/2$ pulses and $\pi/2$ phase shifts are four-quantum selective and decouple groups of spins.

4. – Interaction of radiation and matter.

4˙1. *Two-level system in a quantized field.* – A common question is where are the quanta, the *photons*, in a multiple-quantum experiment? When we say $\Delta M = 2$, are there two quanta really involved or just one off-resonance quantum at double the frequency? What is the relationship to optical multiphoton excitation [108]? To investigate these questions clearly, we would like to study the very simplest case where photons actually appear. In order to do so, it is necessary to bring the radiation field as a legitimate full-fledged partner into the treatment, and not just treat it semi-classically as a time-dependent perturbation on the spins. That way, we can «see» the photons in the field and count them. At the same time this treatment will remind us of Feynman's analogy [109] of any two-level system to a spin $I = 1/2$. Later we shall encounter generalizations of this analogy, *e.g.*, a three-level system to a spin $I = 1$, and so on. The treatment also reminds us of the analogies, often unhappily neglected, between NMR and optical spectroscopy.

Before we treat two-quantum phenomena, let us review the simplest one-quantum problem, a two-level system interacting with a single mode of a cavity (the Jaynes-Cummings model [110]), described by a time-independent Hamiltonian

$$(4.1) \qquad H = H_{\text{mat}} + H_{\text{rad}} + H_{\text{int}}.$$

H_{mat} is the matter Hamiltonian, the matter being the system (atom or spin) irradiated and behavior observed. We assume it consists of two states, 1 and 2, with energies $\hbar\omega_1$ and $\hbar\omega_2$, such that

$$(4.2) \qquad \omega_2 - \omega_1 = \omega_0.$$

Assume further that the particles occupying these states are fermions [111] with coupling λ. Since at most one can occupy each state, the possible states are

$$(4.3) \qquad |0, 0\rangle, \quad |0, 1\rangle, \quad |1, 0\rangle, \quad |1, 1\rangle,$$

where $|0, 1\rangle$ is the vacuum for particle 1 and single occupancy for particle 2, etc. We now introduce fermion operators c_1, $c_1^\dagger$, c_2, $c_2^\dagger$ such that

$$(4.4) \qquad \begin{cases} c_1^\dagger |0, k\rangle = |1, k\rangle, \\ c_1^\dagger |1, k\rangle = |0, k\rangle, \end{cases}$$

and similarly for c_2, $c_2^\dagger$ and $|j, 0\rangle$, $|j, 1\rangle$. All other operations yield zero. The fermion operators satisfy the anticommutation rule

$$(4.5) \qquad \{c_1, c_1^\dagger\} = \{c_2, c_2^\dagger\} = 1.$$

The matter Hamiltonian is given in terms of these operators by

$$(4.6) \qquad H_{\text{mat}} = \hbar\omega_1 c_1^\dagger c_1 + \hbar\omega_2 c_2^\dagger c_2.$$

H_{rad} in eq. (4.1) is the radiation Hamiltonian, which we write in terms of the photon creation and destruction operators for a single mode, a and $a^\dagger$, with frequency ω and eigenkets $|n\rangle$ (n = number of photons) [112]. We remember that for a harmonic oscillator we have

$$(4.7) \qquad \begin{cases} a^\dagger |n\rangle = \sqrt{n+1}\,|n+1\rangle, \\ a|n+1\rangle = \sqrt{n+1}\,|n\rangle. \end{cases}$$

The commutation rules are

$$(4.8) \qquad \begin{cases} [a, a] = [a^\dagger, a^\dagger] = 0, \\ [a, a^\dagger] = 1, \end{cases}$$

and the Hamiltonian, ignoring the zero-point energy, is

$$(4.9) \qquad H_{\text{rad}} = \hbar\omega a^\dagger a.$$

Fig. 4.1. – Two coupling events between a spin $I = 1/2$ or fictitious spin $I = 1/2$ and photons in the cavity. A photon is absorbed and the spin is excited, or a photon is emitted and the spin returns to the ground state. This corresponds to the semi-classical rotating-wave approximation.

For H_{int} of eq. (4.1), the interaction between the radiation and the matter, we assume the coupling depicted in fig. 4.1. A photon $\hbar\omega$ is absorbed, transforming particle 1 into 2, or one is emitted, transforming particle 2 into 1; i.e.

$$(4.10) \qquad H_{\text{int}} = \frac{\hbar\lambda}{2}(c_1 c_2^\dagger a + c_1^\dagger c_2 a^\dagger).$$

This coupling conserves energy for $\omega = \omega_0$ and incorporates the semi-classical rotating-wave approximation (note that we have neglected events in which a photon is absorbed and transforms particle 2 into 1, for example). Equations (4.6), (4.9) and (4.10) now can be collected into the total Hamiltonian in eq. (4.1) to give

$$(4.11) \qquad H = \hbar\omega_1 c_1^\dagger c_1 + \hbar\omega_2 c_2^\dagger c_2 + \frac{\hbar\lambda}{2}(c_1 c_2^\dagger a + c_1^\dagger c_2 a^\dagger).$$

4'2. *Fictitious spin $I = 1/2$ operators.* – At this point we introduce the fictitious spin $I = 1/2$ operators [27-29]

$$(4.12) \qquad \begin{cases} I_x = \dfrac{1}{2}(c_1^\dagger c_2 + c_1 c_2^\dagger), \\ I_y = \dfrac{-i}{2}(c_1^\dagger c_2 - c_1 c_2^\dagger), \\ I_z = \dfrac{1}{2}(c_1^\dagger c_1 - c_2^\dagger c_2). \end{cases}$$

Given the relationships of eqs. (4.4) and (4.5), it can be verified that

$$(4.13) \qquad [I_x, I_y] = i I_z,$$

with the eigenstates $|\pm\rangle$ of I_z defined as

$$(4.14) \qquad I_z |\pm\rangle = \pm \frac{1}{2} |\pm\rangle$$

and

(4.15) $$I_\pm = I_x \pm iI_y,$$

the Hamiltonian eq. (4.11) can be written, aside from a commuting operator (remember that the number of particles is conserved), as

(4.16) $$H = -\hbar\omega_0 I_z + \hbar\omega a^\dagger a + \frac{\hbar\lambda}{2}(a^\dagger I_- + aI_+).$$

This Hamiltonian is identical to that of a spin $I = 1/2$ interacting with the cavity mode. Now assume the cavity mode is at resonance ($\omega = \omega_0$), and transform to an interaction picture defined by

(4.17) $$R(t) = \exp[-it(-\omega_0 I_z + \omega a^\dagger a)].$$

This is the analog of the rotating frame for the full quantum problem. As usual, this transformation simplifies considerably the Hamiltonian, which is now given by

(4.18) $$H = \frac{\hbar\lambda}{2}(a^\dagger I_- + aI_+).$$

4'3. *Evolution of the two-level system.* – Imagine the initial state $|\psi(0)\rangle$ being the excited state $|-\rangle$ for the fictitious spin with n photons in the cavity, as depicted in fig. 4.2:

(4.19) $$|\psi(0)\rangle = |-, n\rangle.$$

The evolution of $|\psi\rangle$ is given by

(4.20) $$|\psi(t)\rangle = \exp\left[-\frac{i}{\hbar}Ht\right]\psi(0),$$

which is easily evaluated using the Hamiltonian of eq. (4.18) together with eqs. (4.7) and (4.15) to yield

(4.21) $$|\psi(t)\rangle = |-, n\rangle \cos\sqrt{n+1}\,\frac{\lambda t}{2} - |+, n+1\rangle i\sin\sqrt{n+1}\,\frac{\lambda t}{2}.$$

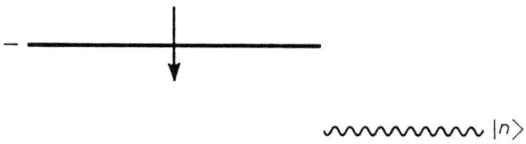

Fig. 4.2. – Initial state. The spin is in the excited state and there are n photons in the field.

This expression describes a quantum Rabi oscillation [113] or nutation that periodically exchanges a photon between the cavity and the spin. The analogy to the usual semi-classical problem of spin $I = 1/2$ interacting with a resonance field of amplitude ω_1 is made by realizing that

(4.22) $$\lambda\sqrt{n+1} \leftrightarrow \omega_1 .$$

Note, however, that, even when there are initially no photons in the cavity ($n = 0$), there is still the evolution, corresponding to spontaneous emission from the excited-spin or two-level system. Note also that, for a 2π pulse

(4.23) $$\omega_1 t = \sqrt{n+1}\,\lambda t = 2\pi ,$$

the ket $|\psi(0)\rangle$ does not come back to itself but instead to $\cos\pi$ times itself, i.e.

(4.24) $$|\psi(2\pi)\rangle = -|\psi(0)\rangle .$$

This effect corresponds to the well-known spinor behavior [114, 115] of a system with an even number of states or of a nonintegral fictitious spin, which we shall discuss further in sect. 9.

4`4. *Evolution off resonance*. – The expression of eq. (4.21) derived for resonance ($\omega = \omega_0$) can be generalized for arbitrary frequency of the mode ω in eq. (4.16). This extension is best made by transforming to a tilted interaction picture in analogy to eq. (3.4). The result, for the initial condition of eq. (4.19), is

(4.25) $$|\psi(t)\rangle = |-, n\rangle [\cos\lambda_e t/2 - i\cos\theta\sin\lambda_e t/2] - |+, n+1\rangle i\sin\theta\sin\lambda_e t/2 ,$$

where

(4.26) $$\lambda_e = (\lambda^2(n+1) + (\omega_0 - \omega)^2)^{1/2} ,$$

(4.27) $$\sin\theta = \sqrt{n+1}\,\lambda/\lambda_e .$$

4`5. *Adiabatic rapid passage*. – The treatment above provides an elegant picture of adiabatic rapid passage for a two-level system or fictitious spin $I = 1/2$. Consider fig. 4.3, which depicts the energy levels of the uncoupled matter and radiation Hamiltonians

(4.28) $$H_0 = H_{\text{mat}} + H_{\text{rad}}$$

as the frequency (or field) is varied through resonance. The resonance between matter and radiation occurs at the level crossings, $\omega = \omega_0$. Adding the interaction Hamiltonian H_{int} causes a mixing of the states near resonance, which gives rise to an avoided crossing as in fig. 4.4. The repulsion between levels at resonance depends on the number of photons in the cavity as $(n+1)^{1/2}\lambda$, the semi-classical field intensity ω_1 [116] from eq. (4.22).

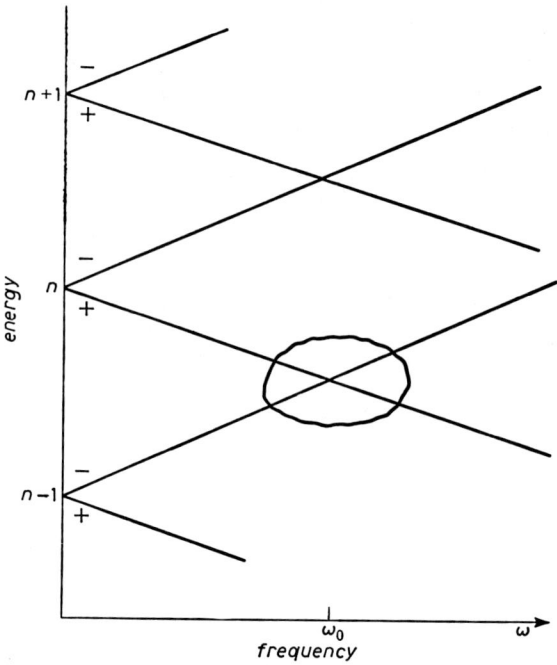

Fig. 4.3. – Energy level diagram for a spin $I = 1/2$ or fictitious spin $I = 1/2$ (two-level system) and a quantized radiation field with photon states $|n\rangle$.

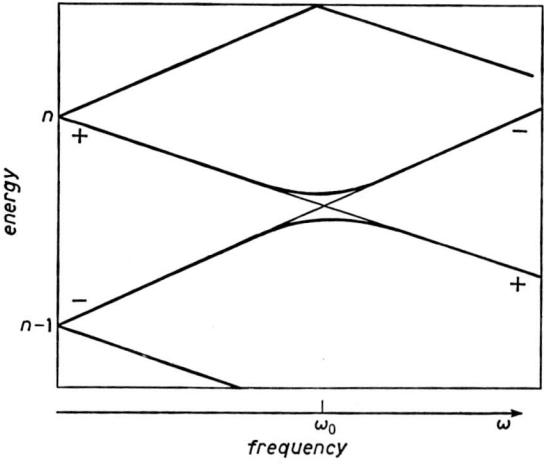

Fig. 4.4. – Coupling between the spin $I = 1/2$ and the radiation causes a level anticrossing at the resonance. From this picture it is easy to visualize how adiabatic rapid passage interchanges the $\pm$ states.

If we begin with the spin (or fictitious spin) in one state away from resonance, e.g. $|-, n\rangle$ on the left of fig. 4.4, and now shift the frequency *adiabatically* [117] through resonance, then we recall from quantum mechanics that the system remains in an eigenstate. Thus it will go to $|+, n-1\rangle$ on the right-hand side. Therefore, sweeping through a resonance line rapidly compared to T_1 but slowly enough so there are no frequency components at ω_0 should invert a resonance line whether we sweep from left to right or right to left. This effect is called population inversion by adiabatic rapid passage, an extremely useful way of making robust π pulses. *The most efficient adiabatic sweeps are not linear (i.e. linear dependence of frequency on time), however, but rather are hyperbolic functions* [118].

4˙6. *Three-level system in a quantized field.* – We adopt the notation of subsect. 4˙3 for three particles interacting with the single mode of a cavity; we assume that only 1 and 2 or 2 and 3 can be directly interconnected by energy-conserving (rotating-wave) absorption or emission of a single photon. Thus the interaction is depicted in fig. 4.5a), with

(4.29) $$\begin{cases} \omega_{12} + \omega_{23} = 2\omega_0, \\ \omega_{12} - \omega_{23} = 2\omega_Q. \end{cases}$$

The full Hamiltonian for this system,

(4.30) $$H = \hbar\omega_1 c_1^\dagger c_1 + \hbar\omega_2 c_2^\dagger c_2 + \hbar\omega_3 c_3^\dagger c_3 + \hbar\omega a^\dagger a + \\ + \frac{\hbar\lambda}{\sqrt{2}}(c_1 c_2^\dagger a + c_1^\dagger c_2 a^\dagger) + \frac{\hbar\lambda'}{\sqrt{2}}(c_2 c_3^\dagger a + c_2^\dagger c_3 a^\dagger),$$

is entirely analogous to eq. (4.11). We will not discuss the general case here, but will instead select simple conditions under which double-quantum behavior emerges [119].

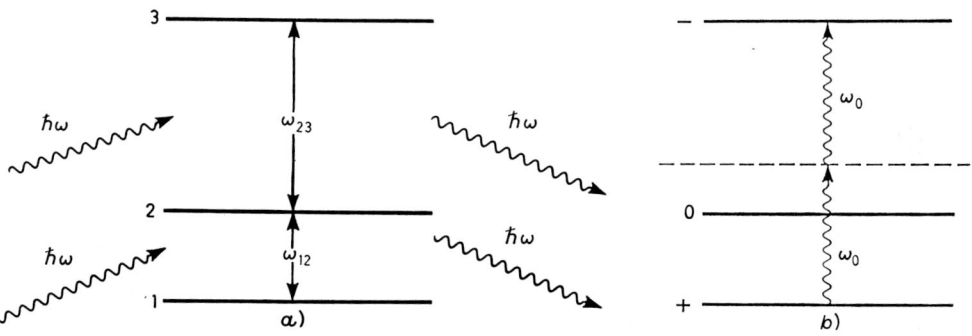

Fig. 4.5. – a) Coupling events between a spin $I = 1$ and photons in a cavity. b) Resonant two-photon process in a three-level spin $I = 1$ system.

4`7. *Fictitious spin $I = 1$ operators.* – First let us introduce the fictitious spin $I = 1$ operators

(4.31)
$$\begin{cases} I_z = c_1^\dagger c_1 - c_3^\dagger c_3 , & Q_z = 2c_2^\dagger c_2 - c_3^\dagger c_3 - c_1^\dagger c_1 , \\ I_+ = c_1 c_2^\dagger + c_2 c_3^\dagger , & Q_+ = c_2 c_3^\dagger - c_1 c_2^\dagger , \\ I_- = c_1^\dagger c_2 + c_2^\dagger c_3 , & Q_- = c_2^\dagger c_3 - c_1^\dagger c_2 , \\ Q_{+2} = c_1 c_3^\dagger , & Q_{-2} = c_1^\dagger c_3 . \end{cases}$$

Here the raising and lowering operators are used directly without going through any intermediate stages involving I_x, Q_x, Q_{xy}, etc. Inserting these spin $I = 1$ operators in eq. (4.30) and ignoring commuting unit operators (again, number of particles is conserved), we can rewrite H as

(4.32) $$H = \hbar\omega_0 I_z + \frac{\hbar\omega_Q}{3} Q_z + \hbar\omega a^\dagger a + \frac{\hbar}{2\sqrt{2}} [(\lambda - \lambda')Q_+ + (\lambda + \lambda')I_+] a +$$
$$+ \frac{\hbar}{2\sqrt{2}} [(\lambda - \lambda')Q_- + (\lambda + \lambda')I_-] a^\dagger .$$

We now specialize to the case in which the two allowed one-quantum absorptions and emissions have the same matrix elements,

(4.33) $$\lambda = \lambda' ,$$

so that

(4.34) $$H = -\hbar\omega_0 I_z + \frac{\hbar\omega_Q}{3} Q_z + \hbar\omega a^\dagger a + \frac{\hbar\lambda}{\sqrt{2}} (I_+ a + I_- a^\dagger) .$$

This Hamiltonian is just the interaction of a quadrupolar spin $I = 1$ with Zeeman frequency ω_0 and quadrupolar frequency ω_Q. Now assume the radiation is at resonance ($\omega = \omega_0$) and that it is weak compared to the quadrupolar splitting,

(4.35) $$\lambda\sqrt{n+1} \ll \omega_Q ,$$

corresponding to the situation in fig. 4.5b). Note that this condition embodies the weaker case in which

(4.36) $$\lambda \ll \omega_Q .$$

4`8. *Double-quantum (two-photon) Hamiltonian.* – In the interaction picture defined by eq. (4.17), the Hamiltonian of eq. (4.34) can be written

(4.37) $$H = H_Q + H_1 = \frac{\hbar\omega_Q}{3} + \frac{\hbar\lambda}{\sqrt{2}} (I_+ a + I_- a^\dagger) .$$

With the condition of eq. (4.36) we now have a situation ideal for a perturbation treatment since the second operator term H_1 in eq. (4.37) is much smaller than the first term H_Q. It is appropriate to retain only the «secular» part of the interaction term, i.e. that part diagonal with respect to the first term. We carry this out using Van Vleck perturbation theory [61] as follows. First, define an infinitesimal transformation

$$(4.38) \qquad H = \exp[-iS] H \exp[iS] \simeq H - i[S, H] + \dots ,$$

where S is an infinitesimal Hermitian operator. Substituting eq. (4.38) into eq. (4.37), we get

$$(4.39) \qquad H = H_Q + H_1 - i[S, H_Q] - i[S, H_1] \dots .$$

Now we demand that the first-order infinitesimal term vanish, so that

$$(4.40) \qquad H_1 - i[S, H_Q] = 0 ,$$

which reduces eq. (4.35) to

$$(4.41) \qquad H' \simeq H_Q + H_1'$$

with the modified interaction Hamiltonian

$$(4.42) \qquad H_1' = \frac{-i}{2} [S, H_1] .$$

After some algebra the S which satisfies eq. (4.40) is found to be

$$(4.43) \qquad S = -i \frac{\lambda}{\omega_Q \sqrt{2}} (Q_+ a - Q_- a^\dagger) .$$

Evaluating the commutator in eq. (4.42) with eq. (4.43), we obtain the effective interaction Hamiltonian (neglecting terms commuting with H_Q)

$$(4.44) \qquad H_1' = \frac{\hbar \lambda^2}{2\omega_Q} (Q_{+2} a^2 + Q_{-2} a^{\dagger 2}) .$$

According to the definitons of $Q_{\pm 2}$ and $a, a^\dagger$ given in eqs. (4.7) and (4.31), this form is a pure double-quantum Hamiltonian with the appealing property that it exhibits the two *photons* directly in the terms a^2 and $a^{\dagger 2}$. These terms represent the destruction and creation of only pairs of photons.

4'9. *Evolution of the system*. – If the system begins in the upper state as shown in fig. 4.6 with n photons in the cavity, i.e.

$$(4.45) \qquad |\psi(0)\rangle = |-, n\rangle ,$$

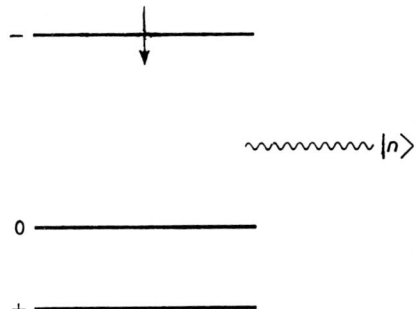

Fig. 4.6. – Initial state. The spin is in the excited state and there are n photons in the cavity.

then under the influence of the Hamiltonian of eq. (4.44) the system evolves as

$$(4.46) \quad |\psi(t)\rangle = |-, n\rangle \cos \sqrt{(n+1)(n+2)}\, \frac{\lambda^{(2)} t}{2} - $$
$$ - |+, n+2\rangle i \sin \sqrt{(n+1)(n+2)}\, \frac{\lambda^{(2)} t}{2}\,,$$

where $\lambda^{(2)}$ is the effective two-photon amplitude

$$(4.47) \quad \lambda^{(2)} = \frac{\lambda^2}{\omega_Q}\,.$$

This result is to be compared with eq. (4.21). *Here two photons are being exchanged with the cavity; a two-photon Rabi oscillation.*

The correspondence to the semi-classical case is given by

$$(4.48) \quad \frac{\lambda^2}{\omega_Q} \sqrt{(n+1)(n+2)} \sim \frac{\lambda^2 n}{\omega_Q} \leftrightarrow \frac{\omega_1^2}{\omega_Q}\,.$$

When we begin with no photons in the cavity, eq. (4.46) corresponds to two-photon spontaneous emission. This process, incidentally, is how metastable He $(1s, 2s)$ decays by optical emission to the ground state.

4˙10. *Double-quantum adiabatic rapid passage.* – The energy levels of the three-level system and cavity photons with no coupling between them appear in fig. 4.7. There are now three relevant crossings indicated in the circle, two corresponding to normal one-quantum resonances and the one we have just described due to a double-quantum resonance at ω_0. With the interaction between matter and radiation, all three crossings are avoided, as shown for the case $\lambda^2((n+1)(n+2))^{1/2} \ll \omega_Q^2$ in fig. 4.8, roughly $\lambda(n+1)^{1/2} \ll \omega_Q$ as stated in eq. (4.35). The levels at the two one-quantum resonances repel by $\sim \lambda(n+1)^{1/2}$,

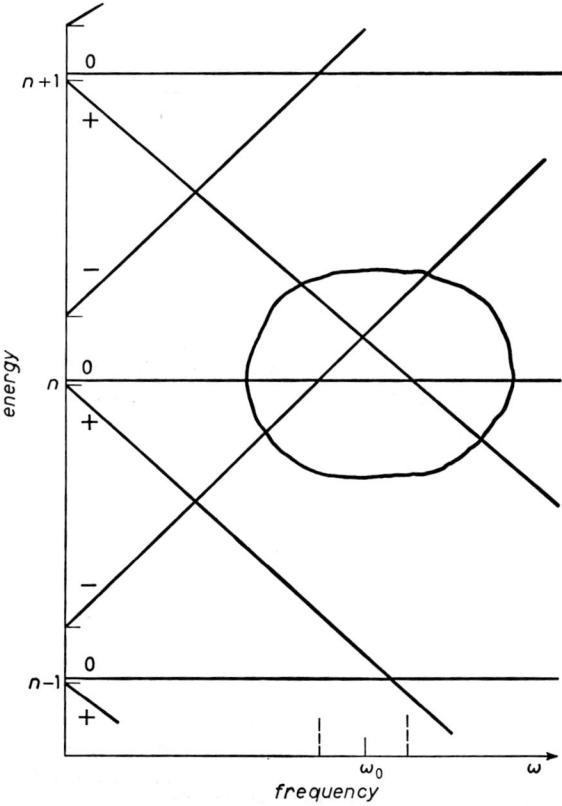

Fig. 4.7. – Energy level diagram for a spin $I = 1$ or fictitious spin $I = 1$ (three-level system) and a quantized radiation field with photon states $|n\rangle$.

whereas those at the double-quantum resonance experience the weaker repulsion $\sim \lambda^2((n+1)(n+2))^{1/2}/\omega_Q$. Thus the adiabatic condition for the double-quantum resonance is considerably more stringent than that for the one-quantum resonances.

Several interesting features are exposed in the diagram of fig. 4.8. The resonances of a three-level system can be inverted by adiabatically sweeping either through all the transitions or only through the double-quantum transition. Individual one-quantum transitions may be inverted by selectively sweeping through them adiabatically. *If we sweep adiabatically to the right through the left resonance and then through the double-quantum resonance, the left line is inverted; if we sweep to the left first through the double-quantum resonance, and then through the left one-quantum resonance, the left line does not invert* [119]. The intensities of each are also affected. This phenomenon was first observed in experiments on spin $I = 1$ nuclei in crystals, where lines would sometimes invert with adiabatic rapid passage in one direction but not in the

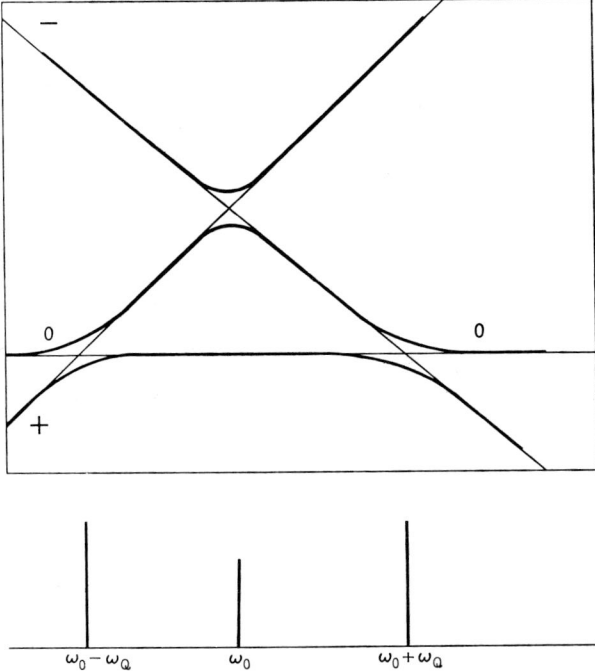

Fig. 4.8. – Coupling between the spin $I = 1$ and the radiation causes level anticrossing. There are two one-photon anticrossings and one two-photon anticrossing. It is possible by adiabatic rapid passage to interchange 0 and −, 0 and + or the ± states. The latter is a two-photon (double-quantum NMR) adiabatic rapid passage.

other. It is, of course, possible to sweep so that passage through the one-quantum transitions is adiabatic, whereas passage through the double-quantum transition is not. If passage through the double-quantum transition is sudden enough compared with the level repulsion, then the system behaves as if the levels cross at ω_0 and there is no double-quantum inversion. Generalizations of these considerations to higher spin and to coupled spins is straightforward.

We note in closing that all arguments concerning adiabatic rapid passage can, of course, be made using traditional semi-classical NMR considerations.

5. – Group theory and dynamics.

In this section we discuss the treatment of dynamic effects in NMR in the presence of symmetry. The two most common motional mechanisms that give rise to such dynamic effects in the solid state are molecular motion (or chemical exchange)[120, 121] and macroscopic reorientational motion of the sample, the latter usually implemented for the purposes of narrowing or simplifying the spectrum.

5‘1. *Molecular motion.* – The effects of symmetry can be appreciated by considering fig. 5.1. A molecule undergoes motion subject to the symmetry G and the motion is investigated by means of an NMR observable characterized by subgroup $g \subset G$. The purpose of the NMR experiment is to provide information about the parameters (for example, the molecular jump rates) of the molecular motion. Now, if the probe has a high enough symmetry, $g = G$, the NMR spectrum is *invariant* to the motion and we can learn nothing further from the spectrum about the parameters of the motion. For the case where the probe has no symmetry subgroup of G except the trivial group C_1, it should be possible to derive, from the NMR spectra, all the parameters of the motion.

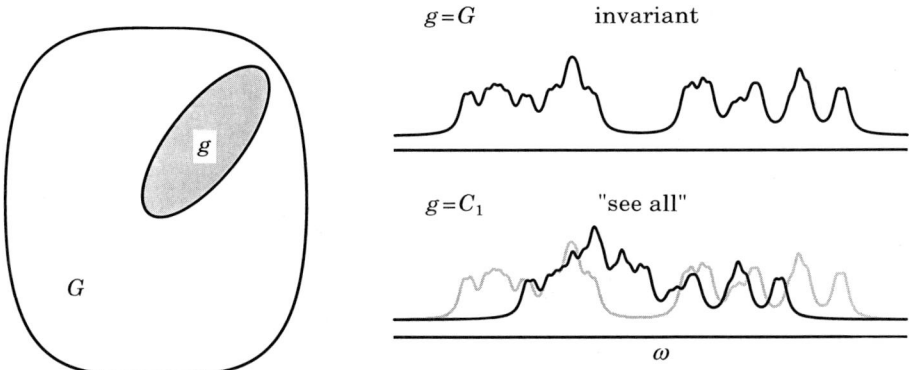

Fig. 5.1. – If a molecule undergoes motion subject to symmetry group G, and the motion is investigated by a probe of symmetry subgroup $g \subset G$, the spectrum is invariant to the motion if $g = G$, but, if the probe contains no symmetry subgroup of G except C_1, it should be possible to derive all the parameters of the motion from the spectrum.

A simple, but realistic example is depicted in fig. 5.2. In this case, the motion consists of 180° flips (with flip rate W_{12}) of a water molecule in a crystal about its C_{2v} axis. If the NMR probe is the dipolar coupling between the two protons, the spectrum is invariant to the motion since the axis of the dipolar interaction is left unchanged by the jump (fig. 5.2a)). In contrast, if the NMR probe has its principal axis roughly along the OH bond as, for example, the chemical shift or deuterium quadrupole coupling, the spectrum displays the classical effects of chemical exchange, collapsing to a motionally averaged spectrum (one line for the chemical shift, and a doublet for the quadrupole coupling) at high temperature, and W_{12} may be extracted from the spectra.

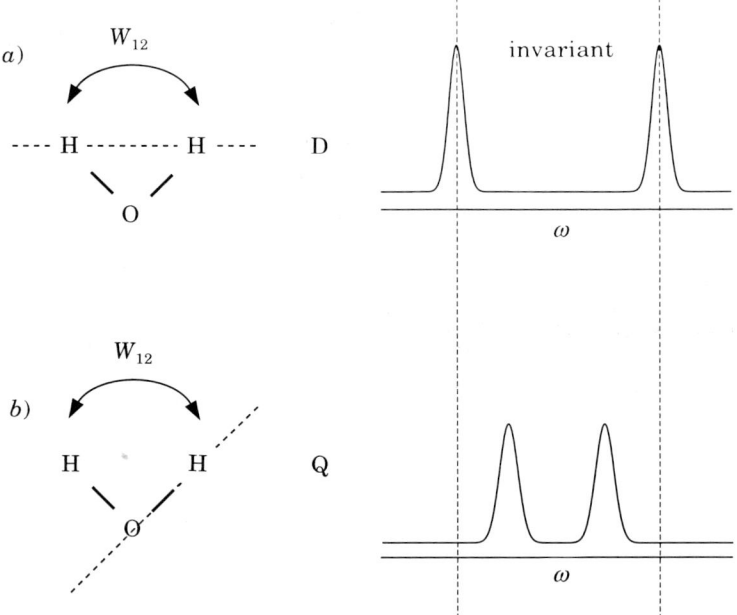

Fig. 5.2. – If the proton dipolar interaction is used to probe the 180° flip about the C_{2v} axis of water of hydration in a crystal, no change in the spectrum is observed since the dipolar interaction is invariant to this motion. On the contrary, if the chemical shift or deuterium quadrupolar interaction were used in place of the dipolar interaction, the motion would have the effect of narrowing the spectrum as illustrated schematically (for the quadrupolar coupling) in the lower part of the figure.

An interesting question is how much can we learn about the motion, how many parameters can be extracted, in the *intermediate* case, where g has a nontrivial symmetry, but less than G [122-124]. Consider, for example, the carbon spectrum of solid benzene. The relevant motion involves 60° jumps (or multiples thereof) about the C_6 axis of the molecule, with jump parameters W_{12}, W_{13}, W_{14},

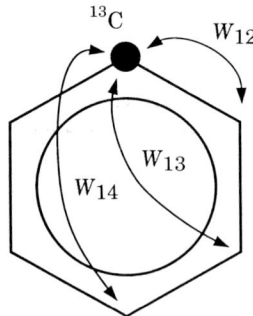

Fig. 5.3. – Jump motions and rates in benzene.

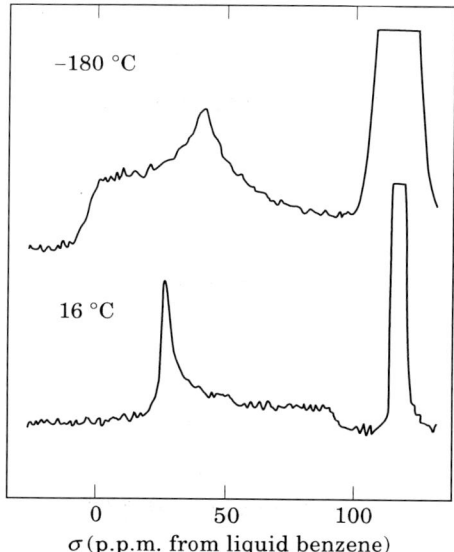

Fig. 5.4. – ^{13}C powder spectra of hexamethyl benzene in the high- and low-temperature limits. The wide peaks are from the ring carbons and the sharp peaks, which have been truncated, are from the methyl carbons. (Courtesy of D. E. WEMMER.)

as illustrated schematically in fig. 5.3 [125]. Here, the chemical-shift tensor has a (nontrivial) symmetry less than that of the molecular jumps, and we expect that the NMR spectra should not provide sufficient information to determine all three parameters. An experimental example of the motional averaging is shown in fig. 5.4.

5˙2. *Group theory and exchange.* – In this subsection we follow the treatment of Alexander, Baram and Luz [123]. In the simplest case, the Bloch equations [126] for the magnetization, y_j, of a spin in site j are

(5.1) $$\alpha_j y_j = P_j ,$$

where

(5.2) $$\alpha_j = i(\omega - \omega_j) - T_{2j}^{-1}$$

and ω and T_{2j} are the Larmor frequency and transverse relaxation time in the j-th site. In the presence of molecular jumps, the magnetization is exchanged between sites [127-129]

(5.3) $$\alpha_j y_j + \sum_k W_{jk}(y_k - y_j) = P_j ,$$

where W_{jk} is the transition rate from j to k. Now consider the effect of symmetry on eqs. (5.1)-(5.3) [130]. Let the symmetry of the molecular dynamics be G. The y_k form the basis for a reducible representation of G, and each W_{jk} may be associated with a group element R in G,

(5.4) $$R_\alpha y_j = y_k ,$$

so that we may write

(5.5) $$W_{jk} = W(R_\alpha) .$$

Moreover, since the $W(R_\alpha)$ are determined by the symmetry of G, *all equivalent elements (those belonging to the same class C) have the same rate W_c*, so we can rewrite eq. (5.3) as

(5.6) $$\alpha_j y_j + \sum_c W_c \sum_{\alpha \in c} (R_\alpha - E) y_j = P_j .$$

We expand the y_j in terms of basis functions $g_{\lambda\mu}$ of the group G

(5.7) $$y_j = \sum_{\lambda\mu} a^j_{\lambda\mu} g_{\lambda\mu} = \sum_\lambda g_{\lambda j} ,$$

where λ labels the irreducible representation and μ is an index of the row in a multidimensional representation. The coefficients $a^j_{\lambda\mu}$ are uniquely determined from G in terms of projection operators formed from the group elements. From group theory, for any $g_{\lambda\mu}$ acted on by the sum of elements in a class [130]

(5.8) $$\sum_{\alpha \in c} R_\alpha g_{\lambda\mu} = n_c \frac{\chi^\lambda_c}{\nu_\lambda} g_{\lambda\mu} ,$$

where n_c is the order of the class C, χ^λ_c is the character, and n_λ the dimensionality of the λ representation. Thus eq. (5.6) becomes

$$\alpha_j y_j - \sum_c W_c n_c \sum_\lambda \left(1 - \frac{\chi^\lambda_c}{\nu_\lambda} \right) g_{j\lambda} = P_j ,$$

which can be simplified to yield

(5.9) $$\alpha_j y_j - \sum_\lambda W_\lambda g_{j\lambda} = P_j$$

by defining W_λ, the rate for the representation, as

(5.10) $$W_\lambda = \sum_c W_c n_c \left(1 - \frac{\chi^\lambda_c}{\nu_\lambda} \right) .$$

Thus we have replaced the summation over individual rates for exchange between sites with rates for the irreducible representations of the symmetry group for the exchange [131].

5˙3. *Relevant representations.* – The symmetry may be used to simplify the analysis even further, by recognizing that not all irreducible representations l

are needed. Because we observe only the total magnetization

(5.11) $$y = \sum_j y_j,$$

which transforms according to the totally symmetric representation A_1 of G, only those representations of G which mix with (couple to) A_1 by the observable are relevant; these *relevant representations* of G are those which, when reduced (decomposed) under g, contain only the totally symmetric representation of g. That is, if

(5.12) $$N_\lambda = \sum_c n_c \chi_c^\lambda \chi_c^{A_1}$$

is not zero, the λ representation is relevant and must be included in the summation. Noting from eq. (5.10) that $W_{A_1} = 0$ in every case, it is clear that *there exists one independent motional parameter for each relevant representation (beyond A_1)*. Indeed, in the case where there is just one relevant representation, Γ beyond A_1, there is a simple analytical solution for the spectral lineshape [125]:

(5.13) $$y(\omega) = \frac{\dfrac{P}{W_\Gamma} \sum_j \dfrac{1}{\alpha_j - W_\Gamma}}{\dfrac{N}{W_\Gamma} + \sum_j \dfrac{1}{\alpha_j - W_\Gamma}}.$$

5'4. *Summary of symmetry considerations.* – To sum up what we have derived above, the effect of motion on the spectrum is determined by the relative

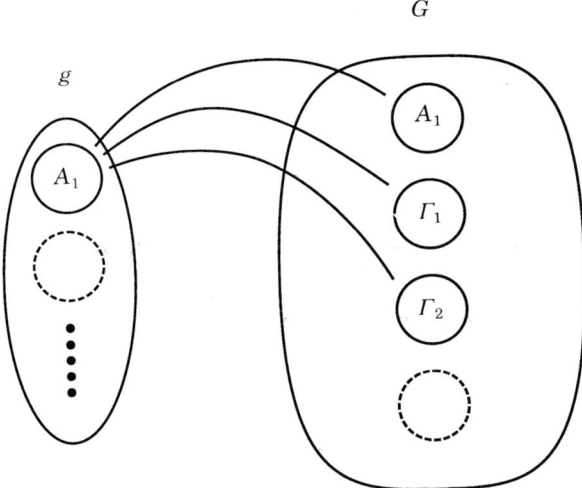

Fig. 5.5. – Relevant representations of G are determined by using character tables to find which elements of G contain the totally symmetric representation of g.

symmetry of the interaction we use to probe the motion (the observable magnetization with symmetry group g) and the symmetry of the dynamics (symmetry group G). If $g = G$ or if g has a higher symmetry than G, then only the A_1 representation of G is relevant, and the spectrum will be invariant to the motion. If g has a lower symmetry than G, then the spectrum will be at least partially affected by the motion, and, if g has no symmetry subgroup of G ($g = C_1$), then the number of relevant representations is equal to the number of classes, and all the W_c can be determined. In general, the approach is to find the relevant representations of G (fig. 5.5) using the character tables for the groups, and then determine the W_λ in terms of the y_j using expression (5.10).

5`5. *Example: dynamics of solid benzene.* – The case of benzene provides an illustrative example of the symmetry considerations [125]. For the molecular symmetry of benzene, with the observable being the anisotropic chemical shift, we can assign $g = D_2$, which has the character table given in table I and contains four irreducible representations A_1, B_1, B_2 and B_3. The symmetry of the molecule in this case can be taken as D_6, with the character table also shown in table I, containing 6 irreducible representations, *only two of which (A_1 and E_2) contain A_1 of D_2*. Thus there are two relevant representations, which corre-

TABLE I. – *Character tables for the D_2 and D_6 groups.* The characters shown in boldface in the D_6 table correspond to the elements of the relevant representations with respect to the D_2 subgroup, namely those elements that do not sum to zero under D_2.

D_2	E	C_2^z	C_2^y	C_2^x
A_1	1	1	1	1
B_1	1	1	-1	-1
B_2	1	-1	1	-1
B_3	1	-1	-1	1

D_6	E	$2C_6$	$2C_3$	C_2	$3C_2'$	$3C_2''$
A_1	**1**	**1**	**1**	**1**	**1**	**1**
A_2	1	1	1	1	-1	-1
B_1	1	-1	1	-1	1	-1
B_2	1	-1	1	-1	-1	1
E_1	2	1	-1	-2	0	0
E_2	**2**	-1	-1	**2**	0	0

spond (from eq. (5.10)) to rates $W_{A_1} = 0$ and $W_{E_2} = W_{12} + W_{13}$. From group theory, we, therefore, expect that there is only one observable rate for the dynamic processes in solid benzene, the sum of the rate parameters for 60° and 120° jumps. This makes sense, because a 180° rotation does not affect the chemical shift (making W_{14} ineffective) and a process which consists of a 60° jump (W_{12}) cannot be distinguished, for example, from one which consists of a 180° jump followed by a $-120°$ jump (W_{13}), hence the sum $W_{12} + W_{13}$. Thus, while the experimental NMR spectra (fig. 5.6) clearly reflect the effects of the sixfold symmetry in the motion, they cannot teach us about 60° vs. 120° jumps. In contrast, in the case of fivefold symmetry, W_{12} (72°) and W_{13} (144°) can be determined separately [132].

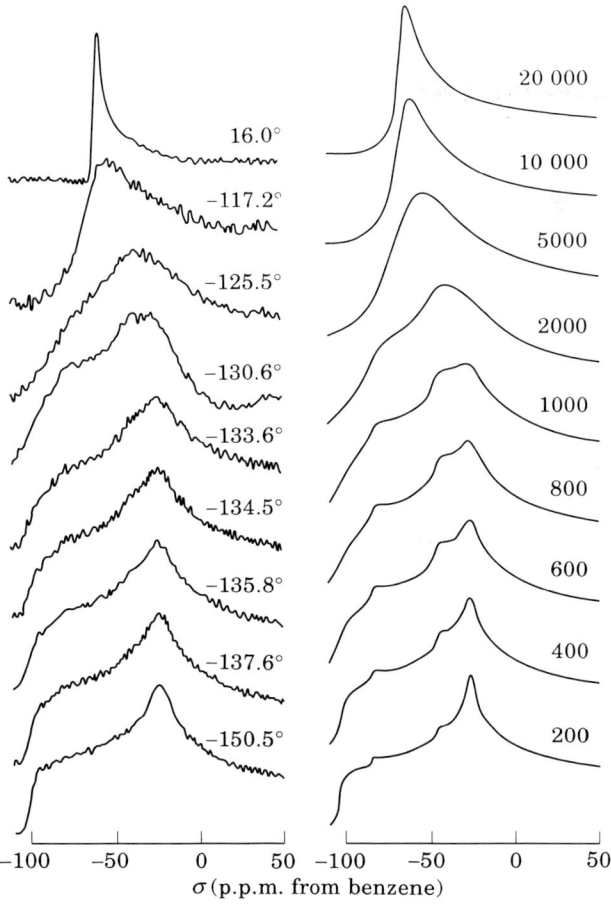

Fig. 5.6. – ^{13}C powder spectra of hexamethyl benzene showing the low-field part of the experimental and theoretical spectra for a number of temperatures. The right-hand side shows the theoretical spectra for a sixfold jump model, demonstrating the features due to motional narrowing present in the experimental spectrum (left). (Courtesy of D. E. WEMMER.)

Based on the principles of this exercise, it is possible to investigate the effects of symmetry in the case of more elaborate motional dynamics. For example, the motion of ethylene absorbed on a surface [133], and a particularly beautiful possibility being the reorientational motion of carbon-60 molecules under icosahedral symmetry [134, 135].

5`6. *Macroscopic motional averaging.* – In this subsection we deal with symmetry considerations in motional averaging of anisotropic interactions. We recall that powder spectra of solids are broad because of the dependence of the frequency on orientation

$$\omega = \omega(\theta, \phi).$$

In principle, by reorienting the sample rapidly and isotropically, simulating, for example, the reorientational motion of molecules in liquids, the anisotropy of the interactions could be made to vanish. Fortunately, it is not normally necessary to implement the full isotropic symmetry of the rotation group SO(3), since the anisotropic interaction has some symmetry, and the dependence of frequency on orientation can be represented as a *finite* sum of components, each irreducible under SO(3)

(5.14) $$\omega = \sum_l \omega_l.$$

The contribution of each component ω_l depends on the $2l+1$ values of the corresponding tensor A_{lm}, which forms the basis of the representation $D^{(l)}$ of SO(3), and on the orientation of the magnetic field in a sample-fixed coordinate system,

(5.15) $$\omega_l = \sum_m A_{lm} D^{(l)}_{m0}(\Omega_{\text{sfc}}(t)),$$

where $D^{(l)}_{m0}(\Omega_{\text{sfc}}(t))$ is an element of the associated Wigner rotation matrix. The question is what is the most economical (perhaps symmetric) set of orientations under which the $l \neq 0$ components of ω vanish.

5`7. *Sample spinning.* – For a single l, for example $l = 2$, one simple answer involves rapid hopping or spinning about an axis. If the sample is undergoing some type of motion on a time scale comparable to the size of the interactions involved (typically a few kHz), then the orientation of the sample-fixed coordinate system will be time dependent and the observed phase of the signal for a given transition at some instant T is proportional to

(5.16) $$\phi_l(T) = \int_0^T \mathrm{d}t\, \omega_l(t) = \sum_m A_{lm} \int_0^T \mathrm{d}t\, D^{(l)}_{m0}(\Omega_{\text{sfc}}(t)).$$

As we saw in subsect. 2`1 and 2`3, stroboscopic sampling, followed by Fourier transform, gives the average frequency of the transition over the interval. The anisotropy described by tensor components $A_{lm \neq 0}$ for a particular l can be averaged away by reorienting the sample so that the field is directed at $N = l + 1$ or more directions to form a cone so that

$$(5.17) \qquad \Omega_{\text{sfc}}^k \equiv \left(\alpha_k = \alpha_0 + \frac{2\pi}{N} k, \beta, \gamma \right)$$

because from eq. (5.16)

$$(5.18) \qquad \sum_{k=1}^{N} D_{m0}^{(l)}(\Omega_{\text{sfc}}) = N d_{m0}^{(l)}(\beta) \delta_{m0} .$$

The remaining term, proportional to A_{l0}, can then be eliminated by a suitable choice of apex angle $2\beta^{(l)}$ such that

$$(5.19) \qquad d_{00}^{(l)}(\beta^l) \equiv P_l(\cos \beta^{(l)}) = 0 .$$

5˙8. *Group theory of motional averaging.* – Now we pose the question of whether motional averaging under a subgroup G of SO(3) can average anisotropic terms in eq. (5.15). Such motion would represent an efficient means of effecting the averaging. We are concerned with the average value of each frequency component over all N symmetry operations R of the group G,

$$(5.20) \qquad \overline{\omega}_l^G = \sum_m A_{lm} \overline{D_{m0}^{(l)}} ,$$

where

$$(5.21) \qquad \overline{D_{m0}^{(l)}} = \frac{1}{N} \sum_{R \in G} D_{m0}^{(l)}(\Omega_R) .$$

The representations of the group SO(3) reduce under the subgroup G as

$$(5.22) \qquad D_{mn}^{(l)}(\Omega_R) = \sum_\lambda a^{(l, \lambda)} D_{mn}^{(\lambda)}(\Omega_R) ,$$

where λ labels the irreducible representation of G. The average of the transformation matrices of eq. (5.21) and, therefore, *the average frequency ω_l can be different from zero only if $D^{(l)}$ contains the totally symmetric representation of G* [136]. The multiplicity $a^{(l, \lambda)}$ of A_1 when $D^{(l)}$ is reduced under G can be evaluated from

$$(5.23) \qquad a^{(l, \lambda)} = \frac{1}{N} \sum_R \chi^{(l)}(R) \chi^{(\lambda)}(R) ,$$

and the characters calculated from

$$(5.24) \qquad \chi^{(l)}(R) = \sum_m D_{mm}^{(l)}(R) = \frac{\sin\left(l + \frac{1}{2}\right) \zeta_R}{\sin \frac{1}{2} \zeta_R} ,$$

where ζ_R is an angle of rotation of the corresponding symmetry operation, related to Euler angles by

(5.25) $$\zeta_R = 2\cos^{-1}\left\{\cos\frac{\beta_R}{2}\cos\frac{\alpha_R+\gamma_R}{2}\right\}.$$

TABLE II. – *Characters of the $D^{(l)}$ in the octahedral and icosahedral point subgroups of SO(3).*

Octahedral (O).

l	E	$8C_3$	$3C_4^2$	$6C_2$	$6C_4$	a^{A_1}
0	1	1	1	1	1	1
1	3	0	−1	−1	1	0
2	5	−1	1	1	−1	0
3	7	1	−1	−1	−1	0
4	9	0	1	1	1	1
5	11	−1	−1	−1	1	0
6	13	1	1	1	−1	1
7	15	0	−1	−1	−1	0
8	17	−1	1	1	1	1

Icosahedral (I), $\rho = (1+\sqrt{5})/2$, $\bar{\rho} = (1-\sqrt{5})/2$.

l	E	$12C_5$	$12C_5^2$	$20C_3$	$15C_2$	a^{A_1}
0	1	1	1	1	1	1
1	3	ρ	$\bar{\rho}$	0	−1	0
2	5	0	0	−1	1	0
3	7	$-\rho$	$-\bar{\rho}$	1	−1	0
4	9	−1	−1	0	1	0
5	11	1	1	−1	−1	0
6	13	ρ	$\bar{\rho}$	1	1	1
7	15	0	0	0	−1	0
8	17	$-\rho$	$-\bar{\rho}$	−1	1	0
9	19	−1	−1	1	−1	0
10	21	1	1	0	1	1

5˙9. *Averaging under cubic and icosahedral symmetry.* – Table II shows the character tables for the octahedral (cubic) and icosahedral groups, with the last column indicating the multiplicity of the totally symmetric representation A_1 for the reduction of $D^{(l)}$ under G. A multiplicity of zero indicates that an anisotropic interaction of rank l will be averaged to zero under the symmetry of group G. For example, under the cubic group, second-rank interactions like chemical-shift anisotropy or first-order quadrupole coupling are averaged to zero. This symmetry may be implemented by reorienting the sample so that the magnetic field lies consecutively along the three orthogonal axes of the sample, a technique known as magic-angle hopping [137, 138]. Magic-angle spinning involves continually reorienting the sample in such a way that, in the sample frame, the magnetic field traces out a cone incorporating the three orthogonal

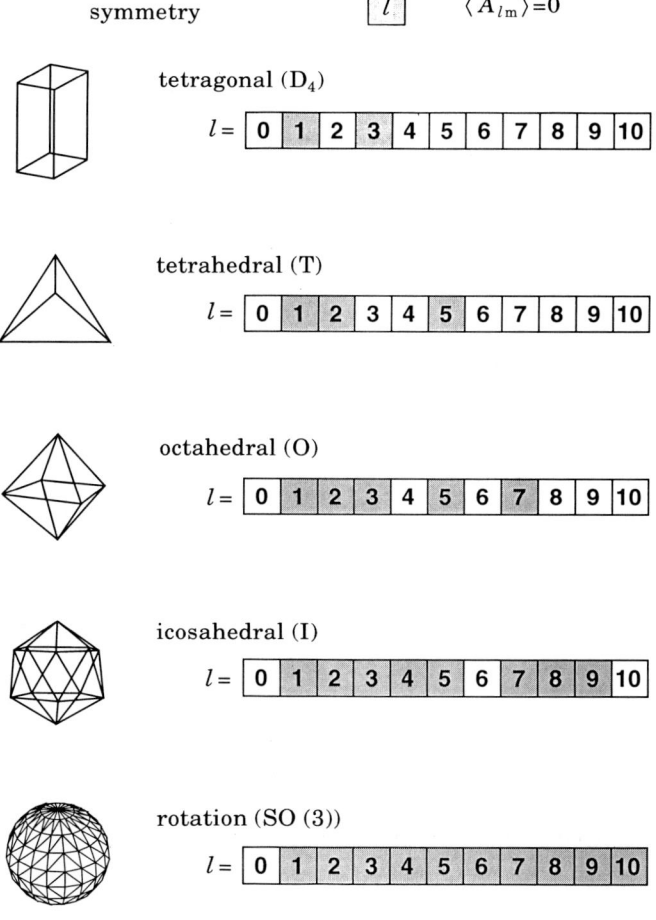

Fig. 5.7. – Averaging of spherical harmonics under subgroups of SO(3).

axes. Thus *magic-angle spinning can be regarded as a continuous implementation of cubic symmetry.*

The central transition for noninteger quadrupole spins is dominated by second-order quadrupole couplings which contain both $l = 2$ and $l = 4$ terms. These terms cannot vanish simultaneously by hopping or spinning about a single axis. However, the character table for the icosahedral group indicates that both these terms should vanish under icosahedral symmetry, which involves, in its most economical form, a motion of the sample amongst six orientations so that the magnetic field lies along the vertices of an icosahedron, a technique dubbed dynamic-angle hopping. A continuous version of this symmetry, dynamic-angle spinning, involves rotation of the sample about two axes so that the magnetic field traces out two cones incorporating the vertices of an icosahedron. *Icosahe-*

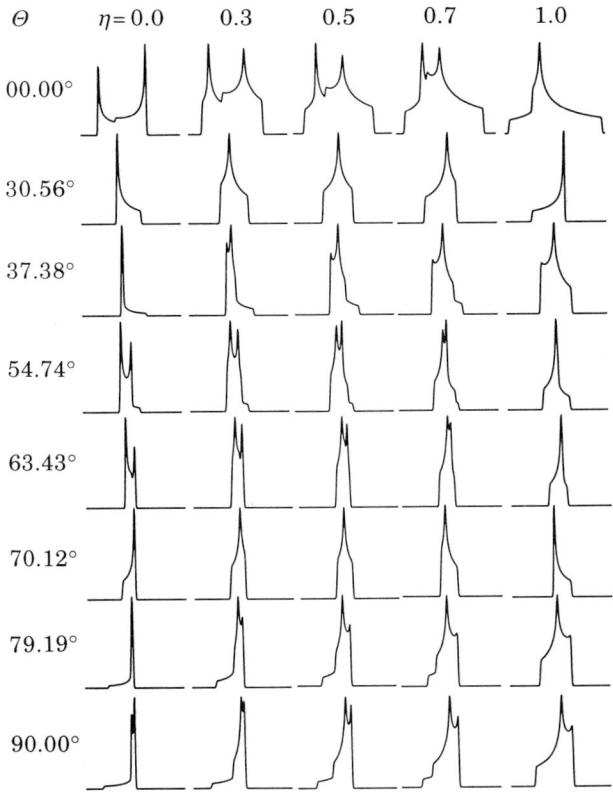

Fig. 5.8. – Simulated second-order powder patterns for the central transition of a spin $I = 3/2$ after motional averaging around a single rotation axis at various angles with respect to the magnetic field. 0° (icosahedral angle), 30.56° (root of $P_4(\cos \Theta)$), 37.38° (icosahedral angle), 54.74° (root of $P_2(\cos \Theta)$), 63.43° (icosahedral angle), 70.12° (root of $P_4(\cos \Theta)$), 79.19° (icosahedral angle). (Adapted from *Solid State Nucl. Magn. Reson.*, **1**, 267 (1992), with permission.)

dral symmetry can be viewed as the next approximation to the sphere for averaging of the higher-rank interactions. In fig. 5.7 we sum up which ranks are averaged to zero by a variety of symmetry groups including tetrahedral, octahedral and icosahedral. Note that the first ranks which are not averaged to zero by icosahedral motion are $l = 6$ and 10.

5˙10. *Dynamic-angle spinning.* – An alternative view of dynamic-angle spinning is suggested by the powder patterns of fig. 5.8 for samples spinning around various axes [139]. Note that some of the lineshapes are in fact mirror images of others, with different scaling factors. For example, the spectra at 0° are scaled mirror images of those at 63.43°, and those spinning at 37.38° are exact images of those for 79.19°. Since reflection occurs about the isotropic-shift value for each crystal orientation, it is possible to rephase the signal of spins from all orientations simultaneously at some time by appropriate choice of spinning angles and evolution times. If evolution for a period τ_1 proceeds at the angle β_1, and subsequently for a period τ_2 at β_2, the accumulated anisotropic

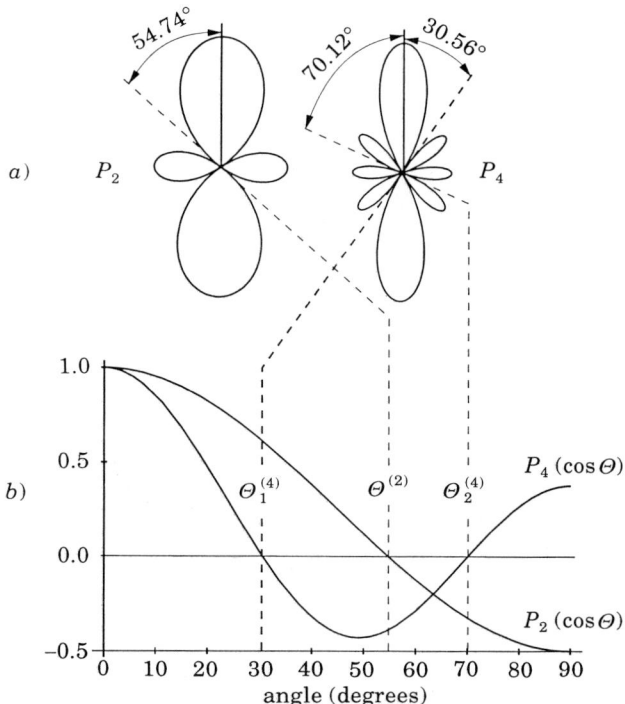

Fig. 5.9. – Plot of the second- and fourth-order Legendre polynomials, $P_2(\cos\Theta)$ and $P_4(\cos\Theta)$, vs. the angle of rotation in variable-angle spinning. *a)* Plot in polar coordinates, *b)* in Cartesian coordinates. The nodes of both functions are indicated by dashed lines. (Adapted from *Solid State Nucl. Magn. Reson.*, **1**, 267 (1992), with permission.)

phase will be

$$(5.26) \quad \phi(T) = \int_0^T \omega(\beta(t))\,dt = \omega(\beta_1)\,\tau_1 + \omega(\beta_2)\,\tau_2\,.$$

The modulation in eq. (5.26) disappears for each interaction of rank n if

$$(5.27) \quad P_n(\cos\beta_1) + P_n(\cos\beta_2)k = 0\,,$$

where $k = \tau_2/\tau_1$ which results in $\phi(\tau_1 + \tau_2) = 0$. In this way the signal recorded

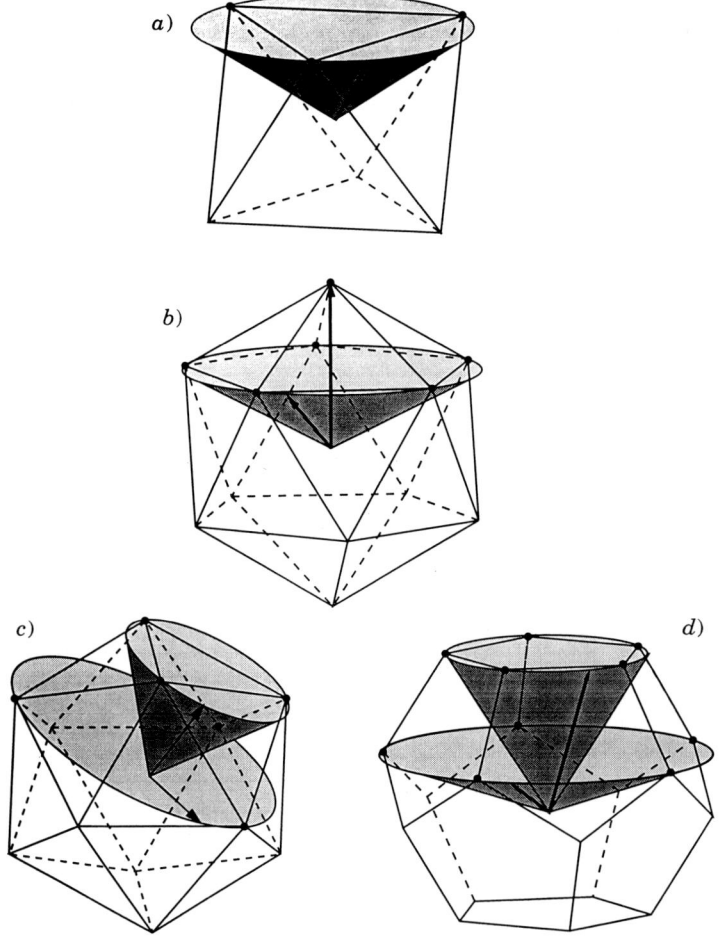

Fig. 5.10. – The external magnetic field viewed in a sample-fixed coordinate frame moves along one or two cones in MAS and DAS, respectively. The DAS trajectory on an icosahedron crosses six vertices for the $k = 5$ case (b)) and the $k = 1$ case (c)). The $k = 1$ case can also be seen as a trajectory crossing ten vertices of a dodecahedron (d)).

as a function of $\tau_1 + \tau_2$ with k constant will be independent of all anisotropic terms, and carry information only about the isotropic shifts.

Many variations of this approach are possible because motion of the spinning axis is not limited to discrete positions and equal probability. Dynamic-angle spinning trajectories have been implemented which follow the symmetry of regular polyhedra. The solutions of eq. (5.27) generally describe surfaces in the three-dimensional space of β_1, β_2 and $\tau_1 + \tau_2$. A simultaneous solution for two equations of the form of eq. (5.27) is obtained at any point on a crossing line of two surfaces, and these points are shown in fig. 5.9. Two solutions, ($\beta_1 = 37.38°$, $\beta_2 = 79.19°$, $k = 1$) and ($\beta_1 = 0.00°$, $\beta_2 = 63.43°$, $k = 5$), describe cones which traverse the vertices of an icosahedron (fig. 5.10), and thus correspond exactly to the icosahedral symmetry of subsect. 5˙9. Figure 5.11a) shows the pulse sequence for a typical DAS experiment and fig. 5.12 shows the resulting spectrum, in this case of the central transitions of the ^{17}O resonances in a series

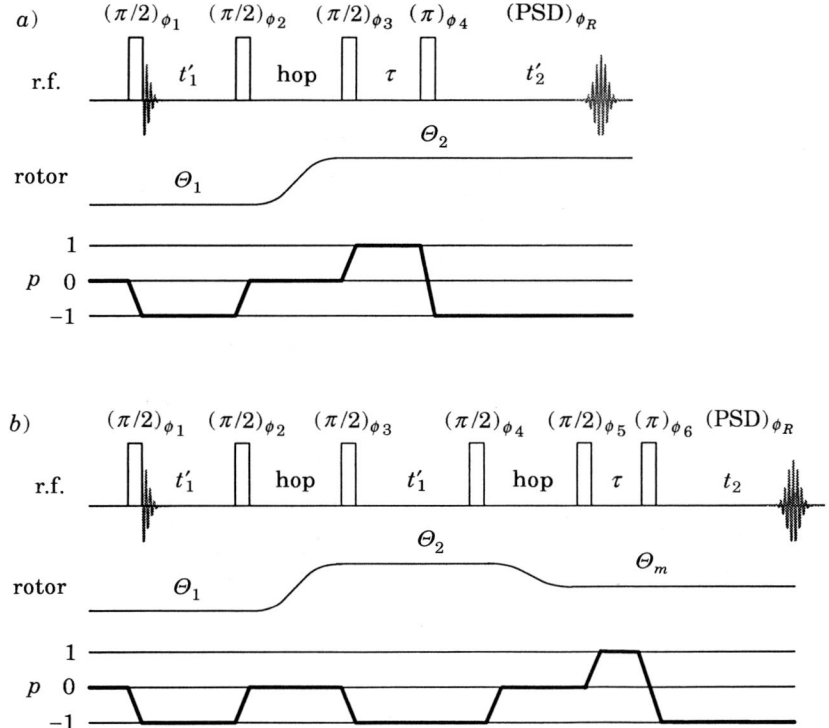

Fig. 5.11. – Pulse sequences for two-dimensional dynamic-angle spinning experiments. a) Conventional sequence to obtain an isotropic-anisotropic correlation, with powder patterns in the anisotropic dimension characteristic of the second spinning angle. b) Sequence including a second hop to allow detection at the magic angle. The phases are cycled so as to select the pathway illustrated below the sequence in order to ensure a pure phase spectrum by selecting the whole echo in t_2. (Courtesy of P. J. GRANDINETTI.)

of crystalline silicates. A simple extension of the sequence, shown in fig. 5.11b), includes an extra hop after t_1 which allows detection of the signal in t_2 at the magic angle. This facilitates analysis of the spectrum since the powder patterns in the anisotropic dimension of the spectrum are free of second-rank contributions and are only due to the second-order quadrupole (fourth-rank) contributions, illustrated for the central transitions of ^{87}Rb in rubidium nitrate in fig. 5.13 [140].

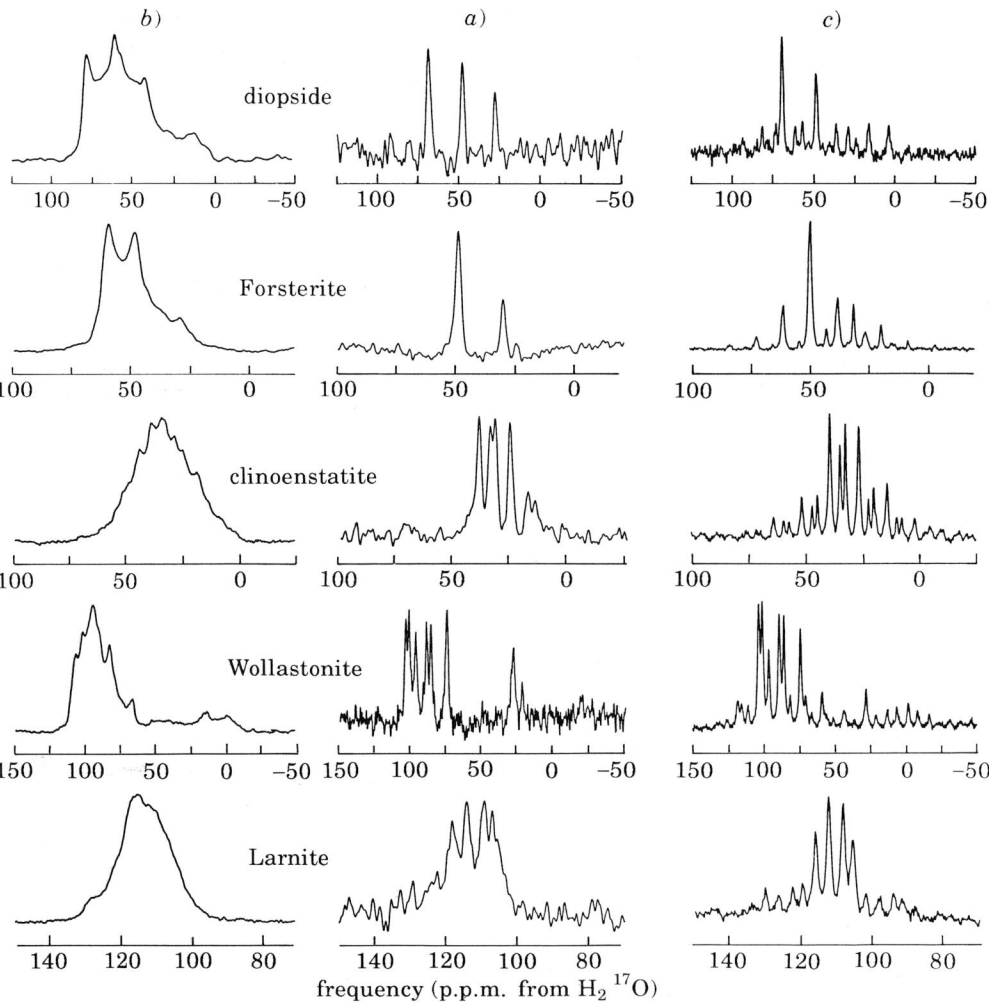

Fig. 5.12. – a) Isotropic ^{17}O dynamic-angle spinning spectra of a series of crystalline silicates at 9.4 T. For comparison we also show b) the magic-angle spinning spectra and c) the double-rotation spectra of the same compounds. (Courtesy of K. T. MUELLER.)

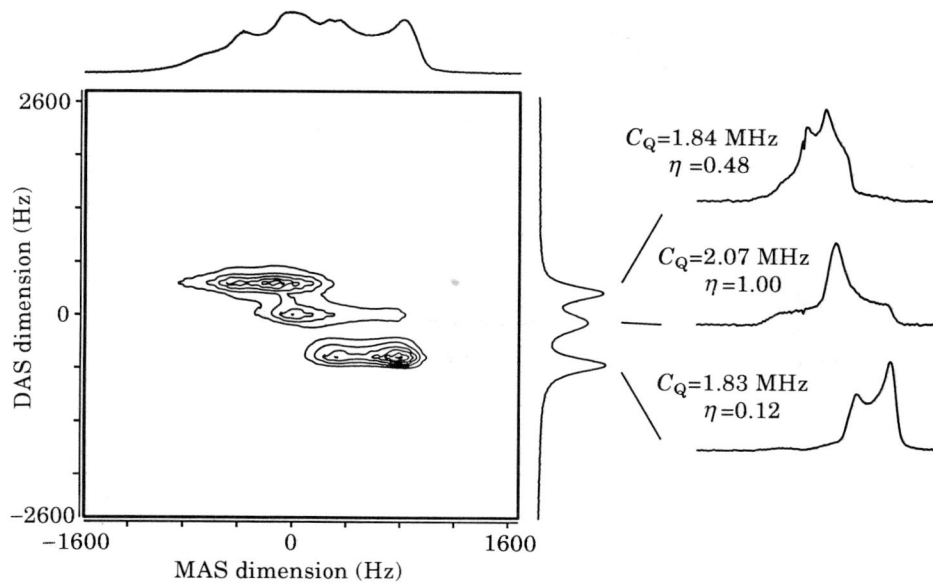

Fig. 5.13. – The two-dimensional dynamic-angle spinning spectrum of the central transition of ^{87}Rb in RbNO$_3$ obtained using the sequence of fig. 5.11b). The isotropic-anisotropic correlation yields powder patterns that are determined purely by fourth-rank interactions, which facilitates the analysis in terms of the quadrupolar coupling constants and asymmetry parameters (Adapted from *J. Am. Chem. Soc.*, **114**, 7489 (1992), with permission).

5˙11. *Isotropic-anisotropic correlation spectra.* – Note that because the isotropic spectrum is detected indirectly in a DAS experiment, according to the pulse sequences shown in fig. 5.11, two-dimensional Fourier transform of the data (followed by a shearing transform) yields a two-dimensional isotropic-anisotropic correlation spectrum. Thus the isotropic resolution is achieved without sacrificing the information contained in the anisotropies, and for each isotropic line we can obtain the corresponding anisotropy[72, 141, 142] which may allow us to assign the isotropic shifts, or in favorable cases gives access to a wealth of chemical information about the material under study. Such a case is shown in fig. 5.14 in which we show the DAS spectrum of a potassium tetrasilicate glass (K$_2$Si$_4$O$_9$)[143]. At first sight the broad line in the isotropic dimension may appear to indicate that the experiment has failed to remove completely the anisotropy. However, closer inspection reveals that this broadening is due to a distribution of isotropic sites in the glass and, moreover, when we examine the anisotropies, we see that they are dependent on the isotropic shift (fig. 5.10). FARNAN *et al.* [143] have shown that this distribution of quadrupolar couplings and asymmetry parameters can be converted into a bond angle distribution for the glass shown in fig. 5.15, thus clearly demonstrating the enormous potential such experiments have for structural studies of complex materials[144].

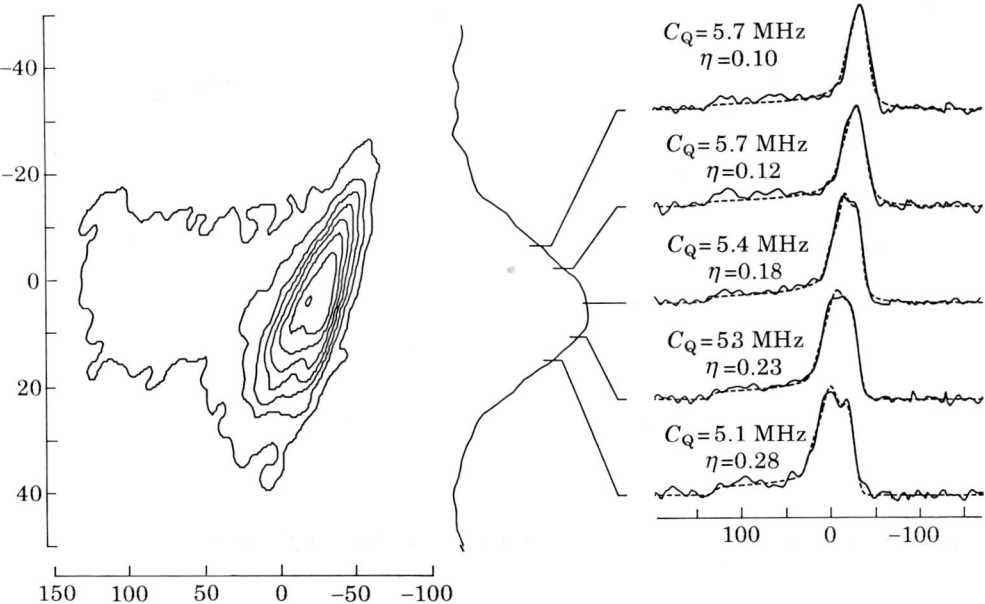

Fig. 5.14. – Two-dimensional ^{17}O DAS NMR spectrum of $K_2Si_4O_9$ glass, showing the resonance of the bridging oxygens. The anisotropic lineshapes are characteristic of spinning at 79.19° with respect to the field, and are due almost entirely to the interaction between the quadrupole moment of ^{17}O and the electric-field gradient at the oxygen sites. Slices through the anisotropic dimension are subjected to multiparameter fits to determine the magnitude and asymmetry of the field gradient across the isotropic line. (Adapted from *Nature (London)*, **358**, 31 (1992), with permission.)

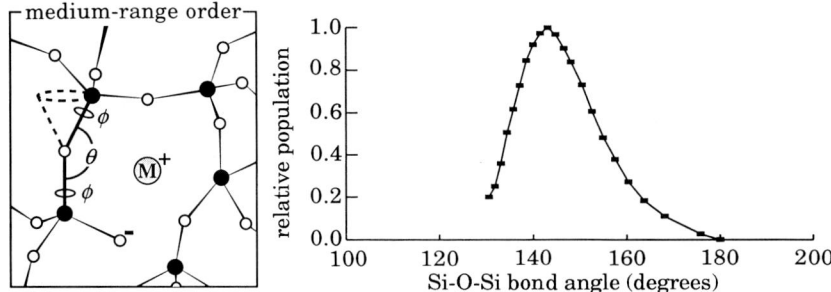

Fig. 5.15. – The quadrupolar-coupling constants and asymmetry parameters extracted from fig. 5.14 can be used to determine the distribution of Si-O-Si bond angles in the $K_2Si_4O_9$ glass which characterizes the degree of medium-range order present in the structure. (Adapted from *Nature (London)*, **358**, 31 (1992), with permission.)

5`12. Variable-angle correlation spectroscopy. – An alternative approach to isotropic-anisotropic correlations in MAS spectroscopy is given by variable-angle correlation spectroscopy (VACSY)[145]. This experiment is representative of a broad spectrum of potential new techniques, dubbed mixed dimension acquisition schemes (MIDAS), for obtaining many kinds of correlation spectra, and it is instructive to look at it in more detail.

From the description of subsect. 5`11 it might seem that either a change in the rotor axis or the application of synchronous pulses is necessary in order to correlate an isotropic (*i.e.* magic-angle spinning) spectrum with a static (*i.e.* spinning at 0°) spectrum. However, we have already seen that continuous versions of the hopping experiments (the transition from magic-angle hopping to magic-angle spinning) yield similar results. Indeed, the VACSY experiment yields two-dimensional isotropic-anisotropic correlation spectra for spin $I = 1/2$ systems experiencing anisotropic chemical shift or heteronuclear dipolar interactions without changing the spinning axis during the sequence. In subsect. 2`2 we derived the chemical-shift Hamiltonian for a spin $I = 1/2$ system undergoing rapid spinning at an angle Θ with respect to the field. Equation (2.20) can be used to derive the precession frequency for a given crystallite orientation in the form

$$(5.28) \qquad \omega(\alpha, \beta, \gamma) = \omega_i + P_2(\cos\Theta)\,\omega_a(\alpha, \beta, \gamma),$$

where the anisotropy is scaled by the second-order Legendre polynomial. Con-

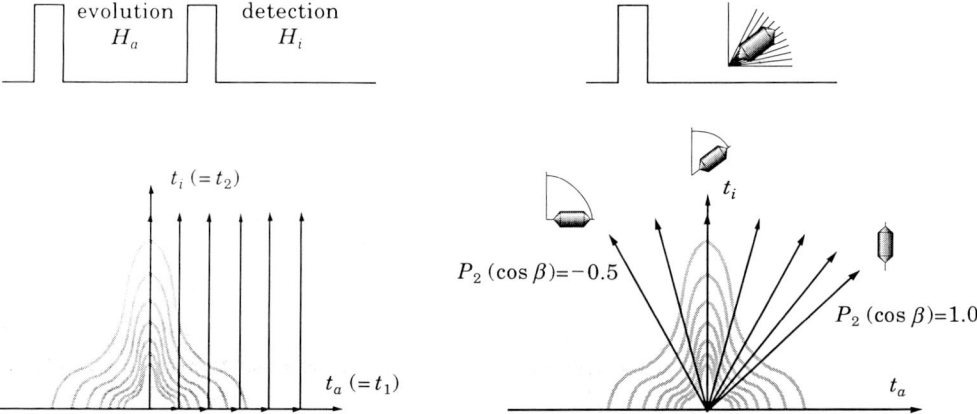

Fig. 5.16. – In «traditional» correlation spectroscopy (left) the time domain is sampled in Cartesian-coordinate fashion by arranging for pure anisotropic and pure isotropic evolution respectively during t_1 and t_2 (perhaps by changing the spinning axis from 0° during t_1 to Θ_m during t_2). In variable-angle correlation spectroscopy (VACSY) (right), the same correlation spectrum is obtained through a non-Cartesian sampling from a series of variable-angle spinning experiments, which progressively scale the anisotropic contribution to the evolution. (Adapted from *J. Chem. Phys.*, **97**, 4800 (1992), with permission.)

sider now the time domain acquisition in a conventional two-dimensional correlation experiment in which the data are sampled in Cartesian coordinates under the pure interactions that are desired in the corresponding dimensions, as illustrated in fig. 5.16. In such an experiment the signal is given by

$$(5.29) \qquad f(t_1, t_2) = \int \int I(\omega_1, \omega_2) \exp[i(\omega_1 t_1 + \omega_2 t_2)] d\omega_1 d\omega_2 .$$

Let us compare eq. (5.29) to the signal acquired in a one-dimensional spinning experiment corresponding to eq. (5.28), for which

$$(5.30) \qquad f(t) = \int I(\omega(\alpha, \beta, \gamma)) \exp[i\omega t] d\omega ,$$

and, if we substitute eq. (5.28) in (5.30), we obtain an expression of the form

$$(5.31) \qquad f(t_i, t_a) = \int \int I(\omega_i, \omega_a) \exp[i(\omega_i t_i + \omega_a t_a)] d\omega_i d\omega_a$$

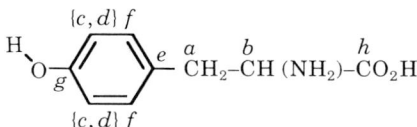

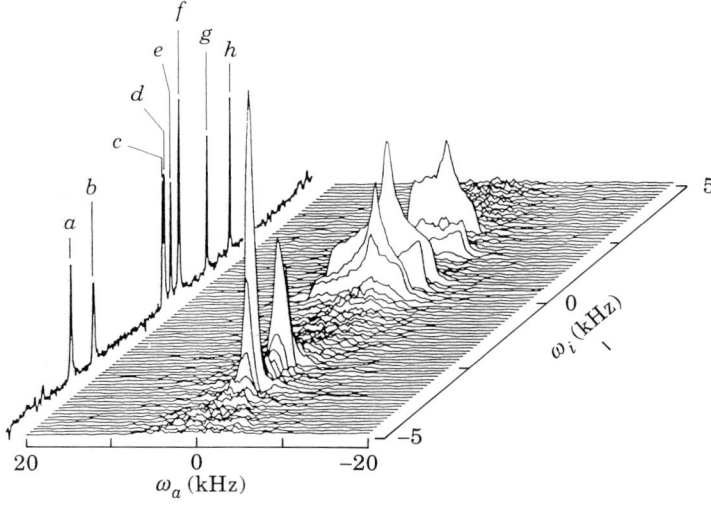

Fig. 5.17. – Two-dimensional isotropic-anisotropic NMR chemical-shift correlation spectrum of L-tyrosine, showing the isotropic projection and the assignment of the peaks to each site in the molecule. A set of 65 signals were digitized at rates of 12 kHz over a range of angles. Experimental data were then interpolated over a regular grid composed of 128 × 256 points and Fourier transformed. (Adapted from *Isr. J. Chem.*, **32**, 161 (1992), with permission.)

with $t_i = t$ and $t_a = P_2(\cos\Theta)t$. This expression has the same form as eq. (5.29) for a normal two-dimensional spectrum and suggests that *by acquiring a series of spectra spinning at different angles we should be able to obtain an isotropic-anisotropic correlation spectrum*. Indeed, by recording such a series of spectra and rearranging the data as shown in fig. 5.16, the spectrum of solid tyrosine, shown in fig. 5.17, was obtained. The great advantage of this technique is that it allows one to obtain spectra by scaling interactions, thus avoiding experiments in which one must switch interactions either by sudden mechanical motion of the sample or the introduction of multiple-pulse irradiation. This analysis turns out to be very general (analogous to imaging by projections instead of Fourier imaging) and it is possible to obtain not only two-dimensional correlations in this way, but also higher-dimensional spectra, the only prerequisite being that the original anisotropic nD spectrum has a form in which the anisotropic contributions scale with $P_2(\cos\Theta)$. The technique has already been extended to the separation of anisotropic two-dimensional exchange spectra[121] according to their isotropic shifts[146].

6. – Cross-polarization.

6˙1. *Spin temperature.* – This section is concerned with cross-polarization, and especially the equilibrium magnetizations observed in cross-polarization (CP) experiments. Using simple models of cross-polarization[147, 148] we may predict the equilibrium magnetizations by observing that in a system containing N_S S spins and N_I I spins, experiencing fields B_I and B_S, the initial I spin magnetization is given by Curie's law

(6.1) $$M_I^0 = \beta_I^0 C_I B_I,$$

where

(6.2) $$\beta_I^0 = 1/kT$$

is the inverse spin temperature, and

(6.3) $$C_I = \gamma_I^2 \hbar I(I+1) N_I / 3$$

is the I spin heat capacity. Two of the most important quantities of the system are the energy

(6.4) $$E_I^0 = -\beta_I^0 C_I B_I^2,$$

and the entropy

(6.5) $$S_I = \text{const} - k(\beta_I^0)^2 C_I B_I^2.$$

If the spins are allowed to come into thermal contact so that they exchange magnetization at a rate characterized by T_{IS}, then after a time $t \gg T_{IS}$ the I and

S spins will come to equilibrium and will have equal spin temperatures

(6.6) $$\beta^{eq} = \beta_I^{eq} = \beta_S^{eq},$$

in which case the energy is now

(6.7) $$E^{eq} = E_I^{eq} + E_S^{eq} = -\beta^{eq}(C_I B_I^2 + C_S B_S^2).$$

Since the Hamiltonian is assumed to be time independent, energy is conserved. By defining $\varepsilon^2 = C_S/C_I$ and $\chi = B_S/B_I$ we obtain

(6.8) $$\beta^{eq} = \beta^0 \frac{1}{1 + \chi^2 \varepsilon^2}$$

and we can predict the S spin equilibrium magnetization

(6.9) $$M_S^{eq} = \frac{\alpha}{1 + \chi^2 \varepsilon^2} \beta_I^0 C_S B_I = \frac{\chi}{1 + \chi^2 \varepsilon^2} M_S^0,$$

where M_S^0 is the normal value of M_S at thermal equilibrium.

6˙2. *Methods for obtaining thermal contact.* – In the normal laboratory frame T_{IS} is comparable to the rate of equilibration with the lattice. In order to achieve rapid cross-polarization we must isolate the I and S spins from the lattice and put them in thermal contact. We discuss two popular methods.

i) *Hartmann-Hahn matching.* The Hartmann-Hahn experiment[149] is a rotating-frame experiment, shown in fig. 6.1, in which the B fields correspond to applied magnetic fields in the transverse plane. T_{IS} is found to be strongly dependent on B_I and B_S and, therefore, on χ. The most rapid cross-polarization occurs under the Hartmann-Hahn (HH) condition

(6.10) $$\gamma_I B_I = \gamma_S B_S.$$

In this case polarization can be transferred by energy-conserving mutual spin flip-flops as shown in fig. 6.1[150]. Normally $N_S \ll N_I$ so that $\varepsilon \ll 1$, in which case eq. (6.9) becomes

(6.11) $$M_S^{eq} = \frac{\gamma_I}{\gamma_S} M_S^0$$

and the HH enhancement is given by the ratio γ_I/γ_S which is 4 for ^{1}H and ^{13}C. We will consider in detail problems associated with HH mismatch ($\chi \neq 1$) below.

ii) *Adiabatic demagnetization in the rotating frame* (ADRF)[147, 151]. A second method for obtaining efficient cross-polarization is shown in fig. 6.2. In this experiment (also performed in the rotating frame) transverse I spin magnetization is first spin locked and then the field B_I is adiabatically allowed to decrease to zero. If we take the most simple example of a two-spin system IS,

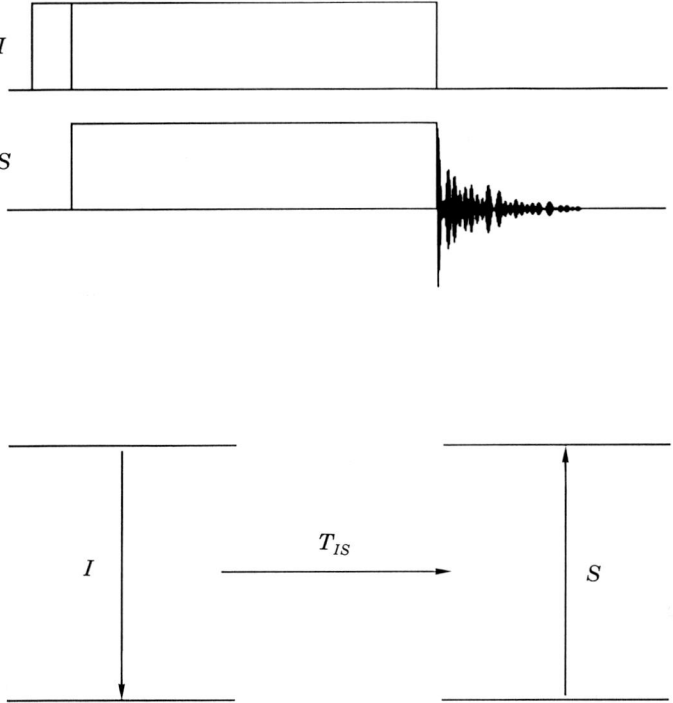

Fig. 6.1. – Pulse sequence for Hartmann-Hahn cross-polarization, in this case with detection of the S-spin free-induction decay. If the magnetic fields B_I and B_S in the rotating frames fulfill the condition of eq. (6.10), then an energy-conserving I-S flip-flop can occur resulting in a transfer of polarization as illustrated schematically in the lower part of the figure. (Adapted from *J. Chem. Phys.*, **59**, 569 (1973), with permission.)

then the initial density matrix in the tilted frame is

(6.12) $$\rho(0) = I_z$$

and the Hamiltonian

(6.13) $$H(t) = \omega_{1I}(t) I_z + b 2 I_x S_x ,$$

where b is the heteronuclear dipolar interaction. If we assume that $\omega_{1I} \gg b$ at the beginning of the experiment, then we obtain

(6.14) $$H_0 \approx \omega_{1I}(0) I_z$$

and we observe that $[H_0, \rho(0)] = 0$. If we turn the field off adiabatically, this corresponds to the condition

(6.15) $$[H(t), \rho(t)] = 0$$

for all values of t. Thus, when the field reaches zero, we obtain a Hamiltonian

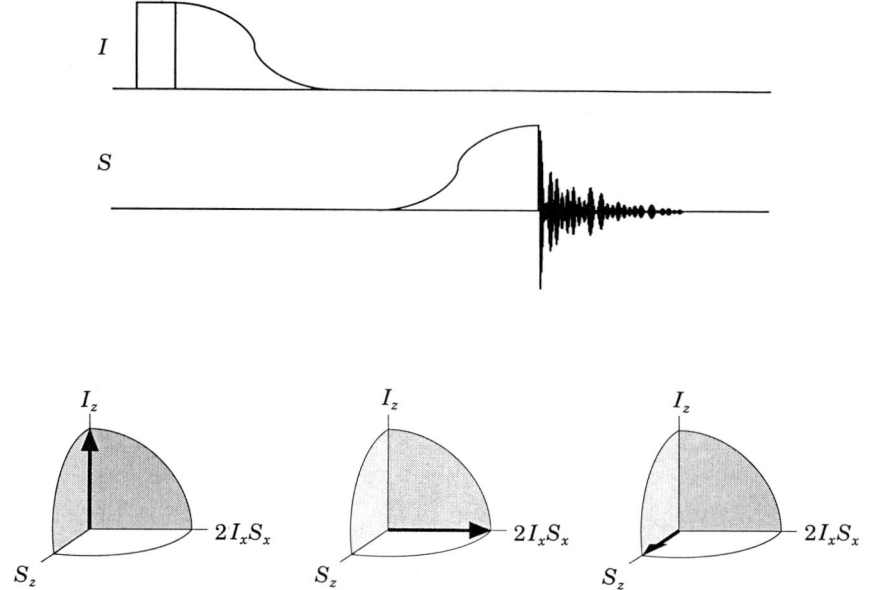

Fig. 6.2. – Pulse sequence for adiabatic demagnetization in the rotating frame. The evolution of the density matrix in a two-spin system is illustrated below. At the start of the experiment (lower left) all the spin order is contained in I_z. After adiabatic demagnetization the order is transferred into the dipolar reservoir, which is in turn transferred to S-spin order after remagnetization at the S-spin frequency (lower right).

and a density operator

(6.16) $$H(t) = b2I_xS_x \,, \qquad \rho(t) = 2I_xS_x \,.$$

Since the change in the density matrix was adiabatic, all the spin order in this two-spin system is transferred from the reservoir I_z into the dipolar reservoir. It can be seen that, if we now adiabatically turn on a strong S spin field, all the order in the dipolar reservoir is transferred to the S spin longitudinal reservoir, so that we can achieve (in the two-spin system) complete transfer of polarization from I to S. The principles of this experiment are illustrated graphically in fig. 6.2. (Note that, if there are two I spins, complete transfer is no longer possible, a point we shall return to in sect. 7.)

6`3. *Statistical picture of cross-polarization.* – An alternative way of looking at CP is to consider the details of the microscopic behavior of the spins [23]. If we consider a system of N_I I spins with N_{I^+} «up» and N_{I^-} «down» coupled to N_S S spins with N_{S^+} «up» and N_{S^-} «down», the I and S polarizations are defined by

(6.17) $$\begin{cases} P_I = N_{I^+} - N_{I^-} \,, \\ P_S = N_{S^+} - N_{S^-} \,. \end{cases}$$

Initially $P_S = 0$ and $P_I = P_I^0$. We will consider briefly a question to which we shall return in the next section: what is the maximum S polarization achievable by the most efficient possible CP process from I to S? Our first consideration is that for maximum efficiency we must conserve entropy (so that the polarization transfer corresponds to a reversible process). The entropy depends on the number of configurations that the system can occupy,

(6.18) $$S = k \ln W,$$

in which W represents the number of ways to place N^+ indistinguishable objects in N positions,

(6.19) $$W = \binom{N}{N^+} = \frac{N!}{N^+!N^-!}.$$

If N is large, we obtain

(6.20) $$W \approx 2^N \exp\left[\frac{-P^2}{2N}\right],$$

so that the entropy is

(6.21) $$S = \text{const} - \frac{P^2}{2N}.$$

If we consider a situation in which entropy is conserved and all the I polarization is transferred to S, then we can equate S^0 with S^{eq} and obtain

(6.22) $$\frac{(P_I^0)^2}{2N_I} = \frac{(P_S^{\text{eq}})^2}{2N_S}.$$

By simple rearrangement we see that the maximum final S polarization is

(6.23) $$P_S^{\max} = P_S^{\text{eq}} = \sqrt{\frac{N_S}{N_I}} P_I^0.$$

Now we recall that the equilibrium polarization is proportional to γ and N, so that

(6.24) $$P_I^0 = \frac{\gamma_I}{\gamma_S} \frac{N_I}{N_S} P_S^0.$$

Inserting eq. (6.24) in eq. (6.23)

(6.25) $$P_S^{\max} = \frac{\gamma_I}{\gamma_S} \sqrt{\frac{N_I}{N_S}} P_S^0,$$

or for the magnetization

(6.26) $$M_S^{\max} = \frac{\gamma_I}{\gamma_S} \sqrt{\frac{N_I}{N_S}} M_S^0.$$

Note that the S magnetization predicted here is a factor $\sqrt{N_I/N_S}$ larger than with HH matching given in eq. (6.11).

6'4. *Hartmann-Hahn mismatch.* – Returning to Hartmann-Hahn experiments (which will form the basis of the rest of the section), we would like to consider in detail the behavior of the system for $\chi \neq 1$. (Similar considerations apply to certain incarnations of the ADRF experiment.) Using eq. (6.10) we can predict the dependence of M_S^{eq} on the parameter $\chi = B_S/B_I$ as is shown in fig. 6.3. Although the treatment of subsect. 6'1 predicts maximum S magnetization for $\omega_S > \omega_I$, experimentally the maximum value of M_S^{eq} is observed at HH match. In order to explain the off-match behavior, we must develop an exten-

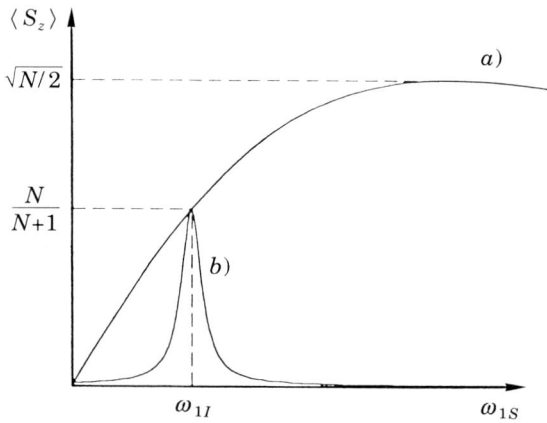

Fig. 6.3. – S-spin magnetization as a function of S-spin r.f. field. The model of subsect. 6'1 predicts curve a), whilst experimental dependences are obtained which follow curve b) and have a peak at HH match. (Adapted from ref. [152], with permission.)

sion to the standard view of spin temperature and thermodynamics which will allow us to predict more accurately the equilibrium positions. Following LEVITT *et al.* [152, 153] we describe the system by an initial density operator $\rho(0)$ which is subjected to a time-independent Hamiltonian H which acts like a unitary transformation on the density operator

(6.27) $$\rho(t) = U\rho(0) U^{-1},$$

where $U = \exp[-iHt]$. From spin thermodynamics we expect that on some time scale $t \gg T_{IS}$ the important observables of the system, such as the magnetizations of the different spins, eventually become time-independent, which is found to be valid in all sufficiently large and complex systems.

In the case of strong r.f. fields (applied along the x-axis) it is most convenient to transform to a picture defined by a rotation about the y-axis

$$A' = R^{-1}AR,$$

with

(6.28) $$R = \exp\left[-i\frac{\pi}{2}\left(\sum_k I_{ky} + S_y\right)\right].$$

At HH match we partition the Hamiltonian into two (noncommuting) reservoirs of spin order,

(6.29) $$\Omega_1 = \frac{1}{\sqrt{N}} \sum_1^N I_{iz},$$

(6.30) $$\Omega_2 = S_z,$$

and

(6.31) $$h_1 = \sqrt{N}\,\omega_{1I}, \qquad h_2 = \omega_{1S},$$

and a perturbation term corresponding to the dipolar interactions,

(6.32) $$V = \sum_1^N b_i 2 I_{ix} S_x + H_{II},$$

in which b_i are the heteronuclear dipolar interactions and H_{II} is the I-I interaction Hamiltonian. The equilibrium position is determined by projection of the initial density operator onto the reservoir terms

(6.33) $$\rho_{eq}^0 = \frac{\langle H^0 | \rho(0) \rangle}{\langle H^0 | H^0 \rangle} H^0,$$

where

(6.34) $$H^0 = h_1 \Omega_1 + h_2 \Omega_2.$$

This has a simple geometric interpretation, demonstrated in fig. 6.4. The total energy of the system,

(6.35) $$\langle H \rangle = \langle H | \rho \rangle = \langle H^0 | \rho^0 \rangle + \langle V | \rho^0 \rangle + \langle H | \rho'(t) \rangle,$$

is conserved since the Hamiltonian is time-independent. Since the excess term ρ' is taken to be orthogonal to H^0, and the energetic influence of the perturbation V is ignored, the energy is contained only in the reservoirs

$$\langle H \rangle = \langle H^0 | \rho^0 \rangle.$$

The reservoir energy is the component of the state vector parallel to the reservoir Hamiltonian, as illustrated in fig. 6.4.

On the contrary, the entropy is split between the reservoirs and the excess term, and in the high-temperature approximation the entropy $S = \rho \ln \rho$ is proportional to the spin order. The spin order is simply

(6.36) $$\langle \rho | \rho \rangle = \langle \rho^0 | \rho^0 \rangle + \langle \rho' | \rho' \rangle.$$

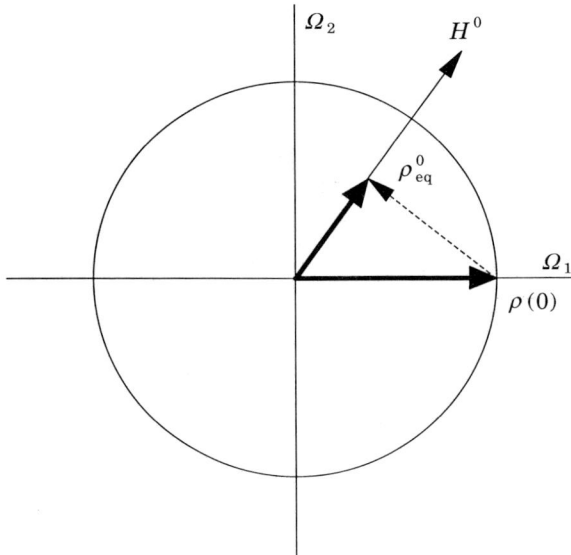

Fig. 6.4. – Geometric picture of equilibrium. In the plane of the reservoir operators the reservoir Hamiltonian is described by a vector. Equilibrium is represented as the projection of the initial density operator $\rho(0)$ onto the Hamiltonian. The circle represents the limit imposed by conservation of spin order.

Since the transformation from ρ^0 to ρ_{eq}^0 is unitary, the *total* entropy must be conserved. The excess term ρ' serves as a repository of entropy but not energy.

The partition of eqs. (6.29)-(6.32) results in the predictions of CP efficiency *vs.* mismatch shown in fig. 6.3. Whilst it may be tempting to attribute this discrepancy between experiment and theory to a dynamic effect, experiments show that even far from match constant values of the S magnetization are obtained fairly quickly, after initial oscillations have decayed (see fig. 6.5)[153].

6˙5. *Thermodynamics of heteronuclear cross-polarization.* – To achieve a description which works off match we must include the effect of broadening of the distribution of I-spin energy levels by dipolar interactions so as to produce a finite width of the HH match. However, we cannot simply replace Ω_1 by $\Omega_1 + H_{II}$ since, although the choice of «reservoirs» and «perturbations» is made arbitrarily, nevertheless each term should either have a harmonic spectrum, or a dense spectrum with a large number of closely spaced eigenvalues. The energy level structure resulting from replacement of Ω_1 by $\Omega_1 + H_{II}$ does not fulfill either of these guidelines. Indeed a combined Zeeman-dipolar reservoir does not attain internal thermal equilibrium rapidly if the applied fields are strong. Thus we are faced with the problem of how to include the energetic effects of H_{II}.

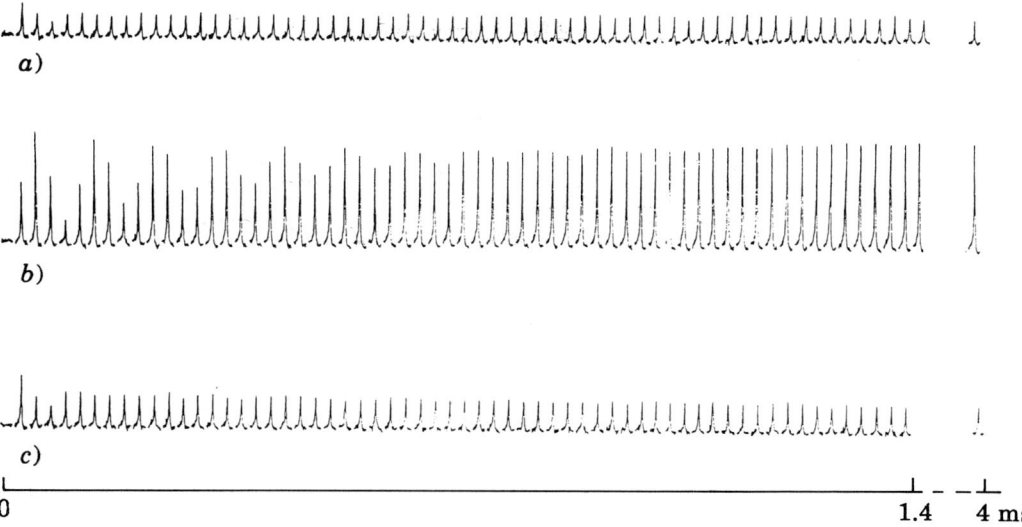

Fig. 6.5. – Experimental evolution of ^{13}C magnetization as a function of cross-polarization time for a single crystal of ferrocene. The experiments were performed at three different r.f. fields where a) has an S-spin field below the Hartmann-Hahn matching condition, b) corresponds to match, and c) is above match. Even away from the Hartmann-Hahn condition an equilibrium state is achieved fairly quickly, and we see that the theoretical behavior predicted by the standard model is not borne out in practice. (Adapted from *J. Chem. Phys.*, **84**, 4243 (1986), with permission.)

The solution is based on the identification of quasi-invariants of the system, operators whose expectation values may be assumed to be conserved during the process of thermal equilibration. In the case of heteronuclear cross-polarization we make the following division into *three* relevant reservoirs: the first two are simply Ω_1 and Ω_2 of eqs. (6.29)-(6.31), and we invoke Ω_3 where

$$(6.37) \qquad h_3 \Omega_3 = -\frac{1}{2} \sum_{jk} d_{jk} \left(2 I_{jz} I_{kz} - \frac{1}{2} (I_j^+ I_k^- + I_j^- I_k^+) \right)$$

(neglecting nonsecular terms), and the perturbation is the heteronuclear coupling term

$$(6.38) \qquad V = \sum_{1}^{N} b_i 2 I_{1x} S_x \,.$$

If this partition is valid (see subsect. 6·9), we may proceed to observe that, since the operator $\sum_k I_{kz} + S_z$ commutes with all the major terms in the Hamiltonian (*i.e.* $I^+ S^-, I^- S^+$), it is plausible to associate a *quasi-invariant operator* Q_1 with it. The expectation value $\left\langle \sum_k I_{kz} + S_z \right\rangle$ then represents the total number of spins with magnetic quantum number $+1/2$ minus the number with $-1/2$, the total angular momentum of the sample, which is clearly conserved by spin pair flips. Thus

$$Q_1 = q_1 \left(\sum_1^N I_{kz} + S_z \right) \tag{6.39}$$

with the normalization

$$q_1 = \frac{\left\langle H^0 \middle| \sum_1^N I_{kz} + S_z \right\rangle}{\left\langle \sum_1^N I_{kz} + S_z \middle| \sum_1^N I_{kz} + S_z \right\rangle} = \frac{N\omega_{1I} + \omega_{1S}}{N+1}. \tag{6.40}$$

Assuming that there are only two quasi-invariants of the system, then the second quasi-invariant Q_2 can be obtained:

$$Q_2 = H^0 - Q_1, \tag{6.41}$$

where Q_2 may be calculated to be

$$Q_2 = \frac{\Delta\omega}{N+1} \left\{ NS_z - \sum_1^N I_{kz} \right\} + h_3 \Omega_3 \tag{6.42}$$

and where $\Delta\omega = \omega_{1S} - \omega_{1I}$.

Note that Q_1 has distinct harmonic eigenvalues, while, for mismatch $\Delta\omega \approx d_{jk}$, Q_2 has a quasi-continuous spectrum. Assuming $N \gg 1$, the quasi-equilibrium density operator is

$$\rho(t_{\text{qe}}) \approx \frac{\langle \rho(0) | Q_1 \rangle}{\langle Q_1 | Q_1 \rangle} Q_1 + \frac{\langle \rho(0) | Q_2 \rangle}{\langle Q_2 | Q_2 \rangle} Q_2, \tag{6.43}$$

$$\rho(t_{\text{qe}}) = \alpha_{0I} \left\{ S_z \frac{\lambda^2}{\lambda^2 + \Delta\omega^2} + \sum_1^N I_{kz} + h_3 \Omega_3 \frac{-\Delta\omega}{\lambda^2 + \Delta\omega^2} \right\}, \tag{6.44}$$

with the CP width λ given by

$$\lambda^2 = \frac{\langle h_3 \Omega_3 | h_3 \Omega_3 \rangle}{\langle I_{kz} | I_{kz} \rangle} = \frac{1}{4} N M_2^I, \tag{6.45}$$

where M_2^I is the second moment of the I-spin resonance line. This process represents a maximization of entropy under the constraints that all $\langle Q_q \rangle$ remain constant. It can be visualized as partial thermal equilibration between the reservoirs H_j^0 which cannot proceed to completion because of the constraint that the invariants $\langle Q_q \rangle$ are conserved, and may be viewed graphically as shown in fig. 6.6.

Equation (6.44) predicts the achievable S-spin polarization is maximum on match, and falls off on either side of match as a Lorentzian function of half-

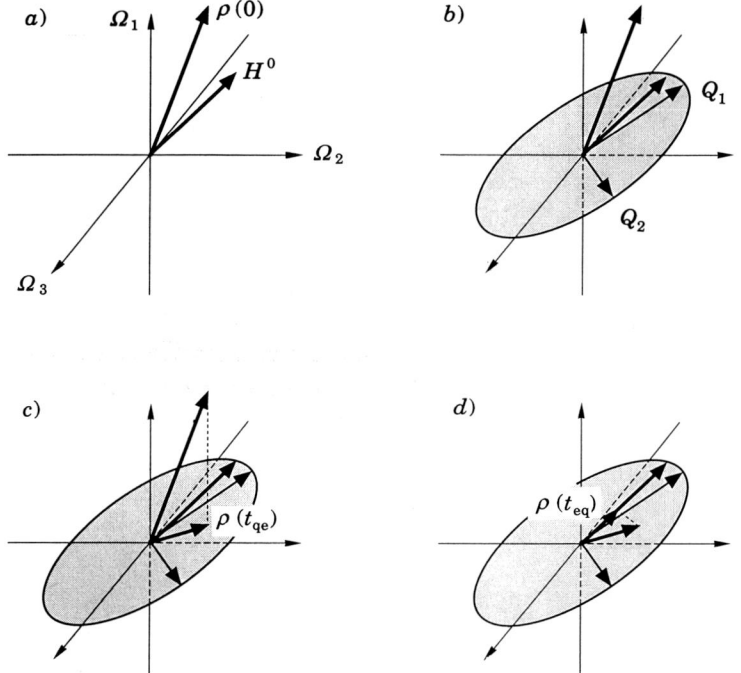

Fig. 6.6. – Given three thermodynamic reservoirs, $\Omega_1, \Omega_2, \Omega_3$, and an initial density operator $\rho(0)$ (a)), then we can define a plane in this space by Q_1 and Q_2, the quasi-invariants of the system, which includes H^0. b) Quasi-equilibrium corresponds to a projection of $\rho(0)$ onto the plane (c)). Only after comparatively long times (longer than the experimental time scale), is the true equilibrium state reached (d)).

width at half-height $\lambda = (1/2)\sqrt{NM_2^I}$. Values of N of around 10 normally produce qualitative agreement with experiment, suggesting that mismatched CP affects strongly the dipolar energies of nearby I spins, but that these effects diffuse only slowly to I spins further away.

Note that this choice of quasi-invariants is only valid around the HH match condition. For other conditions other quasi-invariants must be determined, and indeed one finds that cross-polarization once again becomes efficient at conditions where the S-spin nutation frequency is a multiple of the I-spin frequency, i.e. $N\omega_{1I} = \omega_{1S}$ [154].

6'6. *Resolved heteronuclear coupling.* – As a final step towards an improved determination of equilibrium magnetizations we must recognize that in many organic solids the one-bond heteronuclear dipolar coupling is larger than the I-I dipolar couplings. In many cases it is even resolved and is primarily responsible for the oscillations in the buildup shown in fig. 6.5 [155].

In order to accurately describe such a situation, we must revise the partition of the quasi-invariants. It is convenient to consider the system as a spin pair in a bath of surrounding protons and to partition the tilted-frame Hamiltonian and the density matrix of the $I_1 S$ spin pair into commuting sum and difference terms [153]:

(6.46) $$H_{\text{pair}} = \omega_{1I} I_{1z} + \omega_{1S} S_z + b_1 2 I_{1y} S_y ,$$

(6.47) $$H_{\text{pair}} = H_\Sigma + H_\Delta ,$$

with

$$H_\Sigma = \omega_e^\Sigma I_e^\Sigma , \qquad H_\Delta = \omega_e^\Delta I_e^\Delta ,$$

$$\omega_e^\Sigma = \sqrt{(\omega_{1S} + \omega_{1I})^2 + (b_1)^2} ,$$

$$\omega_e^\Delta = \sqrt{(\omega_{1S} - \omega_{1I})^2 + (b_1)^2} ,$$

$$I_e^\Sigma = I_z^\Sigma \cos\theta^\Sigma + I_x^\Sigma \sin\theta^\Sigma ,$$

$$I_e^\Delta = I_z^\Delta \cos\theta^\Delta + I_x^\Delta \sin\theta^\Delta ,$$

$$I_z^\Sigma = \frac{1}{2}(I_{1z} + S_z), \qquad I_z^\Delta = \frac{1}{2}(I_{1z} - S_z),$$

$$I_x^\Sigma = \frac{1}{2}(I_1^+ S^+ + I_1^- S^-), \qquad I_x^\Delta = \frac{1}{2}(I_1^+ S^- + I_1^- S^+),$$

and where

(6.48) $$\text{tg}\,\theta^\Sigma = \frac{b_1}{\omega_{1I} + \omega_{1S}}, \qquad \text{tg}\,\theta^\Delta = \frac{-b_1}{\Delta\omega} .$$

To explain the observed equilibrium values we assume that ρ_Σ and ρ_Δ, the sum and difference terms in the density operator, reach equilibrium at markedly different rates. Indeed we find that the difference term is isolated from the reservoir of remaining protons, while the sum term comes into contact with the sum term of the bath, as shown qualitatively in fig. 6.7. Note that the separation between energy levels of H_Σ is close to $2\omega_{1I}$ at match, and hence thermal contact between these terms is achieved by *double-quantum spin diffusion*. Therefore, we postulate a quasi-invariant Q_1 whose expectation value is conserved under this process,

(6.49) $$Q_1 = \frac{1}{N+2}(\omega_e^\Sigma + N\omega_{1I})\left(2I_e^\Sigma + \sum_2^{N+1} I_{kz}\right),$$

where N is the number of I spins in contact with the spin pair. Taking the perturbation

(6.50) $$V = H_{II}^{\text{nonsec}} + \sum_2^{N+1} b_k 2 I_{kx} S_x ,$$

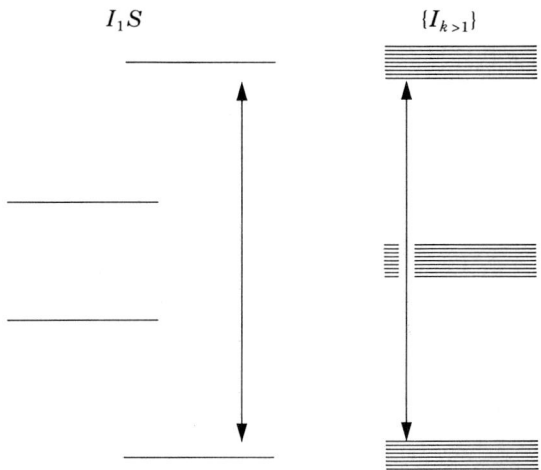

Fig. 6.7. – Schematic representation of the thermodynamic processes responsible for cross-polarization in cases of a resolved heteronuclear coupling. (Adapted from *J. Chem. Phys.*, **84**, 4243 (1986), with permission.)

the remaining quasi-invariants can be derived to be

$$Q_2 = \frac{1}{N+2}(\omega_e^\Sigma + 2\omega_{1I})\left(NI_e^\Sigma + \sum_2^{N+1} I_{kz}\right) + h_3 \Omega_3 , \quad (6.51)$$

$$Q_3 = -\omega_e^\Delta I_e^\Delta . \quad (6.52)$$

Assuming strong r.f. fields and making the approximation $\omega_e^\Sigma \approx \omega_{1S} + \omega_{1I}$, the quasi-equilibrium density operator may be derived from

$$\rho(t_{qe}) \approx \frac{\langle \rho(0)|Q_1\rangle}{\langle Q_1|Q_1\rangle} Q_1 + \frac{\langle \rho(0)|Q_2\rangle}{\langle Q_2|Q_2\rangle} Q_2 + \frac{\langle \rho(0)|Q_3\rangle}{\langle Q_3|Q_3\rangle} Q_3 . \quad (6.53)$$

The expectation value for the equilibrium S magnetization (in the tilted frame) is found to be

$$\langle S_z\rangle = \frac{\hbar\omega_{0I}}{4kT} \frac{1}{2(N+2)} \{N\sin^2\theta^\lambda + (N+2)\sin^2\theta^\Delta\} \quad (6.54)$$

with

$$\operatorname{tg}\theta^\lambda = -\frac{\lambda}{\Delta\omega}\sqrt{\frac{2(N+2)}{N}} . \quad (6.55)$$

In addition the expectation value for dipolar order can be obtained:

$$\langle 2I_{1y}S_y\rangle = \frac{\hbar\omega_{0I}}{4kT}\frac{1}{2}\left\{-\sin\theta^\Sigma\left(1 + \frac{N}{N+2}\sin^2\theta^\lambda\right) + \sin\theta^\Delta\cos\theta^\Delta\right\} \quad (6.56)$$

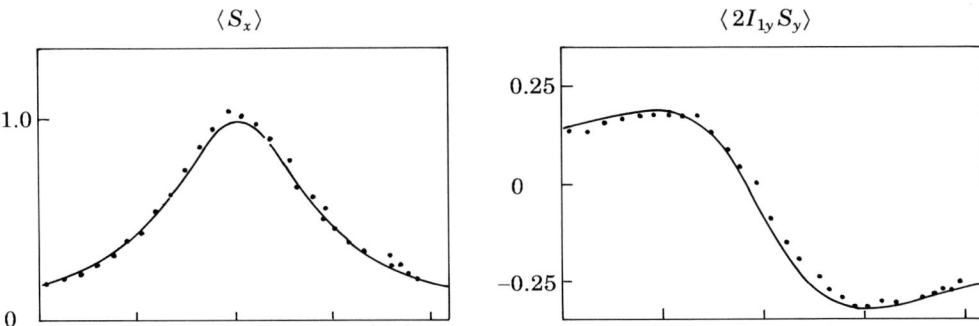

Fig. 6.8. – Dependence of equilibrium S-spin polarization for a single crystal of ferrocene on the S-spin r.f. field. The theoretical curves are predicted using eqs. (6.54) and (6.56), with the dipolar coupling $b_1/2\pi = 10.7\,\text{kHz}$, $N = 10$ and $\lambda/2\pi = 7.0\,\text{kHz}$. (Adapted from J. Chem. Phys., **84**, 4243 (1986), with permission.)

using $\alpha_{0I}\,\text{Tr}\,\{S_x^2\} = \hbar\omega_{0I}/4kT$. As shown in fig. 6.8, these equations give excellent agreement with experiment for the expectation values of both the S magnetization and the dipolar order (which can easily be observed indirectly) in the example of ferrocene. This suggests that the postulated quasi-invariants are a good approximation to reality.

7. – Unitary bounds on spin dynamics.

In this section we shall consider some limits on polarization transfer imposed by the unitary quantum dynamics. Normally, one would expect that the total polarization of the $N\,I$ spins, which amounts to $\sqrt{N}$ in an $I_N S_M$ system, could be transferred to the $M\,S$ spins to which they are coupled. This limit is often referred to as the entropy limit since it is the maximum achievable polarization transfer that one can expect from a thermodynamic approach, as we derived in subsect. 6.3.

In fact *for an isolated system* the amount which can be transferred is always less than the $\sqrt{N}$ limit[156-161]. It has been shown that, if the mechanism for transfer is a unitary transformation, polarization transfer is always incomplete in $I_N S_M$ systems where $M < N$[156]. The origin of this effect is that we do not expect the finite $I_N S_M$ system to behave thermodynamically unless both N and M are large, or the $I_N S_M$ system is part of an ensemble of coupled $I_N S_M$ systems, or the $I_N S_M$ system is coupled to a bath or thermodynamic reservoir.

7.1. *Unitary evolution.* – The evolution of the density matrix ρ in an $I_N S_M$ system in the absence of dissipative behavior such as relaxation is governed by the Liouville-von Neumann equation[2-5]

(7.1) $$d\rho/dt = -i[H(t), \rho(t)],$$

and, as we have seen when H is time-independent, the solution of eq. (7.1) is simply

$$\rho(t) = U\rho(0)\,U^{-1},$$

where U represents a unitary transformation. Even when H is time-dependent, a solution to eq. (7.1) can be found by dividing $H(t)$ into a series of sufficiently small parts $H_1 H_2 \ldots H_n$ such that each part can be considered time-independent. The solution of eq. (7.1) then becomes

(7.2) $$\rho(t) = U_n \ldots U_2 U_1 \rho(0)\, U_1^{-1} U_2^{-1} \ldots U_n^{-1}.$$

Unitary transforms govern the spin dynamics of all NMR experiments, if we neglect relaxation. The evolution of the density operator describes a trajectory in a Liouville space formed by n^2 orthogonal basis operators Ω_k. At any time the density operator $\rho(t)$ corresponds to a point $(a_1, a_2, \ldots, a_{n^2})$ where the a_k are given by

(7.3) $$a_k = \langle \Omega_k | \rho \rangle.$$

7˙2. *The entropy limit.* – The amount of spin order in the spin system is given by $\langle \rho | \rho \rangle$. Since we are considering unitary evolution, this quantity is conserved and evolution satisfies the condition

(7.4) $$\langle \rho(t) | \rho(t) \rangle = \langle \rho(0) | \rho(0) \rangle.$$

Given the initial density operator

(7.5) $$\rho(0) = (r(0)\cos\theta(0),\, r(0)\sin\theta(0),\, 0, \ldots, 0),$$

then

(7.6) $$r(0)^2 = r(t)^2 + \sum_{3}^{n^2} a_k^2(t),$$

which leads to the constraint

(7.7) $$|r(t)| \leq |r(0)|,$$

which is known as the entropy bound. For example, this corresponds to the case of $|r(t)| \leq \sqrt{N}$ in the $I_N S$ system shown in subsect. 7˙7.

7˙3. *Bounds on unitary evolution.* – SØRENSEN [156] has shown that the total amount of achievable polarization transfer in a unitary regime is governed by a theorem about normal matrices [162] and that actual bounds on evolution applicable to unitary behavior depend only on the eigenvalue spectra of the relevant operators. Defining $\Lambda(A)$ as an n-dimensional vector containing the n eigenvalues of an operator A ordered from highest to lowest, we make use of the inequality

(7.8) $$\langle A | B \rangle \leq \Lambda(A) \cdot \Lambda(B),$$

which is related to the Schwartz inequality. Since eigenvalues are conserved by unitary transform, $\rho(t)$ satisfies

(7.9) $$\langle A | \rho(t) \rangle \leq \Lambda(A) \cdot \Lambda(\rho).$$

Equation (7.9) can be summarized as saying that the projection of the density operator onto any operator Ω can never be more than the maximum value obtained by summing products of individual eigenvalues. *This limit is always less than or equal to the entropy limit.* For example, in an I_2S system, the simplest system with $N \neq M$, eq. (7.9) shows that the maximum amount of S polarization achievable using unitary behavior is $1S_z$ rather than the $\sqrt{2}S_z$ which corresponds to the entropy limit.

7˙4. *Redfield's description of polarization transfer.* – REDFIELD [160] has provided an elegant demonstration of the unitary limit. In subsect. 6˙3 we showed how adiabatic demagnetization of the I spin in a two-spin IS system, followed by adiabatic remagnetization of the S spin, resulted in complete polarization transfer. Let us consider the same process in an I_2S system, as illustrated in fig. 7.1. Initially there are two levels with energy approximately $\gamma_I H_I$ ($\hbar = 1$), four with energy zero and two with energy $-\gamma_I H_I$ (left-hand side of fig. 7.1). These sets of levels are not degenerate, being split by the heteronuclear dipolar interaction. We assume that the occupation of the levels is $a - \varepsilon$, a and $a + \varepsilon$, where $a = 1/8$, and their total spin angular-momentum values m_{Ii} are -1, 0 and $+1$, respectively.

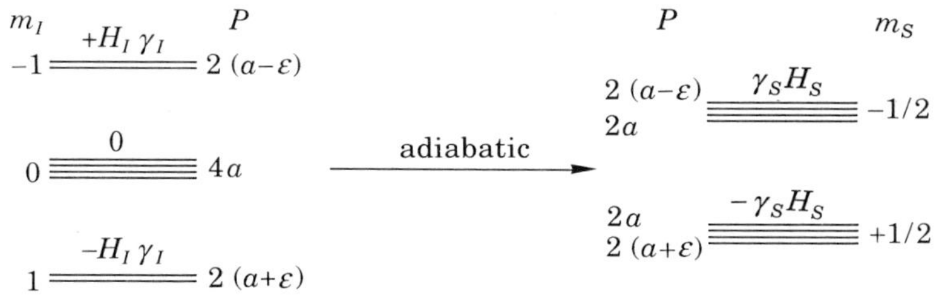

Fig. 7.1. – Energy levels and populations before (left) and after (right) adiabatic demagnetization of the I spins followed by adiabatic remagnetization of the S spin in an I_2S system. Only one unit of I polarization is transferred to the S spin.

This yields an initial polarization, defined as the sum over all eight levels of $\Delta_I = \sum_i P_i m_{Ii}$, which equals 4ε.

After the demagnetization and remagnetization (right-hand side of fig. 7.1), we have two groups of four energy levels separated by energy of approximately $(1/2)\gamma_S H_S$. In this case their total spin angular momentum is $\pm 1/2$, and the populations are determined as described above by retaining the ordering of the populations that existed initially (since the whole process was adiabatic). Thus in the upper group of levels there are two with occupation probability $P_i = a - \varepsilon$ and two with $P_i = a$. Similarly for the lower group, two levels have $P_i = a + \varepsilon$, and two have $P_i = a$. The final polarization, which is the sum $\Delta_S = \sum_i P_i m_{Si}$, is, therefore, equal to 2ε.

Thus there is a net polarization loss by a factor of 0.5, or an efficiency of transfer to single S spin of one unit of initial I-spin polarization. In terms of the norm of the density matrix, this means the polarization has been reduced by a factor $1/\sqrt{2}$ from $P_I = \sqrt{1+1} = \sqrt{2}$ to $P_I = \sqrt{1} = 1$, exactly in agreement with the prediction made in subsect. 7·3.

7·5. *Bounds in $I_N S$ systems.* – We can generalize the results obtained for S polarization in the IS and $I_2 S$ cases to $I_N S$ without too much difficulty by deriving the following formulae for the maximum amplitude of S polarization:

(7.10) $$a^{\max}(N = 2m + 1) = a^{\max}(N = 2m + 2),$$

(7.11) $$a^{\max}(N = 2m + 1) = \binom{2m}{m}(2m + 1) 4^{-m}.$$

Figure 7.2 shows $a^{\max}/\sqrt{N}$ as a function of the number of I spins. Even in the limit of $N \to \infty$ the amount of S polarization does not reach the entropy limit.

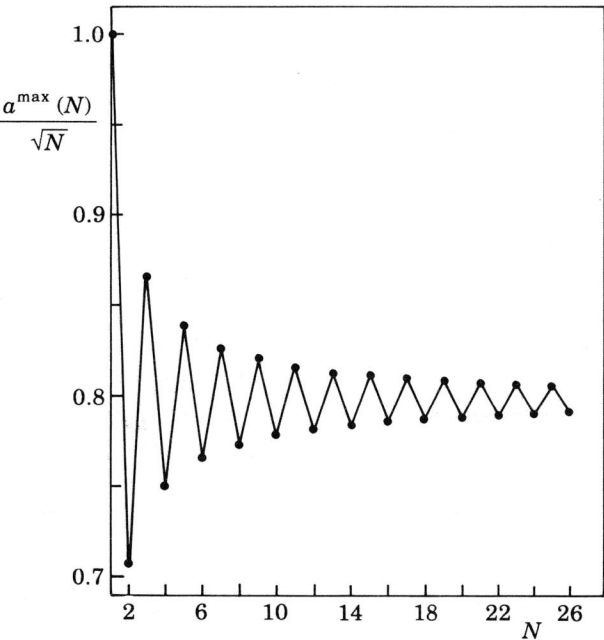

Fig. 7.2. – Illustration of the maximum possible transfer efficiency relative to the efficiency predicted by the entropy bound for the transfer of I-spin polarization to S spins in an $I_N S$ system. (Adapted from *J. Magn. Reson.*, **86**, 435 (1990), with permission.)

Indeed, using Stirling's formula,

$$(7.12) \qquad n! \approx \left(\frac{n}{e}\right)^n \sqrt{2\pi n},$$

which is valid for large n, we can derive a comparison of the two bounds for large N:

$$(7.13) \qquad \frac{a^{\max}}{\sqrt{N}} \approx \sqrt{2/\pi} \approx 0.80.$$

We find that unitary behavior only reproduces the thermodynamic entropy limit when both N and M are large.

7˙6. *The thermodynamic limit.* – By extending Redfield's model (subsect. 7˙4) to larger spin systems we gain more insight into spin temperature, since the final distribution gets closer and closer to the Boltzmann distribution. For example, in an $I_4 S_2$ system, in analogy to fig. 7.1 the uppermost set of four levels has energy roughly $2\hbar\gamma_I H_I$, quantum number $m_{Ii} = -2$, and $P_i = a - 2\varepsilon$, where $a = 1/64$. The next uppermost set of sixteen levels has $m_{Ii} = -1$ and $P_i = a - \varepsilon$. The middle set of 24 levels has $m_{Ii} = 0$ and $P_i = a$, and the lower levels mirror the upper ones. The spin polarization is 64ε. After adiabatic transfer of polarization the two S spins have three sets of levels. The upper and lower set each contain 16 levels with $-m_{Si} = \pm 1$. The populations of the upper set are 4 levels with $P_i = a - 2\varepsilon$ and 12 with $P_i = a - \varepsilon$. The final polarization this time is 40ε, giving a polarization efficiency of $40/64$, which is greater than for the $I_2 S$ case. This is a consequence of the fact that, as the number of I and S spins gets larger, the initial and final distributions of eigenvalues resemble each other and are binomial. Thus Redfield's simple treatment sheds a large amount of light on the connection between spin dynamics of isolated systems and the thermodynamic limit.

To calculate the maximum enhancements for $M > 1$ we remark that the binomial distribution of eigenvalues of the I spins is increasingly well approximated by a normal distribution. Hence the limit $N \to \infty$ was represented by substitution of the binomial for a normal distribution in eq. (7.13). For $M > 1$ we can do this by matching the normal distribution of the I eigenvalues with the binomial distribution of the S eigenvalues, and it can be shown that

$$(7.14) \qquad \frac{a^{\max}(N \to \infty, M)}{a^{\text{entropy}}(N \to \infty, M)} = \sqrt{\frac{8}{M\pi}} \left\{ \sum_{i=0}^{(M-2)/2} \exp\left[-\frac{1}{2}\left(g\left(1 - 2^{-M}\sum_{j=0}^{i}\binom{M}{j}\right)\right)^2\right] \right\},$$

where g is the inverse of the function

$$(7.15) \qquad F(x) = \frac{1}{\sqrt{2\pi}} \int_{-\infty}^{x} \exp[-u^2/2] \, \mathrm{d}u.$$

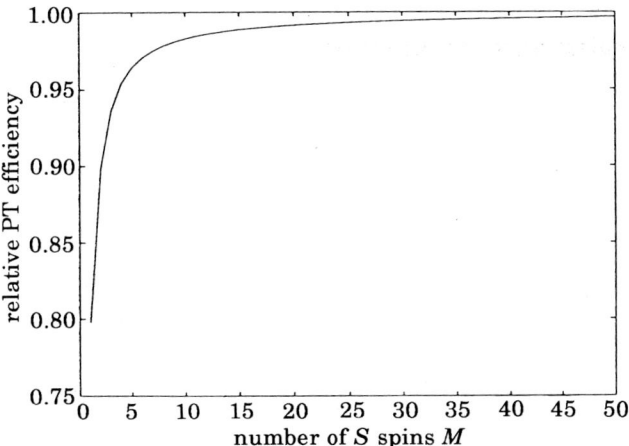

Fig. 7.3. – Relative efficiency of polarization transfer in an $I_N S_M$ spin system as a function of M when N tends to infinity. (Adapted from *J. Magn. Reson.*, **93**, 648 (1991), with permission.)

Equation (7.14) is plotted as a function of M in fig. 7.3, where we see the tendency towards maximum polarization transfer for large M. Indeed, as $M \to \infty$, matching of two normal distributions with standard deviations $\sigma_I = \sqrt{N/2}$ and $\sigma_S = \sqrt{M/2}$ amounts to

(7.16) $\quad a^{\max}(N \to \infty, M \to \infty) =$

$$= \frac{(1/\sigma_I \sqrt{2\pi}) \int_{-\infty}^{\infty} u^2 (\sigma_S/\sigma_I) \exp[-u^2/2\sigma_I^2] \, du}{(1/\sigma_S \sqrt{2\pi}) \int_{-\infty}^{\infty} u^2 \exp[-u^2/2\sigma_S^2] \, du} = \frac{\sigma_I \sigma_S}{\sigma_S^2} = \sqrt{\frac{N}{M}},$$

which is the entropy limit.

7˙7. *Two-dimensional bounds.* – We make use of a graphical technique recently introduced by LEVITT [161], in which the region of Liouville space accessible through unitary transformation is mapped as a function of two operators Ω_1 and Ω_2. For example, an obvious choice of operators might be

(7.17) $\quad \begin{cases} \Omega_1 = \dfrac{1}{\sqrt{N}} I_z \\ \text{and} \\ \Omega_2 = S_z, \end{cases}$

where

(7.18) $$I_z = \sum_{1}^{N} I_{iz}.$$

These two operators define a plane in Liouville space, and we can map the accessible states by evaluating eq. (7.9) for the linear combination

(7.19) $$\Omega_r = \Omega_1 \cos\theta + \Omega_2 \sin\theta$$

for all values of θ. In fig. 7.4a) we see how this leads to a region of accessible space for a two-spin IS system when $\rho(0) = I_{1z}$ for $\Omega_1 = I_z$ and $\Omega_2 = S_z$. In fig. 7.4b) we see the region of allowed space for a three-spin system $I_2 S$, which is the simplest system we can consider with $N \neq M$. In this case $\rho(0) = I_{1z} + I_{2z}$ for $\Omega_1 = I_{1z} + I_{2z}$ and $\Omega_2 = S_z$, and we see that the region of allowed states intersects the S_z axis at a value of 1. In the particular plane considered in fig. 7.4b) the maximum size of the projection of the spin vector onto the plane is continuously getting smaller as θ gets larger. As we are dealing purely with unitary transformations, the norm of the total spin vector (which represents the degree of spin order and is proportional to the entropy of the system) must be conserved, which indicates that other operators must be involved. Indeed, if we consider instead $\Omega_1 = I_{1z}$ and $\Omega_2 = S_z$, again with $\rho(0) = I_{1z} + I_{2z}$, we note that the size changes in a different way (fig. 7.4c)) and there is a combination corresponding to $\Omega_r = I_{1z} + S_z$ which has an amplitude equal to the entropy limit. Thus $\rho(0) = I_{1z} + I_{2z}$ can be completely converted into $\rho(t) = I_{1z} + S_z$. This simple two-dimensional representation can be easily extended to a subspace of three or more dimensions, but we usually find that the two-dimensional projections of fig. 7.4 are the most useful.

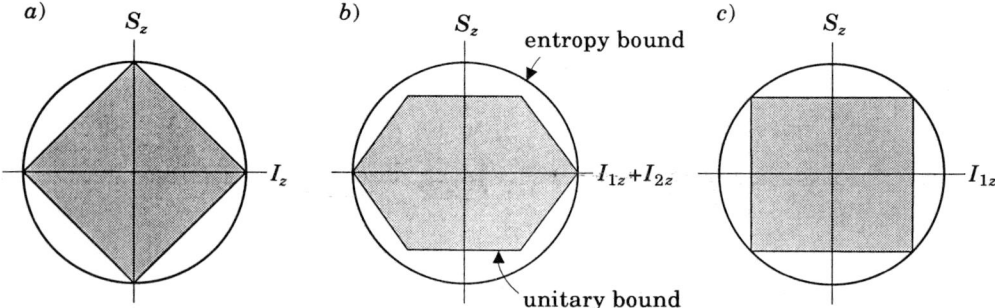

Fig. 7.4. – The shaded regions are projections of the regions of Liouville space accessible within the unitary bound onto planes defined by a) S_z and $I_{1z} + I_{2z}$ in an IS system with $\rho(0) = I_z$ and b) S_z and $I_{1z} + I_{2z}$ for an $I_2 S$ spin system, and c) S_z and I_{1z} for an $I_2 S$ spin system. Both b) and c) have $\rho(0) = I_{1z} + I_{2z}$. The circles correspond to the entropy limit of $\sqrt{N}$ polarization. Note that in both b) and c) the region intercepts the S_z axis at a value of 1.0, but that in c) the region touches the entropy bound for $\Omega_r = I_{1z} + I_z$.

7˙8. *Transfer of basis operators.* – In an I_2S system a total of 32 products of Cartesian operators form a basis set. They can be divided into 8 classes:

(7.20) $\quad\quad\quad I_{1\alpha},\ I_{2\alpha},\ S_\alpha,\ 2I_{1\alpha}I_{2\beta},\ 2I_{1\alpha}S_\beta,\ 2I_{2\alpha}S_\beta,\ 4I_{1\alpha}I_{2\beta}S_\gamma,\ 1,$

where α, β, γ are interchangeably x, y, z. Each of the operators within each class share common eigenvalues. If we take a linear combination of two of these operators according to eq. (7.9), we find that, when Ω_1 and Ω_2 belong to the above set, $\Lambda(\Omega_r)$ *is the same for all combinations* Ω_1 *and* Ω_2 with values of

(7.21) $\quad\begin{cases} \Lambda_{11} = \Lambda_{22} = \cos\theta + \sin\theta, \\ \Lambda_{33} = \Lambda_{44} = \cos\theta - \sin\theta, \\ \Lambda_{55} = \Lambda_{66} = \sin\theta - \cos\theta, \\ \Lambda_{77} = \Lambda_{88} = -\cos\theta - \sin\theta. \end{cases}$

Taking the product $\Lambda(\Omega_r)\cdot\Lambda(\rho(0))$ where $\rho(0) = I_{1z} + I_{2z}$, we find the maximum amplitude of the projection of the spin vector onto the plane $a^{\max} = \sqrt{2}$ for $\theta = 45°$. Thus *any* combination of $\rho(0)$ which consists of a sum of two basis operators can be completely transferred into a sum of *any* other two basis operators, provided the operators within each sum commute with each other. In contrast, two operators cannot be entirely converted into one by unitary behavior. In general we can state the following principle: *Any sum of k commuting basis operators can be entirely converted by unitary transform into a sum of k other commuting basis operators.*

This rule is subject to some restrictions imposed by symmetry in that the sums of operators must belong to the same irreducible representation of the spin Hamiltonian [163]. For example, in an I_2S system the sum $\Omega_r = 2(a_1 I_{1\alpha}I_{2\beta} + a_2 I_{1\alpha}S_\beta + a_3 I_{2\alpha}S_\beta)$ possesses inversion symmetry (*i.e.* it is invariant under the operation of inversion $A' = UAU^{-1}$, where $U = \exp\left[-i\left(\sum_i \pi I_{ix}\right)\right]$ and A is the diagonal matrix of eigenvalues of Ω), and has degenerate eigenvalues, whereas a sum of any other set of three operators in this system, such as $\Omega_r = a_1 I_{1\alpha} + a_2 I_{2\alpha} + a_3 S_\beta$, is antisymmetric under inversion (*i.e.* $A' = -A$) and does not have any degenerate eigenvalues (barring accidental degeneracy if the coefficients are equal). Unitary transformation does not change the symmetry of the density matrix, and cannot, therefore, interconvert these two operators. In order to avoid symmetry-related problems, we shall consider only sums of odd operators (*i.e.* single-spin operators of the type $I_{i\alpha}$), which always belong to the same irreducible representation, in the following examples.

7˙9. *Nonunitary evolution.* – To avoid the restrictions of unitary limits we must change the eigenvalues of the density matrix contained in $\Lambda(\rho)$ by invoking a nonunitary transformation. One mechanism of nonunitary transformation is relaxation, which can be seen as a way of increasing the size of the system through the dipolar interaction. As far as polarization transfer experiments are concerned, we may ignore the origins of the relaxation process, as it is only the form of the relaxation matrix R that is important. The equation of motion of the

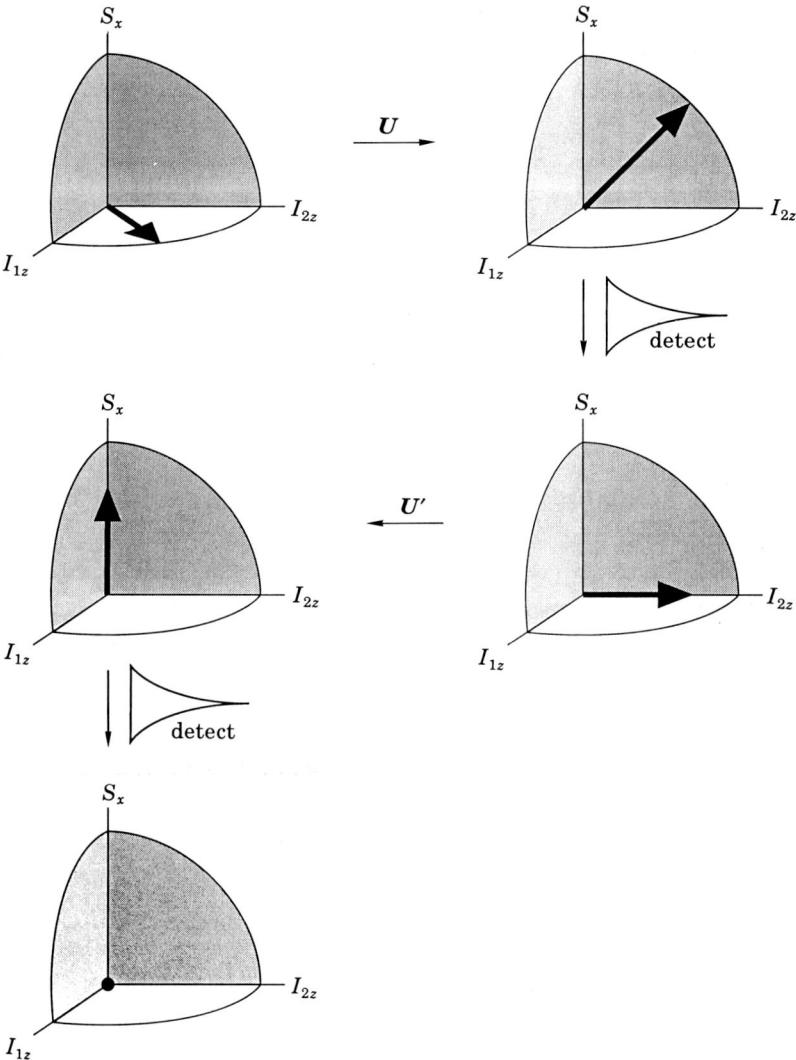

Fig. 7.5. – Schematic representation of an experiment which utilizes transverse relaxation to enhance polarization transfer. In the first step $\rho(0) = I_{1z} + I_{2z}$ is transformed into $\rho = I_{2z} + S_x$ by a unitary transformation U. A signal corresponding to polarization transfer of 1 unit is then acquired as S_x decays to zero through the effect of R. The resulting I_{2z} is then transferred by U' to S_x and a second signal is acquired. The final signal-to-noise ratio is $\sqrt{2}$.

density matrix is modified to become [117, 131, 164]

(7.22) $$d\rho/dt = -i[H(t), \rho(t)] + R(\rho(t) - \rho(\infty)),$$

where R is a tetradyadic with four subscripts and where the element $R_{ab,cd}$ connects the elements ρ_{ab} and ρ_{cd}. If we include only transverse autocorrelation as a relaxation mechanism, R is particularly simple and only contains elements $R_{ab,ab}$ ($a \neq b$) with the result that the off-diagonal elements ρ_{ab} will decay at a rate $-R_{ab,ab}$ which depends on the spectral densities. For our purposes it is convenient to adopt a normal-mode analysis [131] where the normal modes of the system are the $n = 2^{N+M}$ Cartesian basis operators of the $I_N S_M$ system. The evolution of the density matrix under the effect of relaxation alone is now given by

(7.23) $$d\rho/dt = \Gamma \nu(t),$$

where $\nu(t)$ is a vector containing elements $\nu_1 \ldots \nu_n$ corresponding to the n basis operators. In the case under consideration Γ only contains diagonal elements which affect transverse operators. For the moment we wish to observe simply that, if we allow the system to reach equilibrium ($t \to \infty$), the effect of transverse relaxation is to remove all the transverse components of the density matrix.

We can now envisage the *gedanken* experiment shown in fig. 7.5 which is capable of achieving the thermodynamic limit for polarization transfer. For the sake of convenience we have assumed that the time scale of relaxation is much longer than that of U. We note that, if $\rho = S_x + I_{2z}$, then we can detect S_x (or any other observable), which *simultaneously decays to zero* through T_2, whilst I_{2z} (or any longitudinal state) remains unaffected as long as T_1 is sufficiently long. The action of detection corresponds to a nonunitary operator which removes the transverse components of the spin vector whilst at the same time accumulating a signal of size 1. Thus, after the first detection period in the experiment of fig. 7.5, we are left with a density matrix which only contains one operator; by detecting the signal and allowing transverse relaxation to take effect we have changed the numerical values of the eigenvalues of the density matrix. In the experiment of fig. 7.5 there are two detection periods, during each of which we can acquire a signal of size 1. Thus the total resulting signal of 2, coupled with a total random noise level of $\sqrt{2}$, leads to a final signal-to-noise ratio of $\sqrt{2}$, which is the entropy limit. The most reassuring feature of the experiment *is that by changing the eigenvalues of the density matrix nonunitary dissipative behavior can be used to enhance overall accumulated polarization transfer.*

7˙10. *Cross-polarization echoes.* – Before leaving the subject of unitary limits we remark that much of the language in which cross-polarization was discussed in sect. 6 suggests that it is an irreversible process. One speaks loosely of polarization «diffusion» and of projection of the magnetization. However, we have emphasized here that the dynamics are in fact governed by unitary processes. If this is the case, then, although the dynamics are such that forward transfer may be incomplete, reversing the process should be entirely feasible. Therefore, as we saw in subsect. 1`10, we should be able to apply time-reversal techniques to cross-polarization. Indeed, it is quite obvious that time reversal of both H_{II} and H_{IS} would lead to the formation of a cross-polarization echo.

8. – Zero-field NMR.

One of the motivations for doing multiple-quantum NMR described in early sections was to simplify complex spectra. There the complexity of the spectra arose from homogeneous effects, such as the large number of dipolar couplings. But spectra may be intractably complex and broad owing to inhomogeneous broadening as well, which arise from orientational distributions that produce superpositions of inherently simple spectra. This broadening occurs in disordered systems such as polycrystalline solids, amorphous materials, or partially ordered polymers or biological compounds. Because of the anisotropy of the dipolar and quadrupolar interactions, molecules or small groups of spins exhibit different spectra for different orientations; the complete spectrum, therefore, reflects subspectra from all the different orientations, superposed to form a broad, often featureless «powder pattern» from which little information can be obtained.

An example of the proton NMR spectrum of a polycrystalline organic solid, containing four hydrogens per molecule, is shown in fig. 8.1. The situation is similar to that encountered in crystallography by X-ray or neutron diffraction. Although oriented crystals provide diffraction patterns from which structural information can be extracted, a polycrystalline sample yields a considerably less useful powder pattern. We have already considered some methods for removing this broadening in sect. 2 and 5. Another solution to this problem is to perform the NMR in zero field, using principles well known in other forms of magnetic resonance, for example nuclear quadrupolar resonance (NQR) [165] or optically detected electron paramagnetic resonance (EPR). *In the absence of a magnetic field defining an axis in space, all orientations are equivalent and orientationally disordered materials should provide sharp «crystallike» spectra.* The only problem is to overcome the low sensitivity inherent in the low frequencies of zero-field NMR. To take advantage of the high resolution of zero field and the high sensitivity of high field, we employ adaptations of well-known field-cycling methods [166-173].

A diagram of the simplest field cycle and the corresponding apparatus needed is given in fig. 8.2 [174]. Removal of the sample to an intermediate field, followed by a sudden transition to zero field, causes the magnetization carried from the high field to oscillate at frequencies characteristic of local magnetic dipolar or electric quadrupolar interactions. Reapplication of the high field per-

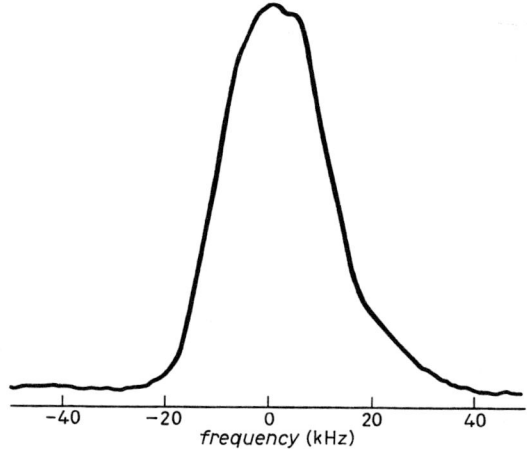

Fig. 8.1. – High-field NMR spectrum of a polycrystalline 1, 2, 3, 4-tetrachloronaphthalene-bis(hexachlorocyclopentadiene) adduct, a four-proton system. As in many dipolar powder patterns, little structure is resolved even though only a small number of spins are strongly coupled together. (Adapted from *J. Chem. Phys.*, **83**, 4877 (1985), with permission.)

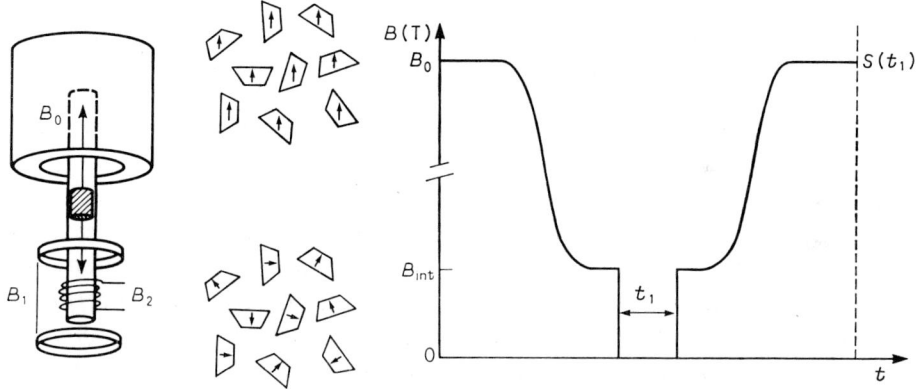

Fig. 8.2. – Zero-field NMR cycle. The applied field is decreased adiabatically by mechanically shuttling the sample out of the bore of a superconducting magnet. The magnetization, preserved in this process, is maintained by an intermediate field, B_1, which is larger than the local internal fields. A second coil produces a pulsed field B_2 that rapidly cancels all other fields and initiates evolution of the spin system in zero field. The local interactions now determine the axis system in zero field and are identical for all crystallites. Reapplication of the intermediate field terminates the zero-field evolution, and the sample is returned to high field, where the magnitude of the signal is measured. The period t_1 is increased incrementally in successive field cycles to produce a time domain signal which produces the zero-field NMR spectrum upon Fourier transformation. (Adapted from *Acc. Chem. Res.*, **20**, 47 (1987), with permission.)

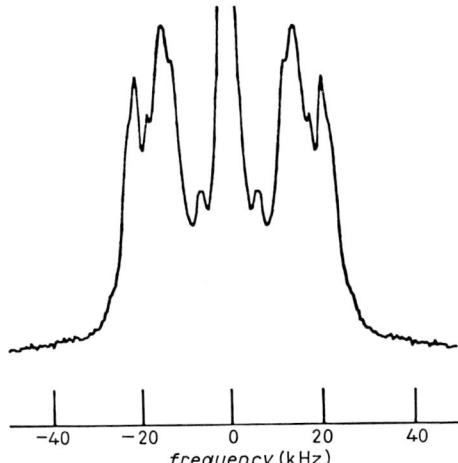

Fig. 8.3. – Zero-field NMR spectrum of the same solid sample shown in fig. 8.1. The sharp peak at zero frequency is truncated for purposes of display. The evolving zero-field magnetization was sampled at 5 µs increments to give an effective zero-field bandwidth of 100 kHz. From the spectrum the configuration of the four-spin central ring could be determined. (Adapted from *J. Chem. Phys.*, **83**, 4877 (1985), with permission.)

mits sensitive detection as a function of the time spent in zero field. Fourier transformation of this time domain signal will produce a zero-field frequency domain spectrum with high resolution and full high-field sensitivity. As an example, the zero-field NMR spectrum of a polycrystalline sample of the four-proton spin system of fig. 8.1 is shown in fig. 8.3 [171]. The relative positions of the hydrogens and the conformation of the central cyclohexane ring can be determined from such a spectrum.

8˙1. *Zero-field NQR of deuterium.* – Frequency domain methods have long been used to observe quadrupolar nuclei ($I \geq 1$), where direct detection of the quadrupolar resonance is possible at high frequency. These methods are of limited applicability, however, when the frequencies are low (< 100 kHz), as, for example, with deuterium. Moreover, direct detection requires the use of radiofrequency irradiation in zero field. Clearly, Fourier transform experiments of the type described in the previous section can avoid many of these problems. As an example, the high-field 55.6 MHz deuterium NMR spectrum of perdeuterated polycrystalline diethylterephthalate is shown in fig. 8.4a) [175]. Only the most prominent singularities of the methyl, methylate and aromatic lineshapes can be resolved, since the deuterium signal is distributed over a wide bandwidth (although in favorable cases it can be «dePaked»). In contrast, the zero-field deuterium spectrum in fig. 8.4b) displays four distinct groups of peaks with sharply resolved fine structure. From such a spectrum resonances from different sites in the molecule can be assigned. In this case, five inequiva-

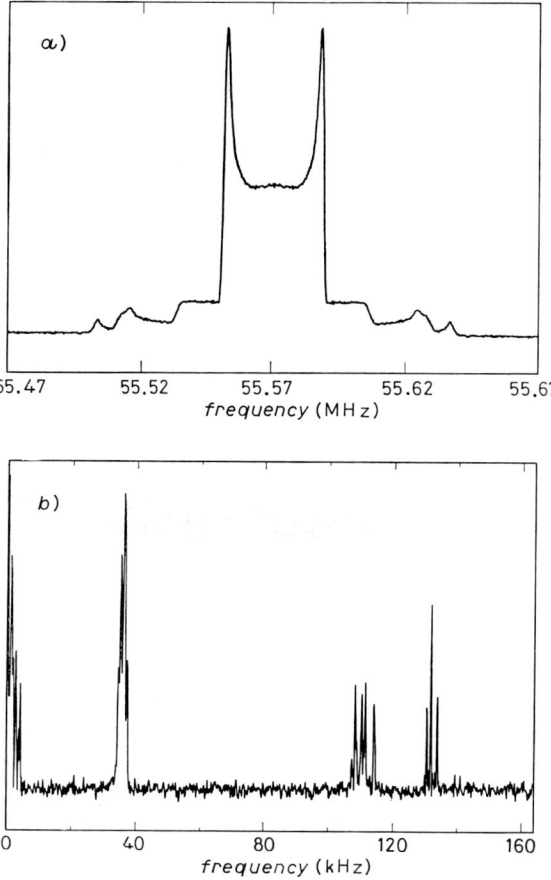

Fig. 8.4. – a) Deuterium high-field NMR spectrum of polycrystalline perdeuterated diethylterephthalate, $CD_3 CD_2 OOCC_6 D_4 COOCD_2 CD_3$. Only the singularities of the methylene and aromatic sites are distinguishable in this broad powder pattern. Three separate quadrupolar sites can be discerned from the overlapping powder lineshapes. b) Zero-field deuterium NQR spectrum of the same sample. Four distinct frequency regions with resolved peaks are evident corresponding to, in order of decreasing frequency, the aromatic, methylene, methyl sites and ν_0 lines. Quadrupolar coupling constants and small asymmetry parameters were established for five inequivalent sites in the molecule. (Adapted from *J. Magn. Reson.*, **69**, 243 (1986), with permission.)

lent sites are established: methyl, two inequivalent methylenes and two inequivalent aromatics. The high resolution of the zero-field experiment permits the measurement of very similar quadrupolar coupling constants and small asymmetry parameters. Figure 8.5 shows a further example with perdeuterated solid dimethoxybenzene [176]. The two doublets around 135 kHz arise from the inequivalence of deuteron sites in the aromatic ring created by the frozen solid-state conformation of the molecule. Many nuclei with low quadrupolar fre-

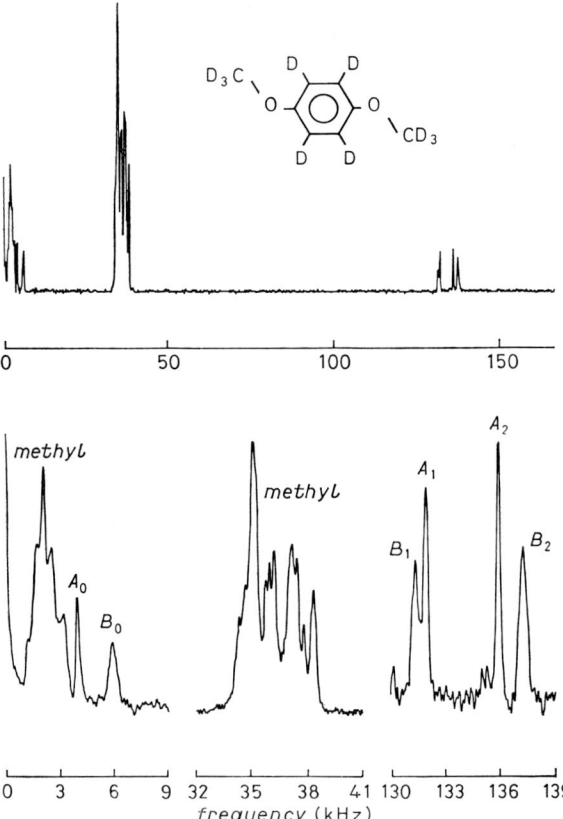

Fig. 8.5. – Fourier transform zero-field deuterium NQR spectrum of polycrystalline perdeuterated 1,4-dimethoxybenzene. The upper plot shows the full spectrum. The lower plot shows expanded views of the three resonance regions assignable to aromatic deuterons, aliphatic deuterons and ν_0 lines. There are two pairs of lines indicating different quadrupole couplings and asymmetry parameters for the aromatic deuterons. The differences are due to the solid-state molecular conformation, which renders pairs of the aromatic positions inequivalent. (Adapted from *J. Chem. Phys.*, **80**, 2232 (1984), with permission.)

quencies are directly accessible by such zero-field NQR studies, and among those studied in our laboratory, in addition to deuterium, are ^{7}Li[175], ^{14}N[177], ^{11}B[178] and ^{27}Al[179].

8·2. *Two-dimensional zero-field* NMR. – Connections between zero-field NMR and NQR transitions (which relate to connections between molecular sites) can be determined by extending the experiment to incorporate two time periods in zero field[174, 180]. The basic idea is to obtain the signal as a function of two independent time variables, t_1 and t_2, and then to Fourier transform

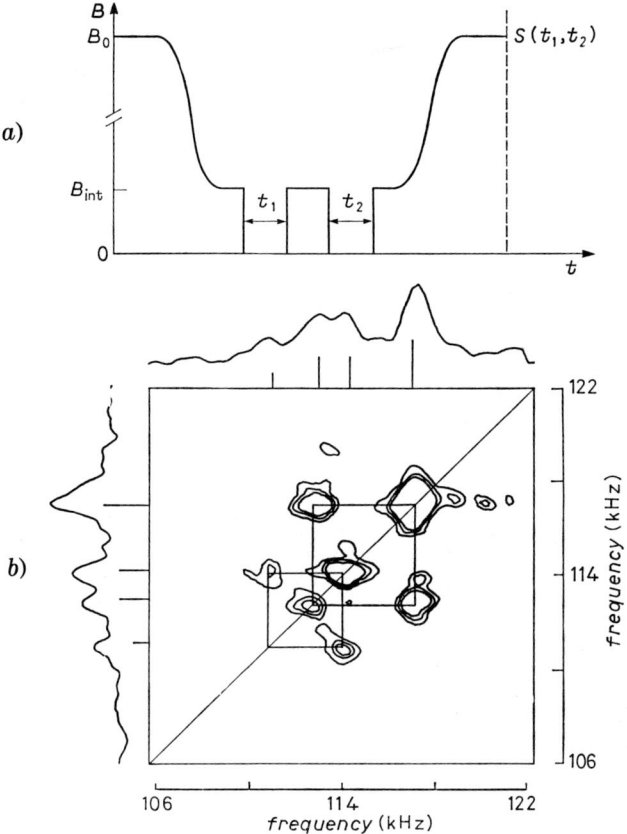

Fig. 8.6. – Two-dimensional zero-field cycle and spectrum of selectively deuterated (CD_2) solid diethylterephthalate. The zero-field spectrum is obtained as a function of two independent time variables, t_1 and t_2, which are separated by application of an intermediate pulsed field. This mixing transfers coherence between the zero-field transitions. As seen in the experimental spectrum, off-diagonal peaks indicate connectivities between the zero-field quadrupolar transitions. In this case, transitions can be assigned to inequivalent deuterium sites in each CD_2 group. (Adapted from *Chem. Phys. Lett.*, **129**, 55 (1986), with permission.)

against both to obtain a two-dimensional zero-field frequency spectrum [5]. An illustration of this experiment is provided in fig. 8.6a), where the field cycle shown employs a pulsed-field mixing period between t_1 and t_2. Magnetization able to oscillate at two possible frequencies in the two time periods, for example the ν_+ and ν_- quadrupolar frequencies of a given site, will produce an off-diagonal peak, or «cross-peak», at the intersection of these two frequencies if the mixing sequence transfers coherence between the transitions. An experimental illustration of the connectivities in a spin $I = 1$ system appears in fig. 8.6b) which shows the two-dimensional zero-field spectrum of the methylene region of a sample of selectively deuterated diethylterephthalate. This result shows that, among the four lines in the CD_2 region of the spectrum, lines 1 and 3 belong to one deuteron and lines 2 and 4 to the other inequivalent deuteron. *With this kind of experiment one can hope to identify sites by their quadrupole couplings and then determine intersite distances through their dipolar couplings in a two-dimensional spectrum.*

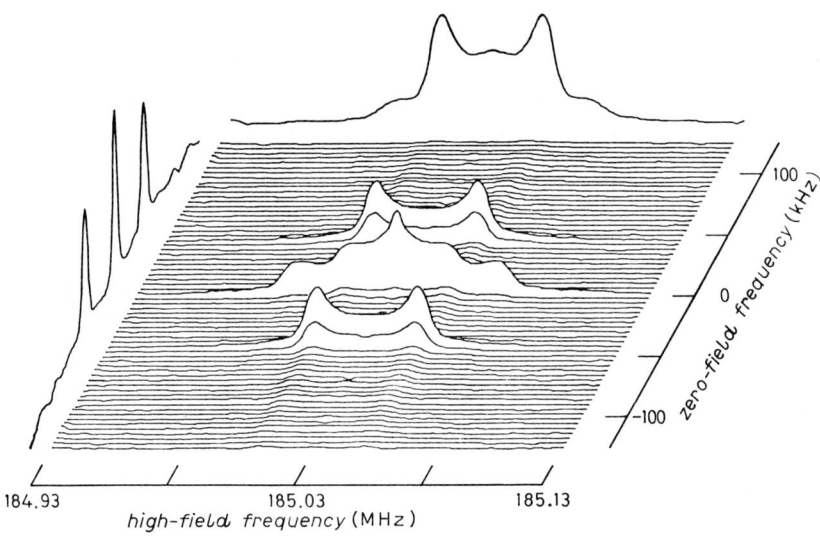

Fig. 8.7. – Two-dimensional zero-field high-field dipolar correlation spectrum of polycrystalline $Ba(ClO_3)_2 \cdot H_2O$. For each of 64 values of t_1, the zero-field interval, the high-field free-induction decay after a solid echo sequence is accumulated and stored. A double real Fourier transform in t_1 and t_2 is applied to the signal $S(t_1, t_2)$. At the left and top are the projections of the zero-field and high-field spectra and in the center the correlations between the two frequency domains. Signals which appear at zero frequency in ω_1 correlate most strongly with signals from orientations of the two-spin system which are near the edges of the high-field powder pattern. Zero-field signals which appear at ~ 42 kHz correlate the orientations which appear near the peaks (at $\sim \pm 21$ kHz) of the high-field powder pattern. (Adapted from *J. Chem. Phys.*, **83**, 4877 (1985), with permission.)

A different class of two-dimensional experiments reserves zero-field evolution for the period t_1 and high-field evolution for the period t_2 in order to correlate high-field and zero-field NMR transitions. An example, shown in fig. 8.7[171] for two water protons in a polycrystalline hydrate, displays the Pake doublet powder pattern in the high-field dimension and a three-line zero-field spectrum in the other dimension. In principle, one can contemplate obtaining high-resolution chemical shifts in the high-field dimension and sharp dipolar couplings between sites in the zero-field dimension.

8'3. *Zero-field pulses.* – The discussion so far has focused on the sudden removal of an applied field to induce zero-field evolution as in fig. 8.2. This approach, however, suffers from two principal disadvantages. The first derives from the requirement that the intermediate field be larger than the local spin interactions so that the Zeeman interaction dominates. For nuclei with small magnetogyric ratios and large quadrupolar coupling constants, this condition requires that a field of a few hundred to a few thousand gauss be applied for some tens of milliseconds—perhaps a difficult task. The other disadvantage comes from the lack of selectivity in the sudden transition that excites evolution of different isotopes and spins (*e.g.*, protons, deuterium, ^{13}C) in zero field. The experiment can be made selective and more flexible, however, by a simple modification of the field cycle to use pulsed d.c. magnetic fields to excite different

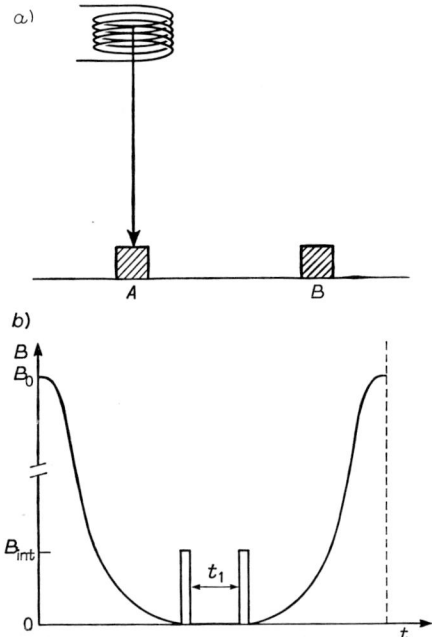

Fig. 8.8. – Pulsed-field cycling with adiabatic demagnetization and remagnetization. *a*) Demagnetized sample A differs from one which has resided on zero field for an extended period B. Both have no magnetization, but A will spontaneously polarize when returned to a field, while B requires a time on the order of the spin-lattice relaxation time, T_1, to polarize. It is this zero-field order of A which is exploited after demagnetizing to zero field by applying d.c. pulses (for example, $\pi/2$ zero-field pulses) to initiate evolution for the time t_1. The evolution is terminated by a second pulse and the sample is returned to high field for sampling of the magnetization. This field cycle provides flexibility in employing large fields for brief periods of time, as well as in selective excitation of different spins (*e.g.*, ^{13}C *vs.* protons), by varying the magnitude, direction and duration of the pulses and level crossings in heteronuclear spin systems. (Adapted from *J. Chem. Phys.*, **83**, 934 (1985), with permission.)

nuclear spins in zero field, as illustrated in fig. 8.8 [174, 177, 181]. The sample is first removed completely to zero field through adiabatic demagnetization in the laboratory frame. Application of a pulsed d.c. field at this point then changes the state of the system and induces evolution in zero field for a time t_1. This evolution may be terminated with a second d.c. pulse. The effect is analogous to pulsed NMR in high field, but here the resonant frequency, and, therefore, the frequency of the pulses, is zero. Finally, the sample is adiabatically remagnetized back to high field for normal high-field NMR detection. Different isotopes and spins can be addressed separately by making the zero-field pulses selective, perhaps by using composite pulses that produce 2π rotations for all isotopes except the one of interest.

An additional advantage of pulsed-field cycles of the sort shown in fig. 8.8 is that they permit level crossings between protons and quadrupolar spins during the adiabatic demagnetization and remagnetization. This possibility allows the zero-field evolution of a quadrupolar spin, say deuterium, to be detected by the effect on the more sensitive proton spins. Indirect detection has long been used in traditional field-cycling NQR experiments, but in the usual procedure the protons are made to absorb low-frequency zero-field irradiation directly, by which a low-frequency signal that obscures the NQR lines is produced. The time domain experiment alleviates such problems by using selective d.c. pulsed fields. An example of a ^{14}N zero-field spectrum obtained by selective pulses in zero field and indirect detection through the protons is shown for polycrys-

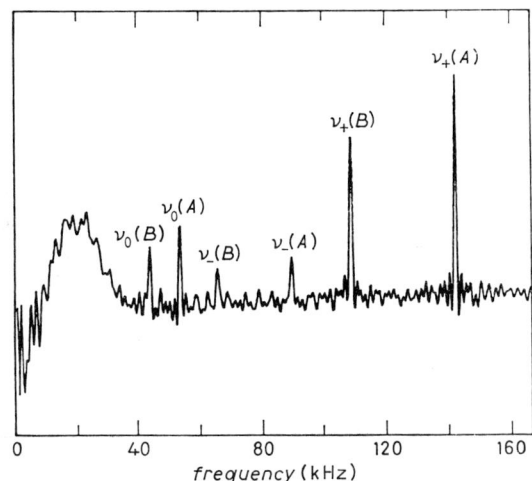

Fig. 8.9. – Pulsed zero-field ^{14}N NQR spectrum of solid $(NH_4)_2SO_4$ using the cycle in fig. 8.8 with selective 2π pulses for the protons. The ^{14}N was detected indirectly by level crossing with the protons. Peaks corresponding to two inequivalent nitrogen sites in the unit cell are labeled A and B. Residual proton signal appears below 40 kHz, but has been reduced sufficiently to allow for resolution of the low-frequency ^{14}N NQR lines. (Adapted from *J. Chem. Phys.*, **83**, 934 (1985), with permission.)

talline ammonium sulfate in fig. 8.9. Such experiments on deuterium and ^{14}N are likely to be useful in the study of biological systems, which are often inherently amorphous or disordered. Recent experiments have demonstrated zero-field analogs to multiple-pulse high-field NMR, including scaling of interactions and time reversal[182-186]. For example, fig. 8.10a) shows the sixteen-pulse

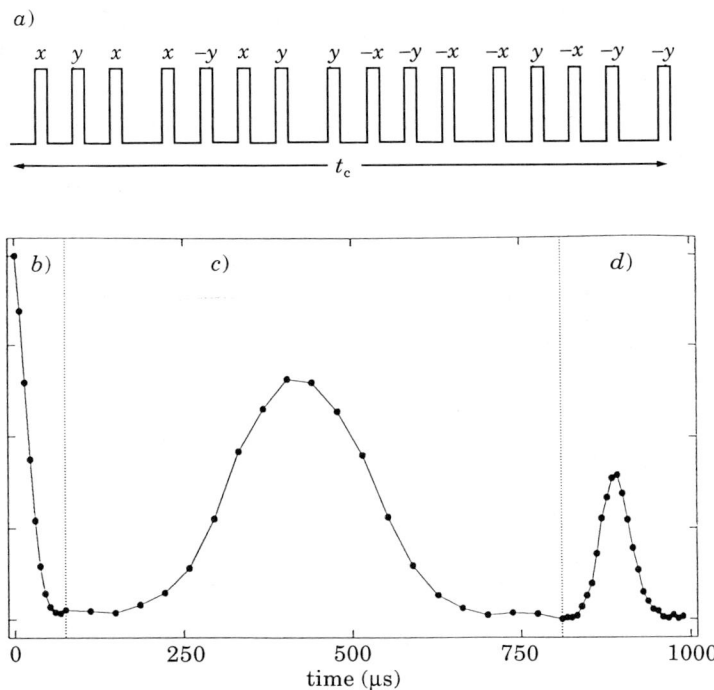

Fig. 8.10. – Isotropic spin echoes for zero-field second-rank interactions. a) The second-rank sixteen-pulse time-reversal sequence. (Small interpulse spacings correspond to τ delays, whilst large spacings correspond to 2τ.) b) The magnetization of the protons in polycrystalline adamantane decays due to the local isotropic dipole-dipole couplings. c) After 74 µs the isotropic time-reversal sequence is applied and the magnetization is retrieved 370 µs later, because the scaling factor is 1/5. d) Free evolution in zero field is resumed at time 810 µs. One sample point is taken per cycle (36.8 µs). The pulse duration is 1 µs. (Adapted from *Phys. Rev. Lett.*, **67**, 1989 (1991), with permission.)

sequence that achieves time reversal of second-rank interactions. The sequence also isotropically scales first-rank interactions by $k = 1/5$. In this process the Hamiltonian is rotated by all of the six $\pi/2$ and the eight $3\pi/2$ rotations of the cubic group with relative weights 2 to 1 (cf. sect. 5). Figures 8.10b)-d) show the result of applying the sequence to a sample of solid adamantane, generating an echo after the zero-field free-induction decay[185].

8'4. Calculation of the zero-field spectrum. – It is instructive to perform a specific calculation of the zero-field signal for a simple case. Consider a molecule or group of spins in a polycrystalline or otherwise disordered sample. The laboratory-based coordinate system is denoted x, y, z as usual, with any high field B_0 along z, and molecule-based axes are labeled xm, ym, zm. We denote by $R(\Omega)$ the operator, or by $\boldsymbol{R}(\Omega)$ the superoperator, that effects the transformation between laboratory and molecular frames. The transformation angles Ω are characterized by a probability distribution $P(\Omega)$ over the sample. $\Omega \equiv (\alpha, \beta)$, where α, β are the two Euler angles, the third angle γ is not required owing to cylindrical symmetry about z in high field and is accordingly set to zero.

In the molecular frame the internal Hamiltonian (dipolar, quadrupolar, etc.) is homogeneous in the sense that it is independent of Ω or position in the sample. For the simple field cycle of fig. 8.2 the initial state of the spin system is given by

(8.1) $$|\varrho(0)\rangle \sim |I_z\rangle,$$

and I_z is detected in high field at the end of the cycle. Thus we can use precisely the formalism of previous sections. The zero-field time domain signal (again we write t instead of t_1) is given by

(8.2) $$f_\Omega(t) = \langle I_z | \boldsymbol{U}_\Omega(t) | I_z \rangle.$$

The subscript Ω reminds us that this signal is for a particular orientation of the molecule or spin system in the laboratory. $\boldsymbol{U}_\Omega(t)$ is the zero-field propagator in the laboratory frame. We then write

(8.3) $$f_\Omega(t) = \langle I_z | \boldsymbol{R}(\Omega) \boldsymbol{U}_\mathrm{m}(t) \boldsymbol{R}^\dagger(\Omega) | I_z \rangle,$$

where

(8.4) $$\boldsymbol{U}_\mathrm{m}(t) = \boldsymbol{R}^\dagger(\Omega) \boldsymbol{U}_\Omega(t) \boldsymbol{R}(\Omega)$$

is the homogeneous propagator in the molecular frame. $|I_z\rangle$ can easily be expressed in terms of operators in the molecular frame through the Wigner rotation matrices

(8.5) $$R(\Omega) I_z R^\dagger(\Omega) = D^{(1)}_{00}(\Omega) I_{zm} - D^{(1)}_{00}(\Omega) \frac{1}{\sqrt{2}} I_{+m} + D^{(1)}_{-10} \frac{1}{\sqrt{2}} I_{+m} =$$

$$= I_{zm} \cos\beta - I_{xm} \sin\beta \cos\alpha + I_{ym} \sin\beta \sin\alpha.$$

Inserting eq. (8.5) into eq. (8.3) we obtain

(8.6) $$f_\Omega(t) = \langle I_{zm} \cos\beta - I_{xm} \sin\beta \cos\alpha + I_{ym} \sin\beta \sin\alpha \cdot$$
$$\cdot | \boldsymbol{X}_\mathrm{m}(t) | I_{zm} \cos\beta - I_{xm} \sin\beta \cos\alpha + I_{ym} \sin\beta \sin\alpha \rangle.$$

8`5. *Average over orientational distribution.* – Expression (8.6) must now be averaged over the distribution $P(\Omega)$ to give the signal for the sample

(8.7) $$f(t) = \int f_\Omega(t) P(\Omega) \, d\Omega.$$

For an isotropic three-dimensional distribution, as in a random powder, all Ω are equally probable, *i.e.*

(8.8) $$P(\Omega) = \text{const},$$

and only three terms in (8.6) survive the integration of (8.7). The result is

(8.9) $$f(t) = \frac{1}{3}(\langle I_{xm} | U_m(t) | I_{xm}\rangle + \langle I_{ym} | U_m(t) | I_{ym}\rangle + \langle I_{zm} | U_m(t) | I_{zm}\rangle).$$

Expanding (8.9) in an eigenbasis of $U_m(t)$ in the manner of sect. **1**, we obtain the final expression

(8.10) $$f(t) = \frac{1}{3} \sum_{jk} (|I_{xmjk}|^2 + |I_{ymjk}|^2 + |I_{zmjk}|^2) \cos \omega_{jk} t,$$

where I_{xmjk} is the (j, k)-th matrix element of I_{xm}. This expression implies that the signal will be linearly polarized along z, and that the Fourier transform spectrum is symmetric around zero frequency. This behavior is expected, of course, because of the axial symmetry around z manifested by eq. (8.8).

8`6. *Dipolar coupled spin $I = 1/2$ pair or quadrupolar spin $I = 1$.* – We can now evaluate (8.10) explicitly for the most basic cases of zero-field NMR and NQR. Consider two spins $I = 1/2$ with an axially symmetric dipolar coupling. With the molecular z axis along the symmetry axis established by the internuclear vector the molecular-frame Hamiltonian in frequency units is

(8.11) $$H_{Dm} = \omega_D (3 I_{zm1} I_{zm2} - \mathbf{I}_1 \cdot \mathbf{I}_2),$$

where the dipolar frequency ω_D is

(8.12) $$\omega_D = \gamma^2 / r^3.$$

The well-known eigenstates of eq. (8.11) are depicted in fig. 8.11a): a triplet 1, 2, 3 (with a degenerate pair 1, 2) and a singlet S. The only nonzero matrix elements in eq. (8.10) are within the triplet manifold. The frequencies are

(8.13) $$\begin{cases} \omega_{12} = 0, \\ \omega_{13} = \omega_{23} = \frac{3}{2} \omega_D. \end{cases}$$

Plugging all this into (8.10) and recalling that we are working with a normalized signal $\langle I_z | I_z \rangle = 1$, as in sect. **1**, we obtain

(8.14) $$f(t) = \frac{1}{3}\left(1 + 2 \cos \frac{3}{2} \omega_D t\right),$$

which predicts a spectrum of lines with equal intensities at 0 and $\pm 3\omega_D/2$, as shown in the lower part of fig. 8.11a). Such a spectrum is indeed observed experimentally for a pair of protons, as we saw previously in fig. 8.7. Figure 8.11b) reminds us that the situation is entirely analogous to a single spin $I = 1$ (cf. the three-level system of subsect. 1˙13) if $\eta = 0$. Remember, however, that here we are in the molecular frame for all Ω. The triplet manifold is analogous to that of the dipolar coupled pair and there is no singlet state.

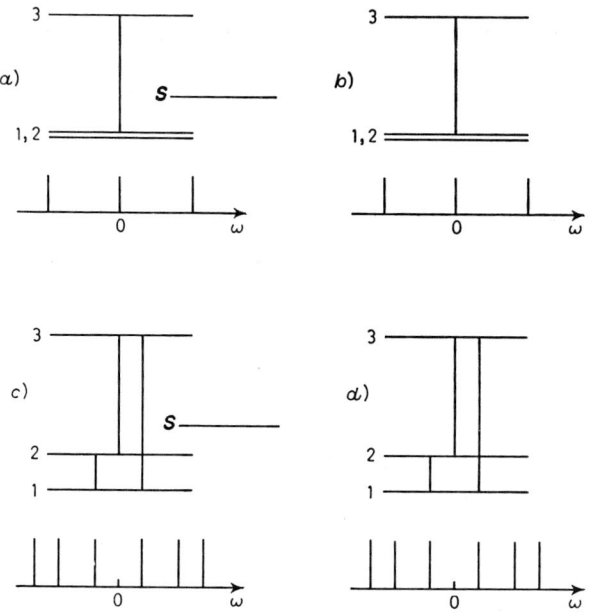

Fig. 8.11. – Energy levels and schematic spectra of two spins $I = 1/2$ and for a spin $I = 1$ in zero field for $\eta = 0$ and for $\eta \neq 0$. Small deviations from local symmetry, or subtle motional effects, lead to small values of η which are easily observable in zero field.

For both the dipolar coupled pair and the single spin $I = 1$, where the asymmetry parameter η is nonzero, the zero-field transitions are split as shown in fig. 8.11c) and d). The molecular-frame Hamiltonian for the dipolar case is

(8.15) $$H_{Dm} = \omega_D[3I_{zm1}I_{zm2} - \boldsymbol{I}_1 \cdot \boldsymbol{I}_2 + \eta(I_{xm1}I_{xm2} - I_{ym1}I_{ym2})],$$

which splits the ω_{12} transition,

(8.16) $$\omega_{12} = \eta\omega_D.$$

For the quadrupolar case, the Hamiltonian is

(8.17) $$H_{Qm} = A[3I_{zm}^2 - \boldsymbol{I}^2 + \eta(I_{xm}^2 - I_{ym}^2)],$$

with the quadrupole coupling constant

$$(8.18) \qquad A = \frac{e^2 qQ}{4}.$$

Thus *the effects of small asymmetry are clearly visible as sharp splittings in the spectrum, although similar effects may be difficult to discern in high-field powder patterns*. This agreeable feature of zero-field NMR was mentioned in subsect. 8·1 and was quite clear in the experimental spectra of fig. 8.4 and 8.5. Consequently, zero-field NMR spectra may provide a useful measure of small-amplitude motions and subtle deviations from local symmetry in disordered systems. Examples of these effects, for example in biaxial smectic phases, have appeared in the literature and might prove useful for biological applications in the future.

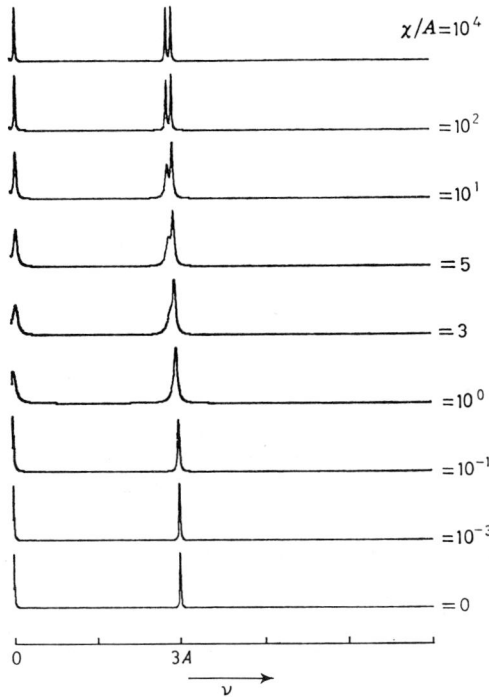

Fig. 8.12. – Simulated zero-field spectra for jumps of symmetry axis of a deuterium spin with $\eta = 0$ between orientations differing by $2\theta = 20°$. Only the positive-frequency halves of the spectra are shown. χ/A, the ratio of the exchange rate to the quadrupolar frequency, varies from the rigid regime (bottom) to the rapid motional limit (top). The ratio of the residual line broadening ($1/T_2$) to the quadrupolar frequency is 0.02. The onset of line splitting due to the motionally induced asymmetry ($\eta \neq 0$) can be seen at large χ/A. The behavior would be similar for the zero-field NMR of two coupled spins $I = 1/2$. (Adapted from *J. Chem. Phys.*, **85**, 4873 (1986), with permission.)

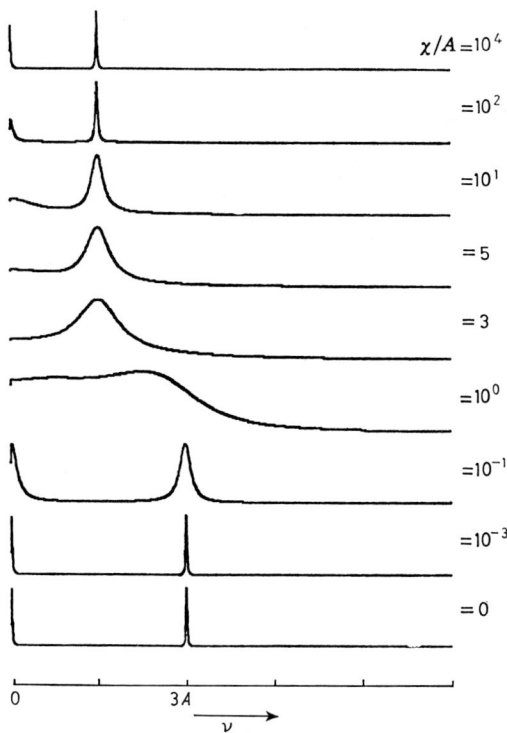

Fig. 8.13. – Same as fig. 8.12 but $2\theta = 90°$. This motion is equivalent to fourfold jumps around an axis, and, therefore, leads to an axially symmetric motionally averaged spectrum. (Adapted from *J. Chem. Phys.*, **85**, 4873 (1986), with permission.)

8`7. *Effects of motion.* – As an example of the effect of introducing a small nonzero asymmetry parameter through motion, consider an axially symmetric dipolar- or quadrupolar-coupling tensor where the symmetry axis jumps randomly through an angle of 2θ. When the jump rate χ increases from zero, the zero-field spectrum should change from the static (axially symmetric) case to the time-averaged case, in analogy to high-field NMR studies of chemical exchange and motion (recall sect. 5)[187].

Examples of spectra simulated for a quadrupolar spin and different relative jump rates (χ/A) are shown in fig. 8.12 for $2\theta = 20°$ and in fig. 8.13 for $2\theta = 90°$; the jumps are equivalent to fourfold jumps around an axis (due to the symmetry of the Hamiltonian under 180° rotations) leading to a time-averaged axially symmetric coupling.

An interesting experimental example of the onset of asymmetry in dipolar couplings occurs in the smectic liquid crystals as shown in fig. 8.14 [188]. The smectic A and B phases are locally uniaxial, and, indeed, the zero-field NMR spectra of methylene chloride (CH_2Cl_2) probe molecules dissolved in these phases display peaks characteristic of axial symmetry. In the smectic E phase,

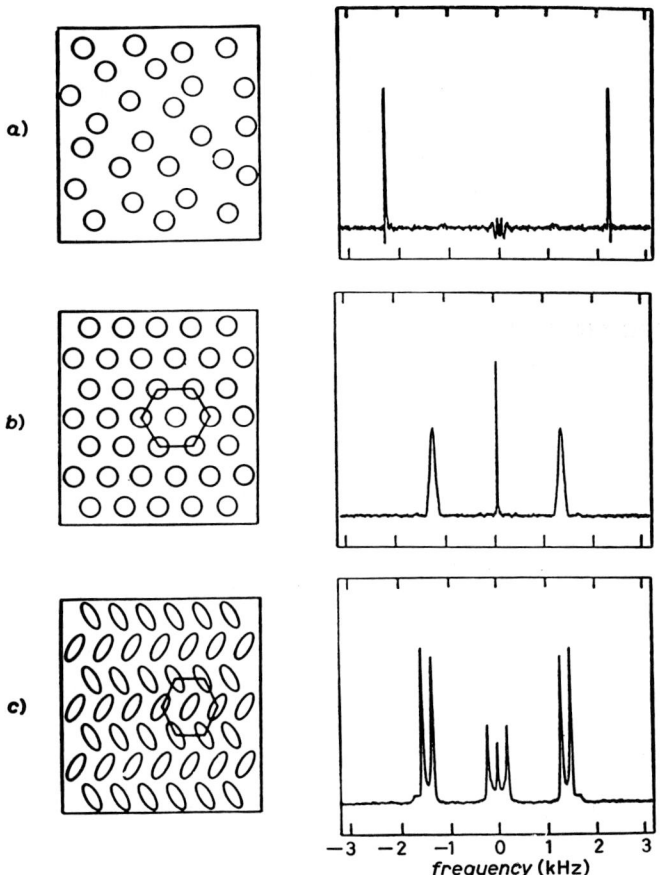

Fig. 8.14. – Zero-field NMR spectra of the proton pairs in methylene chloride (CH_2Cl_2) probe molecules dissolved in smectic a) A, b) B and c) E phases of liquid crystals. The smectic E sample exhibits a line splitting characteristic of an asymmetric dipolar coupling, a consequence of the biaxial environment. (Courtesy of M. LUZAR.)

on the other hand, the line splittings reflect an asymmetry in the dipolar-coupling tensor of the proton spin pair induced by the biaxial environment on the smectic layers. It is useful that such subtle asymmetries can be seen directly in powder samples. Similar effects have been seen in solid carboxylic-acid dimers which exhibit correlated proton jumps in the hydrogen bonds.

8'8. *Magnetic resonance with a* SQUID *detector.* – A recent alternative to magnetic-field cycling is the detection of zero-field and low-frequency spectra using a dc-SQUID (superconducting quantum interference device) detector. Following an early rf-SQUID design by DAY [189], the SQUID has been incorporated into both pulsed- and continuous-wave spectrometers by coupling the nuclear magnetic flux in the sample to the SQUID through a superconducting

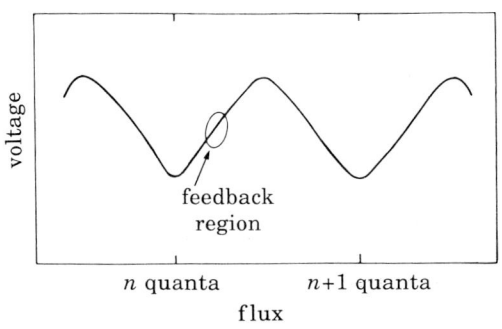

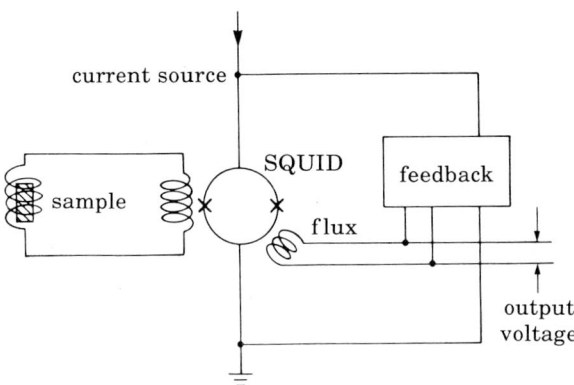

Fig. 8.15. – Schematic representation of a SQUID NMR spectrometer. (Adapted from *Rev. Sci. Instrum.*, **61**, 1059 (1990), with permission.)

transformer as shown in fig. 8.15 [190, 191]. The SQUID spectrometers are quite sensitive and they have been used to detect low-frequency nuclear quadrupole resonances of deuterium, nitrogen-14, boron-10, 11 and aluminum-27 in crystals and glasses, and NMR transitions arising from quantum tunneling of methyl groups at low magnetic field and low temperature shown in fig. 8.16 [191]. Recently a pulsed SQUID spectrometer has been built and used to obtain ^{14}N NQR spectra showing the effects of quantum tunneling of ammonium groups [192].

8'9. *Comment on relationship of spatially selective pulses to zero-field NMR.* – We would like to make a brief comment on the connection between spatial selectivity with a surface coil and the seemingly unrelated area of zero-field NMR. In normal high-field NMR we are accustomed to exciting and detecting different spins (*e.g.*, ^{1}H, ^{13}C, ...) separately and selectively according to their different frequencies. If we wish to work with protons, we may use 400 MHz, for example, and for ^{13}C we would then use 100 MHz. In zero field, however, all

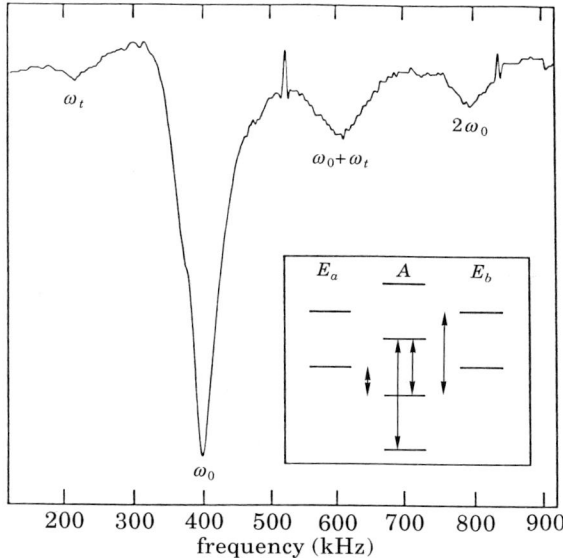

Fig. 8.16. – NMR spectrum arising from quantum tunneling of methyl groups at low magnetic field and low temperature obtained using a SQUID spectrometer. (Adapted from *Rev. Sci. Instrum.*, **61**, 1059 (1990), with permission.)

basic resonance frequencies are zero. So how do we differentiate between different species? The answer is that we use the magnetogyric ratio which is analogous to the r.f. amplitude ω_1 in high-field NMR. If we apply d.c. pulses, the spins respond differently according to their magnetogyric ratios. But is this notion not identical to the idea of spatial selectivity using a surface coil with composite pulses in high-field NMR? The two tubes shown in fig. 1.29 experience different ω_1 fields and this distinction is what is often used to label such regions. Indeed, some of this work in spatial selectivity has been useful in zero-field NMR.

9. – Geometric phases.

9˙1. *Context for geometric phases.* – Nuclear spins provide an ideal arena for the demonstration of a class of phenomena related to the so-called geometric phases [193-197], of which Berry's phase is an important example. The general context for the geometric phase arises from the division of a system into two parts, characterized by coordinates or variables that we shall refer to as external and internal. For example, in the case of a classical object, the external coordinates might be the position of the object, and the internal coordinates might describe its orientation. For a quantum spin, the external coordinates might be the direction of the spin and an internal coordinate might be the phase of the wave function. For a Born-Oppenheimer molecule the external coordinates might be the nuclear coordinates, and the internal coordinates would refer to the electronic wave function.

The separation of the system naturally gives rise to a gauge ambiguity, namely the fact that, for each point of the external-coordinate space, it is possible to choose a coordinate system or reference frame for the internal coordinates, *a choice of gauge*, in different (often an infinite number of) ways. But our observations must be invariant to the choice of gauge, a statement of *gauge symmetry* [198, 199]. Now, a common and important question is, what is the change, the *holonomy*, in the internal variables, when the system undergoes a cyclic evolution in the space of its external coordinates, as shown in fig. 9.1.

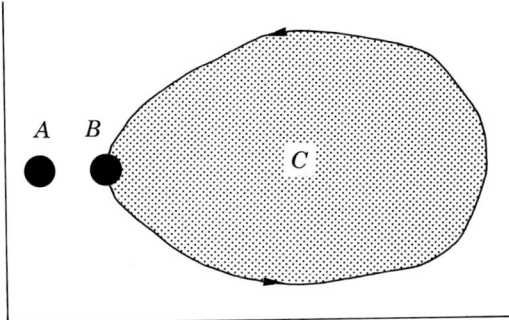

Fig. 9.1. – If a system is characterized by internal and external coordinates, then a change of the internal variables may occur when the system undergoes a cyclic evolution in the external coordinates. For example, if the two twins A and B are initially in the same state and B is taken on a circuit in its external coordinates, the holonomy, or change of internal coordinates of B, can be determined by comparing it with A.

The holonomy (of which the geometric phase is an example) could, of course, be calculated by direct solution (often cumbersome integration) of the appropriate equations of motion, for example the Schrödinger equation. The fact that we can often do better by recognizing the role of geometry and topology is a consequence of gauge symmetry. It is the appreciation of the broad and unifying implications of gauge symmetry that constitutes the basis for the remarkable contributions of Berry's phase and its generalizations. In this section we describe some examples of geometric phases with a view to demonstrating the unification of a diverse range of phenomena, including the phases of spins and of light, the Aharanov-Bohm effect [200, 201], fractional quantum numbers of Jahn-Teller molecules [202-204] and the reorientational kinematics of cats [205-207].

9`2. *Classical holonomy*. – Consider the situation depicted in fig. 9.2; a particle changes its position on the surface of a sphere (the external coordinates)

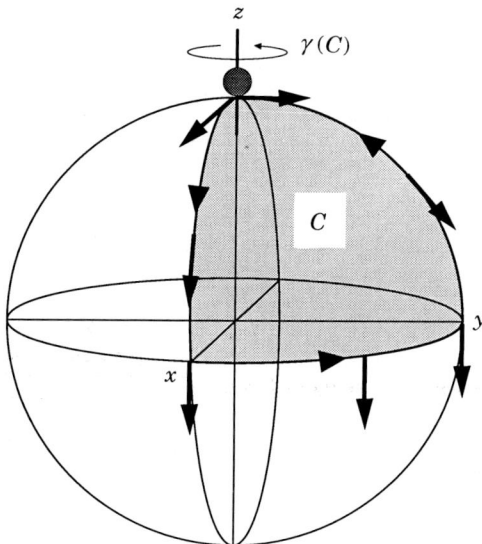

Fig. 9.2. – A simple example of holonomy is obtained by transporting a particle on the surface of a sphere. The direction in which the particle points is indicated by a tangent vector. If the particle is moved around the circuit C in such a way that it maintains a constant angle with the tangent to the path, this corresponds to parallel transport. Upon its return to the north pole, the particle has rotated by an amount $\gamma(C)$, the angle holonomy.

under *parallel transport* [208], namely in such a way that the particle does not rotate locally about its axis, the axis perpendicular to the sphere surface. The orientation of the particle about its axis (the internal coordinate) is indicated by a tangent vector. Under parallel transport on a flat surface, the orientation of the particle would not change; the tangent vector would remain parallel to itself at all times. However, the surface of the sphere is curved, which makes it possible for the orientation of the particle to change under parallel transport. In fact, if the particle undergoes the cyclic trajectory shown in fig. 9.2, starting and ending at the north pole, its orientation changes by [199]

$$(9.1) \qquad \gamma = -\Omega(C),$$

where $\Omega(C)$ is the solid angle subtended by the circuit at the origin of the sphere (the area enclosed by the circuit on the surface of a unit sphere), in this case $\pi/2$. This accumulated global rotation, even though there is no local rotation, is a geometric angle, an example of an *angle holonomy*. A famous example of such a classical holonomy is the rotation of the plane of a Foucault pendulum [209] as the Earth rotates.

9˙3. *Quantum holonomy*. – The classical angle holonomy derives from the curvature of the sphere, a consequence of its nontrivial topology. Similarly, in

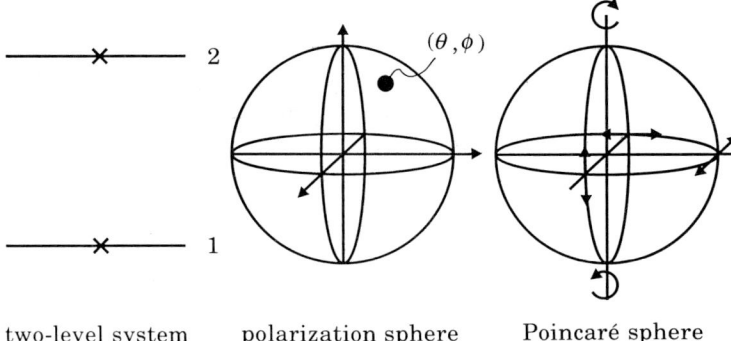

two-level system   polarization sphere   Poincaré sphere

Fig. 9.3. – Representations of two-level systems. A two-level system can be described by θ and ϕ, the external coordinates indicating the direction of the spin, and φ, the internal coordinate indicating the overall phase. The external coordinates define the surface of a sphere. For a spin $I = 1/2$ the poles correspond to «spin up» and «spin down» z polarizations and the equator represents linear combinations of x and y polarizations. On the sphere of light polarization (the Poincaré sphere (right)) the poles correspond to the two helicity states (circular polarization) and the equator describes states of linear polarization with plane angles from 0° to 180°.

quantum mechanics, the state space of the system may also possess a nontrivial topology. In fact, consider one of the simplest quantum systems, a two-level system, for example the states of a spin $I = 1/2$, or the polarization states of a photon (fig. 9.3). Recalling sect. 4, and following FEYNMAN et al. [109, 111, 210], the state of any two-level system can be described by two coordinates indicating the direction of the spin or fictitious spin, and a third coordinate indicating the overall phase of the wave function. Thus any wave function can be written as a product

(9.2) $$|\psi\rangle = \exp[i\varphi]|\bar{\psi}(\theta, \phi)\rangle,$$

where θ and ϕ, the external coordinates, define the surface of a sphere, the polarization sphere (known, for light polarization, as the Poincaré sphere) and φ, the phase, is the internal coordinate. The $|\bar{\psi}\rangle$ constitute a basis or projective space, the space of spin or polarization states without regard to phase. The choice of $|\bar{\psi}\rangle$ implies a reference phase for φ, namely a choice of gauge, under changes of which

(9.3) $$|\bar{\psi}(\theta, \phi)\rangle \to \exp[if(\theta, \phi)]|\bar{\psi}(\theta, \phi)\rangle$$

any observables must be invariant.

Suppose now that, in analogy to the case of classical holonomy in subsect. 9˙2, the spin begins at the north pole (along z), and undergoes the cyclic trajectory of fig. 9.4 (pointing successively along x, y and z), returning to the initial state along z. Such an evolution can be accomplished in a variety of ways

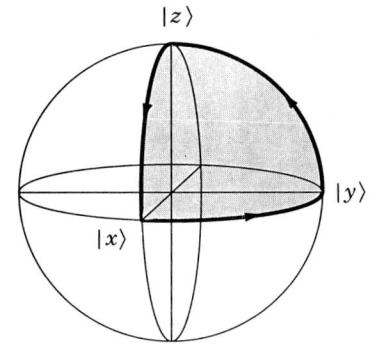

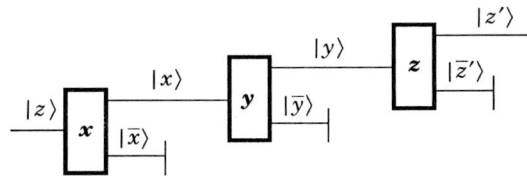

Fig. 9.4. – Quantum (Abelian) holonomy manifests itself as a change in the phase of the wave function when the system undergoes a circuit, in this case on the sphere of polarizations of a spin $I = 1/2$ passing through a series of differently oriented Stern-Gerlach devices.

(for example, by adiabatic following of a magnetic field whose direction changes slowly in a circuit, the situation considered by BERRY[193]). Since the initial and final spin directions are the same, the initial and final wave functions must be the same up to a phase factor:

(9.4) $$|\psi\rangle_{\text{final}} = \exp[i\varphi]|\psi\rangle_{\text{initial}}.$$

As we show in the next subsection, in addition to the familiar $\exp[i\omega t]$ type phase factor (total local phase accumulation)

(9.5) $$|\psi(t)\rangle = \exp[i\omega t]|\psi(0)\rangle,$$

there is a geometric contribution to the phase, arising from parallel transport of the spin state on the curved surface. The geometric part of the phase factor, γ, the *phase holonomy* (the phase change that would be obtained under parallel transport of the spin),

(9.6) $$|\psi\rangle \xrightarrow{C} \exp[i\gamma(C)]|\psi\rangle,$$

is given in analogy to eq. (9.1) by

(9.7) $$\gamma = -m\Omega(C),$$

where the solid angle is *multiplied by the spin quantum number m*, in this case $\pm 1/2$.

9'4. *Equations for the geometric phase.* – In a specific gauge $\{|\tilde{\psi}\rangle\}$, consider a state

(9.8) $$|\psi\rangle = \exp[i\varphi]|\tilde{\psi}\rangle.$$

Differentiation gives

(9.9) $$\frac{d}{dt}|\psi\rangle = i\frac{d\varphi}{dt}|\tilde{\psi}\rangle + \exp[i\varphi]\frac{d}{dt}|\tilde{\psi}\rangle,$$

and from the Schrödinger equation

(9.10) $$\frac{d}{dt}|\psi\rangle = \frac{-i}{\hbar}H|\psi\rangle.$$

Combining eqs. (9.9) and (9.10), followed by multiplication by $\langle\psi|$ and using the fact that $\langle\psi|\psi\rangle = 1$, we arrive at an equation for the evolution of the phase factor:

(9.11) $$d\varphi = \frac{1}{\hbar}\langle\psi|H|\psi\rangle dt - i\langle\tilde{\psi}|d\tilde{\psi}\rangle.$$

The first term in this expression corresponds to the local rotation (phase accumulation) due to the eigenvalue of H, the familiar $\exp[i\omega t]$ term whose integral around a circuit is

(9.12) $$\oint \frac{1}{\hbar}\langle\psi|H|\psi\rangle dt = \overline{\omega}t,$$

where $\overline{\omega}$ is the average eigenvalue of H over the circuit and serves as a clock for the trajectory. The second term, on the other hand, corresponds to the geometric contribution to the overall phase

(9.13) $$\gamma = -i\oint\langle\tilde{\psi}|d\tilde{\psi}\rangle = \oint A.$$

$A = -i\langle\tilde{\psi}|d\tilde{\psi}\rangle$ is a connection [199] (it contains differentially the connections between the $|\tilde{\psi}\rangle$ of the base space, namely the phase factors that define this particular choice of gauge), which behaves exactly like a gauge potential. The geometric phase can be evaluated using Stokes theorem for the integration around a circuit C on a surface s,

(9.14) $$\oint_C A = \int_s dA,$$

where dA is the curvature of the surface. Figure 9.5 illustrates the circuit of $|\tilde{\psi}\rangle$ and the accompanying circuit of $|\psi\rangle$ with the associated holonomy φ. The values

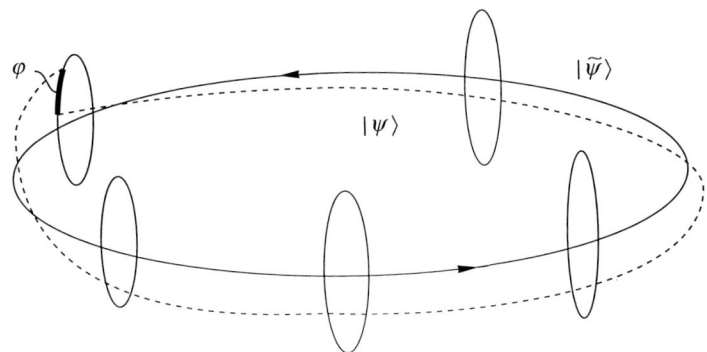

Fig. 9.5. – A circuit in the base space $|\tilde{\psi}\rangle$ and the corresponding circuit of $|\psi\rangle$. At each value of the external coordinate (for each $|\tilde{\psi}\rangle$), the $|\psi\rangle$'s corresponding to different values of the internal coordinates are related to each other by a gauge transformation, eq. (9.8). Since values of the internal coordinates range from 0 to 2π, they are depicted as circles.

of φ range from 0 to 2π and are, therefore, depicted as circles. This construction is the simplest example of a fiber bundle [194], with a fiber circle depicting the internal coordinates (the phase) corresponding to each value of the external co-ordinates for the base space of $|\tilde{\psi}\rangle$ states. A gauge transformation corresponds to threading the fibers with a different set of $|\tilde{\psi}\rangle$

(9.15) $$|\tilde{\psi}(s(\theta, \dot\phi))\rangle \to \exp[if(s(\theta, \dot\phi))]|\tilde{\psi}(s(\theta, \dot\phi))\rangle,$$

where $s(\theta, \dot\phi)$ represents the coordinates in the base space. The geometric phase of eq. (9.13) is now given by

(9.16) $$\gamma' = -i\oint \langle \tilde{\psi} \exp[-if(s)] | \mathrm{d}(\exp[-if(s)]\tilde{\psi})\rangle \, \mathrm{d}s.$$

Integration by parts from 0 to T leads to

(9.17) $$\gamma' = -i\oint \langle \tilde{\psi} | \mathrm{d}\tilde{\psi}\rangle \, \mathrm{d}s + \langle \tilde{\psi} | \tilde{\psi}\rangle (f(T) - f(0)),$$

so, as expected, γ is gauge invariant since $|\tilde{\psi}\rangle$ is single valued.

The case we have described above refers to an Abelian holonomy (the geometric phase is a single real number). However, geometric phases are not limited to be Abelian, and in general the holonomy may be represented by a transformation matrix of higher dimension operating on the vector $|\psi\rangle$ [197, 211-213]

(9.18) $$|\psi\rangle \xrightarrow{C} \exp[i\boldsymbol{R}]|\psi\rangle.$$

9˙5. *Explicit calculation for spin $I = 1/2$.* – As a base space for spin $I = 1/2$, we use the two-component spinors with a choice of gauge corresponding to [214, 215]

$$|\tilde{\psi}\rangle_+ = \begin{pmatrix} \cos\frac{\theta}{2}\exp[i\phi] \\ \sin\frac{\theta}{2} \end{pmatrix}. \tag{9.19}$$

Note that this gauge is ill defined at $\theta = \pi$, and is, therefore, termed the south-pole gauge; it does not represent an appropriate choice for circuits that pass through the south pole. We can evaluate the gauge potential, eq. (9.13):

$$A = -i\langle\tilde{\psi}|\mathrm{d}\tilde{\psi}\rangle = -i\left(\cos\frac{\theta}{2}\exp[-i\phi], \sin\frac{\theta}{2}\right)\mathrm{d}\begin{pmatrix} \cos\frac{\theta}{2}\exp[i\phi] \\ \sin\frac{\theta}{2} \end{pmatrix}, \tag{9.20}$$

$$A = \frac{1}{2}(1+\cos\theta)\,\mathrm{d}\phi. \tag{9.21}$$

This is precisely the gauge potential of a magnetic monopole of charge $m = 1/2$, a point to which we shall return shortly. The geometric phase can now be evaluated using Stokes theorem (eq. (9.14)) and eq. (9.13) to yield

$$\gamma = -\frac{1}{2}\int_S \sin\theta\,\mathrm{d}\theta \wedge \mathrm{d}\phi = -\frac{1}{2}\Omega(C), \tag{9.22}$$

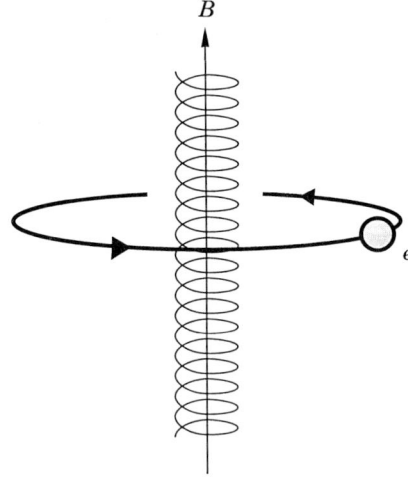

Fig. 9.6.– The Aharonov-Bohm effect involves the change in phase of the wave function of a charged particle transported in a circuit around an (infinitely long) solenoid. Even though there are no forces on the particle, because the magnetic field is zero outside the solenoid, the interaction of the particle charge with the (nonvanishing) electromagnetic gauge potential outside the solenoid gives rise to a phase shift of the particle wave function. The Aharonov-Bohm phase is an example of a geometric phase.

which justifies eq. (9.7) for $m = 1/2$. For a spin I in state m, expression (9.7) is generally valid.

9'6. *The Aharonov-Bohm effect.* – BERRY recognized that what is embodied in expressions (9.8)-(9.18), namely that the geometric phase is the integral of the gauge potential around the circuit, is reminiscent of the phase shift in the famous Aharonov-Bohm effect[109, 200], depicted schematically in fig. 9.6. A

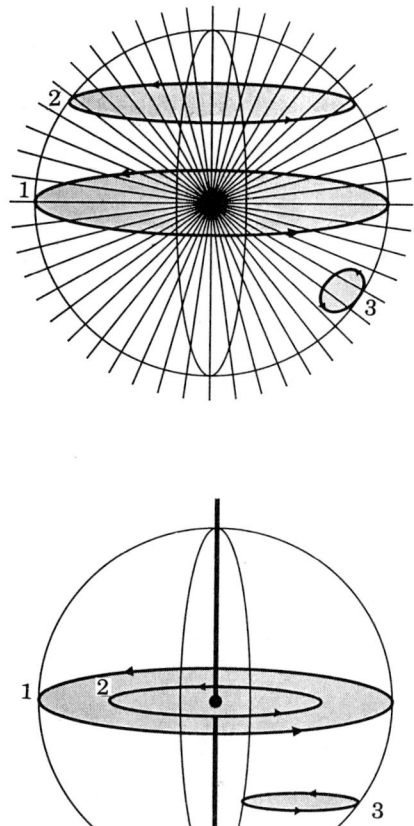

Fig. 9.7. – For a Dirac monopole of strength $m = 1/2$ with flux lines emanating isotropically, a 2π circuit (circle 1) has a geometric phase of $-\pi$ giving rise to the spinor sign change. In the upper figure the phase is proportional to the amount of flux that passes through the circuit $\gamma(C) = -(1/2)\Omega(C)$. The lower picture corresponds to gathering together the lines of flux emanating from the monopole above and below the plane into a flux tube of strength $m = 1/2$. In this case the phase is $\gamma(C) = -(1/2)2\pi = -\pi$ if the circuit encompasses the tube, circles 1 and 2, and $\gamma(C) = 0$ if the circuit does not include the tube, circle 3. (Courtesy of K. T. MUELLER.)

particle of electric charge e is transported in a circuit around an infinitely long solenoid inside which there is magnetic flux m and outside which the electromagnetic field vanishes. Classically, there are no forces on the particle, so there should be no effects due to the solenoid. However, in quantum mechanics, we must consider the electromagnetic gauge (vector) potential A, which does not vanish outside the solenoid (it is only the field, the curl dA of the potential, that vanishes). The interaction of the particle charge gives rise to a quantum phase shift of the particle wave function given by

$$\gamma = \oint_C A,$$

which is proportional to the flux of the magnetic field through the circuit and the particle charge,

(9.23) $$\gamma = -m2\pi e.$$

The relationship to the geometric phase can now be made clear by considering the transport of a particle of unit charge in the vicinity of a Dirac monopole [214-216] of strength m, as depicted in fig. 9.7. The phase shift is again given by the integral of the potential around the circuit which is equal to the flux of the (isotropic) magnetic field through the circuit

$$\gamma = \oint_C A = -m\Omega(C),$$

which is precisely the same as eq. (9.7) for $e = 1$. *Thus the geometric phase for a spin looks exactly like the Aharonov-Bohm effect for the circuit of a charged particle near a magnetic monopole.* The curvature of the spin state space is analogous to the magnetic field, and it is the flux of the curvature or field through the circuit that is responsible for the holonomy or geometric phase.

9'7. *Geometric phase in* NMR *interferometry*. – In order to measure the geometric phase in a two-level system, an additional level is required, to provide the phase reference (the twin or copy of fig. 9.1). The simplest system that can be used is a three-level system. For example, in the experiment of fig. 9.8 the holonomy of a circuit C in the 1-2 two-level system appears as a phase shift in the spin echo of a coherent superposition in the 2-3 system [217, 218]. Once again the geometric phase observed in this experiment is given by eq. (9.7). As we mentioned above, the circuit in the 1-2 system is not limited to be adiabatic, it can be nonadiabatic or even nonunitary (see subsect. 9'9), as long as we know what the circuit is. If the trajectory is not a closed circuit, we can complete it with a geodesic line, which does not affect the resulting geometric phase. Once again we may note some special cases (illustrated with the results shown in fig. 9.9). If the 1-2 transition undergoes a circuit with area 2π, we obtain from

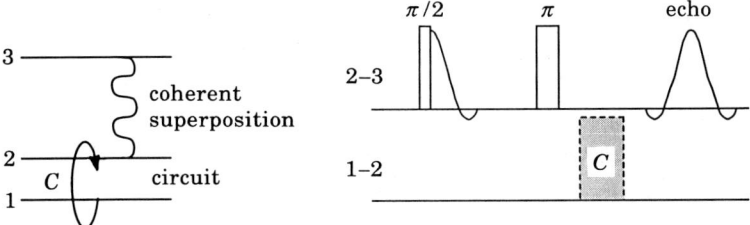

Fig. 9.8. – In NMR interferometry the holonomy of a circuit C in the 1-2 two-level system appears as a phase shift in the spin echo of a coherent superposition in the 2-3 system. (Courtesy of D. SUTER.)

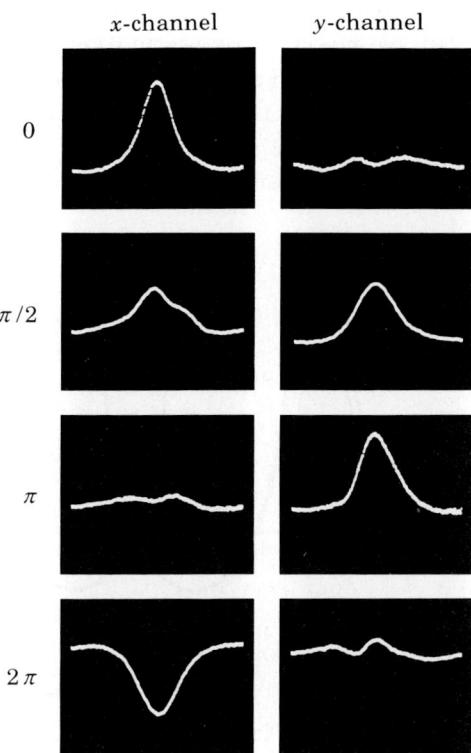

Fig. 9.9. – Holonomy in a two-level system (fig. 9.3) observed in the phase shift of a spin echo. The NMR echo is detected in orthogonal channels of a phase-sensitive detector. The numbers to the left indicate the solid angle of a circuit on the sphere of polarizations for the 1-2 subsystem. The echo arising in the 2-3 superposition moves from one channel to the other as the geometric phase increases from 0° to 90°. (Adapted from *Phys. Rev. Lett.*, **57**, 242 (1986).)

eq. (9.6), for $\exp[-i\gamma(C)] = -1$,

(9.24) $$|\psi\rangle \xrightarrow{C} -|\psi\rangle,$$

which is the well-known spinor sign change; one would need a circuit with an area of 4π for the 1-2 transition to leave the system unchanged. *In fact, we see that the famous spinor sign change is just a special case of the geometric phase for a circuit with area 2π.* Similarly, if the 1-2 spin system undergoes a circuit with area π, we obtain

(9.25) $$|\psi\rangle \xrightarrow{C} i|\psi\rangle$$

and the signal is shifted into the imaginary channel of the phase-sensitive detector.

9˙8. *Fractional quantum numbers.* – As shown in fig. 9.10, the molecule Na_3 undergoes a Jahn-Teller distortion in which the electronic degeneracy (corresponding to the molecular symmetry E representation of the electronic wave functions) is lifted by vibronic interaction with the vibrational degeneracy (corresponding to the e representation of the nuclear wave functions). This is entirely analogous to the lifting of a spin-1/2 degeneracy in the presence of a magnetic field [197, 204]. The distorted triangle can undergo an isoenergetic pseu-

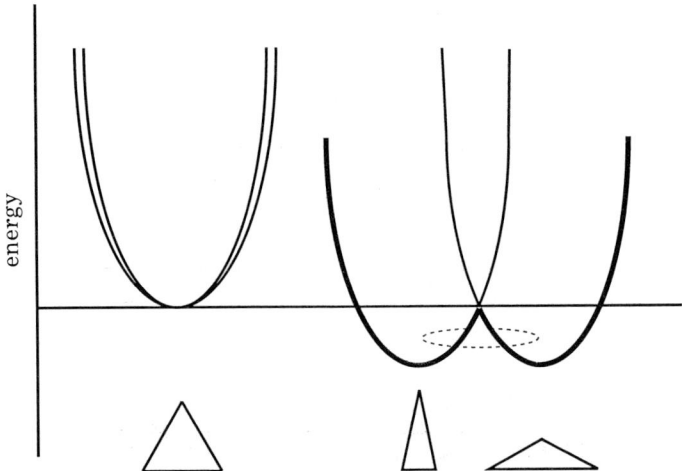

Fig. 9.10. – Born-Oppenheimer potentials for the Jahn-Teller $E \otimes e$ system, shown by taking a slice through the axially symmetric surfaces. Energy is plotted as a function of displacements in the degenerate nuclear vibration. On the left are the surfaces with no vibronic interaction; the degeneracy of these surfaces is lifted, as shown on the right, at all but a single point. The two surfaces diverge linearly from one another at the origin (the point of electronic degeneracy), with the lower surface showing stabilization for certain distorted geometries. The conical intersection at the degeneracy is the source of the geometric phase for the evolution of adiabatic states. (Adapted from *Annu. Rev. Phys. Chem.*, **41**, 601 (1990).)

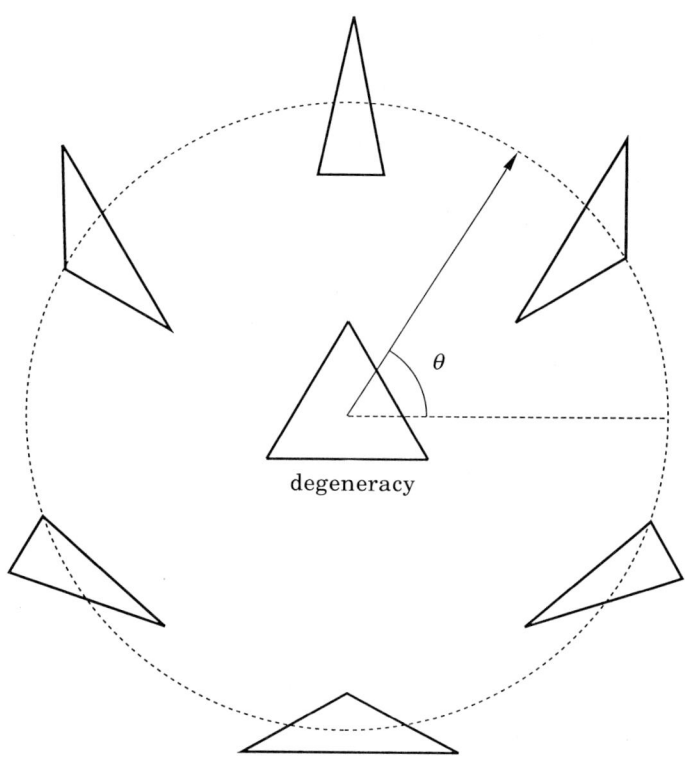

Fig. 9.11. – Pseudorotation in a trimer. The electronic degeneracy occurs at the symmetric configuration, shown as the equilateral triangle in the center. The lower-energy shapes are distorted triangles. Low-lying vibronic states are transported around the degeneracy by the pseudorotation, which results in a closed circuit in parameter space (the space of nuclear shapes) and a corresponding Berry's phase, giving rise to fractional pseudorotation quantum numbers. (Adapted from *Annu. Rev. Phys. Chem.*, **41**, 601 (1990).)

dorotation as shown in fig. 9.11. Within the Born-Oppenheimer approximation, the wave function can be written

(9.26) $$|\psi\rangle = |\psi_{\text{electronic}}\rangle \otimes |\psi_{\text{nuclear}}\rangle$$

with $|\psi_{\text{electronic}}\rangle$ following the nuclear configuration adiabatically:

(9.27) $$|\psi_{\text{electronic}}\rangle = |\psi_{\text{electronic}}(\theta)\rangle,$$

where θ is the pseudorotation angle. The pseudorotation angle can be considered the external coordinate, and the phase of the electronic wave function the internal coordinate, in analogy to the quantum mechanics of a spin $I = 1/2$ adiabatically following a magnetic field. The geometry is that of a 2π circuit in the plane about a monopole (since we are restricted to the plane) of strength $1/2$ and associated geometric phase factor -1. *Thus the nuclear pseudorotation must be quantized over a 4π circuit,* leading to a pseudorotation quantum number J which is half-integer rather than the normal integer quantum number associated with rotation. This observation is borne out by more closely spaced lines in the optical spectroscopy of Na_3 [204]

9˙9. *Nonunitary behavior, quantum projection*. – As we mentioned in subsect. 9˙7, the evolution of $|\psi\rangle$ does not necessarily have to be unitary to give rise to a geometric phase. Indeed, PANCHARATNAM [219] considered two states $|\psi_1\rangle$ and $|\psi_2\rangle$ in which, for example, $|\psi_2\rangle$ may be obtained from $|\psi_1\rangle$ by means of some projection or filtering. A natural way to compare the phases of $|\psi_1\rangle$ and $|\psi_2\rangle$ is by means of the inner product:

$$\langle \psi_1 | \psi_2 \rangle = r \exp[i\alpha]. \tag{9.28}$$

The vectors $|\psi_1\rangle$ and $|\psi_2\rangle$ are said to be in phase or parallel if $\alpha = 0$, so $\langle \psi_1 | \psi_2 \rangle$ is real. SAMUEL and BHANDARI [220] point out that $|\psi_1\rangle$ and $|\psi_2\rangle$ are parallel if $|\psi_2\rangle$ is obtained from $|\psi_1\rangle$ (aside from any nonunitary shrinkage) by transport of the associated density operator along a geodesic in the projective space, the Pancharatnam connection. In the limit of unitary evolution this connection reduces to the adiabatic connection.

As an example of the Pancharatnam phase consider a sequence of (perhaps quantum) filtering measurements on spin $I = 1/2$ particles, as illustrated schematically in fig. 9.4. A beam of particles originally polarized along z with initial state vector $|z\rangle$ is split into a reference channel and a second beam. The second beam enters a Stern-Gerlach apparatus oriented along x and splits into two orthogonal components ($|x\rangle$ and $|\bar{x}\rangle$):

$$|\psi_{\text{initial}}\rangle = |z\rangle = |x\rangle\langle x|z\rangle + |\bar{x}\rangle\langle \bar{x}|z\rangle. \tag{9.29}$$

The $|\bar{x}\rangle$ component is discarded, leaving the filtered component $|x\rangle\langle x|z\rangle$ which is in phase with the original state ($\langle z|x\rangle\langle x|z\rangle = 1/2$ is real). Thus the filtering or projection from $|z\rangle$ to $|x\rangle$ is consistent with parallel transport (together with a shrinkage by $1/2$ and ignoring any dynamical phases) along a geodesic from z to x of the type shown in fig. 9.4. Following the geometry of the geodesic connections in fig. 9.4, the beam now enters a Stern-Gerlach apparatus oriented along y (in which the $|\bar{y}\rangle$ component is discarded) and finally a Stern-Gerlach [221] apparatus oriented along z (in which the $|\bar{z}'\rangle$ component is discarded). The final beam has the same polarization as the original beam (and $1/8$ the intensity), and its state is given by

$$|\psi_{\text{final}}\rangle = |z'\rangle\langle z'|y\rangle\langle y|x\rangle\langle x|z\rangle. \tag{9.30}$$

Although $|z\rangle$ is in phase with $|x\rangle$, $|x\rangle$ is in phase with $|y\rangle$, and $|y\rangle$ is in phase with $|z'\rangle$, it is clear that $|z\rangle$ and $|z'\rangle$ are not in phase; they differ by an holonomy

$$|z'\rangle = \exp[i\gamma(C)]|z\rangle, \tag{9.31}$$

where

(9.32) $$\gamma(C) = -\frac{1}{2}\Omega(C) = -\frac{\pi}{4}.$$

The Pancharatnam phase is thus just Berry's phase for the geodesic triangle and it can be measured by interference with the original beam (whose phase is on record in the reference channel). Note that the limit of a continuous (densely spaced circuit) of quantum filtering measurements is equivalent to unitary evolution along the circuit, and in this limit the Pancharatnam phase reduces to the unitary geometric phase of subsect. 9˙4.

9˙10. *Geometry of light*. – The same result as that described for the Stern-Gerlach-type experiments can be obtained from a circuit of polarizations of light. As shown in fig. 9.12, a beam of light which is initially circularly polarized is passed through a polarizer so that it is subsequently linearly polarized along z, a second polarizer polarizes the light along x, and a final polarizer returns the light to circular polarization. As can be appreciated from the Poincaré sphere of fig. 9.3, such a circuit of polarizations is exactly analogous to the example of the Stern-Gerlach experiments given above, and gives rise to a phase shift of the light corresponding to a monopole $m = 1/2$ given by the expression in eq. (9.32).

A slightly different situation applies to a circuit of *directions* of light (fig. 9.12). For example, a circuit in the directions of propagation of light guided by mirrors or an optical fiber. In this case the light behaves as a particle of spin $I = 1$, and parallel transport of, say, polarized neutrons around, for example, an optical fiber along one of the trajectories shown in fig. 9.12 results in an accumulated geometric phase [222, 223]

$$\gamma(C) = -\Omega(C),$$

corresponding to a monopole $m = 1$.

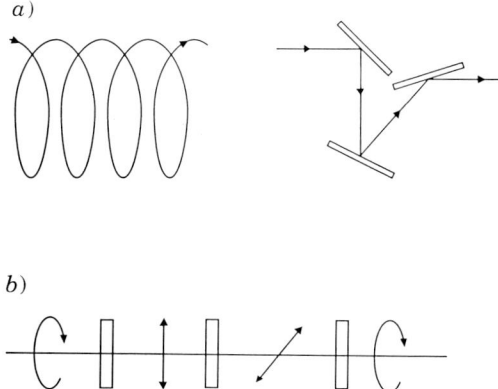

Fig. 9.12. – Circuits in the space of directions (*a*)) and polarizations (*b*)) of light.

9'11. *Rotation of cats.* – The acrobatics of a falling cat may seem a far cry from the preceding discussions of spins, light and molecules, but they share a lot in common and present a beautiful example of holonomy and gauge symmetry. Consider the falling cat of fig. 9.13 beginning with its feet up and maneuvering to land with its feet down. Since there are no external torques, angular momentum must be conserved and the cat cannot just «turn itself upside down»; its only recourse is to change its shape. In fact, if the cat begins and ends with the same shape, the «canonical cat shape», it has executed a circuit in its space of (unoriented) shapes. The question is, what is the angle of rotation of the cat given the circuit of shapes? The question is entirely analogous to those we have treated in previous sections: if we expand the cat shape in spherical harmonics, the coefficients can be considered external coordinates, the space of shapes is the base space, the orientation of the cat comprises the internal coordinates, and the gauge potential connects reference orientations for each of the shapes. SHAPERE and WILCZEK[206] and MONTGOMERY[207] have shown that for a deformable body the angle of rotation is the holonomy of the gauge potential over the circuit of shapes. In general the holonomy is non-Abelian, but a

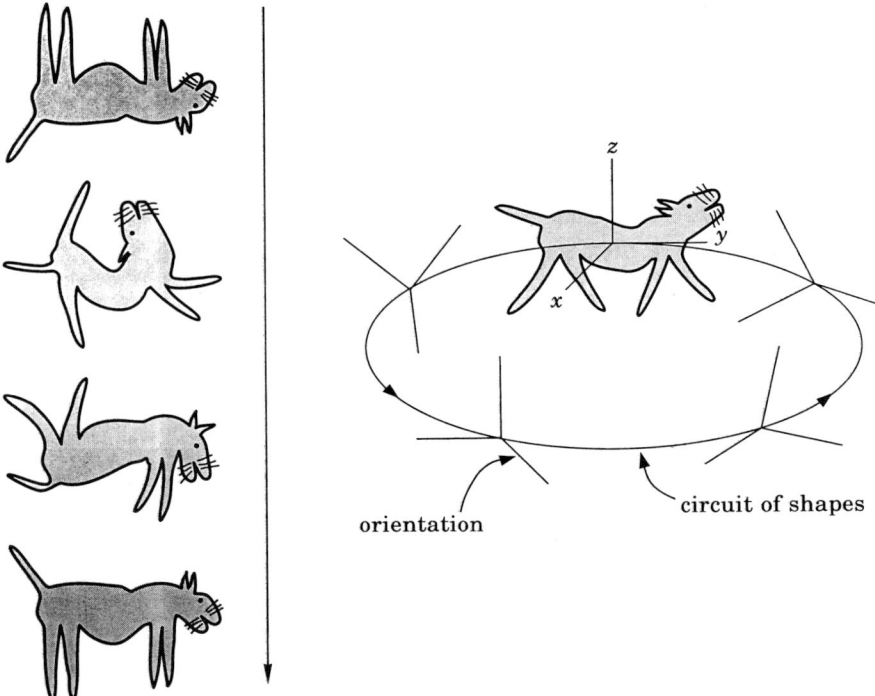

Fig. 9.13. – The reorientational dynamics of a falling cat can be treated by dividing the system into the external coordinates determining the shape of the cat, and the internal coordinates determining the orientation of the cat. In the absence of external torques, the cat can only land on its feet by reorientating through a sequence of shapes. The angle of rotation is the holonomy over the circuit of shapes.

simple Abelian example, the binary cat (following BHANDARI [205]), illustrates the essence of the problem.

The simplest approximation to the cat shape is a rigid sphere or cylinder, which obviously cannot reorient. The next approximation is the binary cat, shown in fig. 9.14, two cylinders joined with muscle fibers allowing *two degenerate bends*; the shape space of this construction corresponds to the surface of a sphere, as shown in fig. 9.14, and the problem is identical to the case of classical holonomy of subsect. 9˙2. The angle of rotation is equal to the area enclosed by the circuit swept out on the surface of a unit sphere by a vector on one of the binary-cat cylinders.

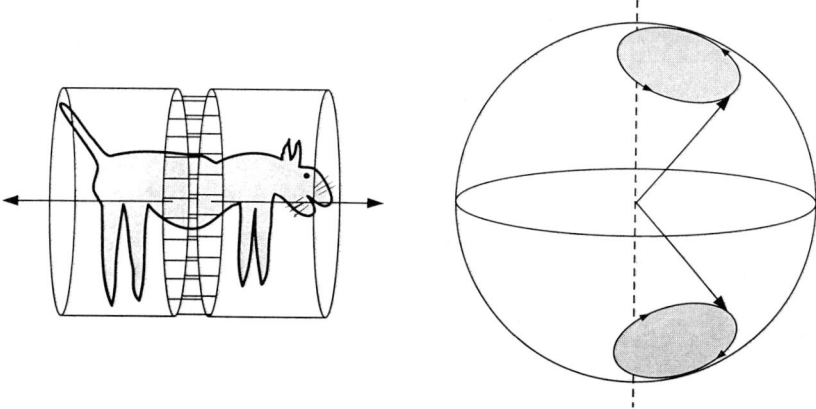

Fig. 9.14. – The binary cat, approximated by two rigid cylinders allowing two degenerate bends, has two external coordinates corresponding to the surface of a sphere. The angle of rotation of the binary cat is the solid angle subtended by the circuit of the cylinder vector at the origin, in analogy to the quantum holonomy of a spin $I = 1$ (Adapted from *Phys. Lett. A*, **133**, 1 (1988).)

A quantum version of the binary cat that corresponds to the classical behavior can be modeled by taking each half (cylinder) of the cat as a spin $I = 1$. Adopting a three-component spinor representation of the spin $I = 1$ [224], we have in the pole gauge for the base space

$$(9.33) \qquad |\tilde{\psi}\rangle_+ = \begin{pmatrix} \cos^2 \frac{\theta}{2} \exp[-i\phi] \\ \sqrt{2} \sin \frac{\theta}{2} \cos \frac{\theta}{2} \\ \sin^2 \frac{\theta}{2} \exp[+i\phi] \end{pmatrix},$$

from which we calculate the gauge potential

$$(9.34) \qquad A = -i\langle \tilde{\psi} | d\tilde{\psi} \rangle = \cos\theta \, d\phi$$

and the curvature

(9.35) $$dA = -\sin\theta\, d\theta \wedge d\phi.$$

The geometric phase is the angle of rotation of the cat and is given by

(9.36) $$\gamma = \oint_C dA = -\Omega(C)$$

exactly equal to the area of the circuit, as mentioned in subsect. 9˙2 and corresponding to the case of a monopole of strength $m = 1$ in subsect. 9˙6. Indeed, binary cats are appreciative of the gauge kinematics, and they tend to execute essentially circular trajectories in the space of shapes in order to maximize the flux of curvature through a given length of circuit.

* * *

We should like to thank the group in Berkeley for providing the stimulating atmosphere in which these lectures were prepared. We are grateful to Ph. J. Grandinetti, A. C. Kolbert, Y. K. Lee, J. R. Sachleben, J. Shore and K. Schmidt-Rohr for comments on the second edition manuscript. We should also like to thank Prof. M. H. Levitt and A. G. Redfield for their comments. We are also grateful to P. Carlson and D. Carmichael for invaluable assistance in the preparation of the manuscript and figures. This work was supported by the Miller Institute for Basic Research in Science, and the Director, Office of Energy Research, Office of Basic Energy Sciences, Materials Sciences Division of the U.S. Department of Energy under Contract No. DE-AC03-76SF00098.

REFERENCES

[1] A. Pines: in *Proc. S.I.F.*, Course C (North-Holland, Amsterdam, 1988), p. 43.
[2] A. Abragam: *Principles of Nuclear Magnetism* (Clarendon Press, Oxford, 1961).
[3] C. P. Slichter: *Principles of Nuclear Magnetic Resonance*, 3rd edition (Springer-Verlag, New York, N.Y., 1990).
[4] M. Mehring: *High Resolution NMR in Solids* (Springer-Verlag, New York, N.Y., 1983).
[5] R. R. Ernst, G. Bodenhausen and A. Wokaun: *Principles of Nuclear Magnetic Resonance in One and Two Dimensions* (Clarendon Press, Oxford, 1987).
[6] G. E. Pake: *J. Chem. Phys.*, **16**, 327 (1948).
[7] F. Creuzet, A. McDermott, R. Gebhard, K. Vander Hoef, M. B. Spijker-Assiak, J. Herzfeld, J. Lugtenberg, M. H. Levitt and R. G. Griffin: *Science*, **251**, 783 (1991).
[8] J. H. Ok, R. G. S. Spencer, A. E. Bennet and R. G. Griffin: *Chem. Phys. Lett.*, **197**, 389 (1992).
[9] A. Saupe: *Z. Naturforsch., Teil A*, **20**, 572 (1965).

[10] P. DIEHL and C. L. KHETRAPAL: *NMR Basic Principles and Progress*, **1**, 1 (1969).
[11] W. S. WARREN, D. P. WEITEKAMP and A. PINES: *J. Chem. Phys.*, **73**, 2084 (1980).
[12] S. YATSIV: *Phys. Rev.*, **113**, 1522 (1959).
[13] W. A. ANDERSON, R. FREEMAN and C. A. REILLEY: *J. Chem. Phys.*, **39**, 1518 (1963).
[14] J. I. MUSHER: *J. Chem. Phys.*, **40**, 983 (1964).
[15] G. BODENHAUSEN: *Prog. NMR Spectrosc.*, **14**, 137 (1981).
[16] D. WEITEKAMP: *Adv. Magn. Reson.*, **11**, 111 (1983).
[17] M. MUNOWITZ and A. PINES: *Science*, **233**, 525 (1986).
[18] M. MUNOWITZ and A. PINES: *Adv. Chem. Phys.*, **57**, 1 (1987).
[19] J. JEENER: in *Ampère International Summer School*, Basko Polje, Yugoslavia, 1971.
[20] W. G. BREILAND, C. B. HARRIS and A. PINES: *Phys. Rev. Lett.*, **30**, 158 (1973).
[21] J. B. MURDOCH, W. S. WARREN, D. P. WEITEKAMP and A. PINES: *J. Magn. Reson.*, **60**, 205 (1984).
[22] U. FANO: *Rev. Mod. Phys.*, **29**, 74 (1957).
[23] R. BALIAN: *From Microphysics to Macrophysics* (Springer-Verlag, Paris, 1991).
[24] J. JENEER: *Adv. Magn. Reson.*, **10**, 2 (1982).
[25] F. DYSON: *Phys. Rev.*, **75**, 486 (1949).
[26] I. J. LOWE and R. E. NORBERG: *Phys. Rev.*, **107**, 46 (1957).
[27] S. VEGA and A. PINES: *J. Chem. Phys.*, **66**, 5624 (1977).
[28] A. WOKAUN and R. R. ERNST: *J. Chem. Phys.*, **67**, 1752 (1977).
[29] S. VEGA: *J. Chem. Phys.*, **68**, 5518 (1978).
[30] O. W. SØRENSEN, G. W. EICH, M. H. LEVITT, G. BODENHAUSEN and R. R. ERNST: *Prog. NMR Spectrosc.*, **16**, 163 (1983).
[31] K. J. PACKER and K. M. WRIGHT: *Mol. Phys.*, **50**, 797 (1983).
[32] F. J. M. VAN DER VEN and C. W. HILBERS: *J. Magn. Reson.*, **54**, 512 (1983).
[33] H. KESSLER, M. GEHRKE and C. GRIESINGER: *Angew. Chem., Int. Ed. Engl.*, **27**, 490 (1988).
[34] B. C. SANCTUARY: *J. Chem. Phys.*, **64**, 4352 (1976).
[35] R. G. GORDON: *Adv. Magn. Reson.*, **3**, 1 (1968).
[36] E. L. HAHN: *Phys. Rev.*, **80**, 580 (1950).
[37] J. BAUM, M. MUNOWITZ, A. N. GARROWAY and A. PINES: *J. Chem. Phys.*, **83**, 2015 (1985).
[38] D. N. SHYKIND, J. BAUM, S.-B. LIU, A. PINES and A. N. GARROWAY: *J. Magn. Reson.*, **76**, 149 (1988).
[39] G. DROBNY, A. PINES, S. SINTON, D. P. WEITEKAMP and D. WEMMER: *Faraday Symp. Chem. Soc.*, **13**, 49 (1979).
[40] G. BODENHAUSEN, R. L. VOLD and R. R. VOLD: *J. Magn. Reson.*, **37**, 93 (1980).
[41] H. HATANAKA, T. TERAO and T. HASHI: *J. Phys. Soc. Jpn.*, **39**, 835 (1975).
[42] H. HATANAKA and T. HASHI: *J. Phys. Soc. Jpn.*, **39**, 1139 (1975).
[43] E. L. HAHN: in *Ampère International Summer School II, Pulsed Magnetic and Optical Resonance*, Basko Polje, Yugoslavia, 1971.
[44] R. MCNAMARA, C. H. WU and S. J. OPELLA: *J. Magn. Reson.*, **100**, 559 (1992).
[45] E. L. HAHN: *NMR Basic Principles and Progress*, **13**, 31 (1976).

[46] M. Gochin, K. V. Schenker, H. Zimmerman and A. Pines: *J. Am. Chem. Soc.*, **108**, 6813 (1986).
[47] G. Celebre, G. DeLuca, M. Longeri and J. W. Emsley: *J. Phys. Chem.*, **96**, 2466 (1992).
[48] L. DiBari, M. Persico and C. A. Veracini: *J. Chem. Phys.*, **96**, 4782 (1992).
[49] B. F. Chmelka, J. G. Pearson, S. B. Liu, R. Ryoo, L. C. de Menorval and A. Pines: *J. Phys. Chem.*, **95**, 303 (1991).
[50] J. Baum, K. K. Gleason, A. Pines, A. N. Garroway and J. Reimer: *Phys. Rev. Lett.*, **56**, 1377 (1986).
[51] D. H. Levy and K. K. Gleason: *J. Phys. Chem.*, **96**, 8125 (1992).
[51bis] S. Lacelle: *Adv. Magn. Opt. Res.*, **16**, 173 (1991).
[52] M. Munowitz, M. Mehring and A. Pines: *J. Chem. Phys.*, **86**, 3172 (1987).
[53] B. E. Scruggs and K. K. Gleason: *Chem. Phys.*, **166**, 367 (1992).
[54] B. E. Scruggs and K. K. Gleason: *J. Magn. Reson.*, **99**, 149 (1992).
[55] D. P. Weitekamp, J. R. Garbow, J. B. Murdoch and A. Pines: *J. Am. Chem. Soc.*, **103**, 3578 (1981).
[56] M. Gochin, D. P. Weitekamp and A. Pines: *J. Magn. Reson.*, **63**, 431 (1985).
[57] J. R. Garbow, D. P. Weitekamp and A. Pines: *Chem. Phys. Lett.*, **93**, 504 (1982).
[58] A. Bax: *J. Magn. Reson.*, **53**, 517 (1983).
[59] U. Haeberlen: *High Resolution NMR in Solids: Selective Averaging, Advances in Magnetic Resonance, Supplement 1* (Academic, New York, N.Y., 1976).
[60] W. Magnus: *Commun. Pure Appl. Math.*, **7**, 649 (1954).
[61] J. H. Van Vleck: *Phys. Rev.*, **33**, 467 (1929).
[62] N. Bloembergen, E. M. Purcell and V. Pound: *Phys. Rev.*, **73**, 679 (1948).
[63] M. Goldman, P. J. Grandinetti, A. Llor, Z. Olejniczak and J. R. Sachleben: *J. Chem. Phys.*, **97**, 8947 (1992).
[64] M. H. Levitt: *Prog. NMR Spectrosc.*, **18**, 61 (1986).
[65] A. J. Shaka and J. Keeler: *Prog. NMR Spectrosc.*, **19**, 47 (1987).
[66] R. Tycko: *Phys. Rev. Lett.*, **51**, 775 (1983).
[67] R. Tycko, J. Guckenheimer and A. Pines: *J. Chem. Phys.*, **83**, 2775 (1985).
[68] J. S. Waugh: *J. Magn. Reson.*, **49**, 517 (1982).
[69] Y. S. Yen and A. Pines: *J. Chem. Phys.*, **78**, 3579 (1983).
[70] H. Y. Carr and E. M. Purcell: *Phys. Rev.*, **94**, 630 (1954).
[71] E. R. Andrew, A. Bradbury and R. G. Eades: *Nature (London)*, **182**, 1659 (1958).
[72] E. O. Stejskal, J. Schaefer and R. A. McKay: *J. Magn. Reson.*, **25**, 569 (1977).
[73] J. S. Waugh, L. M. Huber and U. Haeberlen: *Phys. Rev. Lett.*, **20**, 180 (1968).
[74] C. H. Wang and J. D. Ramshaw: *Phys. Rev. B*, **6**, 3253 (1972).
[75] P. Mansfield: *Phys. Lett. A*, **32**, 485 (1970).
[76] P. Mansfield: *J. Phys. C*, **4**, 1444 (1971).
[77] W. K. Rhim, D. D. Elleman and R. W. Vaughan: *J. Chem. Phys.*, **58**, 1772 (1972).
[78] D. P. Burum, M. Linder and R. R. Ernst: *J. Magn. Reson.*, **44**, 173 (1981).
[79] D. P. Burum and W. K. Rhim: *J. Chem. Phys.*, **71**, 944 (1979).
[80] D. G. Cory: *J. Magn. Reson.*, **94**, 526 (1991).
[81] C. A. Fyfe: *Solid State NMR for Chemists* (CFC Press, Guelph, Ont., 1983).
[82] B. C. Gerstein and C. R. Dybowski: *Transient Techniques in NMR of Solids* (Academic Press, Orlando, Fla., 1985).

[83] M. M. MARICQ and J. S. WAUGH: *J. Chem. Phys.*, **70**, 3300 (1979).
[84] S. F. DEC, R. A. WIND, G. E. MACIEL and F. E. ANTHONIO: *J. Magn. Reson.*, **70**, 355 (1986).
[85] J. ACKERMAN, R. ECKMAN and A. PINES: *Chem. Phys.*, **42**, 423 (1979).
[86] H. J. JAKOBSEN, J. SKIBSTED, H. BILDSØE and N. C. NIELSEN: *J. Magn. Reson.*, **85**, 173 (1989).
[87] R. K. HARRIS and P. JACKSON: *Chem. Rev.*, **91**, 1427 (1991).
[88] A. JURKIEWICZ, C. E. BRONNIMAN and G. E. MACIEL: *Fuel*, **69**, 804 (1990).
[89] S. GANAPATHY, S. SCHRAMM and E. OLDFIELD: *J. Chem. Phys.*, **77**, 4360 (1982).
[90] S. GANAPATHY, J. SHORE and E. OLDFIELD: *Chem. Phys. Lett.*, **169**, 301 (1990).
[91] B. F. CHMELKA and A. PINES: *Science*, **246**, 71 (1989).
[92] A. LLOR and J. VIRLET: *Chem. Phys. Lett.*, **152**, 248 (1988).
[93] A. SAMOSON, E. LIPPMAA and A. PINES: *Mol. Phys.*, **65**, 1013 (1988).
[94] B. Q. SUN, J. H. BALTISBERGER, Y. WU, A. SAMOSON and A. PINES: *Solid State NMR*, **1**, 267 (1992).
[95] Y. WU, B. F. CHMELKA, A. PINES, M. E. DAVIS, P. J. GROBET and P. A. JACOBS: *Nature (London)*, **346**, 6284 (1990).
[96] J. S. WAUGH: *J. Magn. Reson.*, **50**, 30 (1982).
[97] M. H. LEVITT and R. FREEMAN: *J. Magn. Reson.*, **43**, 502 (1981).
[98] M. H. LEVITT, R. FREEMAN and T. FRENKIEL: *J. Magn. Reson.*, **47**, 328 (1982).
[99] A. J. SHAKA, J. KEELER, T. FRENKIEL and R. FREEMAN: *J. Magn. Reson.*, **52**, 335 (1983).
[100] A. J. SHAKA, J. KEELER and R. FREEMAN: *J. Magn. Reson.*, **53**, 313 (1983).
[101] L. C. SNYDER and S. MEIBOOM: *J. Chem. Phys.*, **58**, 5096 (1973).
[102] A. PINES, D. J. RUBEN, S. VEGA and M. MEHRING: *Phys. Rev. Lett.*, **36**, 110 (1976).
[103] C. COHEN-TANNOUDJI, B. DIU and F. LALOË: *Quantum Mechanics* (Wiley, Paris, 1977).
[104] R. C. HEWITT, S. MEIBOOM and L. C. SNYDER: *J. Chem. Phys.*, **58**, 5089 (1973).
[105] J. W. EMSLEY, J. C. LINDON and J. M. TABONY: *J. Chem. Soc. Faraday Trans.*, **269**, 10 (1973).
[106] A. PINES, S. VEGA and M. MEHRING: *Phys. Rev. B*, **18**, 112 (1978).
[107] R. J. WITTEBORT, M. G. USHA, D. J. RUBEN, D. E. WEMMER and A. PINES: *J. Am. Chem. Soc.*, **110**, 5668 (1988).
[108] L. ALLEN and J. H. EBERLY: *Optical Resonance and Two-Level Atoms* (Wiley, New York, N.Y., 1975).
[109] R. P. FEYNMAN, F. L. VERNON and R. W. HELLWARTH: *J. Appl. Phys.*, **28**, 49 (1957).
[110] E. T. JAYNES and F. W. CUMMINGS: *Proc. I.R.E.*, **51**, 89 (1963).
[111] J. J. SAKURAI: *Modern Quantum Mechanics* (Benjamin-Cummings, Menlo Park, Cal., 1985).
[112] W. H. LOUISELL: *Quantum Statistical Properties of Radiation* (Wiley, New York, N.Y., 1973).
[113] I. I. RABI, J. R. ZACHARIAS, S. MILLMAN and P. KUSH: *Phys. Rev.*, **53**, 318 (1938).
[114] Y. AHARONOV and L. SUSSKIND: *Phys. Rev.*, **158**, 1237 (1967).
[115] M. P. SILVERMAN: *Eur. J. Phys.*, **1**, 116 (1980).

[116] S. HAROCHE: *Ann. Phys. (N.Y.)*, **6**, 327 (1971).
[117] A. G. REDFIELD: *Adv. Magn. Reson.*, **1**, 1 (1965).
[118] J. BAUM, R. TYCKO and A. PINES: *Phys. Rev. A*, **32**, 3435 (1985).
[119] M. ALEXANDER: private communication.
[120] J. C. S. JOHNSON: *Adv. Magn. Reson.*, **1**, 223 (1965).
[121] A. HAGEMEYER, K. SCHMIDT-ROHR and H. W. SPIESS: *Adv. Magn. Reson.*, **13**, 85 (1989).
[122] P. RIGNY: *Physica*, **59**, 707 (1971).
[123] S. ALEXANDER, A. BARAM and Z. LUZ: *Mol. Phys.*, **27**, 441 (1974).
[124] A. BARAM, Z. LUZ and S. ALEXANDER: *J. Chem. Phys.*, **64**, 4321 (1976).
[125] D. E. WEMMER: Ph. D. Thesis, University of California, Berkeley, 1978.
[126] R. K. WANGSNESS and F. BLOCH: *Phys. Rev.*, **89**, 728 (1953).
[127] H. S. GUTOWSKY, D. W. MCCALL and C. P. SLICHTER: *J. Chem. Phys.*, **21**, 279 (1953).
[128] H. S. GUTOWSKY and C. HOLM: *J. Chem. Phys.*, **25**, 1228 (1956).
[129] H. M. MCCONNELL: *J. Chem. Phys.*, **28**, 430 (1958).
[130] M. HAMMERMESH: *Group Theory and Its Application to Physical Problems* (Dover, New York, N.Y., 1962).
[131] L. G. WERBELOW and D. M. GRANT: *Adv. Magn. Reson.*, **9**, 189 (1977).
[132] D. E. WEMMER, D. J. RUBEN and A. PINES: *J. Am. Chem. Soc.*, **103**, 28 (1981).
[133] J. WANG and P. D. ELLIS: *J. Am. Chem. Soc.*, **115**, 212 (1993).
[134] R. D. JOHNSON, D. S. BETHUNE and C. S. YANNONI: *Acc. Chem. Res.*, **25**, 169 (1992).
[135] S. E. BARRETT and R. TYCKO: *Phys. Rev. Lett.*, **69**, 3754 (1992).
[136] P. J. STEINHARDT, D. R. NELSON and R. RONCHETTI: *Phys. Rev. B*, **28**, 784 (1983).
[137] A. BAX, N. M. SZEVERENYI and G. MACIEL: *J. Magn. Reson.*, **52**, 147 (1983).
[138] Z. H. GAN: *J. Am. Chem. Soc.*, **114**, 8307 (1992).
[139] K. T. MUELLER, A. SAMOSON, B. Q. SUN, G. C. CHINGAS, J. W. ZWANZIGER, T. TERAO and A. PINES: *J. Magn. Reson.*, **86**, 470 (1990).
[140] J. H. BALTISBERGER, S. L. GANN, E. W. WOOTEN, T. H. CHANG, K. T. MUELLER and A. PINES: *J. Am. Chem. Soc.*, **114**, 7489 (1992).
[141] A. BAX, N. M. SZEVERENYI and G. MACIEL: *J. Magn. Reson.*, **55**, 494 (1983).
[142] R. TYCKO, G. DABBAGH and P. A. MIRAU: *J. Magn. Reson.*, **85**, 265 (1989).
[143] I. FARNAN, P. J. GRANDINETTI, J. H. BALTISBERGER, J. F. STEBBINS, U. WERNER, M. A. EASTMAN and A. PINES: *Nature (London)*, **358**, 31 (1992).
[144] R. DUPREE and R. F. PETTIFER: *Nature (London)*, **308**, 523 (1984).
[145] L. FRYDMAN, G. C. CHINGAS, Y. K. LEE, P. J. GRANDINETTI, M. A. EASTMAN, G. A. BARRALL and A. PINES: *J. Chem. Phys.*, **97**, 4800 (1992).
[146] L. FRYDMAN, Y. K. LEE, L. EMSLEY, G. C. CHINGAS and A. PINES: *J. Am. Chem. Soc.*, **115**, 4825 (1993).
[147] M. GOLDMAN: *Spin Temperature and Nuclear Magnetic Resonance in Solids* (Clarendon Press, Oxford, 1970).
[148] J. JEENER: *Adv. Magn. Reson.*, **3**, 205 (1968).
[149] S. R. HARTMANN and E. L. HAHN: *Phys. Rev.*, **128**, 2042 (1962).
[150] A. PINES, M. G. GIBBY and J. S. WAUGH: *J. Chem. Phys.*, **59**, 569 (1973).
[151] A. G. ANDERSON and S. R. HARTMANN: *Phys. Rev.*, **128**, 2023 (1962).
[152] M. H. LEVITT: in *Pulsed Magnetic Resonance: NMR, ESR, and Optics*, edited by D. M. S. BAGGULEY (Clarendon Press, Oxford, 1992), p. 184.
[153] M. H. LEVITT, D. SUTER and R. R. ERNST: *J. Chem. Phys.*, **84**, 4243 (1986).

[154] J. R. Franz and C. P. Slichter: *Phys. Rev.*, **148**, 287 (1966).
[155] L. Müller, A. Kumar, T. Baumann and R. R. Ernst: *Phys. Rev. Lett.*, **32**, 1402 (1974).
[156] O. W. Sørensen: *Prog. NMR Spectrosc.*, **21**, 503 (1989).
[157] N. C. Nielsen, H. Bildsøe, H. J. Jakobsen and O. W. Sørensen: *J. Magn. Reson.*, **85**, 359 (1989).
[158] O. W. Sørensen: *J. Magn. Reson.*, **86**, 435 (1990).
[159] O. W. Sørensen: *J. Magn. Reson.*, **93**, 648 (1991).
[160] A. G. Redfield: *J. Magn. Reson.*, **92**, 642 (1991).
[161] M. H. Levitt: *J. Magn. Reson.*, **99**, 1 (1992).
[162] G. Strang: *Linear Algebra and its Applications*, 3rd edition (Harcourt Brace Jovanovich, San Diego, Cal., 1988).
[163] P. L. Corio: *Structure of High-Resolution NMR Spectra* (Academic Press, New York, N.Y., 1966).
[164] R. L. Vold and R. R. Vold: *Prog. NMR Spectrosc.*, **12**, 79 (1978).
[165] T. P. Das and E. L. Hahn: *Solid State Phys. Suppl.*, **1**, 18 (1958).
[166] N. F. Ramsey and R. V. Pound: *Phys. Rev.*, **81**, 278 (1951).
[167] R. L. Strombotne and E. L. Hahn: *Phys. Rev. A*, **133**, 1616 (1964).
[168] R. Blinc: *Adv. Nuc. Quad. Reson.*, **2**, 71 (1975).
[169] D. T. Edmonds: *Phys. Rep.*, **29**, 233 (1977).
[170] D. P. Weitekamp, A. Bielecki, D. Zax, K. Zilm and A. Pines: *Phys. Rev. Lett.*, **50**, 1807 (1983).
[171] D. Zax, A. Bielecki, K. Zilm, A. Pines and D. P. Weitekamp: *J. Chem. Phys.*, **83**, 4877 (1985).
[172] J. W. Hennel, A. Birczynski, S. F. Sagnowski and M. Stachurowa: *Z. Phys. B*, **56**, 133 (1984).
[173] R. Blinc and J. Seliger: *Z. Naturforsch.*, **47**, 333 (1992).
[174] A. M. Thayer and A. Pines: *Acc. Chem. Res.*, **20**, 47 (1987).
[175] J. M. Millar, A. M. Thayer, H. Zimmerman and A. Pines: *J. Magn. Reson.*, **69**, 243 (1986).
[176] A. Bielecki, J. B. Murdoch, D. P. Weitekamp, D. B. Zax, K. W. Zilm, H. Zimmerman and A. Pines: *J. Chem. Phys.*, **80**, 2232 (1984).
[177] J. M. Millar, A. M. Thayer, A. Bielecki, D. B. Zax and A. Pines: *J. Chem. Phys.*, **83**, 934 (1985).
[178] C. Connor, J. W. Chang and A. Pines: *J. Chem. Phys.*, **93**, 7639 (1990).
[179] D. Zax, A. Bielecki, A. Pines and S. W. Sinton: *Nature (London)*, **312**, 351 (1984).
[180] A. M. Thayer, J. M. Millar and A. Pines: *Chem. Phys. Lett.*, **129**, 55 (1986).
[181] R. Kreis, D. Suter and R. R. Ernst: *Chem. Phys. Lett.*, **118**, 120 (1985).
[182] R. Tycko: *J. Magn. Reson.*, **75**, 193 (1987).
[183] R. Tycko: *Phys. Rev. Lett.*, **60**, 2734 (1988).
[184] R. Tycko: *J. Chem. Phys.*, **192**, 5776 (1990).
[185] A. Llor, Z. Olejniczak, J. Sachleben and A. Pines: *Phys. Rev. Lett.*, **67**, 1989 (1991).
[186] A. Llor, Z. Olejniczak and A. Pines: *Phys. Rev. B*, in press (1993).
[187] Y. A. Serebrennikov and M. I. Majitov: *Chem. Phys. Lett.*, **157**, 462 (1989); Y. A. Serebrennikov: *Adv. Magn. Reson.*, **17**, 47 (1992).
[188] P. Jonsen, M. Luzar, A. Pines and M. Mehring: *J. Chem. Phys.*, **85**, 4873 (1986).
[189] E. P. Day: *Phys. Rev. Lett.*, **29**, 540 (1972).

[190] J. CHANG, C. CONNOR, E. L. HAHN, H. HUBER and A. PINES: *J. Magn. Reson.*, **82**, 387 (1989).
[191] C. CONNOR, J. CHANG and A. PINES: *Rev. Sci. Instrum.*, **61**, 1059 (1990).
[192] M. D. HURLIMAN, C. H. PENNINGTON, N. Q. FAN, J. CLARKE, A. PINES and E. L. HAHN: *Phys. Rev. Lett.*, **69**, 684 (1992).
[193] M. V. BERRY: *Proc. R. Soc. London, Ser. A*, **392**, 45 (1984).
[194] B. SIMON: *Phys. Rev. Lett.*, **51**, 2167 (1983).
[195] Y. AHARONOV and J. ANANDAN: *Phys. Rev. Lett.*, **58**, 1593 (1987).
[196] A. SHAPERE and F. WILCZEK: *Geometric Phases in Physics* (World Scientific, Singapore, 1989).
[197] J. W. ZWANZIGER, M. KOENIG and A. PINES: *Annu. Rev. Phys. Chem.*, **41**, 601 (1990).
[198] K. MORIYASU: *An Elementary Primer for Gauge Theory* (World Scientific, Singapore, 1983).
[199] N. NASH and S. SEN: *Topology and Geometry for Physicists* (Academic Press, London, 1983).
[200] Y. AHARONOV and D. BOHM: *Phys. Rev.*, **115**, 485 (1959).
[201] R. P. FEYNMAN, R. B. LEIGHTON and M. SANDS: *The Feynman Lectures on Physics* (Addison-Wesley, Reading, Mass., 1963).
[202] G. HERZBERG and H. C. LONGUET-HIGGINS: *Dicuss. Faraday Soc.*, **35**, 77 (1963).
[203] C. A. MEAD and D. G. TRUHLAR: *J. Chem. Phys.*, **70**, 2284 (1979).
[204] G. DELACRÉTAZ, E. R. GRANT, R. L. WHETTEN, L. WÖSTE and J. W. ZWANZIGER: *Phys. Rev. Lett.*, **56**, 2598 (1986).
[205] R. BHANDARI: *Phys. Lett. A*, **133**, 1 (1988).
[206] A. SHAPERE and F. WILCZEK: *J. Fluid Mech.*, **198**, 557 (1987).
[207] R. MONTGOMERY: *Commun. Math. Phys.*, **120**, 269 (1988).
[208] C. W. MISNER, K. S. THORNE and J. A. WHEELER: *Gravitation* (W. H. Freeman and Co., New York, N.Y., 1973).
[209] H. GOLDSTEIN: *Classical Mechanics* (Addison-Wesley, Reading, Mass., 1950).
[210] G. BAYM: *Lectures on Quantum Mechanics* (Benjamin-Cummings, Reading, Mass., 1973).
[211] F. WILCZEK and A. ZEE: *Phys. Rev. Lett.*, **52**, 2111 (1984).
[212] A. ZEE: *Phys. Rev. A*, **38**, 1 (1988).
[213] R. TYCKO: *Phys. Rev. Lett.*, **58**, 2281 (1987).
[214] C. N. YANG: *Ann. N.Y. Acad. Sci.*, **294**, 86 (1977).
[215] S. COLEMAN: *Les Houches 1981: Session 37 Gauge Theories in High Energy Physics* (North-Holland, Amsterdam, 1983).
[216] P. A. M. DIRAC: *Proc. R. Soc. London, Ser. A*, **133**, 60 (1931).
[217] M. E. STOLL, A. G. VEGA and R. W. VAUGHAN: *Phys. Rev. A*, **16**, 1521 (1977).
[218] D. SUTER, A. PINES and M. MEHRING: *Phys. Rev. Lett.*, **57**, 242 (1986).
[219] S. PANCHARATNAM: *Proc. Ind. Acad. Sci.*, **44**, 247 (1956).
[220] J. SAMUEL and R. BHANDARI: *Phys. Rev. Lett.*, **60**, 2339 (1988).
[221] W. GERLACH and O. STERN: *Ann. Phys. (Leipzig)*, **74**, 673 (1924).
[222] R. CHIAO and Y. WU: *Phys. Rev. Lett.*, **57**, 933 (1986).
[223] R. CHIAO, A. ANTARAMIAN, K. M. GANGA, H. JIAO, S. R. WILKINSON and H. NATHEL: *Phys. Rev. Lett.*, **60**, 1214 (1988).
[224] C. BOUCHIAT and G. W. GIBBONS: *J. Phys. Paris*, **49**, 187 (1988).

A Guided Tour through Double-Resonance Phenomena.

M. MEHRING

2. Physikalisches Institut, Universität Stuttgart - D 70550 Stuttgart 80, B.R.D.

1. – Introduction.

Soon after the discovery of the magnetic-resonance phenomenon it was realized that double resonance could enhance the resolution and sensitivity of magnetic-resonance spectroscopy tremendously[1,2]. It is, therefore, not surprising that most of the fascination of nuclear magnetic resonance (NMR), electron spin resonance (ESR) and optical detection of magnetic resonance (ODMR) derives from the ingenious invention of various double-resonance techniques. A general discussion on magnetic-resonance experiments involving double resonance can be found in the standard textbooks by ABRAGAM and SLICHTER [3, 4]. I will make extensive use of the spin temperature concept and the spin density matrix formalism. Both are discussed very thoroughly in the book of Goldman[5]. Part of the discussion presented here can be found in more detail in [6].

The basic principle is rather simple. Consider a multilevel quantum system as shown in fig. 1 which may be generated from a single high-spin system, *e.g.* due to quadrupolar interaction or a coupled spin system, consisting of a number of coupled spins 1/2. Double resonance in general involves the application of two different radiofrequency (r.f.), microwave (mw) or light fields (lf), with frequencies close to specific quantum transitions. In fig. 1 I have depicted three different connected levels for this purpose labelled $|1\rangle$, $|2\rangle$ and $|3\rangle$. The transition frequencies ω_{12} and ω_{23} with $\omega_{12} \neq \omega_{23}$ correspond to $\Delta m = 1$ (allowed) transitions, whereas ω_{13} would correspond to a $\Delta m = 2$ (forbidden) transition. Since we are discussing magnetic-resonance phenomena here, let us interpret these level splittings as caused by a static magnetic field B_0 which leads, by introduc-

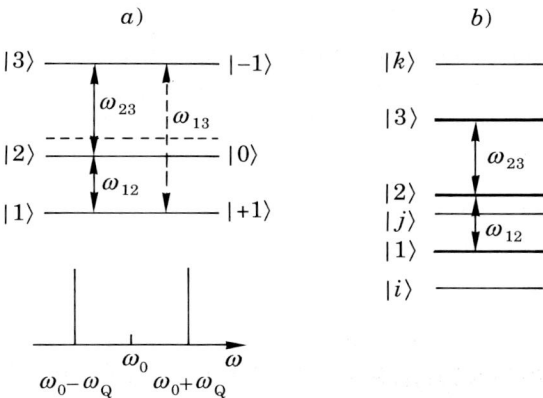

Fig. 1. – a) Three-level system and the resulting allowed (solid arrows) and forbidden (dashed arrows) transitions. The expected two-line spectrum is shown at the bottom, where ω_0 is the centre of the spectrum and the splitting is caused by an interaction of strength ω_Q discussed in the text. b) Multilevel system with three particular levels bolded.

ing effective gyromagnetic ratios γ_{eff}, to the resonance frequencies

$$(1) \qquad \omega_{12} = \gamma_{1\text{eff}} B_0 \quad \text{and} \quad \omega_{23} = \gamma_{2\text{eff}} B_0 \,.$$

Since $\omega_{12} \neq \omega_{23}$ energy is not conserved when mutual flip-flop transitions occur. The «transition rate», *i.e.* the Rabi precession frequency caused by the radiation field, will be labelled $\omega_1 = \gamma_{\text{eff}} B_1$ throughout, where the proper rotating component of the field is chosen which must be close to the transition frequency. In the laboratory frame the radiation field will, therefore, be expressed as

$$(2) \qquad \omega(t) = \omega_1 (\exp[-i\omega t] + \exp[+i\omega t]) \,,$$

where ω is considered to be close to some transition frequency ω_{jk}. Counterrotating components and far off-resonance fields will usually be neglected. Double-resonance phenomena are conveniently discussed in the appropriate rotating reference frames, as will be done in the following.

The motion of spins (or equivalently the coherence of two-level subsystems) under the action of fields represented by the Hamilton operator $\mathscr{H}$ will be described by the Liouville-von Neumann equation

$$(3) \qquad \frac{d\rho}{dt} = -i[\mathscr{H}, \rho] \,,$$

where the spin density matrix ρ in thermodynamic equilibrium is expressed as [5]

$$(4) \qquad \rho = \frac{\exp[-\beta\mathscr{H}]}{\text{Tr}\{\exp[-\beta\mathscr{H}]\}} \qquad \text{with } \text{Tr}\{\rho\} = 1$$

and $\beta = \hbar/k_B T$, which is called the inverse spin temperature but can be viewed equivalently as imaginary time. The expectation value of any operator A can be calculated from

$$\langle A \rangle = \text{Tr}\{A\rho\} \quad \text{with } \text{Tr}\{A\} = 0 \text{ for spin operators}. \tag{5}$$

For the spin systems to be discussed here it is legitimate to invoke the «high-temperature approximation» (HTA), *i.e.*

$$\langle A \rangle \simeq \frac{\text{Tr}\{A(1 - \beta \mathcal{H} + ...)\}}{\text{Tr}\{1\}}, \tag{6}$$

where $\text{Tr}\{1\}$ equals the number of quantum states or to be specific $\text{Tr}\{1\} = (2I+1)^{N_I}(2S+1)^{N_S}$ if N_I I-spins and N_S S-spins are involved. Because of the vanishing trace of spin operators (eq. (5)) I will use in the following the truncated traceless spin density matrix:

$$\rho = -\beta \frac{\mathcal{H}}{\text{Tr}\{1\}}. \tag{7}$$

Since $\mathcal{H}$ in eq. (7) consists of a sum of spin operators, I will loosely speak of the spin density matrix even if only the spin operator in the density matrix is quoted and the prefactors are ignored. Let us consider as an example the Boltzmann equilibrium of N_I I-spins, with the Zeeman Hamiltonian $\mathcal{H} = -\omega_0 I_z$

$$\rho_B = \frac{\beta \omega_0}{(2I+1)^{N_I}} I_z \sim I_z \tag{8}$$

which is the initial density matrix for any magnetic-resonance experiment involving I spins in large magnetic fields.

After having established these prerequisites, we can begin discussing some double-resonance experiments.

2. – Double resonance in three- and multi-level systems.

In this section I will closely follow the treatment outlined in [7]. Let us first consider the three-level system sketched in fig. 1a). Its origin may be a nuclear spin $I = 1$ or an electron spin $S = 1$ (triplet state) which can be expressed by the Hamiltonian

$$\mathcal{H}_0 = -\omega_0 I_z + \omega_Q (I_z^2 - I(I+1)/3), \tag{9}$$

where the allowed transitions ($\Delta m = 1$) can be readily worked out and result in a two-line spectrum at $\omega_{12} = \omega_0 - \Delta\omega$ and $\omega_{23} = \omega_0 + \Delta\omega$. Since we are about to discuss the double-resonance dynamics of any three-level system picked out from a multilevel system as sketched in fig. 1b), I propose to use the fictitious spin-1/2 operators as introduced in [3, 8, 9]. Here we use the following notation:

$$ (10a) \qquad I_x^{r\text{-}s} = \frac{1}{2}\{|r\rangle\langle s| + |s\rangle\langle r|\}, $$

$$ (10b) \qquad I_y^{r\text{-}s} = \frac{-i}{2}\{|r\rangle\langle s| - |s\rangle\langle r|\}, $$

$$ (10c) \qquad I_z^{r\text{-}s} = \frac{1}{2}\{|r\rangle\langle r| - |s\rangle\langle s|\}. $$

The corresponding commutators and related rules can be found in [8, 9] and in [6]. In terms of these fictitious spin-1/2 operators the Hamiltonian in eq. (9) can be expressed in the rotating frame (at frequency ω) as

$$ (11) \qquad \mathcal{H}_0 = -(\omega_0 - \omega)(I_z^{1\text{-}2} + I_z^{2\text{-}3}) + \frac{2}{3}\omega_Q(I_z^{1\text{-}2} - I_z^{2\text{-}3}), $$

which must be complemented by the radiation Hamiltonian

$$ (12) \qquad \mathcal{H}_1 = -\sqrt{2}\,\omega_1(I_{x,y}^{1\text{-}2} + I_{x,y}^{2\text{-}3}) $$

in the same rotating frame, where the phase of the radiation can be chosen appropriately in order to irradiate in the x- or y-direction or any other direction of the rotating frame. To be more specific, let us irradiate near the 2-3 satellite transition at frequency $\omega = \omega_0 + \omega_Q - \delta\omega$. If $\delta\omega, \omega_1 \ll \omega_Q$, it can be shown that the total Hamiltonian in the rotating frame can be expressed as

$$ (13) \qquad \mathcal{H} = \mathcal{H}_0 + \mathcal{H}_1 = -\delta\omega I_z^{2\text{-}3} - \sqrt{2}\,\omega_1 I_{x,y}^{2\text{-}3}. $$

The factor $\sqrt{2}$ in eqs. (12), (13) is specific to a spin $I = 1$. It would be different in another multilevel system and is given simply by the transition matrix element. In order to be more general I will use $\mathcal{H}_1 = -\omega_{1e} I_{x,y}^{2\text{-}3}$ for the effective Rabi frequency with $\omega_{1e} = \sqrt{2}\,\omega_1$ in case of a selective spin-1 transition. We are now well equipped to perform selective spin rotation experiments by the application of pulses. The acquired rotation angle after the application of the pulse of duration t_p is given by $\alpha = \omega_{1e} t_p$.

2˙1. *Population transfer.* – Consider a Boltzmann population as sketched in fig. 2a). The population of a level $|j\rangle$ of a multilevel system can be expressed as

$$ (14a) \qquad p_j = \langle j|\rho_B|j\rangle, $$

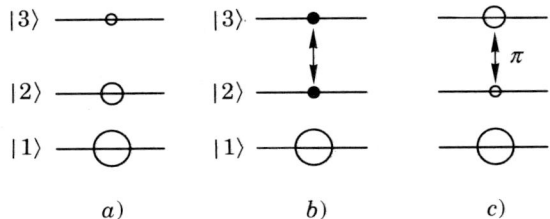

Fig. 2. – Different population scenarios of the three-level system as discussed in the text.

which results in

$$(14b) \qquad p_j = \frac{1}{3}(1 - \beta E_j) \qquad \text{with } p_1 + p_2 + p_3 = 1$$

in the case of a three-level system. The signal intensity at a particular transition r-s is proportional to the z-magnetization

$$(15a) \qquad M_z^{r\text{-}s} = \gamma_{\text{eff}}^{r\text{-}s} \hbar \langle I_z^{r\text{-}s} \rangle = \gamma_{\text{eff}}^{r\text{-}s} \hbar \frac{p_r - p_s}{2},$$

which is readily obtained as

$$(15b) \qquad M_z^{1\text{-}2} = \gamma_{\text{eff}}^{1\text{-}2} \hbar \frac{\beta_{12} \omega_{12}}{6},$$

$$(15c) \qquad M_z^{2\text{-}3} = \gamma_{\text{eff}}^{2\text{-}3} \hbar \frac{\beta_{23} \omega_{23}}{6},$$

in the case of a three-level system.

2˙1.1. *Saturation of the 2-3 transition.* As a very simple double-resonance experiment we saturate the 2-3 transition, *i.e.* we make $p_2 = p_3$ or equivalently $\beta_{23} = 0$ (infinite spin temperature) resulting in $M_z^{2\text{-}3} = 0$. At the same time the signal at the 1-2 transition increases to the value

$$(16a) \qquad M_z'^{1\text{-}2} = \gamma_{\text{eff}}^{1\text{-}2} \hbar \frac{\beta_{12}}{6} \left(\omega_{12} + \frac{\omega_{23}}{2} \right),$$

which can be expressed as

$$(16b) \qquad M_z'^{1\text{-}2} = M_z^{1\text{-}2} \left(1 + \frac{\omega_{23}}{2\omega_{12}} \right),$$

and where the enhancement factor reflects the enhancement in inverse spin temperature of the 1-2 fictitious spin 1/2.

2'1.2. *Inversion of the 2-3 transition.* Figure 2c) shows the population of the levels after the application of a π-pulse at the 2-3 transition. It inverts the population of the levels 2 and 3 with the effect that $M_z'^{2-3} = -M_z^{2-3}$. At the same time the 1-2 signal is enhanced due to the increase in its inverse spin temperature. A simple calculation leads to

$$(17) \qquad M_z'^{1-2} = M_z^{1-2}\left(1 + \frac{\omega_{23}}{\omega_{12}}\right),$$

which corresponds to an enhancement by the factor 2 for $\omega_{23} \sim \omega_{12}$.

2'2. *Spin alignment and pulsed ENDOR experiments.* – The application of a selective π-pulse at the 2-3 transition as demonstrated in the previous subsec-

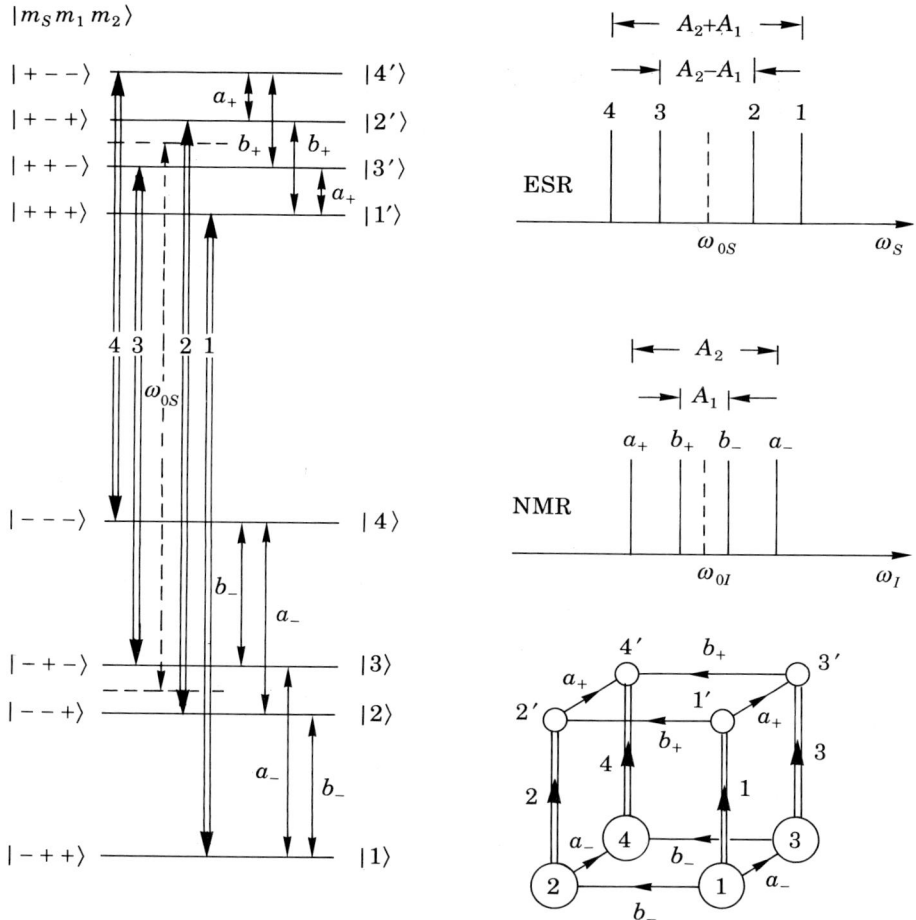

Fig. 3. – Eight-level diagram caused by the two nuclear spins 1/2 coupled to an electron spin 1/2, together with the expected ESR and NMR spectra. The initial «population cube» is drawn schematically in the right-hand corner (according to [10]).

tion created what is called «spin alignment». The total spin polarization is zero after the pulse. Nevertheless there is still a high degree of order in the spin system as will be discussed now. Let us start with the truncated Boltzmann spin density matrix

(18) $$\rho_B = \beta_0 \left\{ 2\omega_0(I_z^{1\text{-}2} + I_z^{2\text{-}3}) + \frac{2}{3}\omega_Q(I_z^{1\text{-}2} - I_z^{2\text{-}3}) \right\}/3,$$

where the dominant term is the Zeeman term because $\omega_0 \gg \omega_Q$ is assumed. We say the system has mostly Zeeman order. The quadrupolar term can be ignored. After the application of a selective π-pulse, however, the density matrix becomes, retaining only the Zeeman term in eq. (18),

(19) $$\rho_\pi = \beta_0 \frac{2}{3}\omega_0 (I_z^{1\text{-}2} + I_y^{2\text{-}3})$$

resulting in a partial transfer of Zeeman order into quadrupolar (or tensorial) order and some off-diagonal order (coherence). It is now straightforward to calculate the 1-2 signal strength $M_z^{1\text{-}2} = \gamma_{\text{eff}} \hbar \langle I_z^{1\text{-}2} \rangle$ by using eq. (19) with the result $M_z'^{1\text{-}2} = 2M_z^{1\text{-}2}$, taking into account that $\text{Tr}\{(I_z^{1\text{-}2})^2\} = 1/2$ and $\text{Tr}\{I_z^{1\text{-}2} I_z^{2\text{-}3}\} = -1/4$.

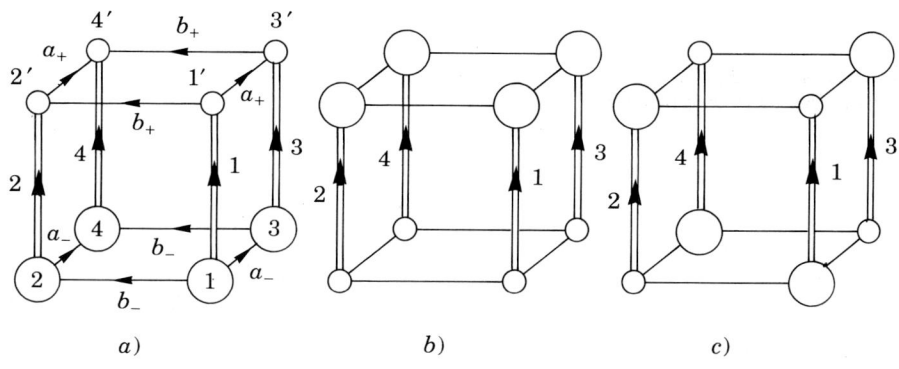

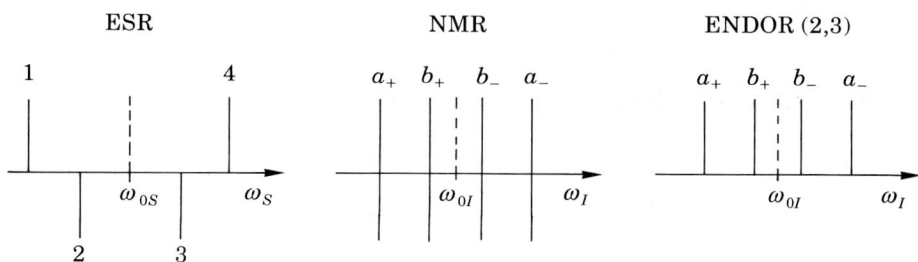

Fig. 4. – Top: Different states of the population cube after application of a hard (b)) and a soft (c)) π-pulse. The ESR, NMR and ENDOR spectra expected after a soft π-pulse are sketched at the bottom (according to [10]).

We note that the selective π-pulse can enhance certain transitions, whereas others are inverted. Nevertheless it can be used to perform a double-resonance experiment in the sense that sublevel transitions are detected by observing the corresponding change in the main-level populations. This has been utilized in electron nuclear double-resonance (ENDOR) experiments[11, 12] as is outlined in the following. For a review on this subject the reader is referred to[10] and the original literature cited there.

Consider an electron spin $S = 1/2$ coupled to two nuclear spins $I = 1/2$. In fig. 3 the level scheme together with population of the different levels is sketched in form of a «population cube» [10]. The vertical transitions with large splittings occur at the ESR frequency (typically 9 GHz) leading to the (four-line) ESR spectrum, whereas the nuclear transitions (smaller splittings, about 5 MHz) are drawn horizontally but with a perspective view. The spheres (circles) represent the population. We start with a higher population in the lower plane (Boltzmann equilibrium), leading to strong ESR, but weak NMR signals. A hard π-pulse (covering the total bandwidth) at the ESR transition inverts the population and the ESR signal, but no change in the NMR signal occurs. Only when a soft π-pulse (selective, *e.g.* at the centre of the ESR spectrum) is applied

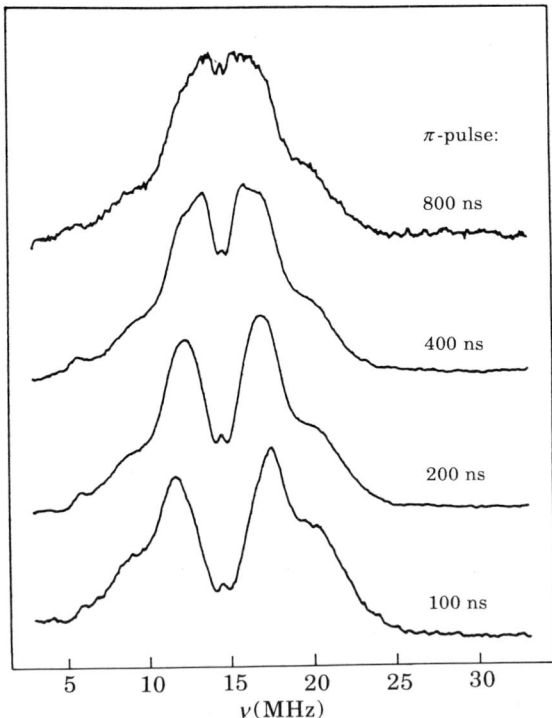

Fig. 5. – ENDOR spectra of polyacetylene obtained after different soft π-pulses of different duration applied at the centre of the ESR transition (according to [10]).

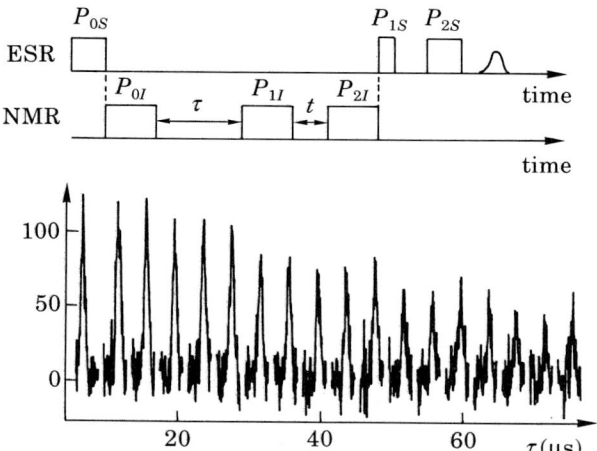

Fig. 6. – Sublevel (NMR) spin echoes detected indirectly at the ESR frequency (according to [10]).

(see fig. 4) the central ESR transitions are inverted and a spin alignment is introduced into the nuclear levels (fig. 4 bottom). Note that their total magnetization is zero, nevertheless a high degree of order exists in the nuclear-level system. If we now sweep with a saturating field of r.f. pulses through the spectral range of the NMR (ENDOR) spectrum, the nuclear-spin alignment is destroyed when we hit an NMR transition, leading to a recovery of the inverted central part of the ESR spectrum (fig. 4 bottom). In this type of ENDOR spectroscopy the increase of the inverted part of the ESR is recorded while sweeping through the NMR frequency range. An example is shown in fig. 5, where a continuum of NMR frequencies is present due to the anisotropy of the hyperfine interaction and the lack of orientation in the sample. We also note that a «blind spot» occurs in the centre of the spectrum depending on the spectral width of the π-pulse. This is readily understood since for weak hyperfine interaction even the soft π-pulse acts as a hard pulse.

A number of pulsed ENDOR double-resonance experiments have been performed by applying these ideas. This includes sublevel coherence spectroscopy demonstrated by exciting sublevel spin echoes [10, 13] with an echo sequence which includes the transfer of population to the ESR transition as shown in fig. 6. In addition multiquantum ENDOR introduced first in [14] was performed by using the rotational symmetry of the multiquantum ENDOR transitions under the action of a phase shift ϕ of the applied r.f. field. This corresponds to the operation $\exp[-in\phi I_z]$ on the n-quantum transition. The corresponding sequence is shown in fig. 7. The response of the electron spin echo on the phase variation is shown in fig. 8 (top). Note the nonharmonic variation. After Fourier transform a multiquantum spectrum (fig. 8 bottom) results.

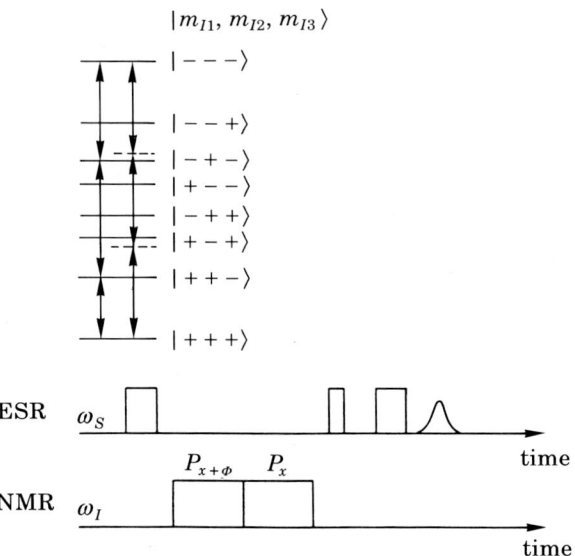

Fig. 7. – Sublevel scheme of three different nuclei coupled to an electron spin, together with a multiquantum ENDOR sequence (according to [14]).

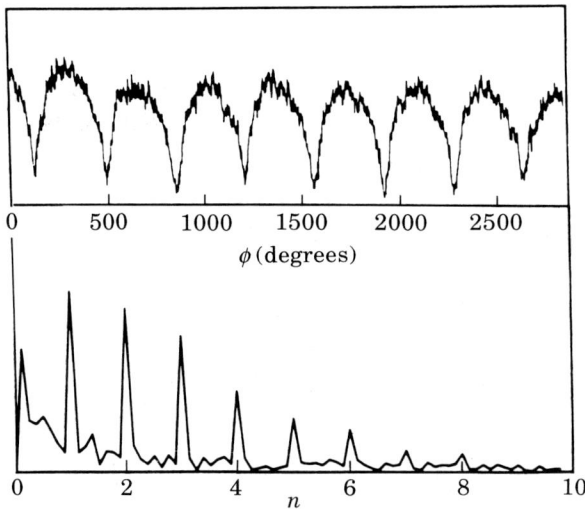

Fig. 8. – Multiquantum ENDOR spectrum (bottom) obtained from the Fourier transform of the phase modulation (top) (according to [14]).

2`3. *Spin echo double resonance* (SEDOR). – The basic pulse sequence for a SEDOR-type double-resonance experiment [15] is sketched in fig. 9a). I would like to discuss its function in the 3-level system again.

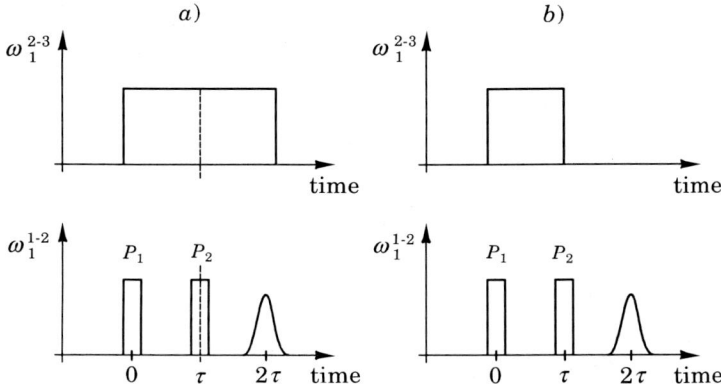

Fig. 9. – SEDOR- and SPINOR-type double-resonance pulse sequences.

2˙3.1. The 1-2 spin echo. We apply the typical spin echo sequence $(\pi/2)$-τ-(π) in the y-direction of the rotating frame selectively at the 1-2 transition, which results in a spin echo at time 2τ in the negative x-direction (negative echo, shown as positive echo in fig. 9). Suppose there is a slight resonance offset by the amount of $\delta\omega_{12}$. We readily calculate the signal response for times $t > \tau$ as

$$(20) \quad S(t > \tau) = \frac{\langle I_x^{1\text{-}2}(t > \tau)\rangle}{\langle I_x^{1\text{-}2}(0)\rangle} = -\cos[\delta\omega_{12}(t - 2\tau)].$$

For a distribution of $\delta\omega_{12}$ values this leads to the (negative) spin echo at $t = 2\tau$.

2˙3.2. SEDOR effect [15]. Let us now take a look at the response of $\langle I_x^{1\text{-}2}(2\tau)\rangle$ when irradiation with strength $\omega_1^{2\text{-}3}$ is performed selectively at the 2-3 transition during the time $0 < t < 2\tau$. By using the transformation [7]

$$(21) \quad \exp[-i\alpha I_y^{2\text{-}3}] I_x^{1\text{-}2} \exp[i\alpha I_y^{2\text{-}3}] = \cos(\alpha/2) I_x^{1\text{-}2} + \sin(\alpha/2) I_x^{1\text{-}3}$$

one arrives for $\delta\omega_{12} = 0$ at

$$(22a) \quad S'(t > \tau) = -\cos[(t - \tau)\omega_1^{2\text{-}3}/2]\cos[\tau\omega_1^{2\text{-}3}/2],$$

$$(22b) \quad S'(t = 2\tau) = -\frac{1}{2}[1 + \cos(\omega_1^{2\text{-}3}\tau)].$$

The SEDOR efficiency can be expressed as

$$(23) \quad \eta = \frac{S(2\tau) - S'(2\tau)}{S(2\tau)} = \frac{1}{2}[1 - \cos(\omega_1^{2\text{-}3}\tau)] = \sin^2(\omega_1^{2\text{-}3}\tau/2),$$

which is maximum ($\eta = 1$) for $\omega_1^{2\text{-}3}\tau = \pi$.

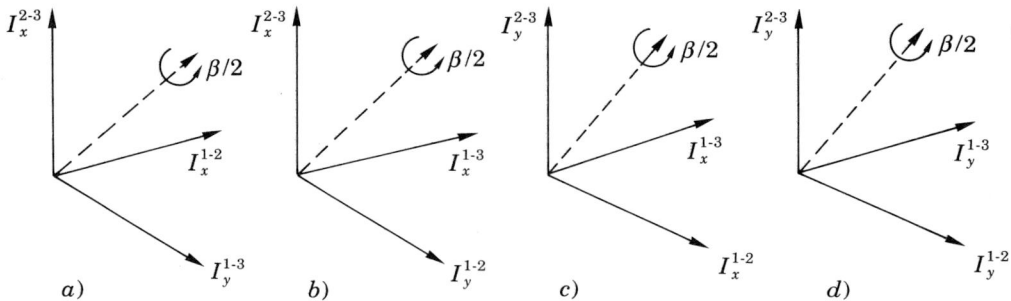

Fig. 10. – Spinor frames in the three-level system as discussed in the text (according to [7]). The different triplet sets of operators transform among themselves as spinors.

Note the $\alpha/2$ variation of the transformed fictitious spin-$1/2$ operators in eq. (21). This reflects the so-called «spinor character» of fermions. For the convenience of the reader I have drawn in fig. 10 a set of fictitious spin-$1/2$ operators which transform among themselves following the properties given by eq. (21). I would like to call them «spinor frames» [7].

2˙3.3. Spinor experiments. A variant of the SEDOR experiment is the SPINOR experiment, first proposed by STOLL, VEGA and VAUGHAN [16], which requires just a slight modification of the standard SEDOR experiment. Suppose irradiation is performed at the 2-3 transition only during half the echo sequence at times $0 < t < \tau$ or $\tau < t < 2\tau$ as is sketched in fig. 9b). Using eq. (21) again one readily obtains [7]

$$(24) \qquad \eta = 1 - \cos\left(\frac{\omega_1^{2\text{-}3}}{2}\tau\right),$$

which is maximum ($\eta = 2$) for $\omega_1^{2\text{-}3}\tau = 2\pi$. Note that the maximum SPINOR efficiency is twice as high as the maximum SEDOR efficiency. Moreover, a 2π-pulse which usually brings a system back to the initial state has the strongest effect. This intriguing behaviour is caused by the spinor character of a spin $1/2$ which also holds by virtue of the Feynman, Vernon, Hellwarth theorem for any two-level quantum system. It can be formulated for a spin $I = 1/2$ as

$$(25) \qquad \exp[-i2\pi \boldsymbol{I} \cdot \boldsymbol{n}] |\psi\rangle = -|\psi\rangle$$

and is expressed in the fictitious spin operator formalism in eq. (21) and shown for different spinor frames in fig. 10. This behaviour was demonstrated in a three-level system for the single- as well as for the double-quantum transition [7]. It has also been applied to pulsed ENDOR spectroscopy [17].

There is, however, a general problem connected with the SPINOR experiment. This is due to the phase shift introduced by any off-resonant irradiation.

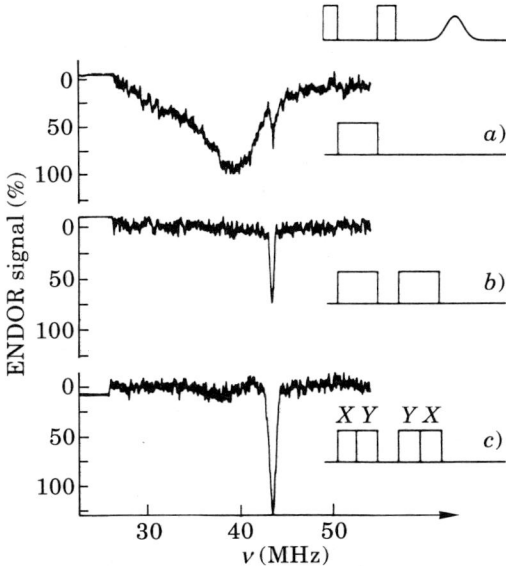

Fig. 11. – SEDOR (b)) and SPINOR ENDOR spectra (a) and c)). Note that the simple spinor ENDOR spectrum (a)) is distorted by Bloch-Siegert shift effects as discussed in the text (according to [17]).

It can be readily calculated that application of a rotating field with frequency ω and field strength ω_1 to a spin $1/2$ with transition frequency ω_0 causes a frequency shift in first order by the amount [6]

$$\omega^{(1)} = \omega_0 \left[\frac{\omega_1^2}{\omega^2 - \omega_0^2} \right]. \tag{26}$$

This corresponds to the Bloch-Siegert shift if the counterrotating field of a linear polarized field is inserted, *i.e.* $\omega = -\omega_0$. In a SPINOR experiment it can introduce a significant phase shift by the amount $\omega^{(1)}\tau$ which reduces for our three-level system to

$$\omega_1^{(1)} \tau = \omega_1^{2\text{-}3} \tau \, \frac{\omega_1^{2\text{-}3}}{4\omega_Q}. \tag{27}$$

In order to remedy this situation a number of pulse sequences were proposed which minimize or cancel this effect [17]. Figure 11 shows such SPINOR-ENDOR experiments for the case of electron-nuclear interactions. Note that in the SEDOR experiment the «Bloch-Siegert» phase shift disappears, if and only if the irradiation at the 2-3 transition (or the secondary transition in general) is performed symmetrically with respect to the π-pulse at the primary transition.

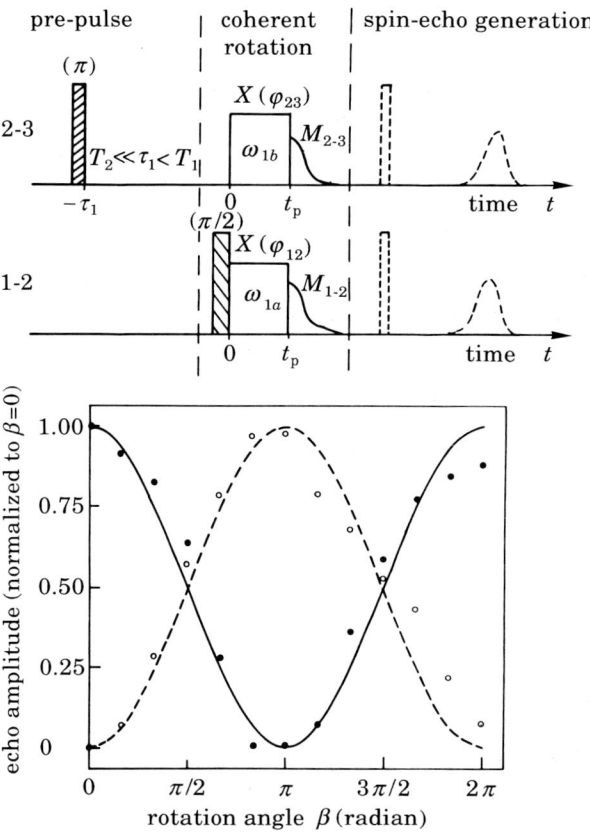

Fig. 12. – Coherence exchange between connected levels in a three-level system by the application of the Pines, Gibby and Waugh sequence [18] (according to [7]).

2˙4. *Cross-polarization in the rotating frame.* – The standard cross-polarization scheme in the rotating frame proposed by PINES, GIBBY and WAUGH [18] consists of a spin-locking sequence applied to the primary spin $I = 1/2$ and irradiation for the same time t_p at the secondary spin $S = 1/2$. In fig. 12 this sequence is sketched with some slight modifications for convenience. Here, we want to deal with fictitious spins $1/2$, *i.e.* the corresponding spin dynamics is discussed within the three-level system. We set $\omega_1^{1\text{-}2} = \omega_1^{2\text{-}3} = \omega_1$ and define the rotation angle $\beta = \omega_1 t_p$. The evolution of the truncated density matrix can be calculated to give [7]

$$(28) \quad \rho(\beta) = \frac{1}{2}(1 + \cos(\beta)) I_x^{1\text{-}2} + \frac{1}{2}(1 - \cos(\beta)) I_x^{2\text{-}3} - \frac{1}{\sqrt{2}} \sin(\beta) I_y^{1\text{-}3}.$$

Figure 12 (bottom) displays the β-dependence of the 1-2 and the 2-3 sublevel magnetization as expected from eq. (28). Note the exchange of magnetization between the two-level systems. This is usually not observable in the standard cross-polarization experiment due to the distribution of frequencies. Here in the three-level system, however, the oscillatory exchange of spin temperature is clearly visible [7]. What is not directly observable is the double-quantum coherence which is contained in the I_y^{1-3} term of the density matrix in eq. (28). However, one should be aware that order is indeed exchanged also with double-quantum coherence [7].

2˙5. *Coherence transfer*. – In this final subsection I want to discuss the possibility of transferring coherence to a transition which is untouched by any irradiation [7]. The experiment is sketched in fig. 13. It begins with a $\pi/2$-pulse at the 1-2 transition, creating a superposition of 1-2 states. This is followed by a β_{13} pulse at the 1-3 transition, which is feasible although it involves a double-quantum transition in our case. These two pulses create a truncated spin density ma-

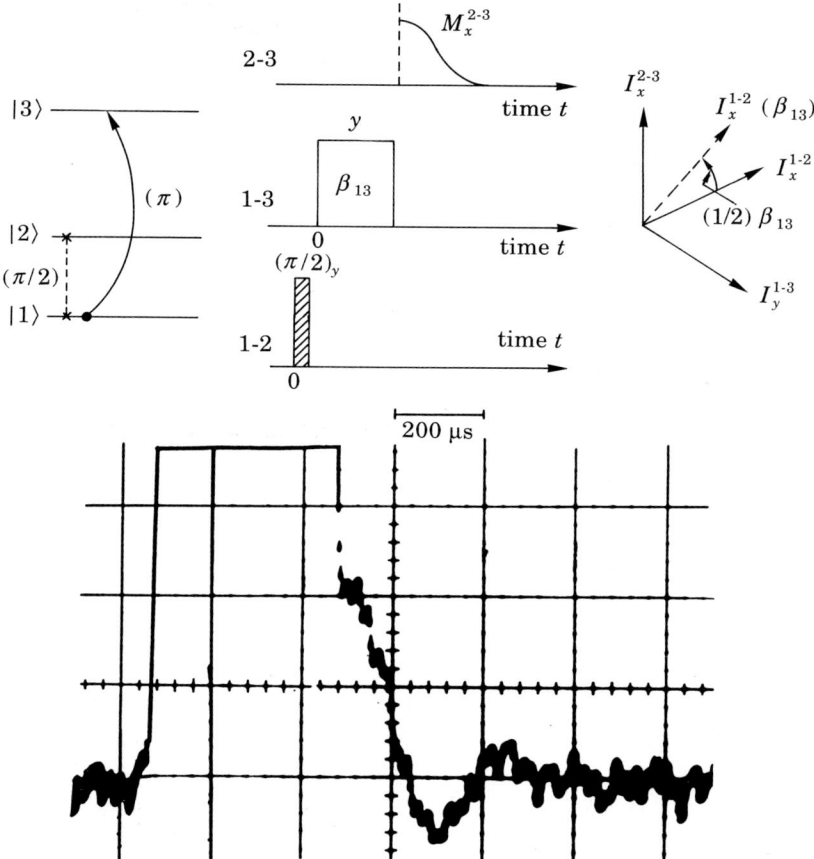

Fig. 13. – Coherence transfer from one two-level system to another one with one level in common (according to [7]).

trix of the following form[7]:

(29) $$\rho(\beta_{13}) = \cos(\beta_{13}/2) I_x^{1\text{-}2} + \sin(\beta_{13}/2) I_x^{2\text{-}3}.$$

Note the spinor behaviour again in eq. (29) as discussed before. The essential aspect here, however, is the occurrence of the term $I_x^{2\text{-}3}$ which reflects coherence at the 2-3 transition. Indeed this is observed as shown by the free-induction decay in fig. 13 which was detected at the 2-3 transition frequency, a frequency which was not present in the r.f. fields. In order to demonstrate that the FID occurs indeed at ω_{23}, phase-coherent detection at the frequency $\omega_{23} = = 2\omega_0 - \omega_{12}$ was employed.

3. – Many-spin systems.

In sect. 2 I have discussed some double-resonance phenomena within a restricted level system. All quantum-mechanical calculations can usually be performed rigorously in such a restricted level system. In many-spin systems, however, this is no longer possible and one has to resort to some sort of approximation. The experiments to be discussed in this section are of the Pines, Gibby, Waugh type[18], sketched in fig. 14. Let us consider two coupled spin systems of N_I I-spins and N_S S-spins. The S spins will be rare so that their mutual inter-

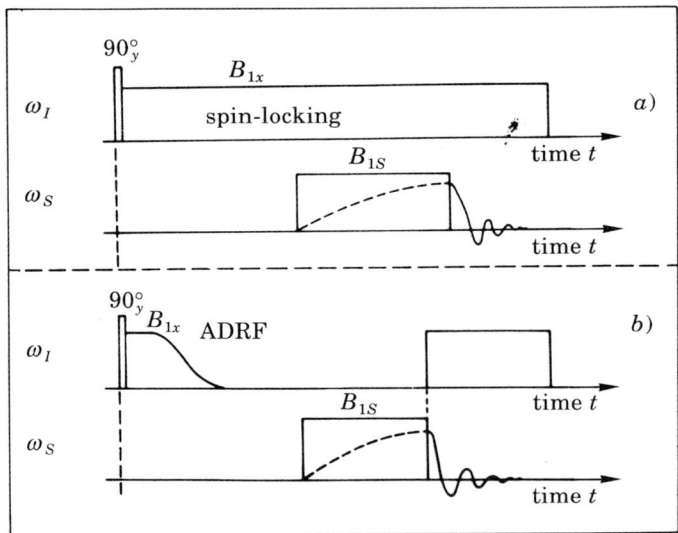

Fig. 14. – Two different cross-polarization schemes involving Hartmann-Hahn condition (a)) and adiabatic demagnetization (ADRF) in the rotating frame (b)).

action can be neglected. The total Hamiltonian can be expressed now as

(30a) $$\mathcal{H} = \mathcal{H}_0 + \mathcal{H}_{\text{int}},$$

(30b) $$\mathcal{H}_0 = \mathcal{H}_I + \mathcal{H}_S,$$

(30c) $$\mathcal{H}_{\text{int}} = \mathcal{H}_{IS} + \mathcal{H}_{II},$$

where in the laboratory frame

(31) $$\mathcal{H}_I = \omega_{0I} I_z \quad \text{and} \quad \mathcal{H}_S = \omega_{0S} S_z.$$

The application of two r.f. fields of the form $2\omega_{1I,S} \cos \omega_{I,S} t$ leads to the following Hamiltonian in the rotating frame:

(32) $$\mathcal{H}_0 = \Delta\omega_I I_z + \omega_{1I} I_{x,y} + \Delta\omega_S S_z + \omega_{1S} S_{x,y}$$

and an interaction Hamiltonian as in eq. (30c), but truncated in the sense that only secular terms in the doubly rotating frame are kept:

(33a) $$\mathcal{H}_{IS} = 2 \sum_{i,j} c_{i,j} I_{zi} S_{zj},$$

(33b) $$\mathcal{H}_{II} = \sum_{i<j} d_{ij} (3 I_{zi} I_{zj} - \mathbf{I}_i \cdot \mathbf{I}_j).$$

3˙1. *Spin temperature and spin density matrix.* – We use here again the truncated spin density matrix in the high-temperature approximation (HTA) as discussed in the introduction. Since we are dealing with spin operators only, we can introduce a simplification by utilizing the fact that $\text{Tr}\{1\} = (2I + 1)^{N_I}(2S + 1)^{N_S}$. In the evaluation of expectation values we will encounter traces of the form $\text{Tr}\{I_z^2\} = N_I(1/3) I(I+1)(2I+1)^{N_I}(2S+1)^{N_S}$. If we define for the following that this trace will be expressed as $\text{Tr}\{I_z^2\} = N_I(1/3) I(I+1)$ and similarly for the S spins, we can further simplify the truncated density matrix in HTA as

(34) $$\rho = -\beta \mathcal{H}$$

with the inverse spin temperature $\beta = \hbar/k_B T$. In a similar way we can express the spin entropy defined as [5]

(35) $$S = -k_B \text{Tr}\{\rho \ln \rho\}$$

by

(36) $$S/k_B = N_I \ln(2I+1) + N_S \ln(2S+1) - \text{Tr}\{\rho \cdot \rho\},$$

which we will truncate to

(37) $$S = -\text{Tr}\{\rho \cdot \rho\}.$$

3˙2. *Spin calorimetry.* – In this subsection we discuss some simple arguments based on spin temperature in order to calculate the possible spin polarization for the S spins. We follow the arguments given in Goldman's book [5]. After the initial $\pi/2$-pulse applied to the I spins the initial density matrix can be written as

$$(38) \quad \rho_0 = -\beta_0(\omega_{0I} I_x + \ldots),$$

where S-spin terms and interaction terms are neglected. The effective Hamiltonian in the doubly rotating frame can be expressed as

$$(39) \quad \mathscr{H}_0 = \omega_{1I} I_x + \omega_{1S} S_x,$$

where interaction terms are neglected, *i.e.* it is assumed that $\omega_{1I}, \omega_{1S} \gg \sqrt{\text{Tr}\{\mathscr{H}_{\text{int}}^2\}}$. Due to the interaction Hamiltonian we assume that both spin systems will approach an equilibrium with the same spin temperature, *i.e.*

$$(40) \quad \rho_f = -\beta_f(\omega_{1I} I_x + \omega_{1S} S_x).$$

Invoking energy conservation, *i.e.* $\text{Tr}\{\mathscr{H}_0 \rho_0\} = \text{Tr}\{\mathscr{H}_0 \rho_f\}$, immediately leads to [5,6]

$$(41) \quad \frac{\beta_f}{\beta_0} = \frac{\omega_{0I}}{\omega_{1I}} \frac{1}{1+\varepsilon'},$$

where

$$(42a) \quad \varepsilon' = \frac{\omega_{1S}^2 N_S S(S+1)}{\omega_{1I}^2 N_I I(I+1)} = \alpha^2 \varepsilon$$

with

$$(42b) \quad \alpha = \frac{\omega_{1S}}{\omega_{1I}} \quad \text{and} \quad \varepsilon = \frac{N_S S(S+1)}{N_I I(I+1)}.$$

From the ratio of the inverse spin temperatures alone the signal enhancement of the S spins may not be readily appreciable. Let us, therefore, also calculate the S-spin magnetization in the rotating frame which is directly proportional to the signal strength. With the definitions

$$(43a) \quad M_{S0} = \gamma_S \hbar \langle S_z \rangle = -\beta_0 \gamma_S \hbar \omega_{0S} \text{Tr}\{S_z^2\}$$

and

$$(43b) \quad M_{Sf} = \gamma_S \hbar \langle S_x \rangle = -\beta_f \gamma_S \hbar \omega_{1S} \text{Tr}\{S_x^2\}$$

one obtains immediately

$$(44) \quad \frac{M_{Sf}}{M_{S0}} = \frac{\gamma_I}{\gamma_S} \frac{\alpha}{1+\alpha^2 \varepsilon}.$$

If the Hartmann-Hahn condition ($\omega_{1S} = \omega_{1I}$, $\alpha = 1$) is fulfilled and ε is very small (rare spins), the signal enhancement is basically given by the ratio γ_I/γ_S. Although this ratio may in certain cases not be very large, cross-polarization is nevertheless very often required either to observe selectively coupled spins or simply when the relaxation rate of the S spins is very small, whereas the I spins relax very rapidly, therefore allowing a short recycle delay of the sequence.

3˙3. *From Hilbert space to Liouville space.* – Here I want to give a brief outline on how to proceed from Hilbert space to Liouville space. This is not new quantum mechanics, but rather a concise formulation of it. I follow here the notation used in [6, 19] where more details can be found. This formulation was also applied by ABRAGAM and GOLDMAN [20] discussing order-disorder phenomena of spins. The basic idea is to switch from wave functions which can be viewed as vectors in Hilbert space to vectors in Liouville space consisting of operators. This way the particular «representation» of a quantum state in terms of wave functions (eigenstates, etc.) becomes irrelevant, *i.e.* the formulation becomes «representation free». Suppose A is any operator (spin operator in our case). When viewed as a vector in an abstract vector space, called Liouville space, A and its adjoint A^+ are written as $|A)$ and $(A|$, respectively. There is a scalar product between two vectors A and B, which is defined as

$$(45) \qquad (A|B) = \mathrm{Tr}\{A^+ B\}.$$

Two operators are called orthogonal if and only if $(A|B) = 0$. The usual associative and distributive laws for combined operations hold similarly as for ordinary vectors [6, 19]. As in ordinary vector spaces there is a superoperator, called Liouvillian, which operates on vectors in Liouville space. This role is usually played by the Hamiltonian which turns into the Liouvillian in Liouville space. It is labelled by a hat on top and it is defined as

$$(46) \qquad \widehat{\mathscr{H}}|A) = |[\mathscr{H}, A]) \quad \text{and} \quad (A|\widehat{\mathscr{H}} = ([\mathscr{H}, A]|\,,$$

which leads to expressions of the following form:

$$(47) \qquad (A|\widehat{\mathscr{H}}|B) = \mathrm{Tr}\{A^+[\mathscr{H}, B]\} = (B|\widehat{\mathscr{H}}|A)^*,$$

where the star labels the complex conjugate. In the following we will also use orthonormalized vectors Q_j in Liouville space, defined as

$$(48) \qquad (Q_i|Q_j) = \delta_{ij}(Q_j|Q_j) \qquad \text{with Kronecker } \delta_{ij}.$$

As an example we define

$$(49) \qquad Q_I = \frac{I_x}{\sqrt{\mathrm{Tr}\{I_x^2\}}}.$$

The total Hamiltonian equation (32) in the rotating frame can now be written as

(50) $$\mathcal{H} = h_I \boldsymbol{Q}_I + h_S \boldsymbol{Q}_S + h_{IS} \boldsymbol{Q}_{IS} + h_{II} \boldsymbol{Q}_{II},$$

where

(51) $$h_k = \sqrt{(\mathcal{H}_k | \mathcal{H}_k)} \quad \text{and} \quad \boldsymbol{Q}_k = \frac{\mathcal{H}_k}{\sqrt{(\mathcal{H}_k | \mathcal{H}_k)}},$$

which results in

(52a) $$h_I^2 = (\mathcal{H}_I | \mathcal{H}_I) = N_I \frac{1}{3} I(I+1) \omega_{1I}^2, \quad h_S^2 = (\mathcal{H}_S | \mathcal{H}_S) =$$

$$= N_S \frac{1}{3} S(S+1) \omega_{1S}^2,$$

(52b) $$h_{IS}^2 = (\mathcal{H}_{IS} | \mathcal{H}_{IS}) = N_S \frac{1}{3} S(S+1) C^2$$

$$\text{with } C^2 = M_{2IS} = \frac{1}{3} I(I+1) \frac{4}{N_S} \sum_{i,j} c_{ij}^2,$$

(52c) $$h_{II}^2 = (\mathcal{H}_{II} | \mathcal{H}_{II}) = N_I \frac{1}{3} I(I+1) D^2 \qquad \text{with } D^2 = \frac{1}{3} M_{2II},$$

and where M_{2IS} and M_{2II} are the second moments of the S and I spins, respectively [6]. Also the spin density matrix can be expanded into orthonormal operators, which in quasi-equilibrium can be expressed as

(53) $$\rho = a_I \boldsymbol{Q}_I + a_S \boldsymbol{Q}_S + a_{IS} \boldsymbol{Q}_{IS} + a_{II} \boldsymbol{Q}_{II} \quad \text{with } a_k = -\beta_k h_k.$$

The dominant part of the spin density matrix after the initial $\pi/2$-pulse applied to the I spins now reads

(54) $$\rho_0 = a_{0I} \boldsymbol{Q}_I \quad \text{with } a_{0I} = -\beta_0 \omega_{0I} \sqrt{N_I \frac{1}{3} I(I+1)}.$$

3'4. *Spin calorimetry in Liouville space*. – In order to consider this time the secular part (in the rotating frame) of the interaction Hamiltonian, namely $-(1/2)\mathcal{H}_{xx}$, as a separate heat bath, we split the Hamiltonian into

(55a) $$\mathcal{H}_0 = \mathcal{H}_I - \frac{1}{2} \mathcal{H}_{xx} + \mathcal{H}_S,$$

(55b) $$\mathcal{H}_{\text{int}} = \mathcal{H}_{IS} + \frac{1}{2} (\mathcal{H}_{zz} - \mathcal{H}_{yy}),$$

where

(55c) $$\mathcal{H}_{kk} = \sum_{i<j} d_{ij} (3 I_{ki} I_{kj} - \boldsymbol{I}_i \cdot \boldsymbol{I}_j) \qquad \text{with } k = x, y, z,$$

or in terms of orthonormal operators

(56a) $$\mathscr{H}_0 = h_I \boldsymbol{Q}_I - \frac{1}{2} h_{II} \boldsymbol{Q}_{xx} + h_S \boldsymbol{Q}_S,$$

(56b) $$\mathscr{H}_{\text{int}} = h_{IS} \boldsymbol{Q}_{IS} + \frac{1}{2} h_{II} (\boldsymbol{Q}_{zz} - \boldsymbol{Q}_{yy}),$$

where

(56c) $$\boldsymbol{Q}_{kk} = \frac{\mathscr{H}_{kk}}{\sqrt{(\mathscr{H}_{kk} | \mathscr{H}_{kk})}},$$

where kk can be any of the indices used in eq. (56a), (56b).

The final density matrix, after equilibrium has been established, is given by

(57) $$\rho_f = a_{If} \boldsymbol{Q}_I - \frac{1}{2} a_{IIf} \boldsymbol{Q}_x + a_{Sf} \boldsymbol{Q}_S.$$

Let us assume that a common spin temperature is reached for all reservoirs, i.e.

$$\rho_f = -\beta_f \mathscr{H}_0.$$

The scenario is sketched in two-dimensional Liouville space in fig. 15, where only the subspace $\boldsymbol{Q}_I$, $\boldsymbol{Q}_S$ is shown. The operator $\boldsymbol{Q}_{xx}$ is orthogonal to $\boldsymbol{Q}_I$ and $\boldsymbol{Q}_S$ and is represented by the out-of-plane component.

3˙4.1. Energy conservation. Let us start with the initial density matrix at a_{0I} along the $\boldsymbol{Q}_I$ axis (open circle in fig. 15). Assuming energy conservation in the rotating frame, i.e. $(\mathscr{H}_0 | \rho_0) = (\mathscr{H}_0 | \rho_f)$, the final density matrix is deter-

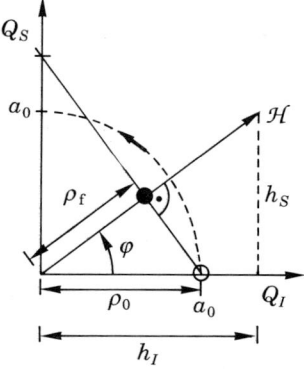

Fig. 15. – Sketch of the 2D Liouville space of the orthogonal operators $\boldsymbol{Q}_I$ and $\boldsymbol{Q}_S$. The Hamiltonian with components h_I and h_S is included. The initial density matrix is represented by a_0 (open circle). Projection onto the Hamiltonian results in the final density matrix under energy conservation. The circle of radius a_0 is the consistent constant entropy.

mined by the projection of the initial density matrix onto $\mathcal{H}_0$. If $\mathcal{H}_{II}$ is neglected, $\mathcal{H}_0$ lies in the ($\boldsymbol{Q}_I$, $\boldsymbol{Q}_S$)-plane and the projection can be performed graphically as drawn in fig. 15. Here we consider the case where $\mathcal{H}_{II}$ plays some role, i.e. $\mathcal{H}_0$ is tilted out of plane. Nevertheless it is straightforward to calculate β_f by the rules given here and one arrives at

$$(58a) \qquad \frac{\beta_f}{\beta_0} = \frac{\omega_{0I}}{\omega_{1I}} \frac{1}{1 + 0.25 D^2/\omega_{1I}^2 + \alpha^2 \varepsilon}.$$

Note that the inverse spin temperature is slightly reduced with respect to the one calculated in sect. 3, where the spin-spin interaction was ignored. Correspondingly is the signal enhancement of the S spins reduced

$$(58b) \qquad \frac{M_{Sf}}{M_{S0}} = \frac{\gamma_I}{\gamma_S} \frac{\alpha}{1 + 0.25 D^2/\omega_{1I}^2 + \alpha^2 \varepsilon}.$$

Note that M_{Sf}/M_{S0} reaches a maximum value as a function of α for $\alpha\sqrt{\varepsilon} = 1 + 0.25 D^2/\omega_{1I}^2$ which can be much larger than γ_I/γ_S. However, this is difficult to achieve, since the time to reach this equilibrium may take much longer than the relaxation times, which are ignored here. Moreover, it has been shown that in some cases a quasi-equilibrium is reached which prevents the system from reaching the final equilibrium on any practical time scale. A sequence called MOIST was, therefore, invented to overcome this bottleneck [21].

3˙4.2. Isentropic mixing. In this type of experiment one tries to keep the entropy constant while performing the mixing of the different heat baths. One way to do this is the sequence sketched in fig. 14b). After an adiabatic demagnetization in the rotating frame (ADRF) of the I spins by reducing the ω_{1I} field adiabatically to zero a r.f. field is suddenly switched on at the S-spin frequency. It would be better, however, if an adiabatic remagnetization (ARRF) of the S spins is performed by turning the ω_{1S} field on slowly up to a maximum value. We will only treat the latter case here. The final density matrix can be expressed as

$$(59) \qquad \rho_f = a_{IIf} \boldsymbol{Q}_{II} + a_{Sf} \boldsymbol{Q}_S \quad \text{with } a_{IIf} = -\beta_f h_{II} \text{ and } a_{Sf} = -\beta_f h_S.$$

Note that the process is not completely adiabatic because of the rapid switching on of the r.f. field at the S-spin resonance, nor is ρ_f the same as discussed under energy conservation (subsect. 3˙4.1) and as is sketched in fig. 15.

Keeping the entropy constant demands $(\rho_0|\rho_0) = (\rho_f|\rho_f)$, i.e. the final state lies on the circle with radius a_{0I} in the 2D Liouville space (fig. 15) when $\mathcal{H}_{II}$ is neglected. This would result in $a_{Sf} = a_{0I}$. Taking $\mathcal{H}_{II}$ into account, however, the final state still lies on the sphere with radius a_{0I} in the 3D Liouville space consisting of the orthogonal operators $\boldsymbol{Q}_I$, $\boldsymbol{Q}_S$, $\boldsymbol{Q}_{II}$. Since $a_{If} = 0$, it lies in the ($\boldsymbol{Q}_S$, $\boldsymbol{Q}_{II}$)-plane. The final inverse spin temperature β_f can be readily calculated

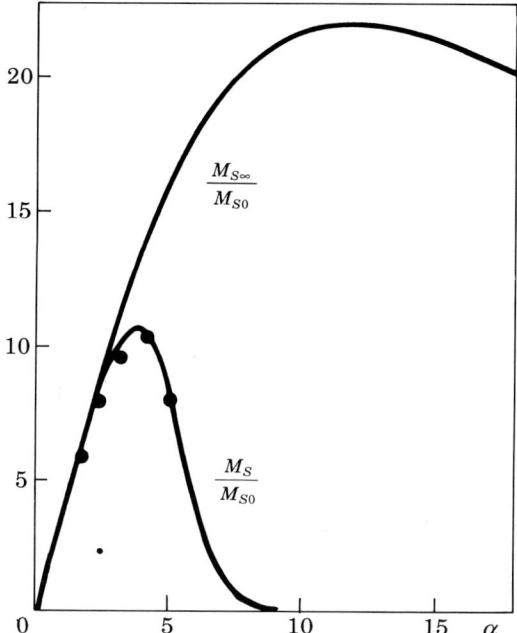

Fig. 16. – Signal enhancement under ADRF conditions with $a = \omega_{1S}/D$ (according to [6]). Note that a sudden switch-on of the ω_{1S} field is performed, however. See the text for a discussion.

and results in

$$(60a) \qquad \frac{\beta_f}{\beta_0} = \frac{\omega_{0I}}{\omega_{1S}} \frac{1}{\sqrt{D^2/\omega_{1S}^2 + \varepsilon}}$$

with the corresponding magnetization (signal strength)

$$(60b) \qquad \frac{M_{Sf}}{M_{S0}} = \frac{\gamma_I}{\gamma_S} \frac{\omega_{1S}}{\sqrt{D^2 + \omega_{1S}^2 \varepsilon}}.$$

Note that the maximum signal attainable is $M_{Sf}/M_{S0} = \gamma_I/\gamma_S \sqrt{\varepsilon}$. Figure 16 shows such an experiment where the maximum signal obtained was much larger than could be achieved under matched Hartmann-Hahn condition [6]. Note, however, that the signal strength in fig. 16 is only half the value expected from eq. (60b) because of the sudden application of the r.f. field used there. In this case energy conservation must be invoked again during the switching-on of the ω_{1S} field. In this case the same expression results for M_S/M_{S0} as in eq. (44), however, with α replaced by ω_{1S}/D.

3'5. SEDOR. – Spin echo double resonance (SEDOR), originally invented by HAHN and co-workers [15], was already discussed in subsect. 2'3 in the case of the simple 3-level system. There the coherent evolution of the sublevel system was demonstrated quite clearly. In a many-spin system, however, there are many coherences and correspondingly many interferences which lead to a loss of oscillatory behaviour. In order to demonstrate the basic effect, let us assume a SEDOR pulse sequence as shown in fig. 9a), with the difference that the observed I spins have the transition frequency ω_I ($\omega^{1\text{-}2}$ in fig. 9a)) and the indirectly detected S spins have the frequency ω_S ($\omega^{2\text{-}3}$ in fig. 9a)). In order to simplify the calculation, we assume that irradiation at ω_S is applied by a short pulse, characterized by the rotation angle β. Furthermore we ignore the coupling among the I spins. Without the pulse at ω_S a spin echo results at time 2τ without destruction.

We will discuss two different variants of the SEDOR experiment:

i) application of a π-pulse at frequency ω_S with variable time delay $0 < T_S < \tau$ after the first pulse,

ii) application of a β-pulse at frequency ω_S symmetrically spaced with respect to the π-pulse at ω_I.

Let us begin with i): The height of the spin echo at ω_I can be expressed by using the Liouville space notation introduced in [19] and applied here as

$$(61) \quad G(2\tau) = \frac{(I_x \mid \exp[-i\tau\mathcal{H}] P_\pi(I) \exp[-i(\tau - T_S)\mathcal{H}] P_\pi(S) \exp[-iT_S\mathcal{H}] \mid I_x)}{(I_x \mid I_x)},$$

where we use $\mathcal{H} = \mathcal{H}_{IS}$ as defined in eq. (33a) and where

$$(62) \quad P_\pi(I) = \exp[-i\pi \hat{I}_y] \quad \text{and} \quad P_\pi(S) = \exp[-i\pi \hat{S}_y].$$

After some algebra one arrives at

$$(63) \quad G_+(2\tau) = -\frac{(I_x \mid \exp[-iT_S \hat{\mathcal{H}}_{IS}] \mid I_x)}{(I_x \mid I_x)}$$

with

$$(64) \quad G(2\tau) = -\frac{1}{N_I} \sum_j \prod_k \cos(T_S c_{jk}).$$

This basically means that the echo amplitude is diminished with increasing T_S, due to the IS interaction. By varying T_S within $0 \leq T_S \leq \tau$ the FID caused by $\mathcal{H}_{IS}$ can be mapped out indirectly. In first order this still holds if II couplings are included. Although the echo height is reduced in this case to begin with, it nevertheless scales with variation of T_S due to $\mathcal{H}_{IS}$ at least for the near-neighbour IS interactions.

Let us now turn to ii): The starting equation for the echo height is now slightly different. It can be formulated as

$$(65) \quad G(2\tau) = \frac{(I_x \mid \exp[-i\tau\hat{\mathscr{H}}] P_\pi(I) P_\beta(S) \exp[-i\tau\hat{\mathscr{H}}] \mid I_x)}{(I_x \mid I_x)},$$

with $P_\beta(S) = \exp[-i\beta \hat{S}_y]$, which results in

$$(66) \quad G(2\tau) = -\frac{1}{N_I} \sum_j \prod_k \left\{ \cos^2\left(\frac{\beta}{2}\right) + \sin^2\left(\frac{\beta}{2}\right) \cos(\tau c_{jk}) \right\}$$

for $I = S = 1/2$.

This opens the possibility to map out the FID of the IS interaction by varying β within $0 \leq \beta \leq \pi$. This second method has the advantage over case i) that, by varying the amplitude ω_{1S} of the pulse at ω_S in fine steps, many points of the FID can be obtained, whereas in case i) the pulse width and the often short time τ limit the number of time increments. On the other hand, case ii) demands a linear r.f. amplitude and a correction for the nonlinear dependence on β.

3˙6. *Double-resonance spin dynamics.* – The lack of space does not allow me to treat the spin dynamics during cross-polarization and decoupling in great detail. I will give only a brief outline here. More details can be found in [6]. We are interested in the time dependence of the parameters

$$(67) \quad \langle \mathscr{H}_I(t) \rangle = (\mathscr{H}_I \mid \rho(t)) = \beta_I (\mathscr{H}_I \mid \mathscr{H}_I) \text{ and } \langle \mathscr{H}_S(t) \rangle = (\mathscr{H}_S \mid \rho(t)) = \beta_S (\mathscr{H}_S \mid \mathscr{H}_S).$$

In order to project the time-dependent density matrix onto the «interesting» part of the operator space, we define the projector

$$(68) \quad P = \frac{\mid \mathscr{H}_I)(\mathscr{H}_I \mid}{(\mathscr{H}_I \mid \mathscr{H}_I)} + \frac{\mid \mathscr{H}_S)(\mathscr{H}_S \mid}{(\mathscr{H}_S \mid \mathscr{H}_S)}.$$

The Liouville-von Neuman equation is now solved starting with

$$(69) \quad \frac{\mathrm{d}}{\mathrm{d}t} \mid \rho(t)) = -i\hat{\mathscr{H}} \{ P \mid \rho(t)) + (1 - P) \mid \rho(t)) \},$$

where $\mathscr{H}$ is the total Hamiltonian. It can be shown [6, 20] that under the conditions given here, where we start with an equilibrium density matrix, the following set of equations can be obtained:

$$(70a) \quad \frac{\mathrm{d}}{\mathrm{d}t} \langle \mathscr{H}_I \rangle = -\int_0^t \mathrm{d}t' \, K_{II}(t-t') \langle \mathscr{H}_I(t') \rangle - \int_0^t \mathrm{d}t' \, K_{IS}(t-t') \langle \mathscr{H}_S(t') \rangle,$$

$$(70b) \quad \frac{\mathrm{d}}{\mathrm{d}t} \langle \mathscr{H}_S \rangle = -\int_0^t \mathrm{d}t' \, K_{SI}(t-t') \langle \mathscr{H}_I(t') \rangle - \int_0^t \mathrm{d}t' \, K_{SS}(t-t') \langle \mathscr{H}_S(t') \rangle$$

with

(71) $$K_{ab}(t) = \frac{(\mathcal{H}_a | \hat{\mathcal{H}}_{IS} S(t) \hat{\mathcal{H}}_{IS} | \mathcal{H}_b)}{(\mathcal{H}_b | \mathcal{H}_b)}$$

and

(72) $$S(t) = \exp[-it(1-P)\hat{\mathcal{H}}] \simeq \exp\left[-it\left(\hat{\mathcal{H}}_0 - \frac{1}{2}\hat{\mathcal{H}}_{xx}\right)\right].$$

A related set of equations is obtained by expressing $\langle \mathcal{H}_a \rangle = -\beta_a (\mathcal{H}_a | \mathcal{H}_a)$ by the inverse spin temperature:

(73a) $$\frac{d}{dt}\langle \beta_I \rangle = -\int_0^t dt' K_{II}(t-t')\langle \beta_I(t') \rangle - \varepsilon' \int_0^t dt' K_{IS}(t-t')\langle \beta_S(t') \rangle,$$

(73b) $$\frac{d}{dt}\langle \beta_S \rangle = -\frac{1}{\varepsilon'}\int_0^t dt' K_{SI}(t-t')\langle \beta_I(t') \rangle - \int_0^t dt' K_{SS}(t-t')\langle \beta_S(t') \rangle.$$

Although these equations can be solved rigorously by using Laplace transform techniques, we want to restrict ourselves here to the short-memory correlation time approximation, where one assumes that the memory functions $K_{ab}(t)$ decay much more rapidly than the functions $\langle \mathcal{H}_a(t) \rangle$ or $\beta_a(t)$. Strictly speaking, this is not obeyed in the double-resonance experiments discussed here. It models, however, somewhat realistically the long-time behaviour of the time evolution of the spin temperature. At short times nonexponential evolution and oscillatory behaviour result if the rigorous treatment is followed [6]. Here it may suffice to treat the long-time behaviour, which can be derived from the above set of equations by invoking symmetry relations of the memory functions and assuming that a final inverse spin temperature is reached under energy conservation. The corresponding differential equations can be expressed as

(74a) $$\frac{d}{dt}\langle \beta_I \rangle = -\varepsilon' \Gamma (\beta_I - \beta_S),$$

(74b) $$\frac{d}{dt}\langle \beta_S \rangle = -\Gamma (\beta_S - \beta_I),$$

where

(75) $$\Gamma = M_{2IS} \frac{1}{2}[J(\omega_{1I} - \omega_{1S}) + J(\omega_{1I} + \omega_{1S})]$$

with

(76) $$J(\omega) = \frac{1}{2}\int_{-\infty}^{\infty} dt\, g(t) \cos \omega t$$

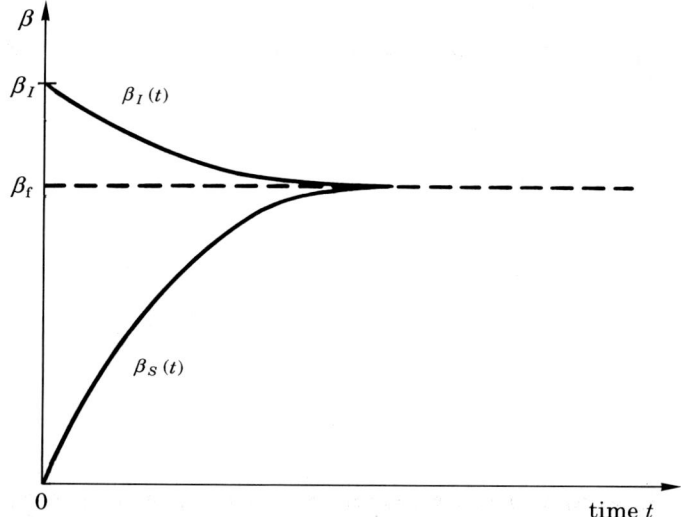

Fig. 17. – Schematic evolution of inverse spin temperatures β_I and β_S under cross-polarization conditions reaching a final spin temperature β_f.

and

(77) $$g(t) = \frac{(\mathcal{H}_{IS}|S(t)|\mathcal{H}_{IS})}{(\mathcal{H}_{IS}|\mathcal{H}_{IS})}.$$

The evolution of the two spin temperatures is schematically drawn in fig. 17. Note that the correlation function $g(t)$ with $g(t = 0) = 1$ is a decaying function where the decay rate involves the second moments of the II and IS couplings [6, 22].

The characteristic time constant for the approach to equilibrium can be calculated from these expressions once $g(t)$ is known. Because of the many-body character, however, a rigorous treatment is not possible. One has to resume to some sort of approximation. One of the most common approximations is that $g(t)$ is Gaussian, i.e. $g(t) = \exp[-N_2 t^2/2]$, which only requires to calculate its second moment [6, 22]:

(78) $$N_2 = \frac{1}{4} \frac{(\mathcal{H}_{IS}|\widehat{\mathcal{H}}_{xx}\widehat{\mathcal{H}}_{xx}|\mathcal{H}_{IS})}{(\mathcal{H}_{IS}|\mathcal{H}_{IS})} = \frac{1}{4} \frac{M_{4ISII}}{M_{2IS}},$$

where the moments M_{4ISII} and M_{2IS} can be determined from the appropriate lattice sums given in [6]. In a similar way one can treat the decoupling problem [6, 23] and the spin temperature oscillations occurring during the cross-polarization process [6] and during spin-locking [24]. Space does not permit, however, to go into details here.

4. – Unconventional double-resonance experiments.

In this section I want to present some unusual double-resonance experiments which either involve an extremely low Larmor frequency for the nuclear spins (less than 1 Hz) which on top is sensitive to rotation (subsect. 4'1) or experiments where only a single frequency is applied but nevertheless a double-resonance effect is achieved (subsect. 4'2 and 4'3).

4'1. Rb-Xe *double resonance and rotating the frame*. – Although space does not permit to go into great detail, I want to touch briefly on an experiment which was performed in our laboratory recently, namely the demonstration that the Larmor frequency of nuclear spins is shifted by the rotational rate with which the whole spectrometer rotates about the magnetic-field axis. This so-called gyro-effect can be used to sense rotational motion. Within a research project for «High Precision Navigation» we have set up an apparatus which is based on a vapour cell containing Rb and Xe atoms, besides N_2 as buffer gas. The Rb atoms are pumped by an i.r. laser diode emitting at the D_1 line of Rb. Via a complex spin exchange process, which has been extensively studied by HAPPER and co-workers[25], the Xe nuclear spins become polarized to a large extent and their FID can be observed either directly or as done in our case indirectly by the change in absorption of the laser light. The double-resonance mechanism is invoked by a rapid field modulation at the Rb sublevel resonance in small magnetic fields (less than 1 mG) (Hanle effect) and field switching in order to initiate an FID of the ^{129}Xe nuclear spins. The corresponding FID and

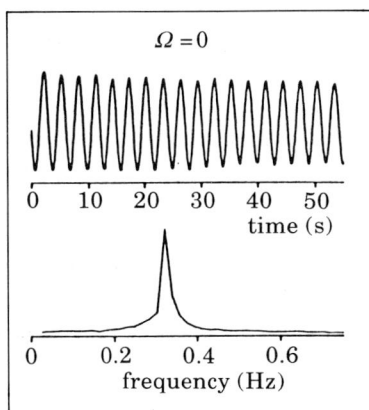 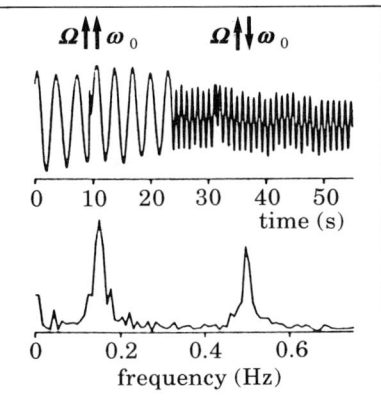

Fig. 18. – ^{129}Xe FID and spectra at a Larmor frequency of less than 1 Hz ($B_0 < 1$ mG). The spectra were obtained by double resonance between ^{85}Rb atoms and ^{129}Xe atoms in the gas phase[26]. Left: Spectrometer without rotation besides Earth rotation (not resolved). Right: With left and right rotation of the whole spectrometer, leading to a shift in Larmor frequency (gyro-effect) (see text).

spectrum are shown in fig. 18 (left). Note that a very small Larmor frequency (less than 1 Hz) has been used in order to reduce field inhomogeneity broadening. The right-hand part of fig. 18 shows what happens if the whole spectrometer is rotated around the B_0 axis [26]. Two different directions of rotation were chosen, one with the Larmor precession of the ^{129}Xe nuclear spins leading to a lower Larmor frequency and one opposite to the Larmor precession of the nuclear spins leading to a higher Larmor frequency. It is obvious that one can determine the sign of the gyromagnetic ratio this way. Besides this the experiment may be useful as a rotational sensor, *i.e.* a nuclear gyroscope. In terms of the theme of this school this is a triple-resonance experiment, employing two spin precession frequencies (Rb, Xe) and the spectrometer rotational frequency.

4˙2. *Hyperfine spectroscopy with correlation to an electron spin* (HYSCORE) [27]. – The experiment to be discussed here involves only single-frequency excitation, namely pulsed excitation at the Larmor frequency of electron spins (typically 9 GHz). Nevertheless a complete two-dimensional sublevel spectrum results, displaying the hyperfine structure of the nuclei coupled to the electron spin as was first demonstrated in [27] and as is shown in fig. 19. The experiment involves a stimulated echo sequence (fig. 19 top) applied at the electron Larmor frequency. It was shown by MIMS [28] that the stimulated echo varies in amplitude if the separation between the second and third pulse is changed. In fact the echo is modulated with the hyperfine interaction of the coupled nuclear spins. When this oscillatory echo envelope modulation is Fourier transformed, a hyperfine spectrum results. This experiment alone shows all the spectroscopic features which are usually only obtained when irradiation at two different frequencies is performed, namely at the main-level and the sublevel transitions. In the electron nuclear case this is called ENDOR.

Coming back to the HYSCORE experiment, we note that a π-pulse is placed between the second and third pulse. This π-pulse inverts the electron spins and, therefore, correlates only those $m_S = 1/2$ with $m_S = -1/2$ hyperfine transitions which belong to the same electron spin. If times t_1, t_2 are varied in a two-dimensional fashion, the two-dimensional hyperfine spectrum shown in fig. 19 results. A closer analysis of this spectrum leads to the distinction of different sets of hyperfine spectra belonging to different electron spins.

The double-resonance mechanism in this case is caused by two facts, i) the large spectral width of the microwave pulses which covers the hyperfine spectral width and ii) the different quantization axis for the nuclear and electron spins. Due to the large anisotropic hyperfine interaction the resulting local field at the nuclear sites points into a different direction than the applied magnetic field B_0 which supplies the quantization axis for the electron spins. In summary in these types of double-resonance experiments a frequency and a time are involved rather than two frequencies.

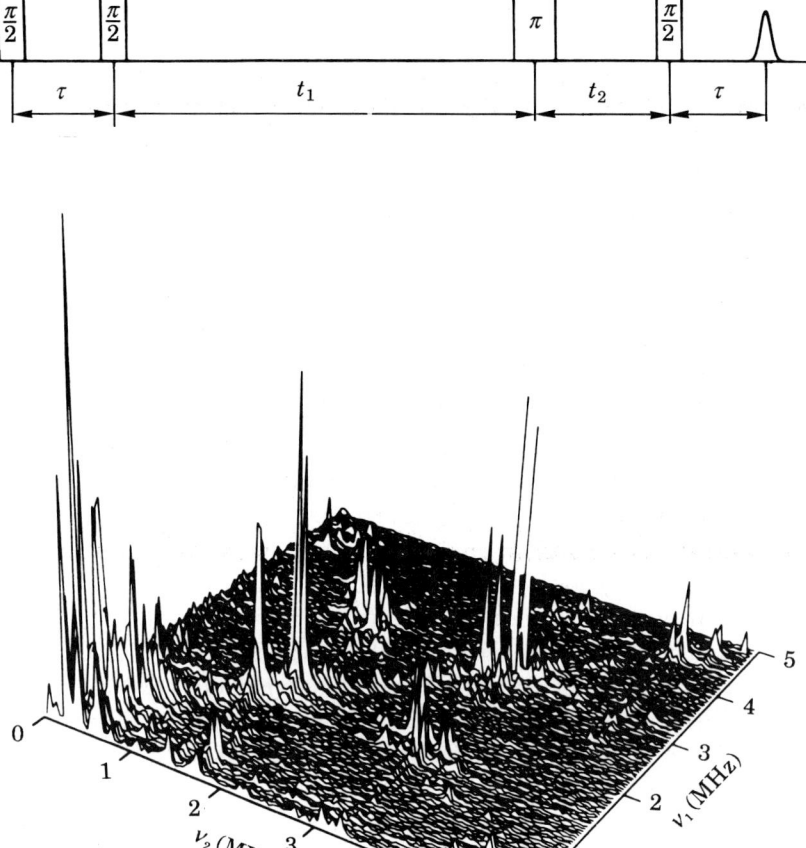

Fig. 19. – Pulse sequence and hyperfine resolved HYSCORE (hyperfine spectroscopy with correlation to an electron spin) experiment on a radical defect in solid squaric acid [27].

4'3. *Optical Zeeman double resonance with a laser diode.* – The final double-resonance experiment to be discussed involves only a single laser diode. The experimental set-up is similar to the one discussed in subsect. 4'1. In subsect. 4'1 besides the laser diode two different frequencies or switched fields were applied in order to achieve the double-resonance effect. Here we restrict ourselves to the use of only a single laser diode emitting near the D_1 lines of ^{85}Rb and ^{87}Rb contained in a gas cell. Tuning of the laser diode through the optical resonance is achieved by temperature and current variation. This way an optical spectrum showing the ^{85}Rb and ^{87}Rb absorption lines is obtained.

How to observe the second resonance, namely the sublevel Zeeman splitting

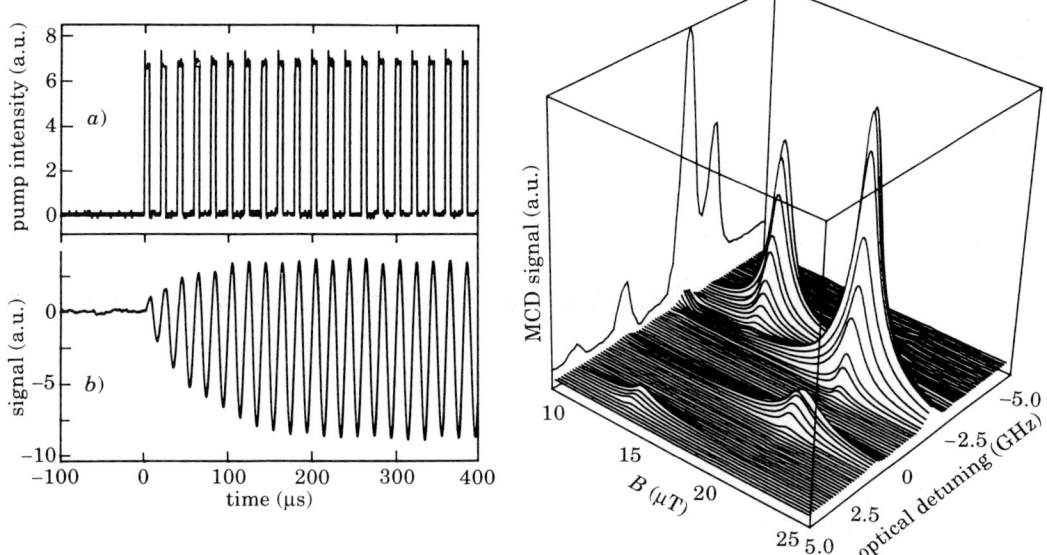

Fig. 20. – Optical Zeeman double-resonance spectroscopy in Rb vapour by using a pulse-modulated infrared laser diode[29]. The laser diode was pulsed with repetition rate $1/\tau$ (τ = pulse spacing) as shown in the left-hand part. When $1/\tau$ corresponds to the Rb sublevel Zeeman splitting an optical response which oscillates with the Larmor frequency is detected (left). If both the wavelength of the diode and the repetition rate of the pulse are varied in a two-dimensional fashion, the corresponding spectrum results (right).

of ^{85}Rb and ^{87}Rb in a small magnetic field? One way to achieve this would be to irradiate the sublevels directly and observe the response as a change in optical absorption (ODMR: optical detection of magnetic resonance)[30]. A more subtle approach is presented in fig. 20[29]. There the laser diode is pulsed with repetition rate $1/\tau$. Part of the continuous laser light has been split off to be used as monitoring light. Under the action of the pulsed light the Rb sublevels are created in a superposition of states leading to coherent oscillation. This oscillation is visible as a change in the absorption of the monitoring light as is demonstrated in the left-hand part of fig. 20. It is to be noted that such a Zeeman oscillation only occurs if the Zeeman splitting corresponds to multiples of the repetition rate $1/\tau$ of the pulse. More details can be found in ref.[29].

The two-dimensional «double resonance» spectrum shown in fig. 20 results if the laser diode detuning is combined with a variation in repetition rate in a two-dimensional fashion. Note the extreme difference in frequency scale of the optical (GHz) and the Zeeman axis (kHz) which arises from the application of very small magnetic fields (mG).

The double-resonance feature of this experiment obviously lies in the application of the optical frequency and the modulation frequency of the light.

5. – Summary.

I have tried to cover a wide range of double-resonance phenomena. Not always was the classical double-resonance schemes applicable, which implies the application of two different frequencies in order to irradiate at two different transitions. We have seen that even the application of a single frequency can result in double-resonance phenomena. What, therefore, is the meaning of «double resonance»? I would like to distinguish the following scenarios of double-resonance experiments:

a) irradiation at two different frequencies ν_1 and ν_2 is applied,

b) pulsed irradiation at only a single frequency ν_1 is applied, but a separate time scale t_2 (pulse separation) is involved,

c) pulsed fields are applied with two different time scales t_1 and t_2.

If we consider any time variation (by Fourier transform) as having frequency components, all these scenarios can still be summarized as «double resonance» experiments.

* * *

I would like to acknowledge many interesting discussions on this subject with A. PINES. I am indebted to G. ZIMMER and K. F. THIER for the careful proofreading of the manuscript. Financial support of part of this work was given by the «Fonds der Chemischen Industrie».

REFERENCES

[1] S. R. HARTMANN and E. L. HAHN: *Phys. Rev.*, **128**, 2042 (1962).
[2] *a*) C. P. SLICHTER and W. C. HOLTON: *Phys. Rev.*, **122**, 1701 (1961); *b*) F. M. LURIE and C. P. SLICHTER: *Phys. Rev.*, **133**A, 1108 (1964).
[3] A. ABRAGAM: *The Principles of Nuclear Magnetism* (Clarendon Press, Oxford, 1961).
[4] C. P. SLICHTER: *Principles of Magnetic Resonance*, 3rd edition (Springer, Berlin, 1989).
[5] M. GOLDMAN: *Spin Temperature and Nuclear Magnetic Resonance in Solids* (Oxford University Press, London, 1970).
[6] M. MEHRING: *Principles of High Resolution NMR in Solids* (Springer-Verlag, Heidelberg, 1983).
[7] M. MEHRING, E. K. WOLFF and M. E. STOLL: *J. Magn. Reson.*, **37**, 475 (1980).
[8] *a*) S. VEGA and A. PINES: *J. Chem. Phys.*, **66**, 5624 (1977); *b*) J. VEGA: *J. Chem. Phys.*, **68**, 5518 (1978).
[9] A. WOKAUN and R. R. ERNST: *J. Chem. Phys.*, **67**, 1752 (1977).

[10] A. GRUPP and M. MEHRING: in *Modern Pulsed and Continuous-Wave Electron Spin Resonance*, edited by L. KEVAN and M. BOWMAN (John Wiley, New York, N.Y., 1990), p. 195.
[11] W. B. MIMS: *Proc. R. Soc. London*, **283**, 452 (1965).
[12] E. R. DAVIES: *Phys. Lett. A*, **47**, 1 (1974).
[13] P. HÖFER, A. GRUPP and M. MEHRING: *Phys. Rev. A*, **33**, 3619 (1986).
[14] M. MEHRING, P. HÖFER, H. KÄSS and A. GRUPP: *Europhys. Lett.*, **6**, 463 (1988).
[15] *a*) D. E. KAPLAN and E. L. HAHN: *J. Phys. Radium*, **19**, 821 (1958); *b*) M. EMSCHWILLER, E. L. HAHN and D. KAPLAN: *Phys. Rev.*, **118**, 414 (1960).
[16] M. E. STOLL, A. J. VEGA and R. W. VAUGHAN: *Phys. Rev. A*, **16**, 1521 (1977).
[17] M. MEHRING, P. HÖFER and A. GRUPP: *Phys. Rev. A*, **33**, 3523 (1986).
[18] A. PINES, M. G. GIBBY and J. S. WAUGH: *J. Chem. Phys.*, **56**, 1776 (1972); **59**, 569 (1973); *Chem. Phys. Lett.*, **15**, 373 (1972).
[19] M. MEHRING: *Principles of High Resolution NMR Spectroscopy in Solids* (Springer-Verlag, Heidelberg, 1976), Appendix E.
[20] A. ABRAGAM and M. GOLDMAN: *Nuclear Magnetism: Order and Disorder* (Clarendon Press, Oxford, 1982).
[21] M. H. LEVITT, D. SUTER and R. R. ERNST: *J. Chem. Phys.*, **84**, 4234 (1986).
[22] D. DEMCO, J. TEGENFELDT and J. S. WAUGH: *Phys. Rev. B*, **11**, 4133 (1975).
[23] *a*) M. MEHRING, G. SINNING and A. PINES: *Z. Phys. B*, **24**, 73 (1976); *b*) M. MEHRING and G. SINNING: *Phys. Rev.*, **15**, 2519 (1977).
[24] U. DEININGHAUS and M. MEHRING: *Phys. Rev. B*, **24**, 4945 (1981).
[25] W. HAPPER: *Rev. Mod. Phys.*, **44**, 169 (1972).
[26] M. MEHRING, S. APPELT, H. LANGEN and G. WÄCKERLE: in *High Precision Navigation 91*, edited by K. LINGKWITZ and U. HANGLEITER (Dümmler Verlag, Bonn, 1992), p. 29.
[27] P. HÖFER, A. GRUPP, H. NEBENFÜHR and M. MEHRING: *Chem. Phys. Lett.*, **132**, 279 (1986).
[28] W. B. MIMS: *Phys. Rev. B*, **5**, 2409 (1972).
[29] G. WÄCKERLE, S. APPELT and M. MEHRING: *Phys. Rev. A*, **43**, 242 (1991).
[30] H. G. DEHMELT: *Phys. Rev.*, **105**, 1924 (1957).

Polarization Transfer Experiments.

B. H. Meier

Laboratorium für Physikalische Chemie, ETH-Zentrum - 8092 Zürich, Switzerland

1. – Introduction.

Polarization transfer processes play an important role in a number of solid-state NMR experiments. Heteronuclear cross-polarization experiments are of particular interest for the signal enhancement in rare-spin, low-γ nuclei[1]. Homonuclear polarization transfer experiments[2-13], often called spin diffusion experiments, are of considerable practical importance because they be can be used to obtain direct information about the geometrical arrangement of the nuclei. Detailed information is most easily available in cases where the resonance lines of the spins participating in polarization transfer can be separated in the solid-state NMR spectrum due to their chemically or crystallographically different environment. The difference in environment may be manifested in the NMR spectrum through a difference in chemical shifts[4], dipolar[14] or quadrupolar interactions[13] or relaxation rate constants[15,16]. The magnetic dipolar interaction («through space» interaction) and the J coupling, an indirect («through bond») interaction mediated by the electrons, that are responsible for polarization transfer processes, are short-range interactions. Polarization transfer methods are, therefore, particularly useful for the investigation of local structure and local order. They can be used to extract geometrical information in solid materials without long-range order which are particularly difficult to investigate by scattering methods. Important classes of materials include polymers, polymer blends, glasses, ceramic materials, microcrystalline and partially disordered crystalline materials.

In this lecture, I shall discuss some of the basic principles of polarization transfer processes. We will restrict ourselves to polarization transfer caused by the dipolar interaction in the presence of a strong field (Zeeman interaction or interaction with a strong spin-lock field in laboratory frame and rotating-frame experiments (*), respectively). The dipolar Hamiltonian can then be written as the sum of secular and nonsecular terms with respect to the dominant interaction with the polarizing field:

(1)
$$H^{(d)} = H^{(d,\text{ sec})} + H^{(d,\text{ ns})}$$

with

(2)
$$H^{(d,\text{ sec})} = \sum_{k<l} A_{20}^{(kl)} T_{20}^{(kl)}$$

and

(3)
$$H^{(d,\text{ ns})} = \sum_{k<l} \sum_{\substack{m=-2 \\ (m \neq 0)}}^{2} (-1)^m A_{2m}^{(kl)} T_{2,-m}^{(kl)} .$$

A spherical-tensor operator notation is used with the basis operators for the spin part,

(4)
$$\begin{cases} T_{20}^{(kl)} = \dfrac{1}{\sqrt{6}} [3 S_{kz} S_{lz} - \mathbf{S}_k \cdot \mathbf{S}_l] = \dfrac{1}{\sqrt{6}} \left[2 S_{kz} S_{lz} - \dfrac{1}{2} [S_k^+ S_l^- + S_k^- S_l^+] \right], \\ T_{2\pm 1}^{(kl)} = \mp \dfrac{1}{2} [S_k^\pm S_{lz} + S_{kz} S_l^\pm], \\ T_{2\pm 2}^{(kl)} = \dfrac{1}{2} [S_k^\pm S_l^\pm], \end{cases}$$

and the spatial part,

(5)
$$\begin{cases} A_{20}^{(kl)} = -\sqrt{\dfrac{3}{2}}\, d_{kl}\, (3 \cos^2 \theta^{(kl)} - 1), \\ A_{2\pm 1}^{(kl)} = \pm \dfrac{3}{2}\, d_{kl}\, \sin 2\theta^{(kl)}\, \exp[\pm i \varphi^{(kl)}], \\ A_{2\pm 2}^{(kl)} = -\dfrac{3}{2}\, (d_{kl}\, \sin \theta^{(kl)})^2\, \exp[\pm 2i \varphi^{(kl)}]. \end{cases}$$

(*) As usual, polarization transfer experiments where the relevant quantization axis is the static magnetic field will be called laboratory frame experiments, while experiments where the applied r.f. field provides the relevant quantization will be referred to as rotating-frame experiments. This classification should not be confused with the frame of reference where the Hamiltonian is set up and the equation of the motion is solved in. Usually, laboratory frame polarization transfer experiments are described in a rotating (or doubly rotating in heteronuclear cases) frame of reference that is an interaction frame with respect to the dominant Zeeman interaction and rotating-frame experiments are described in a frame of reference obtained by going into a further interaction representation, this time with respect to the r.f. field.

The Euler angles $(\theta, \varphi, 0)$ characterize the coordinate transformation leading from the laboratory frame of reference to the principal axis system (PAS) of the dipolar-interaction (spatial) tensor. The dipolar coupling constant d_{kl} (in units of angular frequencies) is defined as

$$d_{kl} = 2 \frac{\mu_0 \gamma_k \gamma_l \hbar}{4\pi r_{kl}^3}, \tag{6}$$

where r_{kl} denotes the internuclear distance between spins k and l, γ_l and γ_k the gyromagnetic ratios and μ_0 the permeability of free space.

For strong fields the nonsecular contributions can be neglected and the flip-flop term of the dipolar Hamiltonian, $[S_k^+ S_l^- + S_k^- S_l^+]$, is the only mechanism that can lead to polarization transfer in laboratory frame as well as in rotating-frame experiments under time-independent Hamiltonians. For time-dependent Hamiltonians, either in spin or in spatial coordinates, flip-flop as well as flop-flop transitions (caused by the $[S_k^+ S_l^+ + S_k^- S_l^-]$ term of the Hamiltonian) can occur and lead to polarization transfer [17, 18]. Flop-flop transitions conserve the difference polarization of the spins system, while the flip-flop terms preserve the sum polarization.

In heteronuclear experiments, the secular part of the dipolar interaction in the usual doubly rotating frame is further truncated because the flip-flop term in $T_{20}^{(kl)}$ is a nonsecular term with respect to the difference of the Zeeman interactions of the two spins and does not influence the time evolution of the spin system. Under r.f. irradiation, for example in cross-polarization experiments close to Hartmann-Hahn match, a secular flip-flop term is reintroduced.

For the description of polarization transfer experiments, it is often convenient to use thermodynamic concepts to characterize the quasi-equilibrium states approached by the spin system after the initial evolution of the density operator. Furthermore, the periodic time evolution, obtained for any system described by a Hamiltonian, can, in many important applications, be approximated by an exponential time evolution characterized by a rate constant. In these cases, the polarization transfer process is often referred to as polarization diffusion or spin diffusion. Later on in this lecture, I shall describe experiments which demonstrate experimentally that it is feasible to reverse the spatial «diffusion» of polarization [19]. The appearance of *polarization echoes* in these experiments demonstrates clearly that care has to be exercised in describing the polarization transfer by a rate constant. Nevertheless, this approach is very useful for a wide range of polarization transfer experiments although it is only justified by its successful predictions but not by first principles.

We start from the description of the spin system by a Hamiltonian and shall introduce the concept of quasi-equilibrium state as we proceed in the description of the time evolution for a polarization transfer experiment. If the total Hamiltonian, H, is time independent, the time evolution of the density operator, as determined by the equation of the motion,

(7) $$\frac{\partial \sigma}{\partial t} = -i[H, \sigma],$$

is given by [20]

(8) $$\sigma(t) = U\sigma(0)U^\dagger,$$

with the unitary propagator

(9) $$U = \exp[-iHt].$$

For our purposes, the time evolution of the density operator $\sigma(t)$ is most conveniently described in Liouville space where the density operator can be represented by a vector $\boldsymbol{b}$ in l-dimensional space:

(10) $$\sigma(t) = \sum_{s=1}^{l} b_s(t) B_s.$$

It is beneficial to use an orthogonal set of basis operators $\{B_s\}$ of the same norm

(11) $$\langle B_i | B_j \rangle = \delta_{ij} \langle B_i | B_i \rangle,$$

where the scalar product is defined as $\langle B_i | B_j \rangle = \mathrm{Tr}\{B_i^\dagger B_j\}$. For n spins 1/2, the dimension of the full Liouville space is $l = 2^{2n}$. For a given initial state $\sigma(0)$ and a given Hamiltonian H, the trajectory describing the time evolution of $\sigma(t)$ is, however, often confined to a subspace of the Liouville space that has a considerably reduced dimension. An example will be presented below.

The norm of the density operator,

(12) $$\|\sigma\| = \sqrt{\langle \sigma | \sigma \rangle},$$

is a constant of the motion under unitary transformations. Therefore, the spin order, defined as

(13) $$\aleph = \langle \sigma | \sigma \rangle = \|\sigma\|^2,$$

is conserved during the time evolution of the density operator under a Hamiltonian operator. In the high-temperature approximation, which is fulfilled in best approximation for temperatures above 1 K, the spin order is related to the entropy of the system by

(14) $$S = -k\,\mathrm{Tr}\{\sigma(\ln \sigma)\} \approx -\frac{k}{2\,\mathrm{Tr}\{1\}} \aleph + \Xi,$$

with $\Xi = k(\ln(\mathrm{Tr}\{1\}))$, where k denotes the Boltzmann constant. Spin order and entropy are both functions of the state in the sense of thermodynamics and differ only by a (negative) proportionality constant and by a different origin of the scale. Therefore, a decrease in spin order corresponds to an increase in entropy. The spin order can be identified with the square of the norm of the density operator.

Here, we consider the time evolution during mixing time for a dipolar coupled spin system in high field where the Zeeman interaction (or the interaction with the r.f. field, in the case of rotating-frame polarization transfer) dominates the laboratory (or rotating-frame) Hamiltonian. The interaction Hamiltonian contains, in the absence of chemical-shielding differences, only dipolar contributions and from eq. (2), eq. (4) and eq. (5) one finds

$$(15) \qquad H = \sum_{i<j} d_{ij} \left(2 S_{iz} S_{zj} - \frac{1}{2} [S_i^+ S_j^- + S_i^- S_j^+] \right)$$

where nonsecular terms have been neglected.

Schemes to achieve such a purely dipolar Hamiltonian, at least approximately, will be discussed later. For the moment, we just note that, for polarization transfer between spins with identical chemical shift, the Hamiltonian in the usual rotating frame has this form.

1˙1. *A simple application: the two-spin system.* – With the initial condition $\sigma(t_1, \tau = 0) = S_{1z}$, the time evolution of the density operator during mixing is easily found to be

$$(16) \qquad \sigma(t) = b_1 S_{1z} + b_2 S_{2z} + b_3 \frac{i[S_1^+ S_2^- - S_1^- S_2^+]}{\sqrt{2}}$$

with

$$(17) \qquad b_1 = \frac{1 + \cos(dt)}{2},$$

$$(18) \qquad b_2 = \frac{1 - \cos(dt)}{2},$$

$$(19) \qquad b_3 = \sqrt{2} \sin(dt).$$

This trajectory is entirely confined to a three-dimensional subspace ($\{S_{1z}, S_{2z}, i[S_1^+ S_2^- - S_1^- S_2^+]/\sqrt{2}\}$) of the full (16-dimensional) Liouville space where it can be geometrically represented by a circle as shown in fig. 1. The initial density operator, $\sigma(0)$, is shown as an emphasized vector in fig. 1. Lighter vectors represent the density operator at later times of the periodic time evolution. The norm of the density operator and, therefore, the length of the vector in fig. 1 remain constant.

In practice, the oscillations described by eqs. (17) to (19) («transient oscillations» [21]) are invariably damped by additional interactions not contained in the simplified Hamiltonian of eq. (15). Examples include interactions to weakly coupled, «extraneous» spins, a spread or time dependence of the dipolar cou-

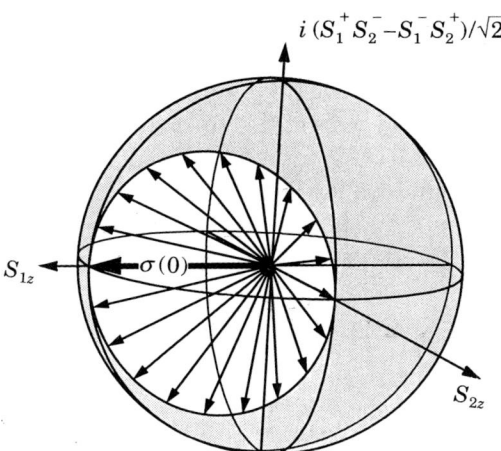

Fig. 1. – Liouville-space representation of the density operator $\sigma(t)$ trajectory for polarization transfer in a two-spin system evolving under a purely dipolar Hamiltonian. The vectors represent the density operators for selected time intervals. The solid line connecting the vectors is the trajectory which falls completely into the depicted three-dimensional subspace of the full 16-dimensional Liouville space.

pling frequencies d_{ij} or additional chemical-shift terms which vary as a function of time or as a function of the position in the sample. In the present context, these interactions are assumed to cause a decay of the transient oscillations without inducing transitions between eigenstates of the unperturbed Hamiltonian. Therefore, the spin system is isolated in the sense that no energy transfer takes place between system and environment. The decay of the off-diagonal elements of the density operator (in the eigenbase of the Hamiltonian) may either be «reversible» or «irreversible» and the classification depends on the degree of approximations made to describe the system. For example, the interaction to extraneous spins is described by a Hamiltonian and is, therefore, in principle, always reversible[19]. Nevertheless, it can be useful and adequate to describe the influence of the extraneous spins approximately by a classical random field leading to an irreversible decay where the spin order decreases as a function of time. Coupling to classical degrees of freedom, for example the modulation of the dipolar interaction by a two-sided chemical reaction, always leads to irreversible behaviour.

After the transient oscillations have decayed irreversibly, the spin system is described by a quasi-equilibrium density operator σ^{qe} which represents the part of the initial density operator that commutes with the Hamiltonian. While this quasi-equilibrium state does not evolve anymore under the simplified Hamiltonian, it is not usually an equilibrium state under the full Hamiltonian (that contains also the terms that have been neglected to obtain the system Hamiltonian) and is not in equilibrium with the lattice. The quasi-equilibrium corresponds to a state with minimum spin order under the constraint that all k constants of the motion $\langle A^{(i)} \rangle$ with

(20) $$[H, A^{(i)}] = 0$$

are conserved:

(21) $$\langle A^{(i)} \rangle = \text{Tr}\{\sigma(0) A^{(i)}\} = \text{Tr}\{\sigma(t) A^{(i)}\} = \text{Tr}\{\sigma^{\text{qe}} A^{(i)}\}, \quad i = 1, \ldots, k.$$

The knowledge about the quasi-equilibrium state is restricted to the expectation values of the constants of the motion. If the solution of the equation of the motion is known, for example for the two-spin system (eqs. (17) to (19)), the quasi-equilibrium density operator can be obtained by simply dropping all time-dependent terms. For the two-spin system, $\sigma^{\text{qe}} = (1/2)(S_{1z} + S_{2z}) = S_z^{(1, 4)}$ is obtained.

The quasi-equilibrium density operator can also be directly evaluated without knowledge of the solution of the equation of the motion. This is of importance for larger spin systems where the solution of the equation of the motion is extremely tedious. The quasi-equilibrium state is obtained by projecting the initial density operator onto the k constants of the motion:

(22) $$\sigma^{\text{qe}} = \sum_{i=1}^{k} P^{(i)} \sigma(0) = \sum_{i=1}^{k} \beta^{(i)} A^{(i)}$$

with

(23) $$P^{(i)} = \frac{|A^{(i)}\rangle\langle A^{(i)}|}{\langle A^{(i)} | A^{(i)} \rangle}$$

and

(24) $$\beta^{(i)} = \frac{\langle A^{(i)} | \sigma(0) \rangle}{\langle A^{(i)} | A^{(i)} \rangle}.$$

The projection coefficients $\beta^{(i)}$ shall be called «inverse spin temperatures» [22].

A particularly simple situation arises in so-called ergodic systems where the total energy (for the high-field case $\langle H^{(\text{Zeeman})} \rangle$) is the only constant of the motion. Ergodicity guarantees that a) the quasi-equilibrium density operator σ^{qe} is proportional to the Hamiltonian, independent of the initial condition, $\lim_{t \to \infty} \sigma(t) = \beta H$, and b) that the time average of the density operator s of any member of the ensemble making up the spin system equals the ensemble average:

(25) $$\lim_{T \to \infty} \frac{1}{T} \int_0^T s(t)\, dt = \sigma^{\text{qe}}.$$

For a dipolar coupled two-spin system (see eq. (15)), two constants of motion $A^{(1)}$ and $A^{(2)}$ exist and can be found as the solutions of the equation $[H, A] = 0$:

$$A^{(1)} = \frac{1}{2}(S_{1z} + S_{2z}) = S_z^{(1,4)}, \tag{26}$$

$$A^{(2)} = \frac{1}{2}[S_1^+ S_2^- + S_1^- S_2^+] = S_x^{(2,3)}, \tag{27}$$

where the single transition operators [23-25]

$$(28) \begin{cases} S_x^{(1,4)} = \frac{1}{2}[S_1^+ S_2^+ + S_1^- S_2^-], & S_y^{(1,4)} = \frac{-i}{2}[S_1^+ S_2^+ - S_1^- S_2^-], \\ S_z^{(1,4)} = \frac{1}{2}[S_{1z} + S_{2z}], & S_x^{(2,3)} = \frac{1}{2}[S_1^+ S_2^- + S_1^- S_2^+], \\ S_y^{(2,3)} = \frac{-i}{2}[S_1^+ S_2^- - S_1^- S_2^+], & S_z^{(2,3)} = \frac{1}{2}[S_{1z} - S_{2z}] \end{cases}$$

have been used.

$A^{(2)}$ is not a member of the relevant subspace $\{S_{1z}, S_{2z}, i[S_1^+ S_2^- - S_1^- S_2^+]/\sqrt{2}\}$ containing the solution of the equation of the motion (eq. (16)) for the initial condition $\sigma(0) = S_{1z}$, and $A^{(1)}$ is proportional to the Hamiltonian. Therefore, the system behaves ergodically. Starting from the initial condition given above, the quasi-equilibrium density operator is easily obtained as

$$\sigma^{qe} = \frac{\text{Tr}\{\sigma(0)H\}}{\text{Tr}\{HH^\dagger\}} H = \frac{\text{Tr}\{\sigma(0)A^{(1)}\}}{\text{Tr}\{A^{(1)}A^{(1)}\}} A^{(1)} = \frac{1}{2}(S_{1z} + S_{2z}). \tag{29}$$

The projection procedure of eq. (29) is illustrated in fig. 2 in a two-dimensional subspace spanned by $\{S_{1z}, S_{2z}\}$. The initial density operator, $\sigma(0)$, is chosen to be proportional to S_{1z}, corresponding to polarization on one spin only. The trajectory of the spin system is required to follow the solid emphasized line in fig. 2 which corresponds to the conservation of the energy as the only relevant constant of the motion. The circle is the so-called spin order bound or entropy bound. As mentioned before, the spin order can never increase during time evolution. Therefore, the trajectory for the density operator is entirely confined within this circular bound. The quasi-equilibrium density operator is obtained by projection of the initial density operator onto the Hamiltonian. Ergodicity guarantees that each particular point on the trajectory is actually passed by each member of the ensemble at some point in time.

If the relevant Hamiltonian of the spin system consists not only of a dipolar term, as assumed above, but, in addition, a chemical-shift difference term is present,

$$H^{(2,3)} = \Sigma S_z^{(1,4)} + \delta S_z^{(2,3)} - d S_x^{(2,3)} \tag{30}$$

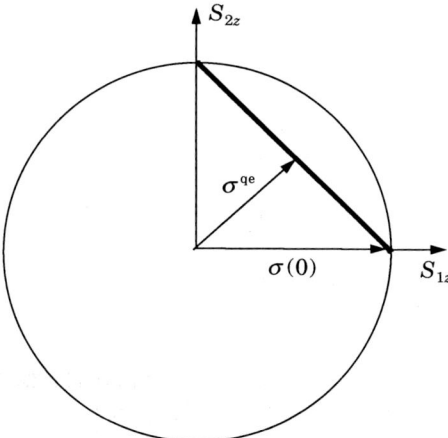

Fig. 2. – Liouville-space representation of the projection onto the Hamiltonian that leads from the initial density operator, $\sigma(0) = S_{1z}$, to the quasi-equilibrium density operator σ^{qe}. The projection of the trajectory onto the shown subspace is indicated by the emphasized line.

with $\Sigma = (\Omega_1 + \Omega_2)/2$ and $\delta = (\Omega_1 - \Omega_2)/2$, where the Ω_i denote the rotating-frame resonance frequencies of the spins. The constants of the motion take the form

$$(31) \qquad A^{(1)} = S_z^{(1,\,4)}, \qquad A^{(2)} = \delta S_z^{(2,\,3)} - d S_x^{(2,\,3)}.$$

Starting from an initial condition $\sigma(0) = S_{1z}$, a quasi-equilibrium density operator of

$$(32) \qquad \sigma^{\text{qe}} = S_z^{(1,\,4)} + \frac{\delta}{d^2 + \delta^2} \left(\delta S_z^{(2,\,3)} - d S_x^{(2,\,3)} \right)$$

is obtained. For $\delta \to 0$, eq. (32) reduces to eq. (29) that describes the purely dipolar case. For $\delta > 0$ the polarizations of the two spins do not equilibrate in the course of the time evolution, but less than half of the initial polarization on spin one is transferred to spin two [22]. Furthermore, the transfer of polarization is now accompanied by a build-up of a dipolar term $S_x^{(2,\,3)}$ [22] and the energy required for the flip-flop transition of the two spins is provided by the dipolar interaction. For $\delta/d = 1$, only half as much polarization as in the degenerate case $\delta = 0$ is transferred. For large chemical-shift differences $\delta \gg d$, the polarization transfer is entirely quenched.

2. – Application to the MOIST experiment.

The graphical interpretation of the polarization transfer in a Liouville state, as introduced above, allows for an intuitive description of the MOIST experi-

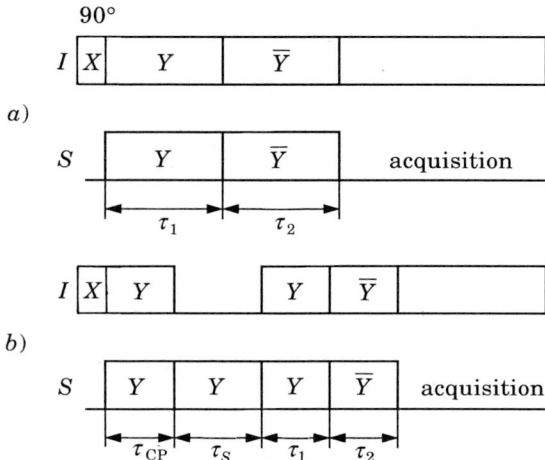

Fig. 3. – Spin-polarization (a)) and depolarization (b)) pulse sequences with simultaneous π phase shifts of the two r.f. fields (MOIST) for $m = 2$. The depolarization sequence starts with cross-polarization during $\tau_{\rm CP}$. τ_S is chosen such that the proton polarization decays to zero.

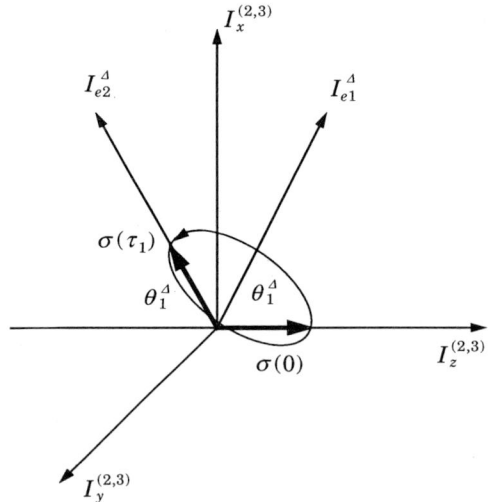

Fig. 4. – Diagram showing the evolution of the density operator in the zero-quantum subspace. Initially aligned along $S_z^{(2,\,3)}$, the evolution of the density operator during τ_1 and τ_2 is described by rotations about the two effective field axes. Here, a special case is shown where, after τ_1, the density operator is aligned along the effective field direction I_{e2}^{Δ}.

ment [26, 27] in systems with resolved dipolar interactions. The aim of the MOIST experiment is to reduce the sensitivity of the cross-polarization efficiency to a mismatch of the Hartmann-Hahn condition. The pulse sequence is shown in fig. 3a). Here we discuss a greatly simplified description of the spin system and take into account explicitly only one S spin and its most strongly coupled I-spin neighbour. All other I spins are included into the «environment» and are taken into account only as a damping mechanism for the transient oscillations between I and S spins. Such a situation is approximately realized in ferrocene, where the heteronuclear dipolar coupling is about a factor of four larger than the homonuclear proton interaction. It is seen from eq. (31) that the sum polarization is a constant of the motion of the simplified spin system. We can, there-

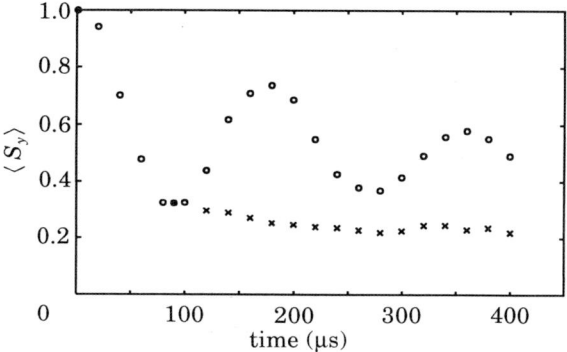

Fig. 5. – Depolarization of ^{13}C in a single crystal of ferrocene obtained with the pulse sequence of fig. 3b) with $\tau_{CP} = 5$ ms and $\tau_S = 1$ ms, $\omega_{1I}/2\pi = 41.66$ kHz and $\omega_{2I}/2\pi = 38.63$ kHz. Circles represent the transient oscillation as a function of τ_1. The signal represented by crosses is obtained after a π phase shift applied at $\tau_1 = 90$ μs corresponding to half a precession period. The transient oscillation is quenched by the phase shift (adapted from ref. [27]).

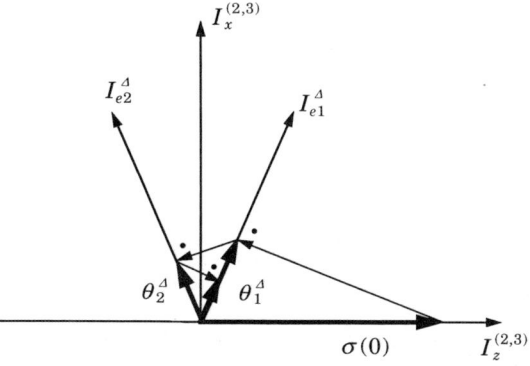

Fig. 6. – Schematic representation of the successive projections of the Hamilton operator caused by phase switching (adapted from ref. [27]).

fore, represent the time evolution in the zero-quantum (2, 3) subspace introduced in eq. (28) as shown in fig. 4. The double-quantum (1, 4) subspace does not contribute to the evolution of the density operator, because $\sigma^{(1,\,4)}(t) = \sigma^{(1,\,4)}(0)$. The evolution during the first cross-polarization period τ_1 is described by a rotation around an effective field axis that is tilted by an angle $\theta_1^{\downarrow}$ from the z-axis of the zero-quantum subspace of eq. (28), where $\mathrm{tg}\,\theta_1^{\downarrow} = d/(\omega_{1I} - \omega_{1S})$. After the r.f. phase on both channels is switched simultaneously by 180°, the orientation of the effective field axes changes to a new tilt angle $\theta_2^{\downarrow} = -\theta_1^{\downarrow}$. In general, a new transient oscillation around that axis will occur during the second cross-polarization time τ_2. In fig. 4, a special situation is shown, where the

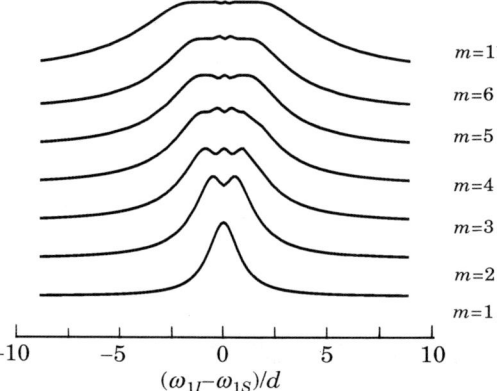

Fig. 7. – Calculated quasi-equilibrium S-spin polarization in a MOIST experiment for an IS spin system as a function of the Hartmann-Hahn mismatch for different numbers of phase switches $m - 1$ (adapted from ref. [27]).

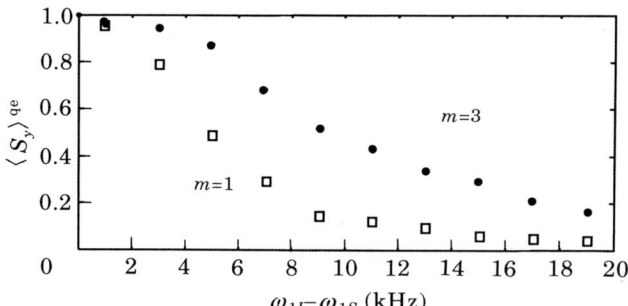

Fig. 8. – Experimental quasi-equilibrium values for the S-spin polarization in a ferrocene single crystal after $m - 1$ phase switches for $m = 1$ (open squares) and $m = 3$ (circles) as a function of the Hartmann-Hahn mismatch. The r.f. field on the proton channel was kept constant at 42 kHz, while the carbon r.f. field was adjusted to achieve the desired mismatch. A total cross-polarization time of 2 ms was applied. The heteronuclear dipole coupling frequency for the chosen crystal orientation amounted to 5.8 kHz. Qualitatively, the behaviour calculated for the isolated two-spin system (see fig. 7) is reproduced [28].

density operator after the first oscillation is aligned in parallel with the effective field axes for the second cross-polarization period. This leads to a quench of the transient oscillation. This behaviour was experimentally verified using a single-crystal sample of ferrocene. For the experimental investigation of the cross-polarization process, it is often beneficial to prepare an initial state of pure S-spin polarization and observe its time evolution during cross-polarization contact to the I spins. This experiment is sometimes referred to as «depolarization» experiment and is schematically shown in fig. 3b). The experimental

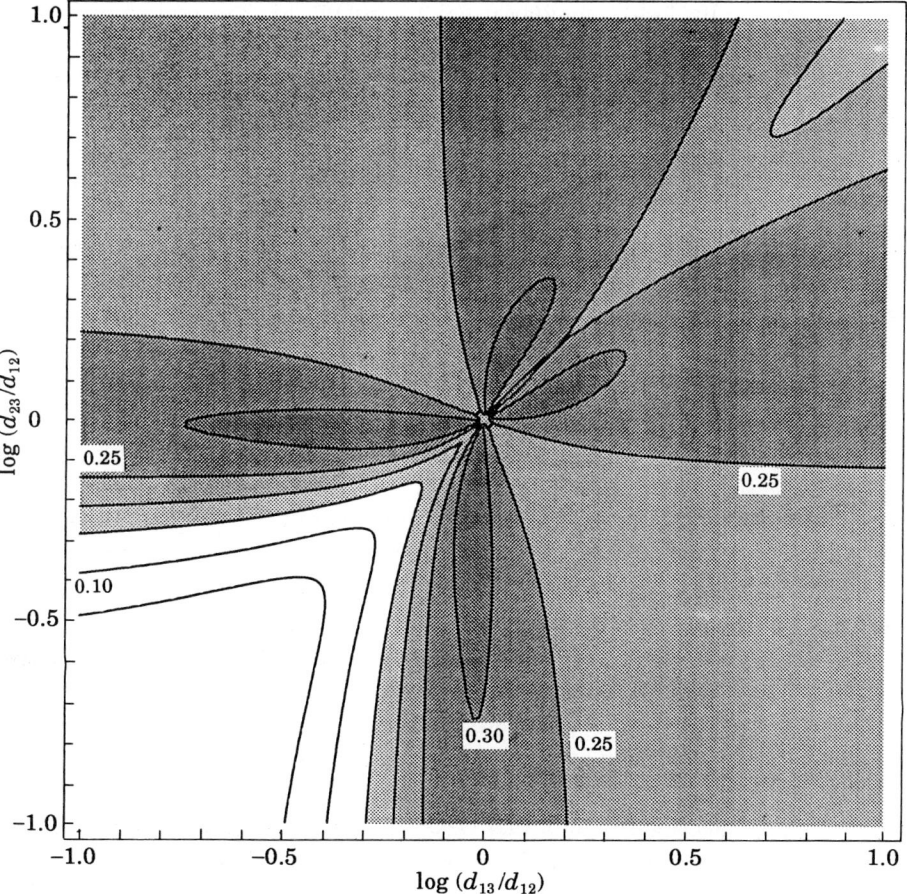

Fig. 9. – Plot of the expectation value of $\langle S_{3z} \rangle$ in the quasi-equilibrium state as a function of the ratios q and r of the dipolar-coupling frequencies in a three-spin-1/2 system. The initial condition is chosen to be $\sigma(0) = S_{1z} + S_{2z}$ and the total polarization $\langle S_{1z} + S_{2z} + S_{3z} \rangle$ is normalized to 1. Darker shaded areas correspond to higher polarization transfer. The maximum value of $\langle S_{3z} \rangle$ is 1/3 corresponding to a uniform distribution of the polarization over all three spins. This value is reached only for special combinations of dipolar-coupling constants. It is, in particular, not reached for the symmetric case $d_{12} = d_{13} = d_{23}$ where $\langle S_{3z} \rangle = 2/9$ is found.

results are given in fig. 5. For the phase-switched experiment (crosses in fig. 5), the predicted quenching of the transient oscillation is verified. If the time period between consecutive switches of the r.f. phase is chosen long enough such that all transient oscillations have decayed, the MOIST experiment with $m - 1$ phase switches is described by consecutive projections of the initial density operator onto the two effective field axes, as shown in fig. 6. After a number of projections, the difference polarization decays and, even for mismatched Hartmann-Hahn conditions, the polarizations of the I and S spins are equilibrated. The larger the mismatch, the more phase switches are required. This behaviour was numerically simulated for a two-spin system and the broadening of the CP spectrum as a function of the number of phase switches $(m - 1)$ is shown in fig. 7. The experimentally determined dependence of the quasi-equilibrium S-spin polarization on the Hartmann-Hahn mismatch is shown in fig. 8 for no phase switches (normal cross-polarization experiment) and for 2 phase switches. Similar results for a different crystal orientation are given in ref.[26]. A qualitative agreement is found. No quantitative comparison shall be attempted because the model used here has been oversimplified. A more adequate description of the spin system has to take into account the neighbouring proton spins and the two-spin approximation has to be released. Attempts for such a treatment can be found in the literature[26, 27].

3. – Larger spin systems.

For more than two spins, several relevant constants of the motion are found even without a chemical-shift difference term or a Hartmann-Hahn mismatch. For a three-spin system, one finds, for general values of the ratios of the dipolar coupling frequencies $r = d_{23}/d_{12}$ and $q = d_{13}/d_{12}$, three constants of the motion and, for the special case $r = q = 1$, four constants of the motion[29]. As an example, I shall shortly discuss the quasi-equilibrium values for the polarization in a homonuclear three-spin system where two spins (spins 1 and 2) are not distinguishable by the chemical shielding. For an initial condition where spins 1 and 2 are polarized, $\sigma(0) = S_{1z} + S_{2z}$, the quasi-equilibrium density operator was calculated by projection of the initial density operator onto the constants of the motion and the expectation value $\langle S_{3z} \rangle$ is plotted as a function of r and q in fig. 9. It is interesting to note that for the degenerate case, $r = q = 1, \langle S_{3z} \rangle = 2/9$ is found, which is less than $1/3$, corresponding to an even distribution over the three spins.

For more complicated spin systems, the evaluation of the constants of the motion is a formidable task. Therefore, bounds for the range of possible quasi-equilibrium states as well as for accessible transient states shall be discussed. For simplicity, only spin systems with two inequivalent groups of spins with N and M members, respectively, shall be discussed. In short, we denote these by

$I_N S_M$. Total I and S spin polarization is defined as the expectation values of the operators $F_{1z} = \sum_{i=1}^{N} I_{iz}$ and $F_{2z} = \sum_{i=1}^{M} S_{1z}$, respectively. For the following discussion, we shall assume an initial condition with polarization on the first group of spins only:

$$\sigma(0) = \alpha F_{1z}. \tag{33}$$

The quasi-equilibrium polarization on the second spin group can be written as

$$\langle F_{2z} \rangle = \alpha(1 - \langle F_{1z} \rangle) = \alpha \left(1 - \frac{N}{M+N} - \sum_{i=2}^{k} \frac{(\mathrm{Tr}\{F_{1z} A^{(i)}\})^2}{\mathrm{Tr}\{A^{(i)} A^{(i)}\}} \right), \tag{34}$$

where all the terms under the sum are necessarily positive numbers. Their existence can only decrease the amount of transferred polarization and the transfer is most effective for the ergodic case, where it yields a polarization ratio (for spin 1/2) of

$$\frac{\langle F_{2z} \rangle}{\langle F_{1z} \rangle} = \frac{M}{N}. \tag{35}$$

It is instructive to plot the projection of the trajectory followed by the density operator onto a plane spanned by the two normalized Liouville-space basis operators $\{F_{1z}/\sqrt{N}, F_{2z}/\sqrt{M}\}$. The high-field Hamiltonian in the laboratory frame, $H_0 = F_{1z} + F_{2z}$, is contained in this plane and the projection onto the two axes yields the normalized expectation values $\langle F_{1z} \rangle/\sqrt{N}$ and $\langle F_{2z} \rangle/\sqrt{M}$. Figure 10 shows such a representation for $I_x S$, $I_x S_x$ and IS_x spin systems for a few values of x. In high field, the total polarization is a constant of the motion. Therefore, all possible states lay on a straight line of slope $-\sqrt{N/M}$ and abscissa $\sqrt{N/M}$, shown as a light line in the figure. The segment of this line that corresponds to possible quasi-equilibrium states lies between the initial value (in the figure along the horizontal axis) and the intersection of the line with the Hamiltonian. This region is emphasized by a broader line in fig. 10.

In a spin system with a limited coupling network, not all states allowed by the entropy bound can actually be reached from a given initial condition by unitary transformations. This was pointed out first by SØRENSEN [30], and is also discussed in ref. [31-34]. Reference [34] describes the shape of the «unitary bound» for a number of spin clusters as obtained by a computer search procedure. These bounds are plotted for some IS_x spin clusters in fig. 11 and it is clearly seen that only part of the high-field trajectory allowed by the spin order bound lays within the unitary bound. The allowed segment of the trajectory is distributed asymmetrically around the intersection of the trajectory with the Hamiltonian. Nevertheless, these systems can, if the values of the dipolar coupling parameters are appropriate, achieve a quasi-equilibrium density operator

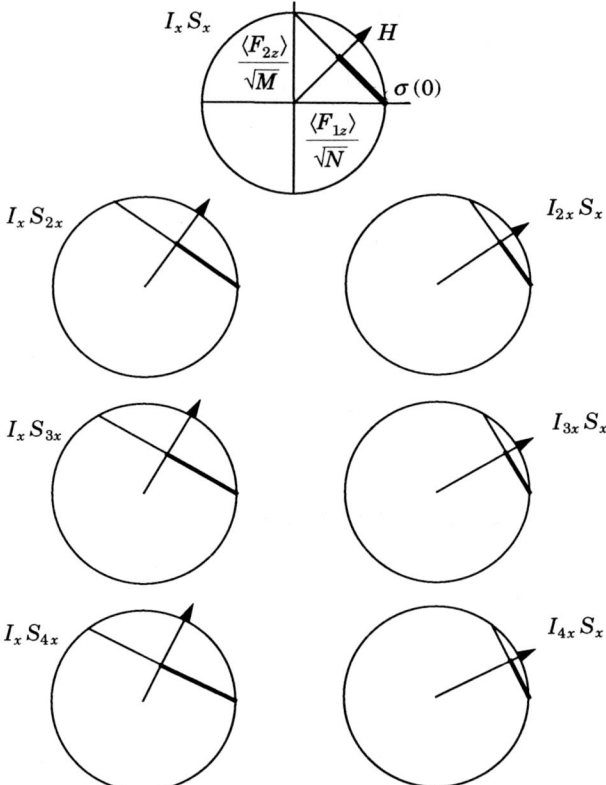

Fig. 10. – Range of possible quasi-equilibrium states (emphasized line) for a number of spin systems of the type $I_N S_M$. The index x stands for an arbitrary integer number. The thin line represents the condition of conservation of the total magnetization as it has to be fulfilled in the high-field approximation. The circle indicates the range of constant spin order. Because the spin order can never increase, all allowed states have to lay within this circle through the origin and the initial condition $\sigma(0)$. The scaled magnetization of the two spin groups (I and S) can be obtained by projection of the quasi-equilibrium state onto the coordinate axes $F_{1z}/\sqrt{N}$ and $F_{2z}/\sqrt{M}$, respectively. The initial condition is always chosen along $F_{1z}/\sqrt{N}$.

proportional to the Hamiltonian. The time average of the trajectory will lead to a quasi-equilibrium value located at the intersection of the trajectory with the Hamiltonian. Obviously, the probability distribution of states along the allowed part of the trajectory can, for these cases, not be uniform.

In summary, it can be said that the unitary bound restricts the accessible states in Liouville state and thereby the maximum transient polarization transfer. It does, however, not influence the quasi-equilibrium polarization transfer under high-field conditions. The maximum quasi-equilibrium transfer is obtained under ergodic conditions which, for the three-spin system discussed above, amounts to $\langle F_{2z} \rangle / \langle F_{1z} \rangle = M/N$.

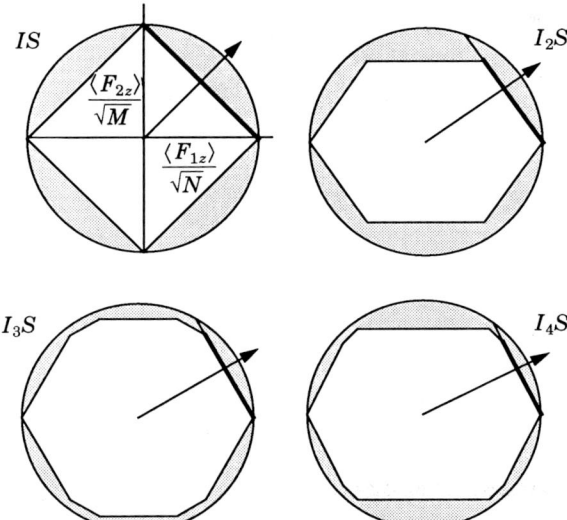

Fig. 11. – Range of allowed transient states (projection onto the $\{F_{1z}/\sqrt{N}, F_{2z}/\sqrt{M}\}$-plane) for a number of spin clusters of the type $I_x S$. States of constant spin order lie on the circle shown in the figure. Allowed states have to lay within this circle («entropy bound»). The shaded areas denote the range allowed by the entropy bound which, however, can never been reached through processes characterized by a unitary transformation. The co-ordinates of the polygons have been taken from [32]. The thin line represents the condition of conservation of the total magnetization as it has to be fulfilled in the high-field approximation. The emphasized line indicates the allowed part of the trajectory.

For large spin systems, it is usually assumed that they are ergodic and the total energy is the only constant of the motion. The description of larger spin systems is, therefore, considerably simpler than the one of intermediate-size spin systems. One should, however, keep in mind that this assumption is not justified in all cases. If, for example, a dominant dipolar coupling exists, leading to a partially resolved spectrum, the assumption of ergodicity is not well justified.

The rate constant for the time evolution towards quasi-equilibrium is usually calculated, using perturbation theory, as $W_{ij} = (\pi/2) d_{ij}^2 f(0)$, where $f(0)$ is the intensity of the normalized zero-quantum spectrum at frequency zero [13, 35].

4. – Time-dependent Hamiltonians.

The concepts for polarization transfer discussed above rely on the assumption that the Hamiltonian is explicitly time independent. The case of explicit time independence includes random time dependences characterized by a correlation time or a time constant for an exchange process as they appear in the relaxation superoperator $\widehat{\varGamma}$ and in the exchange superoperator $\widehat{K}$, respectively.

Excluded, however, are terms which contain the time as a variable, for example time-dependent r.f. irradiation in multiple pulse sequences or time-dependent chemical shifts and dipolar coupling constants in magic-angle spinning (MAS) experiments.

For explicitly time-dependent Hamiltonians, the solution of the equation of the motion (eq. (7)) is no longer given by eq. (8) but by the more complicated form

$$\sigma(t) = U(t)\,\sigma(0)\,U(t)^{-1} \tag{36}$$

with

$$U(t) = T\exp\left[-i\int_0^t H(t')\,\mathrm{d}t'\right], \tag{37}$$

where T denotes the Dyson time-ordering operator [36, 37].

An additional complication arises because the thermodynamic picture of quasi-equilibrium states, as has been extensively used above, is no longer valid. Thermodynamical arguments are restricted to conservative, time-independent systems [38, 39].

In the following, we restrict the discussion to the practically important case of a periodic time dependence with a cycle time t_c much shorter than the inverse rate constant for the polarization transfer. The periodic Hamiltonian, $H(t + t_c) = H(t)$, can then be approximately replaced by a time-independent Hamiltonian $\overline{H}$ and the solution of the equation of the motion is again of the form

$$\sigma(t) \approx \exp[-i\overline{H}t]\,\sigma(0)\exp[i\overline{H}t]. \tag{38}$$

An average Hamiltonian, $\overline{H}$, can be obtained after transformation into an appropriate interaction frame by dropping the nonsecular or time-dependent parts of the interaction Hamiltonian [40]. An alternative and more general way to treat the situation is to apply Floquet theory [25, 41, 42].

As an illustrative application of the interaction frame approach, we consider ^{13}C-^1H cross-polarization under MAS in the «fast-spinning» limit where the MAS frequency, $\omega_r/2\pi$, exceeds the proton-proton interactions [18]. The basic phenomenon was demonstrated by WAUGH et al. [43]. The Hartmann-Hahn condition [44], as known from CP dynamics in static samples, $\omega_{1I} = \omega_{1S}$, is split into sidebands appearing at the Hartmann-Hahn match plus or minus integer multiples of the MAS frequency, $\omega_{1I} = \omega_{1S} \pm n\omega_r$ [45, 46], with $\omega_{1I} = -\gamma_I B_{1I}$ and $\omega_{1S} = -\gamma_S B_{1S}$. For sufficiently fast spinning, the sidebands for $n = \pm 1$ and ± 2 dominate the CP spectrum. Here and in the following, we refer to the observed S-spin polarization in a CP experiment as a function of the strength of the S-spin irradiation ω_{1S} as the «CP spectrum». The thermodynamics in this system is altered by the absorption or emission of energy by the mechanical rotation of

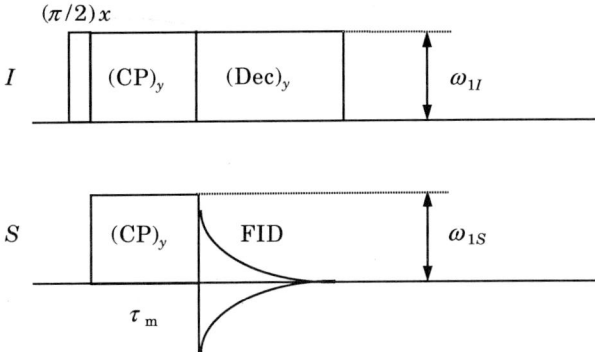

Fig. 12. – Standard CP pulse sequence. The adamantane FID was acquired as a function of the carbon r.f. irradiation strength ω_{1S} in the form of a two-dimensional data matrix. After Fourier transformation of the time domain data, the magnetization transferred from protons to ^{13}C is obtained as a function of ω_{1S} for each of the two carbon resonances separately. Typically 256 to 2048 ω_{1S} increments were recorded. The spectra presented in fig. 13 and 14 are traces along the ω_{1S} axis at the position of a ^{13}C resonance. The experiments were performed on a Bruker MSL-400 spectrometer equipped with a selective excitation unit (SEU) and a linear power amplifier on the carbon channel (Amplifier Research Type 100LM8). The SEU was used because it has a high resolution in the ω_{1S} domain (12 bits). It is, however, neither linear in attenuation (as a function of the nominal digital input value) nor phase constant as a function of the attenuation. The spectrum has, therefore, to be corrected for both effects based on a calibration measurement. The field strength on the proton channel was set to 12.5 kHz. (Figure adapted from ref. [18].)

the sample. This behaviour does not only result in a number of modified «Hartmann-Hahn conditions» for the heteronuclear zero-quantum (flip-flop) transitions, but it will be demonstrated in the following that additional «Hartmann-Hahn conditions« appear where the heteronuclear double-quantum transition becomes an energy-conserving mechanism. For the double-quantum CP, the expectation value of the difference polarization $\langle \sum_k I_{kz} - S_z \rangle$ becomes a constant of the motion in contrast to the sum polarization $\langle \sum_k I_{kz} + S_z \rangle$ in the zero-quantum CP experiment. In a static CP experiment, only zero-quantum CP is allowed.

CP experiments using the standard pulse sequence shown in fig. 12 have been performed on a powder sample of adamantane. The intensity of the CH$_2$ resonance as a function of the carbon r.f. field strength ω_{1S} at constant proton field strength («CP spectrum») for a MAS frequency of 5 kHz is shown for mixing times τ_m of 1 ms and 16 ms in fig. 13. The behaviour for the CH resonance, not shown, is similar. At 1 ms mixing time, the ± 1 and ± 2 sidebands clearly dominate the CP spectrum, the ± 1 sidebands being more intense. At 16 ms mixing, the ± 1 and ± 2 sidebands grow to nearly the same height. The centreband, only weakly present in 1 ms mixing, reaches nearly the same intensity as

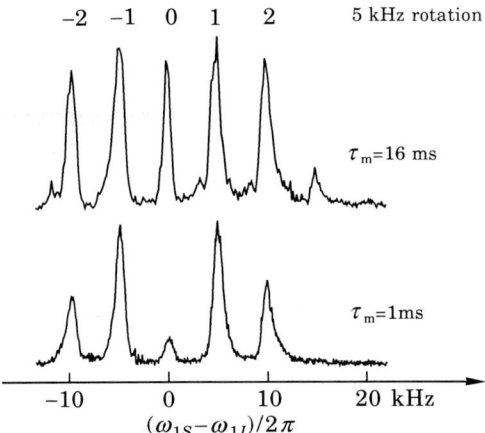

Fig. 13. – Intensity of the CH_2 carbon signal in adamantane after a CP experiment as a function of the carbon r.f. field strength for two different mixing times. The experimental details are given in the caption to fig. 12. The MAS frequency was 5 kHz. The «sidebands» are located at the Hartmann-Hahn condition plus or minus multiples of the MAS frequency. For the shorter mixing time, the ± 1 and ± 2 sidebands are dominant. The centreband and the higher sidebands appear only for longer mixing times. (Figure adapted from ref.[18].)

the discussed sidebands. Higher sidebands remain weak. The spectrum is, within experimental error, symmetric about the static Hartmann-Hahn condition $\omega_{1S} = \omega_{1I}$.

In fig. 14, the intensity of the CH_2 resonance is shown for a MAS frequency of 10 kHz. All other parameters are the same as for fig. 13. Because $2\omega_r > \omega_{1I}$, the − 2 sideband would appear in this CP spectrum at negative ω_{1S} fields. In fact, the sideband is «folded back» and appears at positive ω_{1S} field but surprisingly with negative intensity. Again, the absolute intensities of the ± 1 and ± 2 sidebands tend to equalize with long mixing times.

In the following, we shall briefly sketch the simplest model of CP thermodynamics for time-independent Hamiltonians to introduce the notation. The Hamiltonian is separated into a sum of commuting, orthogonal reservoir terms H_i^0 and a perturbation term V which does not commute with the reservoir terms:

(39) $$H = H^0 + V$$

with

(40) $$H^0 = \sum_i H_i^0.$$

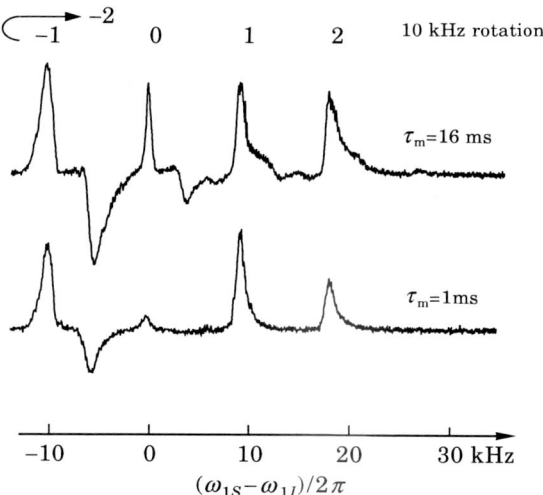

Fig. 14. – Intensity of the CH_2 carbon signal in adamantane after a CP experiment as a function of the carbon r.f. field strength for two different mixing times. The experimental details are given in the caption to fig. 12. The MAS frequency was 10 kHz. The ± 1 and ± 2 sidebands as well as the centreband behave the same way as at 5 kHz rotation frequency (see fig. 13), but the -2 sideband, which would be positioned at a value corresponding to a negative carbon field strength, appears with negative intensity at a frequency position folded back at the left margin of the spectrum. (Figure adapted from ref. [18].)

This decomposition of the Hamiltonian is somewhat arbitrary. To ensure that reservoir terms can establish internal equilibrium quickly, each of the reservoir Hamiltonians has to possess either an equally spaced spectrum or a dense spectrum [26].

In the simplest description, the reservoirs are identified by the two Zeeman Hamiltonians [47, 48]

(41) $$H_1^0 = H_I ,$$

(42) $$H_2^0 = H_S ,$$

and the perturbation by the dipolar couplings

(43) $$V = H_{IS} + H_{II} .$$

For a static sample, this description is only valid close to the Hartmann-Hahn condition [26]. In the CP experiment (see fig. 12), the density operator after the initial 90° pulse is, in the doubly rotating tilted frame, given as

(44) $$\sigma(0) = \alpha \sum_k I_{kz} ,$$

with $\alpha = (-\hbar\omega_{0I})/kT\,\text{Tr}\,\{1\}$, where T is the lattice temperature and $\omega_{0I}/2\pi$ is the Larmor frequency of the I spins.

The quasi-equilibrium density operator $\sigma^{\text{qe}} = \sigma(t \to \infty)$ is obtained, according to eq. (21), by projecting $\sigma(0)$ onto the constants of the motion. We approximate the system Hamiltonian by the reservoir term only. In this approximation, the system is ergodic, the reservoir Hamiltonian itself is the only constant of the motion. Therefore, we find

$$\sigma^{\text{qe}} = \frac{\text{Tr}\,\{\sigma(0)H^0\}}{\text{Tr}\,\{H^0 H^0\}} H^0, \tag{45}$$

and

$$\sigma^{\text{qe}} = \alpha \left(\frac{N\omega_{1I}^2}{N\omega_{1I}^2 + \omega_{1S}^2} \sum_k I_{zk} + \frac{N\omega_{1I}\omega_{1S}}{N\omega_{1I}^2 + \omega_{1S}^2} S_z \right), \tag{46}$$

leading to an S-spin polarization as a function of the r.f. field strength as shown in fig. 15. It should, however, be pointed out that H^0 is an appropriate constant of the motion only if the Hartmann-Hahn condition is fulfilled. At exact Hartmann-Hahn match, $\langle S_z \rangle = (-\hbar\omega_{0I}/4kT)(N/(N+1))$ is found.

It is clearly seen from fig. 15 that the naive application of eq. (46) predicts

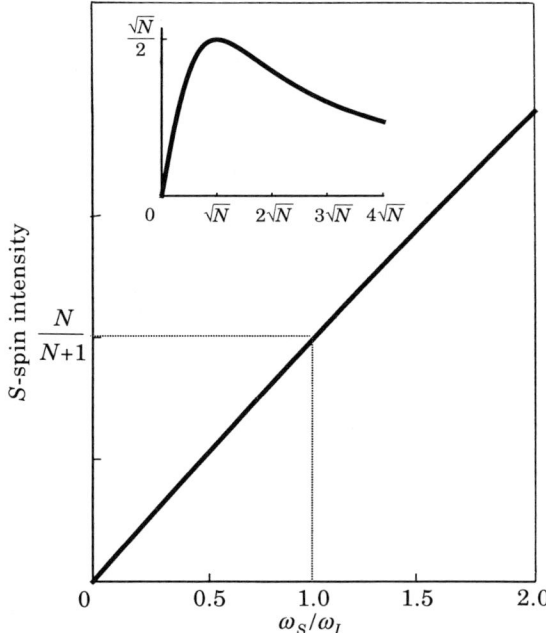

Fig. 15. – S-spin polarization derived from the quasi-equilibrium density operator of eq. (46). The assumption used to derive eq. (46) is, however, valid only if the two field strengths ω_1 and ω_2 match.

sideband intensities that contradict the experimental findings. Namely, it is predicted that the sideband intensities at quasi-equilibrium are proportional to the S-spin field strength ω_{1S}, while, experimentally, they are found to be ω_{1S} independent.

To be able to apply thermodynamics properly to the MAS case, we replace the time-dependent Hamiltonian by a time-independent Hamiltonian $\overline{H}$. We choose an appropriate interaction frame where the nonsecular parts of the interaction Hamiltonian can be neglected. We can then calculate the thermodynamic quasi-equilibrium state at the sideband positions $\omega_{1S} - \omega_{1I} = n\omega_r$ of the CP spectrum.

If we describe the CP-MAS experiment in the doubly rotating frame used for the static CP, the reservoir terms remain time independent under MAS but the perturbation V, containing the heteronuclear and homonuclear dipole couplings, becomes time dependent. For the description of the centreband of the CP spectrum ($n = 0$), this frame is an adequate choice. The perturbation Hamiltonian $V(t)$ has frequency components at ω_r and $2\omega_r$. For fast spinning, these do not coincide with any transition of the reservoir term. The secular-part Hamiltonian is given by

$$\overline{H} = H_1^0 + H_2^0 . \tag{47}$$

The thermodynamic description developed for the static solid remains valid for the Hartmann-Hahn match, $\omega_{1S} = \omega_{1I}$. However, the CP rate is predicted to vanish in zeroth order because the average of the perturbation $V(t)$ vanishes. Because $V(t)$ does not commute with itself at different times, higher-order average Hamiltonian corrections introduce a nonvanishing time-independent coupling term which, for long CP times, establishes the equilibrium condition:

$$\langle S_z \rangle = \frac{-\hbar \omega_{0I}}{4kT} \frac{N}{N+1} . \tag{48}$$

For the sidebands of the CP spectrum with $n = -2, -1, 1, 2$, the transition frequencies of a heteronuclear flip-flop (or flop-flop) process $\omega_{1S} - \omega_{1I}$ (or $\omega_{1S} + \omega_{1I}$, respectively) is matched by a frequency component of $V(t)$ and the doubly rotating frame is not a valid interaction frame. We transform, therefore, to a different doubly rotating coordinate system, where the resonant part of $V(t)$ becomes time independent [40]. We approximate the perturbation by the heteronuclear part only, which is permissible for fast-spinning $\omega_r \gg \|H_{II}\|$:

$$V(t) = H_{IS} = \sum_k b_k(t) 2I_{ky} S_y , \tag{49}$$

where the dipolar-coupling frequency $b_k(t)$ contains Fourier components at frequencies ω_r and $2\omega_r$. To render the resonant part (with frequency $n\omega_r$) time independent, we transform into an interaction frame (in spin space) where the I- and S-spin frames of reference rotate around the z-axis with a frequency difference of $n\omega_r$. One possible choice for the unitary operator U is

(50) $$U = \exp[iH^\# t]$$

with

(51) $$H^\# = n\omega_r S_z \,.$$

In the interaction frame, the Hamiltonian is again written as a sum of two reservoir terms and a perturbation term:

(52) $$\tilde{H} = \tilde{H}_1^0 + \tilde{H}_2^0 + \tilde{V},$$

with

(53) $$\tilde{H}_1^0 = H_1^0 = \omega_{1I} \sum_k I_{kz} \,,$$

(54) $$\tilde{H}_2^0 = (\omega_S - n\omega_r) S_z = \tilde{\omega}_S S_z \,.$$

The time-dependent part of the perturbation $\tilde{V}_1(t)$ contains frequencies at integer multiples of ω_r and there is no resonance with transitions of the new reservoir Hamiltonian. The thermodynamic equilibrium can, therefore, be calculated analogously to the static case. We find that the quasi-equilibrium density operator is the same as for the static case in agreement with the experimental spectrum and in disagreement with the naive interpretation of fig. 15.

An interesting situation arises if $\tilde{\omega}_{1S}$ and ω_{1I} have different signs. This is the case if a sideband of the CP spectrum moves beyond the origin of the ω_{1S}-axis in the CP spectrum ($n\omega_r > \omega_{1I}$). The expectation value for the S-spin polarization is now of opposite sign but still of identical absolute value. The elementary step is a heteronuclear double-quantum transition and one can speak of double-quantum CP. This behaviour is found in the experimental spectrum of fig. 14.

5. – Spin diffusion.

In homonuclear experiments, the flip-flop process is obviously not energy conserving (in the laboratory frame of reference) if the two involved species have different resonance frequencies. Therefore, an additional source of energy («heat bath») has to be present to allow for energy-conserving transitions. We shall classify the basic polarization transfer experiments according to the nature of the energy source as

	energy source	
coupling mechanism	coherent	incoherent
spin	r.f. driven	proton driven
spatial	rotor driven	motionally driven

Fig. 16. – Polarization transfer experiments classified according to the description of the properties of the energy source (coherent vs. incoherent) and the coupling mechanism (coupling through the spatial or spin part of the Hamiltonian).

proton-driven polarization transfer (proton-driven spin diffusion), if the energy for the flip-flop transition is provided by the strongly coupled, abundant proton spins;

motionally-driven polarization transfer, if the energy is provided by molecular motion or chemical reactions;

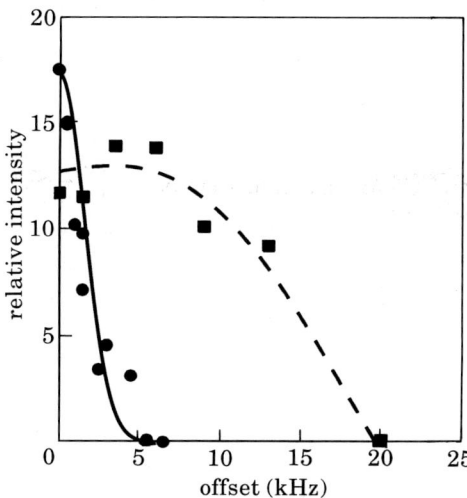

Fig. 17. – Ratio of cross-peak to ^{13}CH diagonal-peak intensity for r.f.-driven spin diffusion between the ^{13}CH and ^{13}CH$_2$ resonances of adamantane (powder sample) as a function of the offset of the r.f. carrier frequency from the centre of the two resonance lines. The circles represent experimental values obtained with a 30 ms c.w. spin-lock. The squares represent experimental values obtained with a 30 ms WALTZ17 spin-lock using a spin-lock pulse $\alpha = \pi/4$. The dashed line is drawn as a guide to the eye. The ^{13}C r.f. field strength was set to 50 kHz. Figure adapted from ref.[11].

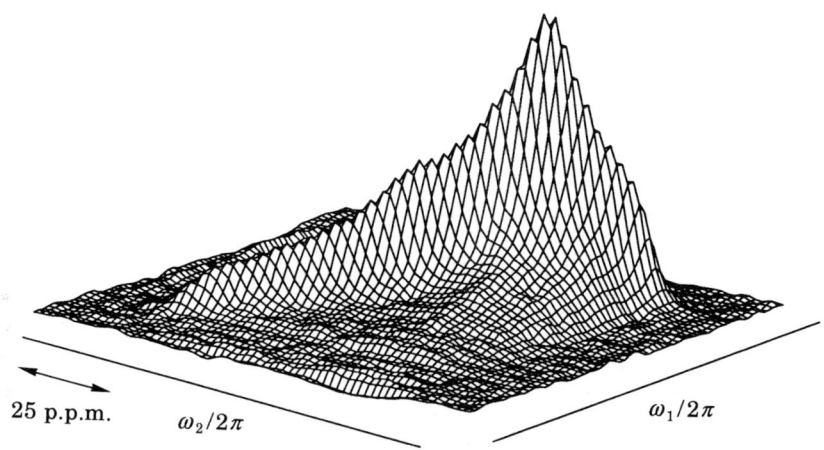

Fig. 18. – R.f.-driven spin diffusion spectrum of a sample of selectively enriched PMMA (see text) using a mixing time of 8 ms. For mixing, a WALTZ17-BLEW12 sequence was used with an r.f. field strength of 78 kHz for carbon and protons [49].

r.f.-*driven polarization transfer*, if the energy is provided by a radiofrequency field, and

rotor-driven polarization transfer, if the energy is provided by the macroscopic sample rotation (*e.g.*, around the magic-angle direction).

The four types of polarization transfer experiments are examples for coherent and incoherent energy sources that interact with either the spatial or the spin part of the Hamilton operator as shown in fig. 16.

Proton-driven spin diffusion uses the broadening of the carbon resonance lines by coupling to the protons to induce energy-conserving flip-flop transitions. For many coupled spins, the rate constant for the flip-flops is given by $W_{ij} = (\pi/2) d_{ij}^2 f(0)$, where $f(0)$ denotes the intensity of the normalized zero-quantum line. The main deficiency of proton- and motionally-driven spin diffusion experiments is the relatively low spin diffusion rate constants typically found. The line broadening which, on the one hand, suppresses the chemical-shielding difference terms in the Hamiltonian reduces, on the other hand, the intensity of the normalized zero-quantum line $f(0)$ resulting in a spin diffusion rate constant that is often too small for practical applications. A considerably more efficient scheme keeps the zero-quantum linewidth narrow by applying proton decoupling and tries to move the effective resonance positions close to zero frequency. The largest spin diffusion rate constant is then obtained for vanishing chemical-shift difference. This can be achieved in an interaction frame under r.f. irradiation. The situation is similar to a Hartmann-Hahn polarization transfer with the important difference that all the involved spins are irradiated by the same r.f. field. It is seen from the experimental results of fig. 17 that contin-

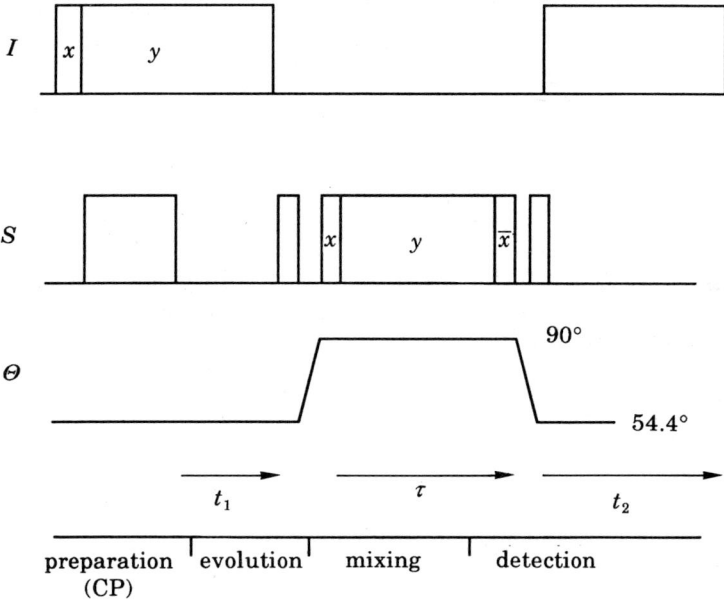

Fig. 19. – Pulse sequence and position of the sample rotation axis with respect to the static magnetic field in a dynamic-angle r.f.-driven spin diffusion experiment. After the evolution period, the magnetization is stored along the z-direction to avoid disturbances of the spin-lock by the mechanical flip of the sample rotation axis. During the mixing time, the magnetization remains spin-locked.

uous-wave irradiation is usually not sufficient to obtain efficient polarization transfer over the spectral range of a typical carbon spectrum It is, therefore, necessary to find pulse sequences with a much broader bandwidth than c.w. irradiation.

The following four points have to be taken into account for the design of improved spin-lock sequences: i) The effective chemical-shift difference δ should be minimized for all spin pairs, ii) the zero-quantum line shape function $G^{(2,3)}(\omega)$ should optimize $G^{(2,3)}(\delta) = f(0)$, iii) the dipolar-coupling constant d should be scaled as little as possible by the spin and spatial rotations applied to fulfil i) and ii), and iv) the sum polarization should not decay. The design considerations are related to the design of optimum pulse sequences for the TOCSY transfer in liquid-state NMR [50]. However, points iii) and iv) put additional constraints to be fulfilled for the r.f.-driven experiments which are not relevant in TOCSY experiments.

Figure 17 shows the improvement in bandwidth by the use of a WALTZ17 irradiation instead of c.w. irradiation. On resonance, the same dipolar scaling factor is found for c.w. and WALTZ irradiation. The off-resonance behaviour is much improved with respect to both a large dipolar interaction and an efficient

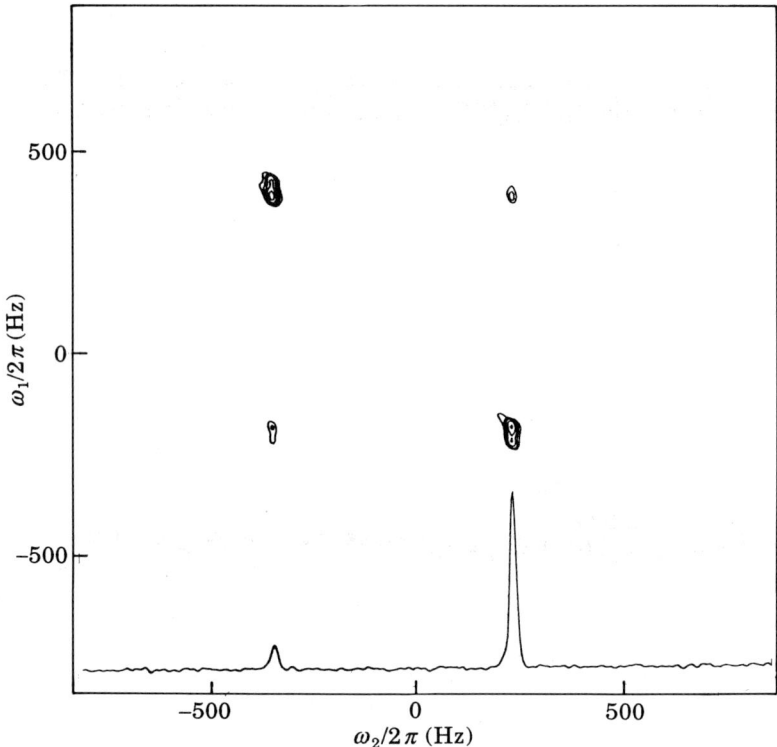

Fig. 20. – R.f.-driven spin diffusion spectrum of a powder sample of adamantane rotating at 1900 Hz using the pulse sequence of fig. 19. A mixing time of 80 ms was used. The experiment was performed on a modified CXP-300 spectrometer using a home-built MAS probe head that allows rapid changes in the orientation of the sample rotation axis. Proton and carbon r.f. fields were matched at 40 kHz. 30 ms were allowed for the flip of the rotation axis. The 1D slice insetted into 2D plot is a horizontal slice through the cross-peak at approximately − 250 Hz.

suppression of chemical-shift difference terms. It should be pointed out that the application of an effective spin-lock pulse (17th pulse in the present example) is mandatory to avoid a quick decay of the sum polarization. In fig. 18, a two-dimensional powder spin diffusion spectrum of a 95% enriched sample of O-$^{13}CH_3$ isotactic, amorphous PMMA (polymethylmethacrylate) is shown. Strong cross-peaks are found for a mixing time of 8 ms. The spectrum was acquired using a WALTZ17/BLEW12 spin-lock sequence with an effective spin-lock pulse of $\pi/4$. From two-dimensional spin diffusion spectra, the relative orientation of the chemical-shielding tensors of spins in spatial neighbourhood can be extracted [6, 7, 10, 51]. R.f.-driven spin diffusion is also observed between carbon spin in natural-abundance samples [11]. It is then particularly important to avoid cross-talk between the carbon spins and the proton dipolar reservoir as this provides an efficient mechanism for $T_{1\rho}$ relaxation [29]. As mentioned above, the simultaneous application of WALTZ17 (for the carbons) and BLEW12 (for the protons) provides a means to obtain sufficiently long $T_{1\varrho}$ values.

The observation of r.f.-driven polarization transfer under magic-angle sample spinning is not straightforward. MAS at frequencies exceeding the ^{13}C homonuclear dipolar couplings, that, for natural-abundance samples, are of the order of a few ten hertz, quench r.f.-driven spin diffusion. An obvious solution

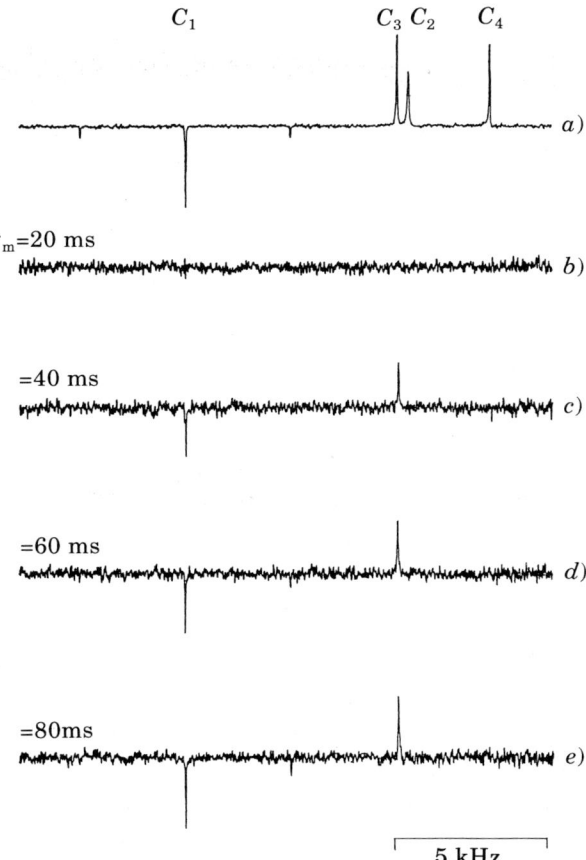

Fig. 21. – Experimental ^{13}C spin diffusion spectra of a powder sample of threonine. The top spectrum was recorded without a mixing time ($\tau_m = 0$). The C_1 resonance was selectively inverted using a DANTE π pulse. Magic-angle sample spinning at 3965 Hz, corresponding to half of the difference in the resonance frequencies of C_1 and C_3, was applied. Spectra b)-e) were recorded as difference spectra as described in the text. Signals in spectra b) to e) are indicative for spin diffusion and the signal amplitude is a measure for the corresponding rate constant. For spectrum a), 64 scans were accumulated, for the other spectra 1024 scans. The amplitudes of the spectra were normalized by the number of scans. The plot scale of spectra b) to e) is 10 times enhanced, compared to a). After 80 ms mixing time, the spin diffusion signals reach 10% of the signal intensity in the spectrum without mixing (a)). (Figure adapted from ref.[12].)

is to perform dynamic-angle spinning experiments, where the sample rotation axis during mixing is switched to an angle other than the magic angle. For a NMR probe design using a stator-fixed solenoid coil, where the r.f. field is coaxial with the sample rotation axis, no r.f. pulses can be applied at $\theta = 0$. For these purely technical reasons, it is, therefore, useful to switch the rotation axis to a position with $\theta = \pi/2$ where the r.f. field is perpendicular to the static field. The dipolar interactions at $\theta = \pi/2$ are scaled by a factor of $-1/2$ in comparison to $\theta = 0$. The experimental scheme is shown in fig. 19. An experimental r.f.-driven polarization transfer from a powder sample of adamantane rotating at 1.9 kHz is shown in fig. 20. C.w. irradiation was employed. The application of multiple pulse spin-locking at $\theta = \pi/2$ still suffers from the modulation of the Hamiltonian by MAS. It would, therefore, be beneficial to use either a solenoid coil which is fixed with respect to the laboratory frame (does not move with the MAS stator) or a stator-fixed Helmholtz coil to be able to apply pulses at $\theta = 0$.

An alternative method to obtain fast polarization transfer between rare spins is rotor-driven spin diffusion [12, 52], where the energy for the flip-flop process is provided by the coupling to the MAS rotation. Experimental results for a powdered sample of threonine (at natural abundance of ^{13}C) are shown in fig. 21. Rotor-driven spin diffusion is a selective experiment. Only spins with a chemical-shift difference of $\delta = n\omega_r$, where ω_r is the MAS frequency and $n = \pm 1, \pm 2, \pm 3...$, participate in spin diffusion. An enhancement for the spin diffusion rate constant of more than two orders of magnitude is found in threonine, compared to the proton-driven experiments.

6. – Polarization echoes.

Usually, the evolution of locally perturbed polarization is described by a dissipative process approaching a quasi-equilibrium state in an exponential process characterized by a diffusion rate constant. For large spin systems with many dipolar couplings of comparable size, the quasi-equilibrium state corresponds usually to an equal distribution of the polarization over all coupled spins. This simplified description is adequate for a number of experiments including most of the spin diffusion and cross-polarization experiments I have discussed above. In some experiments, however, the deterministic nature of the polarization transfer process becomes apparent.

Here, we discuss an experiment that reverses the spatial «diffusion» of polarization after spatially selective excitation in a homonuclear spin system. A schematic picture is shown in fig. 22. Proton spin diffusion in ferrocene is used as an experimental example. The pulse sequence is shown in fig. 23. After cross-polarization (during t_c) and after a time t_S to allow the proton coherence, present after the spin-lock pulse, to decay, the protons directly bound to a (nat-

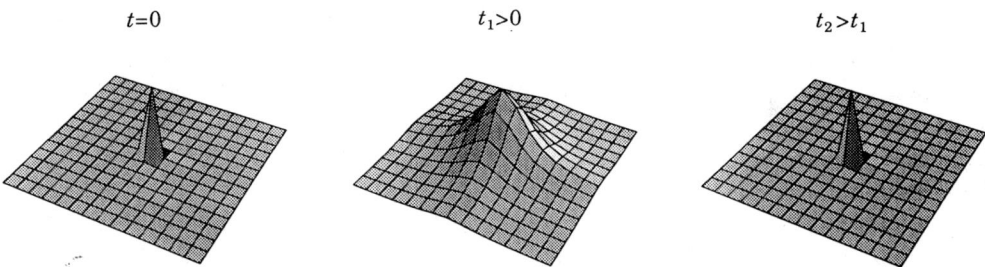

Fig. 22. – Schematic representation of a polarization echo. Initially, a spatially localized polarization is prepared, that, by flip-flop processes, loses its spatial localization. However, after application of a time reversal, the spatial localization is restored at the time of the polarization echo.

ural-abundance) ^{13}C spin are selectively repolarized during the short cross-polarization period $t_{\rm d}$. At this point in time, a spatially localized proton polarization is found which evolves, during the time period τ_1, in the rotating frame under a dipolar Hamiltonian. The polarization is then transferred into the laboratory frame by a $\pi/2$ pulse. During τ_2, laboratory frame polarization transfer takes place, again under a dipolar Hamiltonian. At the end of τ_2, the polarization is transferred to the carbon spins, by a spatially selective cross-polarization period, where it is detected.

Experimental results are shown in fig. 23. The crosses indicate the time evolution of the polarization of the ^{13}C-bonded I spins as a function of τ_1 in an experiment where τ_2 was set to zero. The decay towards an asymptotic value of 0.2 is nearly exponential and can be interpreted as unhindered spin diffusion among the five protons of one of the five-membered rings in ferrocene. The system seems to reach a quasi-equilibrium state among all protons in the same ring. Spin diffusion to protons on the other rings is slower and does not become evident on the time scale of the experiment. If, however, the experimental final I-spin polarization is measured as a function of τ_2 (indirectly through the carbon spins) keeping τ_1 fixed at 160 µs, the spin diffusion process, which takes place during τ_1, is not continued during τ_2, but an echo formation is observed. The experimental results are indicated by circles in fig. 23.

The echo is based on the fact that the dipolar interaction in the laboratory frame, $H_{\rm d}$, has an identical form but a different sign than the dipolar interaction in the rotating frame, $H_{\rm d}'$. The laboratory frame dipolar interaction that is truncated with respect to the Zeeman field becomes, in the rotating frame, truncated once more by the interaction with the strong r.f. field:

(55) $$H_{\rm d}' = -\frac{1}{2} H_{\rm d}.$$

This property was already by exploited RHIM et al. [53, 54] in the famous «magic-echo» experiments and by SCHMIEDEL and co-workers [55].

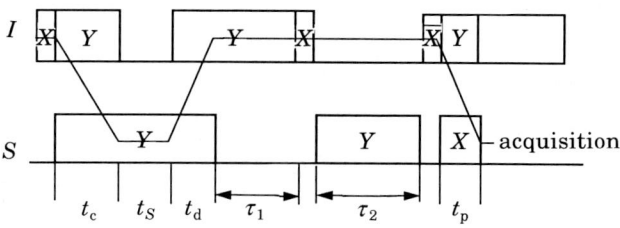

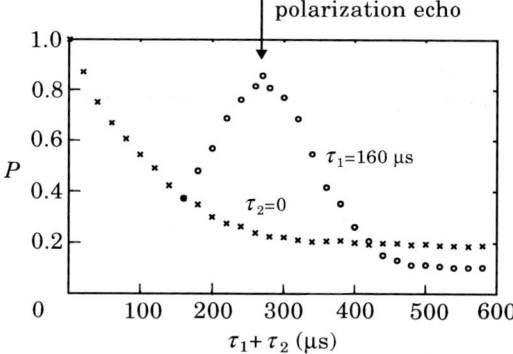

Fig. 23. – Polarization echoes: Pulse sequence and experimental time evolution of the polarization P of those protons that are in spatial neighbourhood of a ^{13}C spin. (Figure adapted from ref. [19].)

The density operator at time $\tau_1 + \tau_2$ is given by

(56) $\quad \sigma(\tau_1, \tau_2) = \exp\left[i(\pi/2)F_x\right]\exp\left[iH_d \tau_2\right]\exp\left[-i(\pi/2)F_x\right]\exp\left[-iH_d' \tau_1\right] \times$

$\quad\quad\quad \times \sigma(0) \exp\left[iH_d' \tau_1\right]\exp\left[i(\pi/2)F_x\right]\exp\left[(-i)H_d \tau_2\right]\exp\left[-i(\pi/2)F_x\right].$

Using eq. (55), the formula for the final density operator can be simplified to

(57) $\quad \sigma(\tau_1, \tau_2) = \exp\left[-iH_d'(\tau_2 - \tau_1/2)\right]\sigma(0)\exp\left[H_d'(\tau_2 - \tau_1/2)\right]$

and, therefore, at the time $\tau_2 = \tau_1/2$, we find

(58) $\quad\quad\quad\quad\quad\quad \sigma\left(\dfrac{\tau}{2}, \tau\right) = \sigma(0),$

and an echo is expected. In practice, the echo is slightly delayed from the time calculated above because proton spin diffusion is already taking place during the cross-polarization period t_d.

* * *

The concepts and experimental results presented here have been inspired by numerous discussions with and experimental findings by M. COLOMBO, TH.

GISLER, P. ROBYR, M. TOMASELLI and SHANMIN ZHANG. The author is particularly indebted to Prof. R. R. ERNST for his help and advice. Financial support by the «Schweizerischer Nationalfonds» and the ETH Zürich is gratefully acknowledged.

REFERENCES

[1] A. PINES, M. G. GIBBY and J. S. WAUGH: *J. Chem. Phys.*, **59**, 569 (1973).
[2] R. A. ASSINK: *Macromolecules*, **11**, 1233 (1978).
[3] J. VIRLET and D. GHESQUIERES: *Chem. Phys. Lett.*, **73**, 323 (1980).
[4] P. CARAVATTI, J. A. DELI, G. BODENHAUSEN and R. R. ERNST: *J. Am. Chem. Soc.*, **104**, 5506 (1982).
[5] P. CARAVATTI, G. BODENHAUSEN and R. R. ERNST: *J. Magn. Reson.*, **55**, 88 (1983).
[6] P. M. HENDRICHS and M. LINDER: *J. Magn. Reson.*, **58**, 458 (1984).
[7] M. LINDER, P. M. HENDRICHS, J. M. HEWITT and D. J. MASSA: *J. Chem. Phys.*, **82**, 82 (1985).
[8] P. CARAVATTI, P. NEUENSCHWANDER and R. R. ERNST: *Macromolecules*, **18**, 119 (1985).
[9] D. L. VANDERHART: *J. Magn. Reson.*, **72**, 13 (1987).
[10] P. ROBYR, B. H. MEIER and R. R. ERNST: *Chem. Phys. Lett.*, **187**, 471 (1991).
[11] P. ROBYR, B. H. MEIER and R. R. ERNST: *Chem. Phys. Lett.*, **162**, 417 (1989).
[12] M. G. COLOMBO, B. H. MEIER and R. R. ERNST: *Chem. Phys. Lett.*, **146**, 189 (1988).
[13] D. SUTER and R. R. ERNST: *Phys. Rev. B*, **32**, 4905 (1985).
[14] A. SCHMIDT and S. VEGA: *Chem. Phys. Lett.*, **157**, 539 (1989).
[15] M. GOLDMAN and L. SHEN: *Phys. Rev.*, **144**, 321 (1966).
[16] T. T. P. CHENG: *J. Chem. Phys.*, **76**, 1248 (1982).
[17] J. BAUM, M. MUNOWITZ, A. N. GARROWAY and A. PINES: *J. Chem. Phys.*, **83**, 2015 (1985).
[18] B. H. MEIER: *Chem. Phys. Lett.*, **188**, 201 (1992).
[19] S. ZHANG, B. H. MEIER and R. R. ERNST: *Phys. Rev. Lett.*, **69**, 2149 (1992).
[20] R. R. ERNST, G. BODENHAUSEN and A. WOKAUN: *Principles of Nuclear Magnetic Resonance in One and Two Dimensions* (Clarendon Press, Oxford, 1987).
[21] L. MULLER, A. KUMAR, T. BAUMANN and R. R. ERNST: *Phys. Rev. Lett.*, **32**, 1402 (1974).
[22] M. GOLDMAN: *Spin Temperature and Nuclear Magnetic Resonance in Solids* (Clarendon Press, Oxford, 1970).
[23] R. R. ERNST, G. BODENHAUSEN and A. WOKAUN: *Principles of Nuclear Magnetic Resonance in One and Two Dimensions* (Clarendon Press, Oxford, 1987).
[24] A. WOKAUN and R. R. ERNST: *J. Chem. Phys.*, **67**, 67 (1977).
[25] S. VEGA: *J. Chem. Phys.*, **58**, 5518 (1978).
[26] M. H. LEVITT, D. SUTER and R. R. ERNST: *J. Chem. Phys.*, **84**, 4243 (1986).
[27] S. ZHANG, B. H. MEIER, S. APPELT, M. MEHRING and R. R. ERNST: *J. Magn. Reson. A*, **101**, 60 (1993).
[28] SHANMIN ZHANG: unpublished results.
[29] B. H. MEIER: in *Advances in Magnetic and Optical Resonance*, edited by W. S. WARREN (Academic Press, New York, N.Y. (in press)).

[30] O. W. SØRENSEN: *J. Magn. Reson.*, **86**, 435 (1990).
[31] O. W. SØRENSEN: *J. Magn. Reson.*, **92**, 642 (1991).
[32] O. W. SØRENSEN: *J. Magn. Reson.*, **93**, 648 (1991).
[33] M. H. LEVITT: *J. Chem. Phys.*, **94**, 30 (1991).
[34] N. C. NIELSEN and O. W. SØRENSEN: *J. Magn. Reson.*, **99**, 449 (1992).
[35] A. ABRAGAM: *The Principles of Nuclear Magnetism* (Clarendon Press, Oxford, 1961).
[36] F. DYSON: *Phys. Rev.*, **75**, 486 (1949).
[37] F. DYSON: *Phys. Rev.*, **75**, 1736 (1949).
[38] W. WEIDLICH: *Thermodynamik und statistiche Mechanik* (Akademische Verlagsgesellschaft, Wiesbaden, 1976).
[39] M. M. MARICQ: *Phys. Rev. B*, **31**, 127 (1985).
[40] A. G. REDFIELD: *Science*, **164**, 1015 (1969).
[41] J. H. SHIRLEY: *Phys. Rev.*, **138** B, 979 (1965).
[42] S. VEGA, E. T. OLEJNICZAK and R. G. GRIFFIN: *J. Chem. Phys.*, **80**, 4832 (1984).
[43] E. O. STEJSKAL, J. SCHAEFER and J. S. WAUGH: *J. Magn. Reson.*, **28**, 105 (1977).
[44] S. R. HARTMANN and E. L. HAHN: *Phys. Rev.*, **128**, 2042 (1962).
[45] S. SARDASHTI and G. E. MACIEL: *J. Magn. Reson.*, **72**, 467 (1987).
[46] R. A. WIND, S. F. DEC, H. LOCK and G. E. MACIEL: *J. Magn. Reson.*, **72**, 136 (1988).
[47] M. MEHRING: *Principles of High Resolution NMR in Solids*, 2nd edition (Spinger, Berlin, 1983).
[48] B. C. GERSTEIN and C. R. DYBOWSKI: *Transient Techniques in NMR of Solids* (Academic Press, Orlando, Fla., 1985).
[49] J. STRAKA: unpublished results.
[50] L. BRAUNSCHWEILER and R. R. ERNST: *J. Magn. Reson.*, **53**, 512 (1983).
[51] R. TYCKO and G. DABBAGH: *J. Am. Chem. Soc.*, **113**, 5392 (1991).
[52] D. P. RALEIGHT, M. H. LEVITT and R. G. GRIFFIN: *Chem. Phys. Lett.*, **146**, 71 (1988).
[53] W. K. RHIM, A. PINES and J. S. WAUGH: *Phys. Rev. Lett.*, **25**, 218 (1970).
[54] W. K. RHIM, A. PINES and J. S. WAUGH: *Phys. Rev. B*, **3**, 684 (1971).
[55] H. SCHNEIDER and H. SCHMIEDEL: *Phys. Lett. A*, **30**, 298 (1969).

Introduction to Two-Dimensional NMR in Liquids.

R. FREEMAN

Department of Chemistry, Cambridge University - Cambridge, U.K.

1. – Introduction.

Double resonance has always played an important role in high-resolution NMR of liquids because it provides a powerful method for assignment, indicating which nuclei are connected by spin-spin coupling. In a coupled two-spin system IS, selective irradiation of the I spins causes the collapse of the multiplet structure of the S spins, providing direct evidence of the connectivity[1]. Similarly, if the I and S spins are in slow chemical exchange, saturation of the I spins is carried over to the S spins at a rate determined by the rate constant for the exchange process[2]. A third application of double resonance is to determine the nuclear Overhauser enhancement, a measure of the proximity of two nuclei within the molecule[3]. All three techniques were available in the 1960's, but it was the advent of multidimensional spectroscopy[4, 5] that made them the common currency of all high-resolution spectroscopists. One can make a good case that the early double-resonance experiments were essentially two-dimensional, since it was usually necessary to search one frequency dimension with the irradiation field B_2 while investigating the normal spectroscopic dimension with the observing field B_1. Indeed, theoretical calculations of double-resonance spectra sometimes employ a two-dimensional «stacked-trace» display of the intensity as a function of the two frequency parameters[6] closely reminiscent of the ubiquitous «2D» displays that have become so familiar in recent years. Figure 1 shows a two-dimensional representation of the carbon-13 spectrum of a methyl group calculated for a coherent proton decoupling experiment. The observing frequency (B_1) is swept through the carbon-13 frequencies while the decoupling frequency (B_2) is stepped through the proton dimension. This simulation shows the distortion of the peak heights from the expected 1:3:3:1 ratio as a result of differential broadening in the spatially inhomogeneous B_2 field.

True two-dimensional spectroscopy was introduced in a seminar entitled

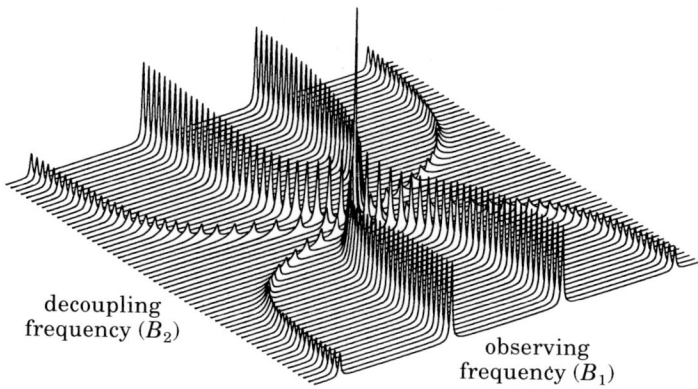

Fig. 1. – Double resonance viewed as a two-dimensional experiment. A computer simulation of coherent decoupling of the carbon-13 quartet of methyl iodide from protons. The spatial inhomogeneity of B_2 gives rise to a preferential broadening of the outer lines of the carbon-13 quartet.

Pulse pair technique in high resolution NMR given by J. JEENER at a Summer School in Yugoslavia in 1971 [4] and subsequently developed by R. ERNST [5], who first appreciated the full generality of the two-dimensional Fourier transform principle and led the way to its exploitation in structural organic chemistry. We may ask what are the essential ingredients that make two-dimensional Fourier transform spectroscopy so much more popular than the «old-fashioned» double-resonance experiments. One factor must be convenience; double resonance requires some rather painstaking manual searches in frequency space at different levels of the irradiation field B_2. Two-dimensional spectroscopy examines the second frequency dimension automatically by scanning the corresponding time dimension t_1. A second factor is the more sophisticated mode of display that is forced upon us for two-dimensional spectroscopy. Intensity contour plots are very convenient for displaying two-dimensional correlation spectra. This direct pictorial representation of the connectivities is much appreciated by the structural organic chemist. As more complex molecules are studied, this comprehensive display mode becomes essentially indispensable, and for protein studies automatic analysis by computer is taking over from manual interpretation methods.

The fundamental difference between two-dimensional Fourier transformation and double resonance is the adoption of time domain pulse experiments as opposed to continuous-wave methods. This has opened the door to an endless variety of «spin gymnastics» where zero-, single- and multiple-quantum coherences are employed in a complex coherence transfer pathway to filter, edit, simplify, enhance, correlate or assign the high-resolution spectrum. It is perhaps in these offshoots of the two-dimensional concept that the real importance of the method lies. Ideas derived from two-dimensional spectroscopy have proved a fertile source of new experiments, whereas the double-resonance methods tended to remain an end in themselves.

2. – *Pseudo*-two-dimensional spectroscopy.

Nevertheless, pulsed-double-resonance ideas can be useful for understanding the classic two-dimensional correlation experiment COSY [5, 7]. Consider the high-resolution spectrum of a two-spin (IS) system coupled by a scalar coupling J_{IS}. It consists of two doublets, centred at the chemical-shift frequencies δ_I and δ_S with splittings J_{IS}. Suppose a radiofrequency pulse is applied that is so selective in the frequency domain that it only affects one of these four transitions at any given time, leaving the other three unperturbed. This implies that the intensity during the pulse is weak compared with J_{IS}, and that the pulse duration is relatively long. Suppose that the pulse flip angle on resonance is 90°.

The double-resonance experiment is designed to test whether I and S are coupled by a resolvable scalar coupling J_{IS}. The pulse sequence is

soft 90°(x)-hard 90°(x)-acquire

and the frequency of the soft pulse (F_1) is incremented in small steps over a range of frequencies that covers the chemical shifts δ_I and δ_S. We thus explore this frequency dimension directly, whereas the two-dimensional experiment derives this information through Fourier transformation. When F_1 is far from all resonances, the soft pulse has no effect and the conventional unperturbed spectrum is obtained. When the frequency is exactly on resonance for the first I-spin transition (I_{12}), the soft pulse converts M_z into M_y, and then the hard pulse converts this into $-M_z$. This corresponds to a population inversion (fig. 2) and there are now changes in the population differences across the two S transitions. One has an enhanced population difference (S_{24}) and the other has equal populations (S_{13}).

We may decompose the hard 90° pulse into a «cascade» of two soft 90° pulses,

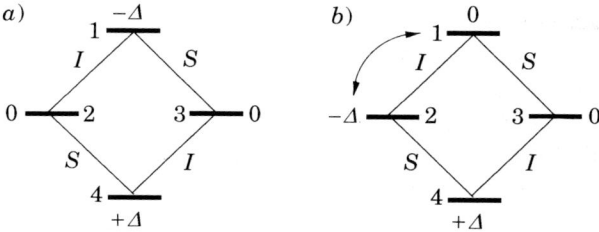

Fig. 2. – Energy level diagram of a two-spin system *a*) with Boltzmann populations and *b*) after a selective 180° pulse on the 1-2 transition.

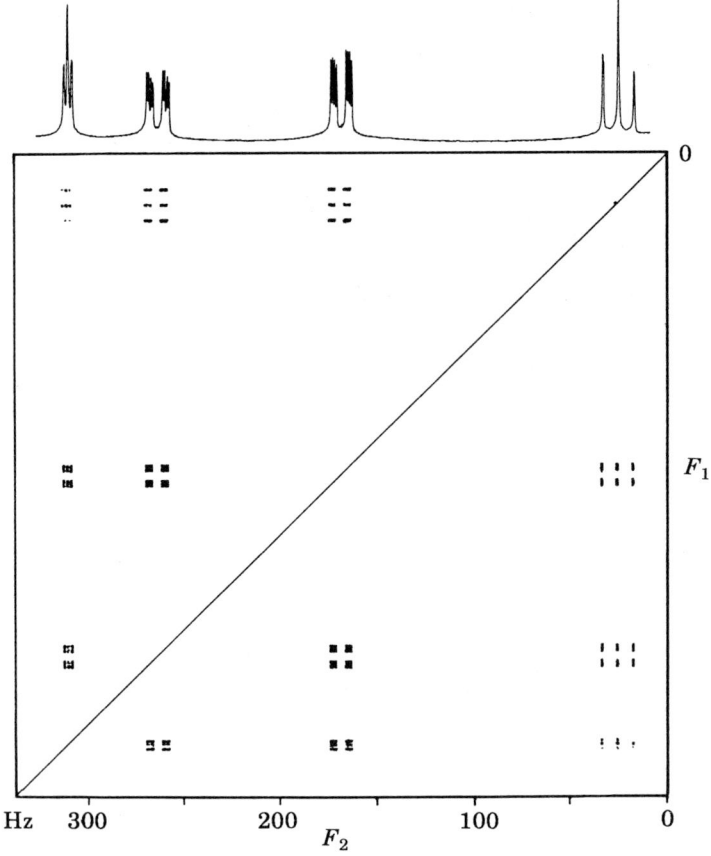

Fig. 3. – Pseudo-two-dimensional correlation spectrum of metabromonitrobenzene obtained by stepping the frequency of a soft 90° pulse through the F_1 dimension (see text). Dispersion-mode responses have been suppressed by a difference method, giving negligible signals on the principal diagonal. Otherwise this has the same form as a COSY spectrum.

the first applied to the I spins, the second to the S spins. The latter acts as a «read» pulse, converting the population disturbances into changes in signal intensity. When the spectrum is recorded as a function of the frequency F_2, one S transition doubles its intensity while the other falls to zero. Subtraction of this response from the unperturbed S spectrum gives the characteristic «up-down» pattern of absorption mode signals. When the frequency of the soft pulse reaches the second I-spin transition (I_{34}), there is a similar transfer of polarization but in the opposite sense, giving a «down-up» pattern for the S-spin response. The overall result is a square pattern of lines for the S spin, alternating in sense in both the F_1 and F_2 frequency dimensions, identical to the pattern observed for a cross-peak in a two-dimensional COSY spectrum. This frequency-stepped double-resonance investigation is, therefore, essentially equivalent to the two-dimensional coherence transfer experiment so familiar to present-day spectroscopists.

The spectra may be displayed by stacking them one behind the other to give the impression of a surface in three dimensions—a graph of intensity plotted as a function of two frequency parameters. Alternatively we may represent this *pseudo*-two-dimensional spectrum as an intensity contour diagram as illustrated in fig. 3 for the proton spectrum of metabromonitrobenzene. This experiment was performed in a slightly different manner using a «spin pinging» sequence [8] for the soft pulse. This has the practical advantage of suppressing the dispersion-mode signal that would normally appear on the principal diagonal. The only appreciable responses are «cross-peaks» centred at the frequency coordinates (δ_I, δ_S) and (δ_S, δ_I), indicating that there has been a transfer of polarization from I to S or from S to I during the experiment. The metabromoni-

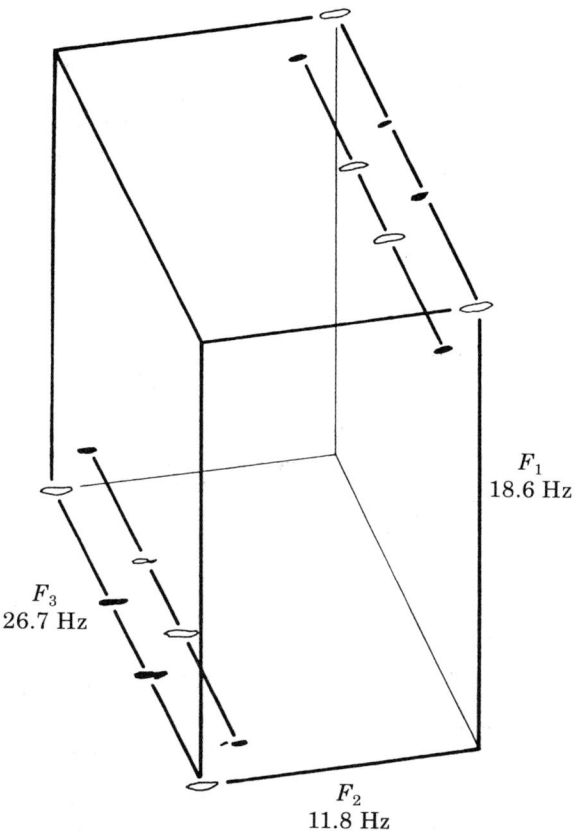

Fig. 4. – A cross-peak from a three-dimensional correlation spectrum investigated by stepping selective 90° pulses through the F_1 and F_2 dimensions. The F_3 axis has been obtained by Fourier transformation of a free-induction signal.

trobenzene correlation spectrum shows that each spin is coupled to the other three, giving twelve cross-peaks in all.

We can now appreciate the essential equivalence between sweeping the frequency F_1 and incrementing the evolution time t_1. Both modes require the same overall spectrometer time, but the two-dimensional experiment benefits from a higher sensitivity through the multiplex advantage in t_1. On the other hand, the frequency sweep mode has the advantage that the increments in F_1 need not be regular, and the experiment can be designed to examine the detailed fine structure of cross-peaks with only sparse sampling between the main responses. An extreme example is provided by a «triple-resonance» version [9], where the frequencies F_1 and F_2 are incremented independently through carefully chosen narrow frequency ranges while the third dimension (F_3) is obtained by Fourier transformation of a free-induction signal. This allows the examination of a «3D» cross-peak with very high digital resolution (fig. 4). The corresponding three-dimensional correlation experiment would have required an extremely large data table and a very protracted experiment, although, of course, it would also have delivered a great deal of useful information about the remainder of the spectrum.

3. – Correlation spectroscopy (COSY).

Jeener's original two-pulse experiment is now universally known as correlation spectroscopy (COSY). The pulse sequence is very simple:

$$90°(x)\text{-}t_1\text{-}90°(x)\text{-acquire }(t_2).$$

It may be formally analysed by the density operator method or by its shorthand equivalent, the product operator treatment. For didactic purposes it is probably simpler to treat it as a polarization transfer experiment in which the disturbance of the spin populations is crucial. No signals are detected during the evolution period t_1, and the information about the behaviour of the spins during t_1 is only obtained indirectly, being coded into a perturbation of the spin populations.

The problem is symmetrical with respect to the I and S spins, so we may concentrate on the transfer of coherence from I to S. The first (hard) 90° pulse may be thought of as exciting two separate components of I-spin magnetization, represented by two vectors initially aligned along the $+y$-axis of the rotating frame of reference. Free precession during t_1 leaves these vectors at angles

(1) $$\alpha_1 = (2\pi\delta_I + \pi J_{IS})t_1, \qquad \alpha_2 = (2\pi\delta_I - \pi J_{IS})t_1,$$

with respect to the $+y$-axis. When $t_1 = 0$, these angles are zero and the second (hard) 90° pulse simply combines with the first to give a 180° rotation, corre-

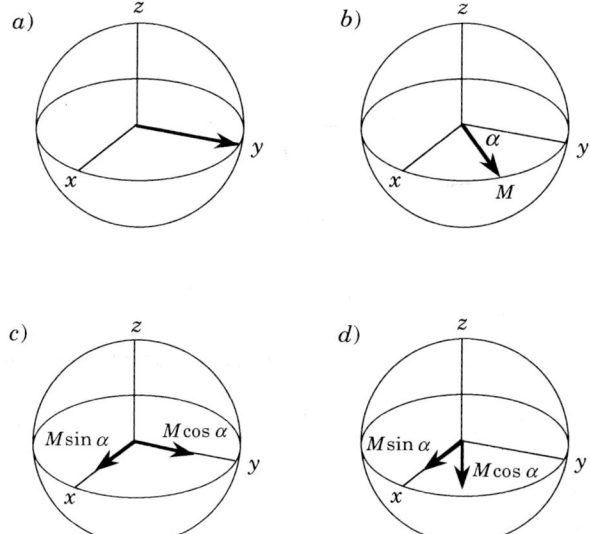

Fig. 5. – Motion of a representative vector M in a two-dimensional correlation experiment: a) Initial 90° pulse creates y-magnetization, b) free precession during the evolution period t_1, c) decomposition into two component vectors, d) after the second hard 90° pulse.

sponding to a spin population inversion across both I-spin transitions. Population differences across the S transitions are not affected since populations on both the upper level and the lower level change by identical amounts.

For polarization to be transferred from I to S there must be a *differential* effect on the two I transitions, which implies differential precession during t_1. If a representative vector M precesses through an angle α during the evolution time (fig. 5b)), it may be resolved into two orthogonal components, $M \sin \alpha$ along the x-axis and $M \cos \alpha$ along the y-axis (fig. 5c)). The second (hard) 90° pulse may be decomposed into a cascade of two soft 90° pulses, the first affecting the I spins while the second acts as a read pulse for the S spins. The soft I-spin pulse converts $M \cos \alpha$ into $-z$ magnetization (fig. 5d)). The two I-spin vectors are affected differently, giving different degrees of population inversion:

(2) $\qquad \Delta_{12} = M \cos (2\pi \delta_I + \pi J_{IS})$, $\qquad \Delta_{34} = M \cos (2\pi \delta_I - \pi J_{IS})$.

These disturbances are distributed equally between the two levels involved (fig. 6). Through a standard trigonometrical identity, the differential effect on the S transitions may be written

(3) $\qquad (\Delta_{34} - \Delta_{12})/2 = M \sin (2\pi \delta_I t_1) \sin (\pi J_{IS} t_1)$.

The second soft 90° pulse of the cascade converts these population changes into observable S-spin signals. One S transition gains intensity and the other loses intensity, and the profile in the F_2 dimension is that of an antiphase doublet cen-

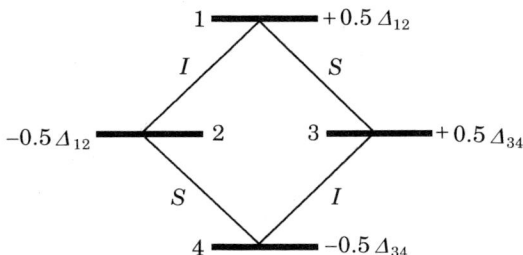

Fig. 6. – Disturbance of the I-spin populations on the energy levels of an IS spin system by the first 90° pulse of a COSY experiment. The S-spin intensities are only affected if $\Delta_{12} \neq \Delta_{34}$.

tred at δ_S with a splitting J_{IS}. Fourier transformation as a function of t_1 gives a profile in the F_1 dimension that is also an antiphase doublet, centred at δ_I with a splitting J_{IS}. The cross-peak is centred at (δ_I, δ_S) and consists of a square pattern of four absorption-mode lines, alternating in intensity in both frequency dimensions. There is a symmetry-related cross-peak at (δ_S, δ_I) denoting transfer from S to I.

Diagonal peaks arise from I magnetization which remains at the I-spin frequency during both t_1 and t_2 and is *not* transferred to S. It corresponds to the x-component of magnetization ($M \sin \alpha$) generated at the end of the evolution period and not affected by the second 90°(x) pulse. There are two components which may be summed together to give

$$\text{(4)} \quad M \sin[(2\pi\delta_I + \pi J_{IS})t_1] + M \sin[(2\pi\delta_I - \pi J_{IS})t_1] = \\ = 2M \sin(2\pi\delta_I t_1) \cos(\pi J_{IS} t_1).$$

In the F_2 dimension the two lines are in phase and have the dispersion-mode lineshape (M_x). Similarly the profile in the F_1 dimension is an in-phase doublet (the cosine term) of splitting J_{IS} centred at the I-spin chemical-shift frequency. The diagonal peak is, therefore, a square pattern of four lines

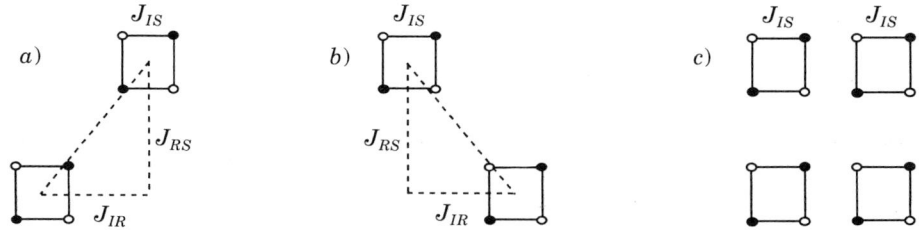

Fig. 7. – Schematic representation of the IS cross-peak of a COSY spectrum in the presence of a passive spin R: a) COSY-45, like signs; b) COSY-45, opposite signs; c) COSY-90. If the second pulse is 45°, two square patterns are observed, disposed according to the relative signs of J_{IR} and J_{RS}. If the second pulse is 90°, there are four square patterns arranged as shown in c).

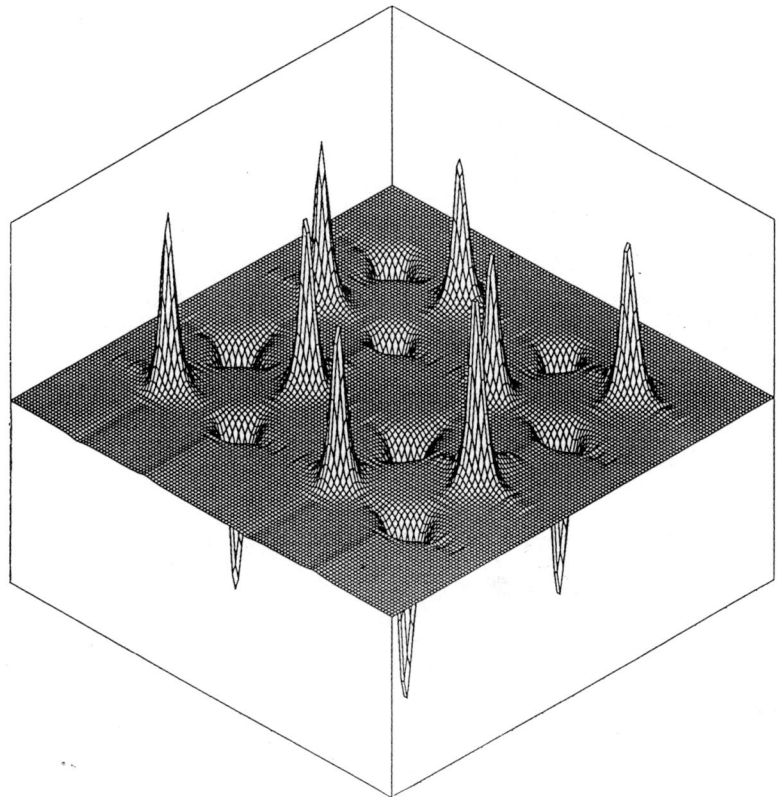

Fig. 8. – Computer simulation of a COSY-90 cross-peak for a three-spin system, showing four square patterns with intensities alternating in both frequency dimensions.

all in the same sense and all having the two-dimensional dispersion lineshape (M_x).

Correlation spectra of real molecules are, of course, more complicated than this two-spin case, but are readily understood by introducing the concept of active and passive spins. The two active spins (I and S) are those involved in the polarization transfer. It is sufficient here to consider a single passive spin R; further passive spins merely complicate the picture without introducing any new physics. The passive spin may be considered as a spectator, generating small local fields at the I and S sites and creating effective chemical shifts

$$\delta_I \pm J_{IR}/2 \quad \text{and} \quad \delta_S \pm J_{SR}/2.$$

If the flip angle of the second pulse of the COSY experiment is kept small (45° is usually small enough), then the R spin can be assumed not to change its state during the experiment and the COSY spectrum is simply the superposition of two basic IS spectra centred at the coordinates

$$(\delta_I + J_{IR}/2, \delta_S + J_{SR}/2) \quad \text{and} \quad (\delta_I - J_{IR}/2, \delta_S - J_{SR}/2)$$

corresponding to the two possible states of the R spin. Each cross-peak is a superposition of two square patterns separated by the «displacement vector» which is the vector sum of J_{IR} in one dimension and J_{SR} in the other. The relative signs of J_{IR} and J_{SR} may be obtained by inspection (fig. 7). If the flip angle of the second pulse is increased to 90°, then the R spin has a 50% chance of changing its spin state at this point, and the cross-peak acquires two further square patterns of side J_{IS}, centred at the coordinates

$$(\delta_I + J_{IR}/2, \delta_S - J_{SR}/2) \quad \text{and} \quad (\delta_I - J_{IR}/2, \delta_S + J_{SR}/2)$$

giving a total of 16 lines. Figure 8 shows a computer simulation of a COSY cross-peak for this case. The information about the relative signs of J_{IR} and J_{SR} is now lost.

COSY spectra are usually run with a phase-cycling scheme [10] or pulsed-field gradients [11, 12] to compensate for pulse imperfections. It is usual to display them as intensity contour maps [7] after precautions have been taken to en-

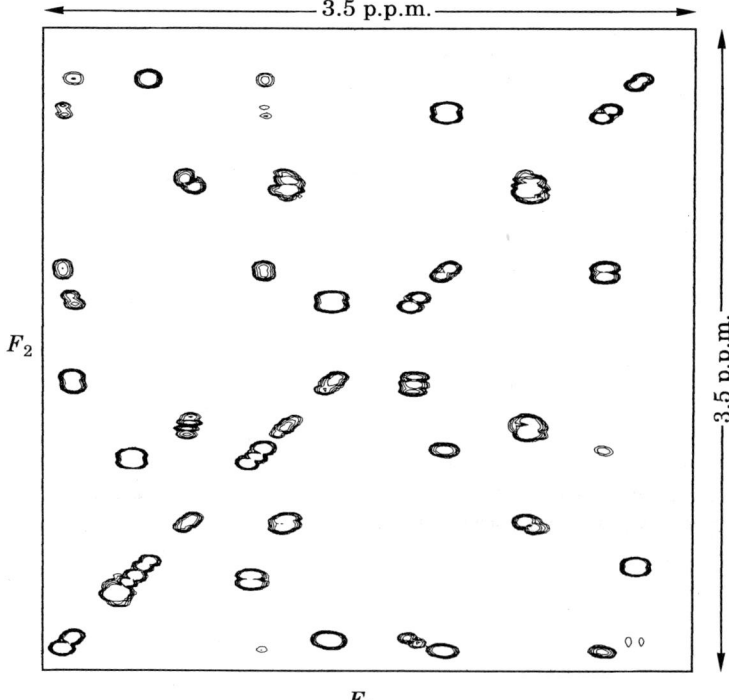

Fig. 9. – Correlation spectroscopy (COSY) on part of the 400 MHz proton spectrum of strychnine.

sure that the spectra are in the pure absorption mode and that frequency discrimination in the F_1 dimension has been properly implemented[13, 14]. The relatively intense diagonal peaks are normally de-emphasized by filtering the signals through double-quantum coherence in the widely used modification called DQ-COSY[15]. A typical COSY spectrum is shown in fig. 9.

4. – Total-correlation spectroscopy.

The strength of the COSY experiment is that only a single stage of coherence transfer is permitted; observation of a pair of cross-peaks is unequivocal evidence that the two spins in question are related by a scalar coupling. Sometimes it can be advantageous to lift this restriction and allow unrestricted flow of coherence throughout the entire coupling network. This has been called total-correlation spectroscopy (TOCSY)[16] although a more descriptive name is the homonuclear Hartmann-Hahn (HOHAHA) experiment[17]. Transfer of coherence is achieved by applying an intense spin-lock field for a period τ (several tens of milliseconds) so that all pairs of coupled spins experience essentially equal effective fields in the rotating frame of reference. Since the precession

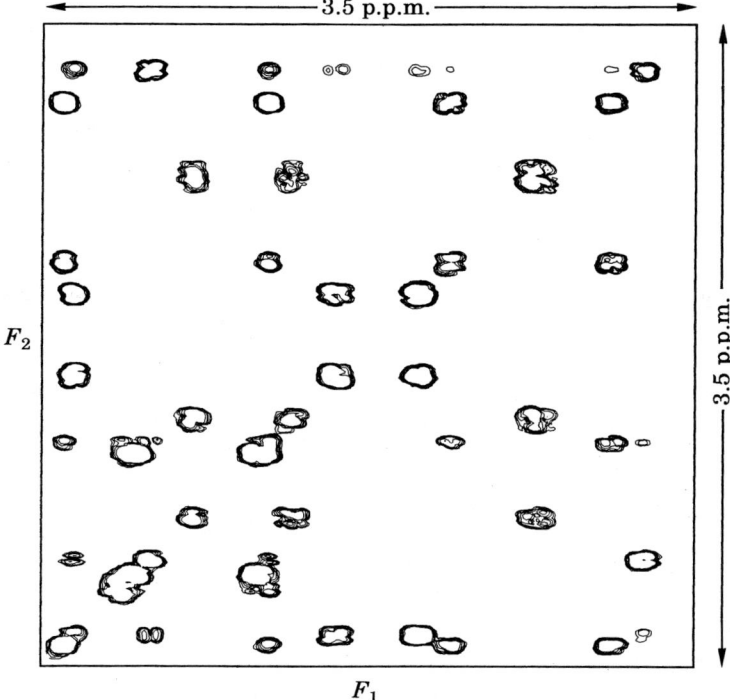

Fig. 10. – Total-correlation spectroscopy (TOCSY or HOHAHA) on part of the 400 MHz proton spectrum of strychnine.

frequencies in the rotating frame are all equal, they behave like two coupled pendulums, exchanging coherence back and forth between them at a rate determined by the scalar coupling. The coherence can be propagated throughout the entire coupling network. A typical HOHAHA spectrum is shown in fig. 10. The effective bandwidth may be increased by replacing the continuous spin-lock field with one of the periodic pulse sequences normally used for broad-band decoupling. We shall see later that interesting experiments may be performed by reversing this trend, employing a very selective spin-lock field.

5. – Chemical-exchange spectroscopy.

Any experiment which «passes on» spin population disturbances, magnetization, or coherence from one site to another can form the basis for correlation spectroscopy. The familiar double-resonance experiment of Forsén and Hoffman[2], where a population disturbance (saturation) is transferred between proton sites by slow chemical exchange, was soon adapted to two-dimensional NMR[18]. After labelling the various magnetizations according to their precession frequencies in t_1, the experiment returns these vectors to the z-axis (spin populations) allowing them a short period τ for chemical exchange before reconversion into detectable y-magnetization. This is very useful for studying the mechanisms of exchange, since a two-stage migration $I \to S \to R$ will show up as such, with cross-peaks between I and S and between S and R, but with no cross-peak between I and R. The beauty of the method is that the two-dimensional spectrum closely mimics the type of diagram that an organic chemist might draw to indicate which sites are related by a direct chemical exchange.

6. – Nuclear Overhauser effect.

A third type of correlation spectroscopy relies on the intensity changes induced by the nuclear Overhauser effect. Under normal circumstances, saturation of the I-spin signal in an IS-spin system would not affect the S-spin intensity. However, if dipole-dipole interactions dominate the spin-lattice relaxation processes of the S spin, a rearrangement of spin populations occurs and the S-spin intensity is enhanced. There is an initial transient effect, followed by the establishment of a steady state that differs from Boltzmann equilibrium. The phenomenon is best understood by the analogy with Kirchoff's laws governing the voltages, currents and conductances of a complex electrical network. The voltages represent population disturbances from Boltzmann equilibrium and they must be nonzero for a nuclear Overhauser enhancement. The currents correspond to the flux of spins between the energy levels and the conductances represent the relaxation transition probabilities.

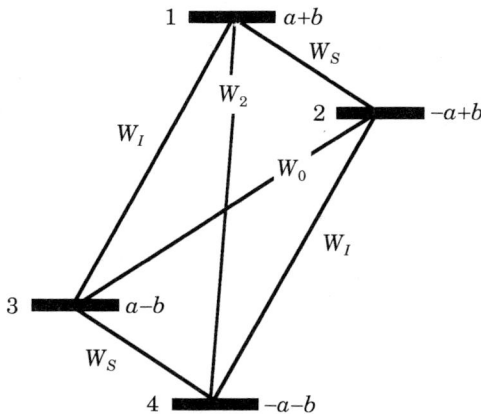

Fig. 11. – Populations and relaxation transition probabilities for a nuclear Overhauser experiment on an IS spin system.

A nuclear spin S may experience a local dipolar field B_{DD} from a neighbouring spin I given by the expression

$$B_{DD} = \pm \mu (3\cos^2 \theta - 1)/r^3 , \tag{5}$$

where μ is the strength of the I dipole, θ is the angle of the internuclear vector with respect to the direction of the applied field B_0 and r is the internuclear distance. In the solid state, the dipole-dipole interaction is a dominant effect and gives rise to structure and broadening of the S resonance. In liquids, the rapid molecular reorientation averages out the local dipolar field since θ is changing rapidly through all possible values. However, the interaction is still responsible for a major contribution to spin-lattice relaxation, proportional to B_{DD}^2 and hence inversely proportional to the sixth power of the internuclear distance.

For the general case of two species I and S with different magnetogyric ratios γ_I and γ_S, the energy level diagram is as shown in fig. 11. The «conductances» are:

W_I: the I-spin transition probability;
W_S: the S-spin transition probability;
W_2: the double-quantum transition probability;
W_0: the zero-quantum transition probability.

The parameters W_2 and W_0 are called cross-relaxation probabilities and correspond to simultaneous I and S transitions, either in the same (W_2) or opposite senses (W_0). The Overhauser effect relies on the fact that W_2 is relatively more important than W_I, W_S and W_0.

The deviations of spin populations from equilibrium may be represented by the numbers a and b where

(6) $$2a = S_Z - S_0 \quad \text{and} \quad 2b = I_Z - I_0$$

with

(7) $$I_0/S_0 = \gamma_I/\gamma_S .$$

Relaxation creates a flux of spins proportional to the population differences multiplied by the appropriate relaxation transition probabilities. By writing the rate equations for the inflow and outflow from energy levels 1 and 2, and by setting the condition for a steady state we can derive the expression

(8) $$-a/b = (W_2 - W_0)/(2W_S + W_2 + W_0).$$

The Overhauser effect involves I-spin saturation ($I_Z = 0$) giving $2b = -I_0$. The Overhauser enhancement factor is then

(9) $$1 + \eta = \frac{S_Z}{S_0} = 1 + \frac{W_2 - W_0}{2W_S + W_2 + W_0} \frac{\gamma_I}{\gamma_S} .$$

If the mechanism for spin-lattice relaxation is purely dipolar, and if the molecular motion is fast enough that $\omega\tau_c \ll 1$, then the relative importance of the transition probabilities is known to be in the ratio

(10) $$W_2 : W_S : W_0 :: 12 : 3 : 2 .$$

This gives the simple result

(11) $$1 + \eta = 1 + 0.5 \frac{\gamma_I}{\gamma_S} .$$

For proton-proton systems this means that irradiation of the I protons leads to a 50% enhancement of the intensity of the S protons if the mechanism is exclusively dipolar. If other («leakage») relaxation mechanisms are appreciable, they reduce the observed enhancement, and, if leakage dominates, the Overhauser effect disappears. In practice there is usually a balance between dipolar and leakage mechanisms, favouring the dipolar contribution when the internuclear distance is short, owing to the inverse sixth-power dependence. For larger molecules and more viscous solutions where $\omega\tau_c$ becomes comparable with or greater than unity, the nuclear Overhauser effect decreases and eventually becomes negative even though the relaxation may be predominantly dipolar. It is for these cases that the rotating-frame Overhauser effect, sometimes called «Camelspin» [19], can be useful, since this enhancement is always positive and there is no risk of missing the effect because it just happens to be near the zero-crossing point.

7. – Nuclear Overhauser spectroscopy (NOESY).

The early nuclear Overhauser enhancements were measured by performing double-resonance experiments, observing the S-spin intensity with and without a saturating radiofrequency field B_2 at the I-spin frequency. With the advent of two-dimensional spectroscopy, the measurements may be made in a much more efficient manner, displaying all the Overhauser enhancements at the same time. The two-dimensional spectrum has the same general form as a COSY spectrum but the cross-peaks are made up of all positive signals. The pulse sequence may be written

$$90°(+x)\text{-}t_1\text{-}90°(-x)\text{-}\tau\text{-}90°(+x)\text{-acquire}.$$

The evolution period t_1 serves to label each chemically distinct site by allowing its magnetization to precess freely at its characteristic frequency. These magnetization components are then converted into z-magnetization (population disturbances) by the second pulse and a suitable time τ is allowed for cross-relaxation. Then the transfer of spin population is monitored by the final «read» pulse.

Cross-peaks centred at coordinates (δ_I, δ_S) and (δ_S, δ_I) indicate that I and S are connected through a cross-relaxation mechanism. The intensity of the cross-peak depends on the time allowed for cross-relaxation (τ) and on the degree to which the relaxation is dipolar. It is this latter parameter which is important to the biochemist. Although the detailed balance between dipolar relaxation and leakage mechanisms is unknown, and although reorientational correlation times (τ_c) may not all be the same throughout the molecule, valuable information about internuclear distances can nevertheless be obtained. The key is the inverse sixth-power dependence of dipolar relaxation on the internuclear distance, ensuring that cross-peaks in NOESY spectra are only observed if the two spins are reasonably close in space, say less than about 5 or 6 ångström units. The biochemist relies heavily on these «distance constraints» to refine his proposed three-dimensional structures. Fortunately the problem is normally greatly overdetermined since there are very large numbers of cross-peaks in the NOESY spectra. If the aminoacid resonances have been assigned, just one observation of a nuclear Overhauser cross-peak may be sufficient to establish that there is a loop of a particular size. Quantitative measurements of cross-peak intensities are normally unnecessary.

8. – Forbidden transitions.

Two-dimensional Fourier transformation has another very useful property—it can be used to monitor forbidden transitions. The selection rule for magnetic-resonance experiments is $\Delta m = \pm 1$. Although the other types of transition are formally forbidden, they are not without interest. Double- and triple-quantum transitions were detected in the early continuous-wave spectrometers by increasing the radiofrequency level to compensate for the much lower transition probabilities of multiple-quantum transitions. They may be used to assign transitions to the appropriate energy level diagram and determine the relative signs of coupling constants. But it was not until the advent of two-dimensional Fourier transformation that forbidden transitions really came into their own. The trick is to excite the desired multiple-quantum coherence during the evolution period t_1 when the receiver is inactive, reconverting it into observable transverse magnetization (single-quantum coherence) during the detection period.

Double-quantum coherence may be excited in a two-spin IS system by the sequence

$$90°(x)\text{-}\tau\text{-}180°(x)\text{-}\tau\text{-}90°(x),$$

where the interval τ has been chosen to equal $1/4J_{IS}$. We recognise this as a spin echo experiment with J-modulation (see below), arranged so that the two I-spin vectors are aligned in opposition along the $\pm x$-axes of the rotating frame. Unfortunately this is where the vector model loses its predictive capabilities and the conversion into double-quantum coherence can only be properly described in terms of density matrix (or product operator) theory. If a vector picture has any validity in this context, double-quantum coherence would be represented as a pair of antiphase vectors of identical frequencies, permanently locked in opposition and unable to induce any signal in the receiver coil. Precession occurs at the double-quantum frequency $(\delta_I + \delta_S)$ Hz, which appears in the F_1 dimension.

Double-quantum coherence has some interesting properties. Compared with single-quantum coherence, it is twice as sensitive to radiofrequency phase shifts and to magnetic-field gradients. These features may be put to very good use in separating double-quantum signals from other orders of coherence.

9. – «INADEQUATE».

Double-quantum coherence requires the presence of at least two coupled spins; isolated spins can never support multiple-quantum coherences. We may exploit this to suppress undesirable signals. A most successful application has been to carbon-13 spectroscopy [20, 21] for the study of carbon-carbon coupling. Only one molecule in 10^4 contains two adjacent carbon-13 spins in natural-abundance samples, whereas one molecule in 100 has isolated carbon-13 spins, which are of no interest in this application. This much stronger signal may be suppressed by filtration through double-quantum coherence, leaving the desired carbon-carbon coupling information. This is achieved by a phase cycle that «fol-

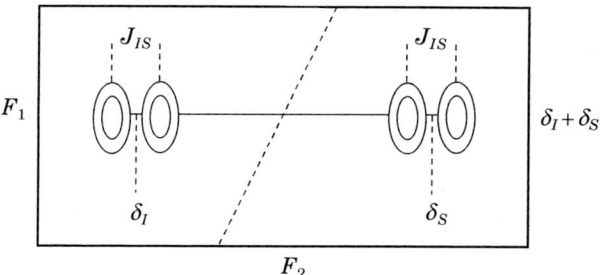

Fig. 12. – Schematic representation of the pattern of lines from two coupled carbon-13 spins in the INADEQUATE experiment.

lows» the double-quantum precession, or by pulsed-field gradients designed to select double-quantum coherence and to suppress the rest.

Now the carbon atoms are the backbone of almost all molecules of interest to the organic chemist and, if we can identify directly bonded carbon atoms, we can hope to build up the entire skeleton one link at a time. It is fortunate that directly bound carbon atoms show coupling constants in the range $(30 \div 40)$ Hz, whereas all longer-range interactions are much weaker. Thus, by suitable choice of the τ delay, double-quantum coherence from directly bound carbon pairs can be excited with reasonable uniformity, with a negligible contribution from more distant carbon pairs. Each coherence is excited independently of the rest, for each carbon pair is in a different molecule. We may, therefore, identify the links of the chain one at a time with no danger of interference effects.

The corresponding two-dimensional («INADEQUATE») spectra are extremely simple but of low sensitivity because of the adverse effect of the natural isotopic abundance. Each pair of coupled carbon spins generates a four-line spectrum in the F_2 dimension just as in conventional NMR spectroscopy (fig. 12). The centre of gravity of the pattern is at $(\delta_I + \delta_S)/2$ Hz. These four-line patterns are dispersed in the F_1 dimension according to the appropriate double-quantum frequency $(\delta_I + \delta_S)$ Hz. Consequently, the centre of gravity of each pattern lies on a skew diagonal of slope 2.

This is a considerable aid to assignment. We may identify at once four important structural features—a straight chain, a branched chain, ring formation and quaternary carbon sites (fig. 13). In principle, one could proceed directly to the assembly of a molecular model based on the INADEQUATE results.

Figure 14 is a tracing of some experimental results from the natural-abundance carbon-13 INADEQUATE spectrum of panamine, one of the *Ormosia* family of alkaloids. We see from this limited extract from the two-dimensional spectrum that site i is directly bonded to sites a, d, e and n. The site i must be a quaternary carbon, a conclusion confirmed by the low relative intensity of this line since it is partially saturated (long T_1). This provides a good starting point for tracing out the remaining carbon-carbon linkages and, despite the presence

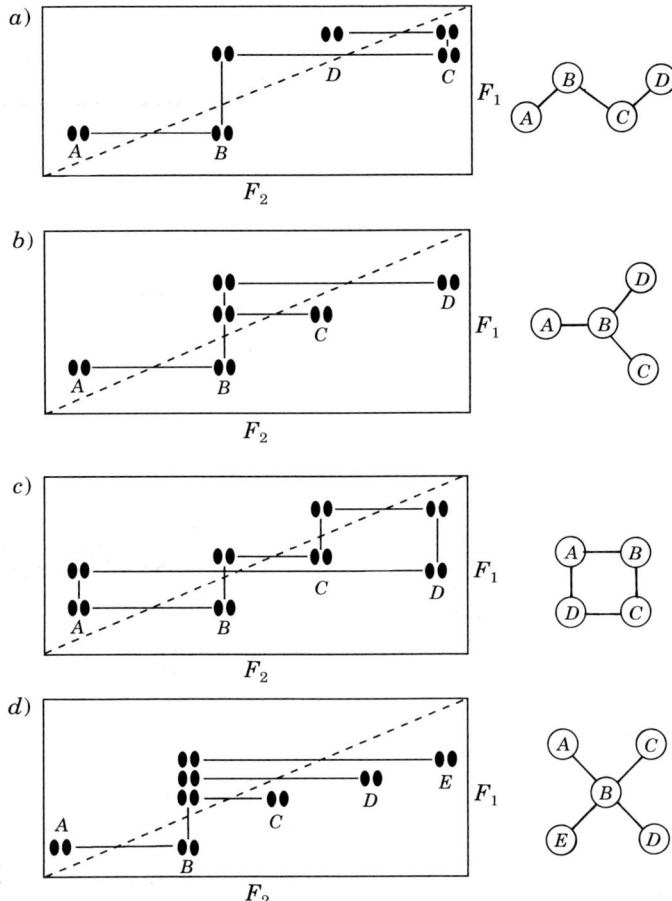

Fig. 13. – Structural features that may be deduced from the topology of the INADEQUATE spectrum: a) straight chain, b) chain branching, c) ring formation, d) quaternary site.

of three nitrogen atoms in the ring structure (interrupting the carbon chains), the overall panamine structure can be successfully deduced [22].

Isotopic enrichment can be used to improve the signal-to-noise of INADEQUATE spectra, although it is preferable to avoid situations where three or more carbon-13 spins reside in the same molecule. An exciting example is provided by the new form of carbon—the Buckminster fullerenes. An enriched sample of the fullerene C_{70} is easily obtained by embedding some carbon-13 in the graphite electrodes used in the preparation [23]. There are five distinct chemical sites in C_{70} and the two-dimensional INADEQUATE spectrum gives direct evidence of their connectivity, providing confirmation that the molecule has the form of an ellipsoid of revolution (a rugby ball).

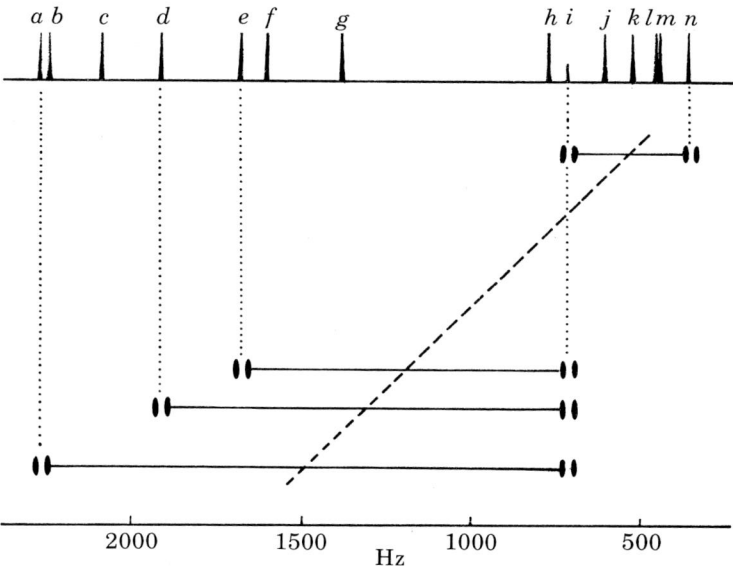

Fig. 14. – Tracing from the experimental INADEQUATE spectrum of panamine showing that carbon i is directly bound to carbons a, d, e and n. The signal from site i is weak because of slow spin-lattice relaxation.

Filtration through higher-order multiple-quantum coherences provides a very effective method for simplifying two-dimensional correlation spectra. The separation is achieved by a suitable phase cycle or a combination of matched pulsed-field gradients with the appropriate ratio of pulse durations. As an example, filtration of a COSY spectrum through four-quantum coherence eliminates all signals that arise from groups of less than four coupled spins—a major simplification in many cases [24]. It has become common practice to filter COSY spectra through double-quantum coherence as a method of reducing the relative importance of a strong water resonance and undesirable diagonal peaks [15].

10. – Spin echoes.

The concept of the spin echo is the basis of many experiments employed in high-resolution NMR, including techniques as disparate as composite radiofrequency pulses and phase-cycling schemes to compensate instrumental shortcomings. HAHN and MAXWELL [25] showed that spin echoes are modulated by scalar spin-spin coupling, provided both coupled spins are «resonant». The process can be visualized in terms of a phase evolution diagram (fig. 15) for a coupled two-spin system IS. The S spins may be represented by two vectors α and β which precess at frequencies determined by the S-spin chemical shift and the coupling constant J_{IS}. Suppose for the moment that the I spin is nonresonant,

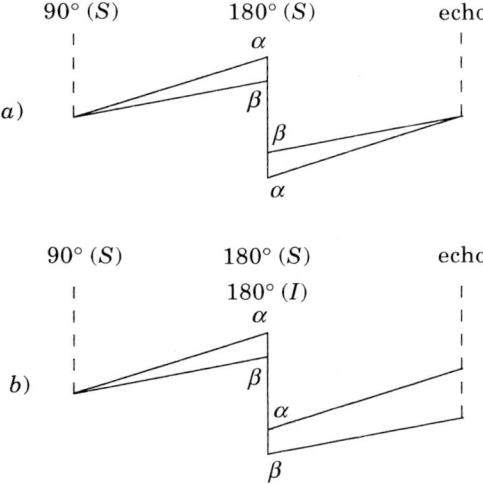

Fig. 15. – Phase evolution diagrams for a spin echo experiment on a coupled IS spin system. a) When only the observed S spin experiences the effect of the 180° pulse, then both chemical-shift and spin-spin coupling effects are brought to a focus at time 2τ. b) When the I spin is also affected, this interchanges the α and β labels and the divergence due to spin-spin coupling persists throughout the 2τ interval.

that is to say, it does not experience the effects of the 180° pulse. For example, it may be a heteronuclear species or the pulse may be frequency selective. After the initial 90° pulse, the S-spin vectors diverge as shown in fig. 15a) during the interval τ and are brought to an exact focus at time 2τ. On the other hand, if both the I and S spins are resonant (a homonuclear system with hard pulses), then the 180° pulse interchanges the α and β labels and the two vectors continue to diverge throughout the entire period 2τ (fig. 15b)). This phase divergence translates into a modulation of the spin echo at frequencies $\pm J_{IS}/2$.

Echo modulation provides a method for distinguishing chemical-shift effects from scalar coupling. Applied to two-dimensional spectroscopy, it offers a complete separation of shifts and coupling constants. For example, an experiment can be devised which refocuses chemical-shift effects during the evolution period, restricting evolution to spin-spin coupling alone, so that the F_1 dimension shows only J-splittings. If the detection period monitors the conventional free-induction decay, then the F_2 dimension comprises both chemical shift and spin-spin coupling information. This is called a two-dimensional «J-spectrum». Figure 16 shows a typical proton J-spectrum obtained in this manner. The spin multiplets are arranged rather like a half-open venetian blind, running at 45° across the two-dimensional display. Each spin multiplet can be examined separately with no overlap from adjacent multiplets. Unfortunately the resonances have the undesirable «phase twist» lineshape, giving sections with dispersion-mode contributions. There is an even more serious consequence. The 45° projec-

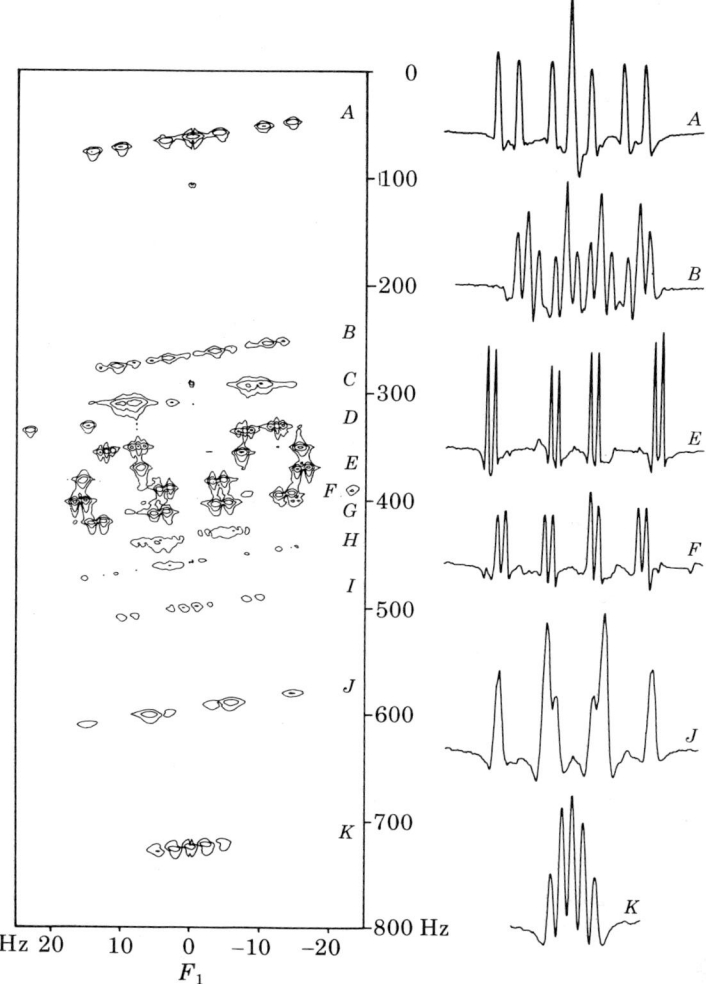

Fig. 16. – Conventional two-dimensional J-spectrum of protons in a tricyclodecanone derivative. For clarity the contour diagram is drawn in the absolute-value mode, while the skew sections through the spin multiplets are displayed as phase-sensitive spectra, illustrating the dispersion-mode tails. Projections in *enfilade* along the multiplets vanish because the negative dispersion tails exactly cancel the positive absorption components.

tion, which would be expected to give a «broad-band decoupled» version of the spectrum, vanishes in practice because the dispersion-mode tails exactly cancel the absorption-mode contribution. The only simple solution appears to be to project the absolute-value mode, which has unacceptably broad lines and is distorted when adjacent lines overlap. The goal of proton spectra without spin-spin splitting appears to be as elusive as ever.

11. – «Broad-band decoupled» proton spectra.

One of the most attractive features of carbon-13 spectroscopy is the simplicity of the spectra when broad-band proton decoupling is used. It would be very useful to have a similar mode of proton spectra—just a single resonance at each chemical-shift value and no spin-spin splittings. This can be achieved with the following pulse sequence [26, 27]:

$$90°(x)-t_1/2-180°(y)-t_1/2-\text{spin-lock } (\pm y)\text{-acquire}.$$

This is a standard modulated echo sequence except for the introduction of the spin-lock pulse just before signal acquisition. It serves to «purge» x-components of magnetization by dispersing them in the inhomogeneous B_1 field, retaining the y-components. This fundamentally changes the character of the spin-spin multiplet structure. Instead of multiplets that run along the 45° diagonal, there is a symmetrical pattern with identical (multiplicative) splittings in both the F_1 and F_2 dimensions (fig. 17). Each individual multiplet acquires the valuable C_4 symmetry property, that is to say, it is invariant with respect to 90° rotations about its centre. Figure 18 shows some typical two-dimensional spin multiplets from the purged J-spectrum of protons in strychnine.

This C_4 symmetry property can be put to very good use as a «symmetry filter» which retains signals only when they are distributed in the appropriate C_4 pattern, and rejects the rest. An algorithm can be written that searches the two-dimensional spectrum to locate the centres of C_4 symmetry and then projects the multiplet structure onto an axis through the centre parallel to the F_1 axis. This displays each multiplet individually and also provides the chemical-shift frequency of that particular site.

The algorithm searches through the entire experimental data matrix with a square «test zone» (typically 50 Hz by 50 Hz) looking for patterns of lines that have the requisite C_4 symmetry (fig. 19a)). Normally one would test for this symmetry feature by rotating the pattern through $\pm 90°$ and checking that it remains invariant. In practice it proves more convenient to employ a folding oper-

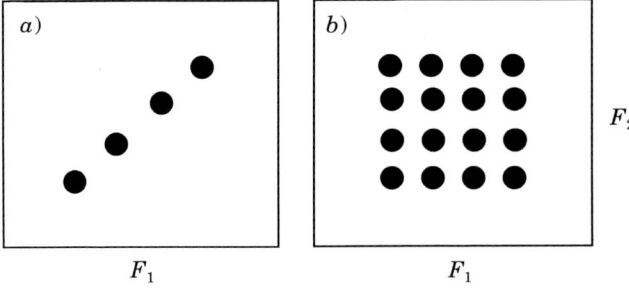

Fig. 17. – Schematic spin multiplets from a) a conventional J-spectrum, b) from a purged J-spectrum. Pattern b) has C_4 symmetry.

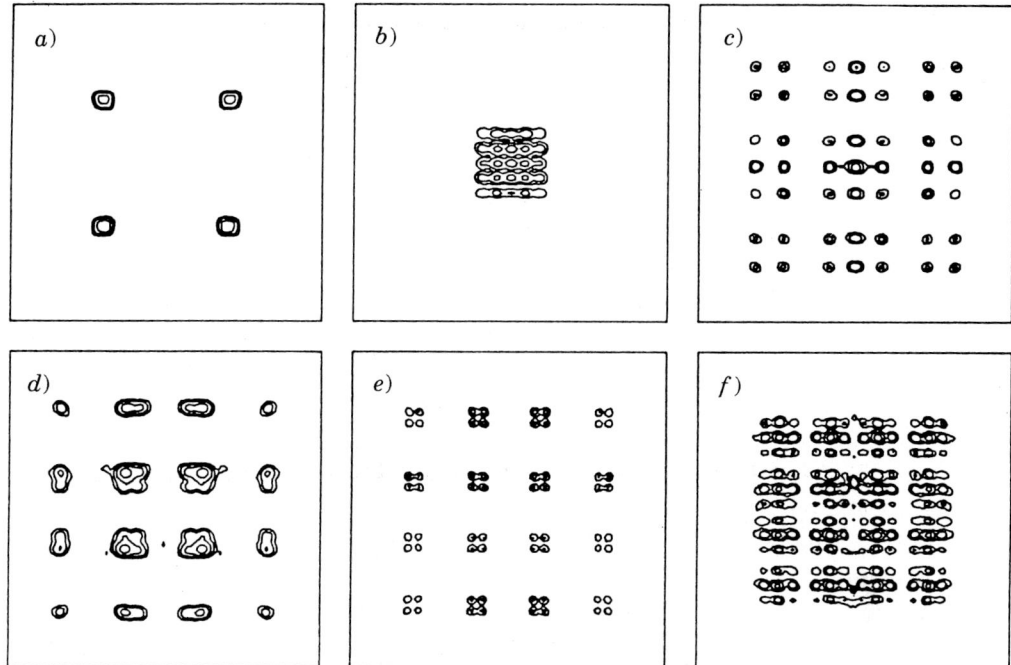

Fig. 18. – Some typical spin multiplets from the purged two-dimensional J-spectrum of protons in strychnine. Note that they all possess C_4 symmetry.

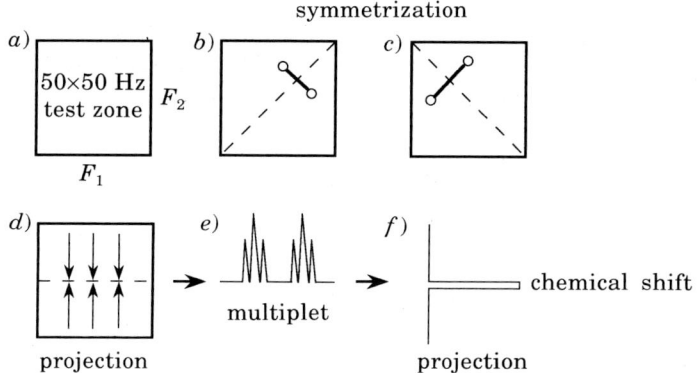

Fig. 19. – Post-acquisition data processing to locate the centres of C_4 symmetry in a purged J-spectrum. A test zone (a)) is stepped through the spectrum. b) At each step, the intensities of signals at symmetrically related positions with respect to the 45° diagonal are compared and replaced by the lower ordinate. c) This is repeated for the 135° diagonal. d) The integral over the test zone reaches a maximum for a centre of symmetry. e) Projection of individual multiplets. f) Projection of the «broad-band decoupled» spectrum.

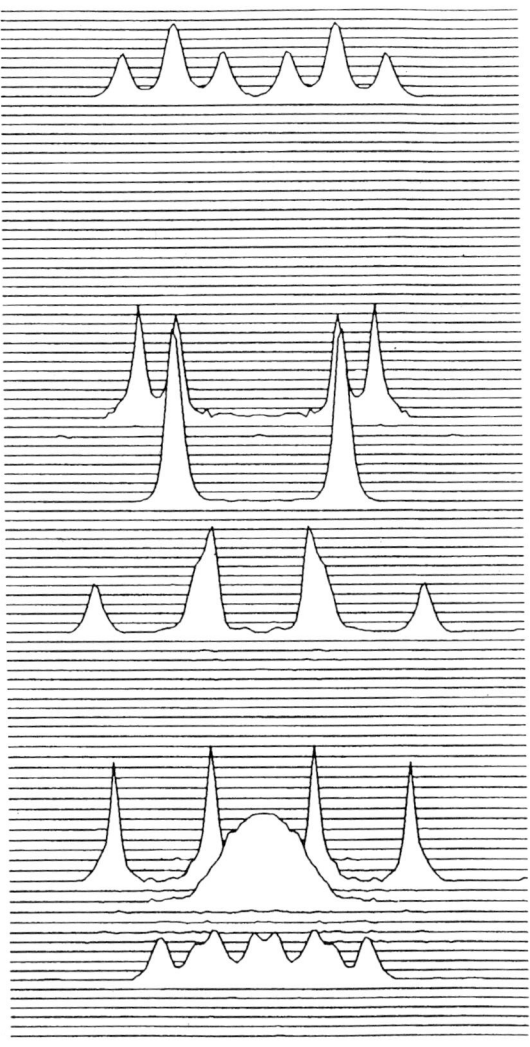

Fig. 20. – Part of the processed two-dimensional purged J-spectrum of strychnine. Only every tenth trace has been plotted.

ation about the 45° diagonal (fig. 19b)), comparing the intensities at symmetrically related points and setting the lower value in both locations. If the two test ordinates are equal, there is no change, but, if this symmetry is lacking, then the stronger signal is eliminated, usually leaving a value near zero. The process is then repeated for the 135° diagonal (fig. 19c)). Thus, where there is no mirror symmetry about the 45° and 135° diagonals, the signals are suppressed, but, where there is symmetry, they are retained. The integral over the 50 Hz by 50 Hz test zone is, therefore, appreciable only for a centre of C_4 symmetry. A

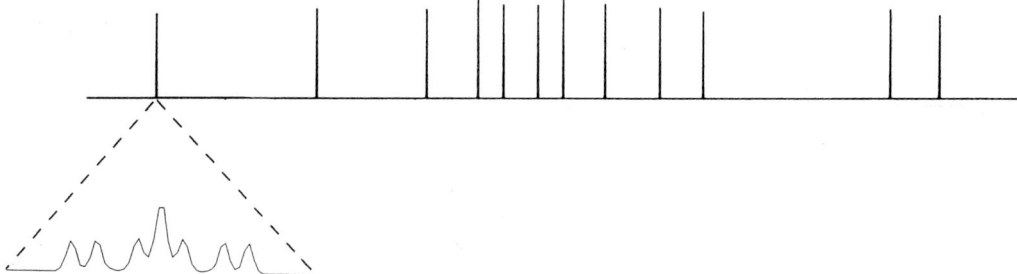

Fig. 21. – Proton spectrum of a tricyclodecanone derivative obtained by processing the purged J-spectrum. Each of the twelve chemically distinct sites gives a singlet response. The spin multiplet structure can be retrieved if necessary (left-hand line).

plot of this integral as a function of F_2 gives a spectrum with peaks at the chemical-shift values.

Once the symmetry centre has been located, the individual spin multiplet pattern for that site may be examined by projecting the intensities within the test zone onto the F_1 axis (fig. 19e)). The lines are all in the pure absorption mode and good resolution is achieved. The same information may be presented in the form of a purged two-dimensional J-spectrum with the multiplet structure confined to the F_1 dimension, just as in a *heteronuclear* broad-band decoupled experiment. Figure 20 shows the result for part of the 400 MHz high-resolution J-spectrum of strychnine. For practical reasons of clarity of presentation, only every tenth trace is plotted. In a single diagram this gives the chemical shifts (in F_2) and the multiplet structure for each individual proton site.

The most exciting result of purged J-spectroscopy is the possibility of recording proton spectra without the complication of spin-spin splitting. Figure 21 shows an example, the 400 MHz proton spectrum of a tricyclodecanone derivative with just one resonance line from each of twelve chemically distinct sites. Note the uniformity of the intensities. The multiplet structure of any given site can be displayed if necessary.

12. – Discussion.

This lecture presents only an outline of the main themes of two-dimensional spectroscopy; many elaborations and embellishments are in common use by high-resolution spectroscopists. Although most two-dimensional experiments may be viewed as being derived from existing double-resonance experiments, by introducing the idea of pulse sequencing they have influenced the entire field of experimental NMR. For example, the new broad-band decoupling techniques that have now superseded noise decoupling can be traced directly to ideas developed for two-dimensional spectroscopy.

Many of the new experiments can be regarded as extensions of basic two-dimensional techniques, using the building-block principle. For example, heteronuclear polarization experiments are now usually performed as double transfers from protons to the X-nucleus and back to protons, in order to preserve the maximum sensitivity. Complexity of pulse sequences is not a practical problem since the entire experiment is governed by a computer program much of which was probably written by the spectrometer manufacturer. Tacking two or more sequences together is normally a simple matter that requires only a minor modification of computer code. Where the early double-resonance experiments often demanded some expertise with a soldering iron, the modern spectroscopist merely needs access to a keyboard and some understanding of the spectrometer software.

* * *

The author is indebted to Dr. $\overline{\text{E}}$. Kupče for the COSY and TOCSY spectra shown in fig. 9 and 10.

REFERENCES

[1] W. A. Anderson: *Phys. Rev.*, **102**, 151 (1956).
[2] S. Forsén and R. A. Hoffman: *J. Chem. Phys.*, **39**, 2892 (1963).
[3] J. H. Noggle and R. E. Schirmer: *The Nuclear Overhauser Effect* (Academic Press, New York, N.Y., 1971).
[4] J. Jeener: *Ampere International Summer School, Basko Polje, Yugoslavia, 1971*.
[5] W. P. Aue, E. Bartholdi and R. R. Ernst: *J. Chem. Phys.*, **64**, 2229 (1976).
[6] R. Freeman, J. B. Grutzner, G. A. Morris and D. L. Turner: *J. Am. Chem. Soc.*, **100**, 5637 (1978).
[7] A. Bax and R. Freeman: *J. Magn. Reson.*, **44**, 542 (1981).
[8] P. Xu, X. L. Wu and R. Freeman: *J. Magn. Reson.*, **84**, 198 (1989).
[9] J. Friedrich, S. Davies and R. Freeman: *Mol. Phys.*, **64**, 691 (1988).
[10] G. Bodenhausen, R. Freeman and D. L. Turner: *J. Magn. Reson.*, **27**, 511 (1977).
[11] P. Barker and R. Freeman: *J. Magn. Reson.*, **64**, 334 (1985).
[12] R. Hurd: *J. Magn. Reson.*, **87**, 422 (1990).
[13] D. J. States, R. A. Haberkorn and D. J. Ruben: *J. Magn. Reson.*, **48**, 286 (1982).
[14] D. Marion and K. Wüthrich: *Biochem. Biophys. Res. Commun.*, **113**, 967 (1983).
[15] U. Piantini, O. W. Sørensen and R. R. Ernst: *J. Am. Chem. Soc.*, **104**, 6800 (1982).
[16] L. Braunschweiler and R. R. Ernst: *J. Magn. Reson.*, **53**, 521 (1983).
[17] A. Bax and D. G. Davis: *J. Magn. Reson.*, **63**, 207 (1985).
[18] J. Jeener, B. H. Meier, P. Bachmann and R. R. Ernst: *J. Chem. Phys.*, **71**, 4546 (1979).

[19] A. A. BOTHNER-BY, R. L. STEPHENS, J.-M. LEE, C. D. WARREN and R. W. JEANLOZ: *J. Am. Chem. Soc.*, **106**, 811 (1984).
[20] A. BAX, R. FREEMAN and S. P. KEMPSELL: *J. Am. Chem. Soc.*, **102**, 4849 (1980).
[21] A. BAX, R. FREEMAN and T. A. FRENKIEL: *J. Am. Chem. Soc.*, **103**, 2102 (1981).
[22] N. S. BHACCA, M. F. BALANDRIN, A. D. KINGHORN, T. A. FRENKIEL, R. FREEMAN and G. A. MORRIS: *J. Am. Chem. Soc.*, **105**, 2358 (1983).
[23] R. D. JOHNSON, G. MEIJER, J. R. SALEM and D. S. BETHUNE: *J. Am. Chem. Soc.*, **113**, 3619 (1991).
[24] A. J. SHAKA and R. FREEMAN: *J. Magn. Reson.*, **51**, 169 (1983).
[25] E. L. HAHN and D. E. MAXWELL: *Phys. Rev.*, **88**, 1070 (1952).
[26] P. XU, X. L. WU and R. FREEMAN: *J. Am. Chem. Soc.*, **113**, 3596 (1991).
[27] P. XU, X. L. WU and R. FREEMAN: *J. Magn. Reson.*, **95**, 132 (1991).

Selective Excitation in High-Resolution NMR.

R. FREEMAN

Department of Chemistry, Cambridge University - Cambridge, U.K.

1. – Introduction.

Double-resonance experiments in high-resolution NMR are inherently frequency selective, being implemented by continuous-wave radiofrequency fields that must be adjusted to the appropriate frequencies. On the other hand, Fourier transform spectroscopy is normally initiated with a «hard» radiofrequency pulse to ensure essentially uniform excitation across the entire spectrum; there is no selectivity. There are, however, several new applications where it is advantageous to lift this restriction on pulsed spectroscopy, in order to achieve selective excitation [1, 2]. This might involve a single transition (a line-selective pulse), a single spin multiplet (multiplet-selective) or a range of chemical shifts (band-selective). Such experiments make it possible to take a closer look at the fine detail of the spectrum, to unravel complicated overlapping features or to simplify multidimensional spectroscopy.

In general, selective excitation involves working with a much weaker radiofrequency field B_1, such that off-resonance spins experience an effective field that is tilted away from the x-axis towards the z-axis where is causes negligible excitation of transverse magnetization. In order to achieve a 90° (or 180°) rotation at resonance, the pulse duration must be correspondingly increased. The simplest approach is to employ a long weak rectangular pulse, and most of the early experiments were performed in this manner. It is even possible to dispense with radiofrequency switching altogether, deriving the soft pulse as an audiofrequency modulation sideband with the modulation field being switched [3]. The centreband is left on continuously but set far away from all resonances in the spectrum. This technique has the advantage that the NMR signal may be observed during the pulse as it nutates about the effective field in the rotating frame. In this manner several types of relaxation studies can be performed on individual proton lines in a high-resolution spectrum. Spin-lattice

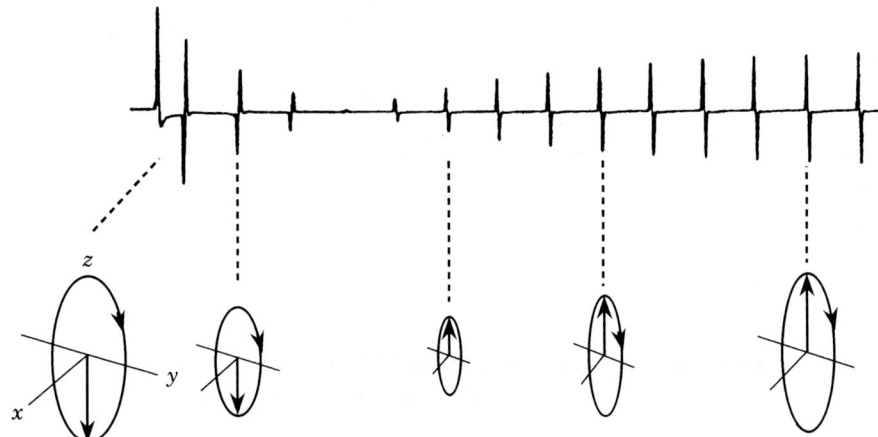

Fig. 1. – Spin-lattice relaxation studied by a selective 180° pulse followed by a train of selective 360° pulses. These were generated as switched audiofrequency modulation sidebands of a continuous carrier frequency, permitting observation of the signal during the pulses.

relaxation measurements are particularly simple, involving the sequence

$$180°\text{-}\tau\text{-}360°\text{-}\tau\text{-}360°\text{-}\tau\text{-}360°\text{-}\ldots .$$

The initial pulse inverts the spin populations across the selected transition while the 360° pulses monitor the signal as it recovers, leaving it free to continue its recovery during the next τ period [3]. Figure 1 shows some typical results.

2. – The DANTE sequence.

In many high-resolution spectrometers there is no provision for switching the transmitter down to a sufficiently low level for soft-pulse operation, nor is it possible to fine-tune the frequency to the accuracy demanded by selective excitation experiments. In this situation the «DANTE» sequence can be useful [4]. This replaces the long, low-level pulse by a regular sequence of hard pulses at high transmitter level, the sequence being continued for the same duration as the soft pulse it replaces. The operation is best visualized in terms of magnetization trajectories in the rotating frame. The soft-pulse trajectory is a smooth curve, representing nutation about the effective field B_{eff}; the equivalent DANTE sequence induces a zig-zag motion corresponding to low-amplitude nutations about the x-axis (the hard pulse) alternating with short periods of free precession about the z-axis. If there is a reasonably large number of pulses in the DANTE sequence, the resulting zig-zag path closely approximates the smooth curve (fig. 2). Consequently the DANTE sequence behaves in a manner very similar to that of a single soft pulse of the same duration. However, there

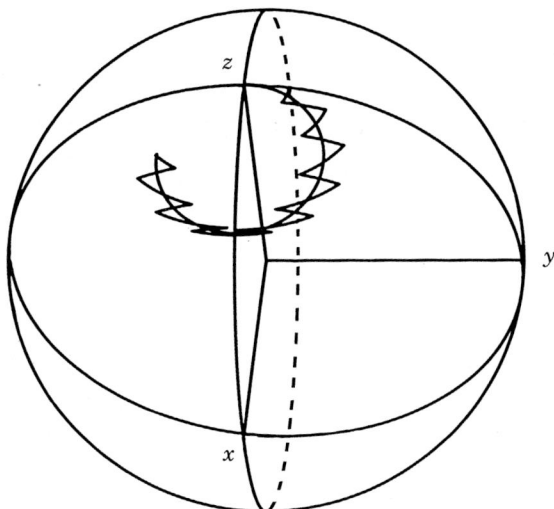

Fig. 2. – Magnetization trajectories of a single soft pulse (smooth curve) and the equivalent nine-pulse DANTE sequence (zig-zag path).

is one crucial difference, the DANTE sequence possesses a set of sideband responses arrayed on either side of the main response at offsets n/τ Hz, where τ is the interval between the hard pulses. For example, the first sideband response corresponds to free precession through 360° during each τ interval. Sideband excitation can be put to very good use to fine-tune the effective irradiation frequency by adjusting the interval τ rather than the spectrometer transmitter frequency.

Two essentially simultaneous selective pulses at different frequencies can be implemented by interleaving two DANTE sequences (A and B) of the same repetition rate [5]. Sequence A would have all pulses of the same phase, whereas in sequence B the transmitter phase would be incremented in small steps δ radians, corresponding to a frequency shift of $\delta/2\pi\tau$ Hz. The idea can be extended to generate several different irradiation frequencies. If the component DANTE sequences are set up with different total durations, their frequency selectivities may be separately adjusted.

Since a DANTE sequence and a single low-level pulse have essentially the same behaviour, in what follows we shall take the term «soft pulse» to apply to either of these implementations.

3. – Shaped pulses.

When one examines the behaviour of a selective pulse at intermediate offsets from resonance, some complications come to light. The excitation does not fall off monotonically with offset but exhibits sidelobe responses which, although they get weaker with offset, nevertheless extend over a considerable range of frequencies where they can cause spurious excitation of adjacent resonances. In the vector picture this corresponds to rotation about an effective field that is still appreciably tilted away from the $+z$-axis of the rotating frame, causing the magnetization to make small circular excursions away from the z-axis. For off-resonance excitation it is a fairly good approximation to say that the frequency domain excitation spectrum is the Fourier transform of the radiofrequency pulse shape. The sidelobes may then be described in terms of a sinc function, the transform of a time domain rectangular pulse.

This suggests that sidelobes may be reduced or eliminated by suitably «rounding off» the pulse envelope. Early experiments used simple shaping functions, and the Gaussian was found to be a very good choice[6] since its frequency domain excitation spectrum falls off rapidly in the tails. There are no sidelobe responses. Alternatively the pulse may be shaped according to a sinc function with a view to obtaining an excitation that is approximately rectangular (the «top-hat» pulse). Owing to the nonlinearity of the NMR response, this Fourier transform relationship is only approximate. Combinations of a Gaussian pulse with a polynomial function («Hermite» pulse) have also proved useful for selective excitation[7].

A compromise is possible. We may retain one sharp edge but round off the other. For example, a pulse shaped according to the first (rising) half of a Gaussian curve but truncated abruptly at the peak generates an approximately Gaussian absorption-mode frequency domain response with no sidelobes. The dispersion-mode response is very broad, but for some coherence transfer applications it is only the absorption-mode component that is important. Dispersion contributions may be «purged» from the response by appending hard $\pm 90°(y)$ pulses immediately after the soft half-Gaussian, returning x-magnetization to the $+z$-axis.

4. – Phase gradients.

These simple shaped pulses would be perfectly acceptable if it were possible to begin signal acquisition near the mid-point of the pulse. In practice, acquisition is usually started at the end of the pulse, and the consequent delay induces a frequency-dependent phase shift across the spectrum. Another way to visualize this effect is to calculate magnetization trajectories as a function of offset from resonance. The larger the tilt of the effective field, the larger the phase shift away from the $+y$-axis, giving increasing dispersion components (fig. 3). In some simple applications these frequency-dependent phase shifts may be corrected after signal acquisition, using existing phase correction routines. In many multipulse experiments, however, these phase gradients are undesirable.

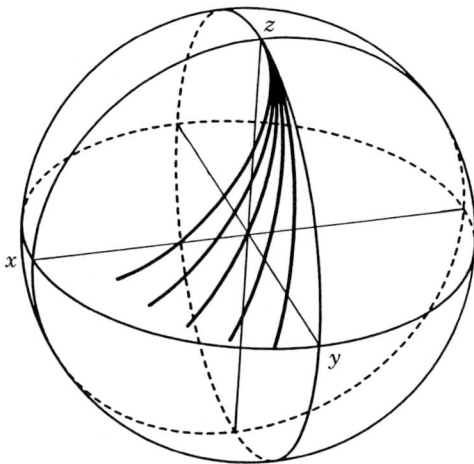

Fig. 3. – Phase gradient generated by a soft pulse. Magnetization trajectories have been calculated for increasing offsets from resonance (increased tilt of the effective field). The phase error is an approximately linear function of offset.

To a good approximation the phase shift is a linear function of offset, allowing refocusing by a hard 180° pulse followed by the appropriate delay.

Extension of this idea suggests that it might be feasible to incorporate a refocusing feature into the soft pulse itself, that is to say, to shape a pulse in such a manner that off-resonance trajectories are brought to a focus on the $+y$-axis, eliminating dispersion contributions. Outside the effective bandwidth the trajectories would be returned to the $+z$-axis (no excited signal). In a narrow transition zone between the two, some dispersion-mode excitation would be tolerated. Band-selective pulses of this type («BURP» pulses) have been designed by the simulated annealing method[8]. They give «pure-phase» spectra, that is to say, all responses within the defined band are of uniform intensity and are in the absorption mode, while the transition regions flanking the excitation band are relatively narrow, with negligible excitation elsewhere. The effective bandwidth is readily scaled up by decreasing the pulse duration. Since these pulses are designed on a refocusing principle, the flip angle at resonance must be carefully calibrated; they are not particularly tolerant of pulse length errors.

5. – «Spin pinging».

This is a new type of selective pulse[9, 10] that is insensitive to pulse width miscalibration, radiofrequency inhomogeneity and timing delays, without sacrificing performance. It also has an acceptable frequency domain excitation profile without pulse shaping, and a very good profile if only simple shaping functions are used, provided they are symmetrical in time. The key is to prepare the magnetization along the $+y$-axis with a hard 90° pulse, followed by a soft 180°(x) pulse. The experiment is then repeated with a soft 180°(y) pulse and the difference spectrum recorded. By calculating trajectories, one can easily show that dispersion components vanish everywhere, while the absorption components are only significant near resonance. If the soft 180° pulses are rectangular, the excitation profile is an approximate sinc^2 function which has only weak sidelobe responses, while a (time-symmetric) triangular pulse envelope gives an approximate sinc^4 excitation profile where the sidelobe responses are essentially negligible. Consequently, sophisticated shaping functions are unnecessary and there is little point in designing self-focusing schemes along the lines of the BURP pulses, since there are no dispersion contributions after a spin-pinging sequence. Furthermore, the performance is little affected by gross missetting of the 180° pulse length; this simply scales down the observed signal with very little degradation of the excitation profile. This suggests applications where the radiofrequency field is seriously inhomogeneous, for example *in vivo* NMR experiments with surface coils. Spin pinging (named after its inventor) is thus a «user-friendly» selective pulse to be used in situations where careful calibration may not be feasible or even appropriate.

6. – Practical implementation.

Single low-level pulses may be shaped by using a pattern generator, a device which converts a table of ordinates (a histogram) into the corresponding stepwise variations of the transmitter level. The steps are made fine enough that the histogram is a good approximation to the ideal smooth-pulse envelope. The pattern generator drives a power amplifier that is sufficiently linear that there is no appreciable distortion of the pulse envelope. DANTE sequences may be shaped by modulating the individual hard-pulse widths according to the appropriate shaping function. Very short pulses may be implemented by a forwards-backwards pulse pair. Since performance is significantly improved by pulse shaping, there seems little point in using rectangular soft pulses in present-day spectrometers.

7. – Resolution enhancement.

The technology for shaped pulses was largely developed in the context of magnetic-resonance imaging, where the essence of the method is selective excitation in an applied field gradient. These ideas may be appropriated by high-resolution spectroscopists for particular applications, for example, resolution enhancement. In principle the resolving power of a high-resolution spectrometer could be increased simply by reducing the sample size, since it is the spatial inhomogeneity over the sample volume that normally determines the linewidth.

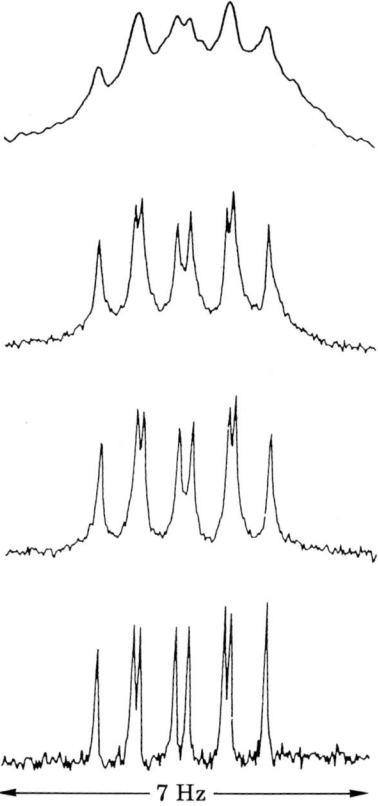

Fig. 4. – Resolution enhancement by selective excitation in an applied field gradient to reduce the effective height of the sample in the z dimension. After excitation the gradient is switched off. As more intense field gradients are applied the resolution improves from 0.6 Hz (top) to 0.08 Hz (bottom).

Unfortunately this is not no so in practice, since, as the sample is reduced in size, the container walls exert an increasingly important magnetic-susceptibility effect, distorting the field across the sample. Tubes of less than about 2 mm diameter usually show poorer resolution than the standard 5 mm tubes.

The problem can be circumvented by reducing the *effective* volume of the sample by selective excitation in a field gradient[11]. Since spinning does nothing to mitigate the effect of the z-gradient, this is where the greatest improvement can be expected. An additional z-gradient is applied during excitation by a selective pulse, but is switched off before signal acquisition. A flat disc-shaped region of the sample is excited, the short dimension being determined by the intensity of the applied gradient and the selectivity of the pulse. Free precession then occurs in the usual static-field inhomogeneity, but now the residual z-gradients have a much smaller influence on the linewidth, while the effects of x-

and y-gradients are reduced by sample spinning. Sensitivity is lost, but resolution is appreciably enhanced.

Figure 4 shows an example from the high-resolution spectrum of one of the protons in furan-2-aldehyde[11]. A selective pulse of 50 ms duration has been employed. As the sample is excited in more and more intense applied field gradients, the resolution improves from 0.6 Hz to 0.08 Hz, while the signal-to-noise ratio is degraded.

8. – Reduction of dimensionality.

Two-dimensional spectroscopy is now in widespread use for the purpose of correlation through scalar coupling, chemical exchange or the nuclear Overhauser effect. The only practical disadvantages are the increased demands on instrumental time and certain limitations on digital resolution. In such situations, the two-dimensional experiment may profitably be reduced to the equivalent one-dimensional investigation by the use of a selective pulse. Suppose we are interested in obtaining shift correlation information of the kind provided by

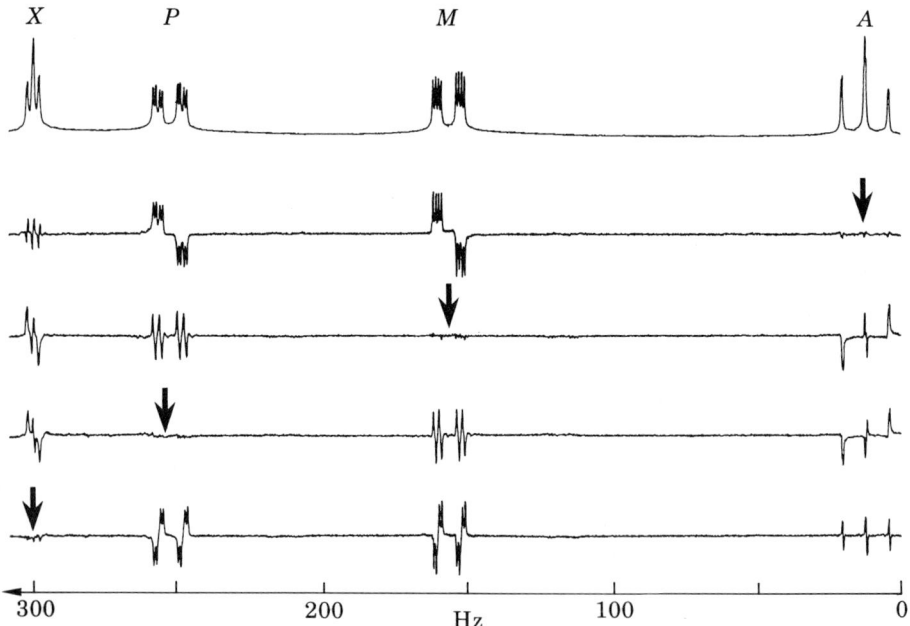

Fig. 5. – Selective one-dimensional shift correlation experiments on the four protons of metabromonitrobenzene initiated by a spin-pinging sequence. The arrows indicate the approximate excitation frequencies; in fact a single transition was chosen for excitation in each case. Correlation responses have the familiar «up-down» pattern, but the excited proton shows only a minimal response since all dispersion-mode signals have been cancelled.

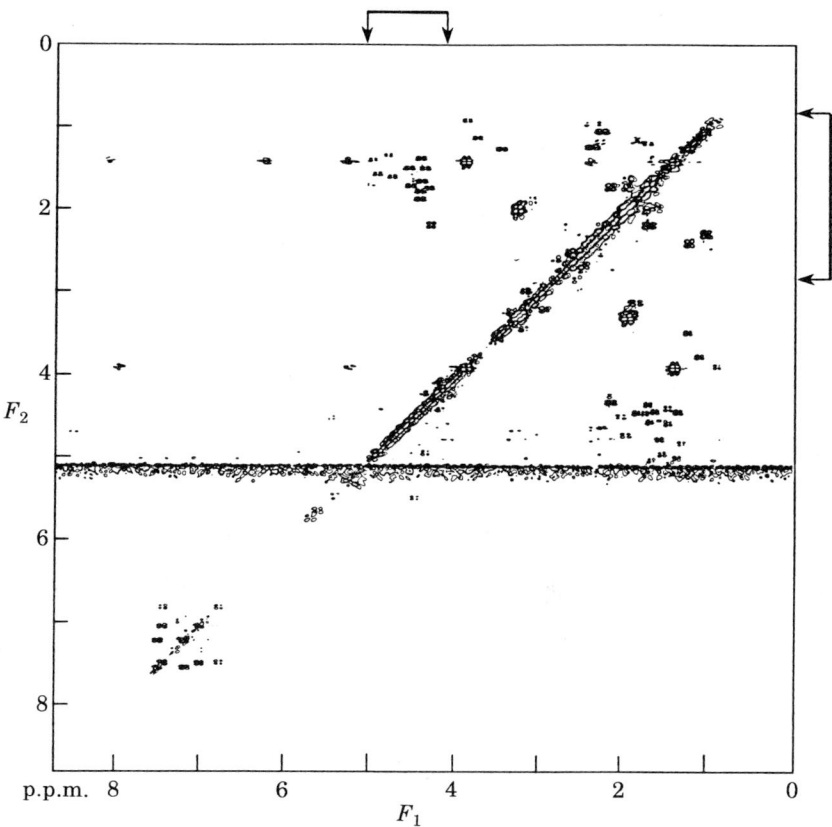

Fig. 6. – The conventional 400 MHz COSY spectrum of protons in ribonuclease A. The rectangular 0.9 p.p.m. by 2 p.p.m. region indicated by the arrows is reinvestigated with band-selective spin pinging in fig. 7.

the COSY spectrum, but in a situation where the *entire* connectivity pattern may not be required. A line-selective pulse may be used to seek out the appropriate transition from a chosen proton site while correlation responses are recorded from all the coupled sites, taking advantage of exactly the same coherence transfer mechanism as the two-dimensional COSY analog. For example, we might choose the spin-pinging sequence[9, 10] as the mode of excitation since it performs well without pulse shaping and it has the valuable property of eliminating dispersion-mode signals. Figure 5 shows an example from the 400 MHz proton spectrum of metabromonitrobenzene, a system of four coupled protons. The excited proton shows a negligible response (since dispersion components are suppressed) while the coupled sites show the familiar antiphase pattern of lines for the active splitting. Four one-dimensional measurements identify all the coupling partners.

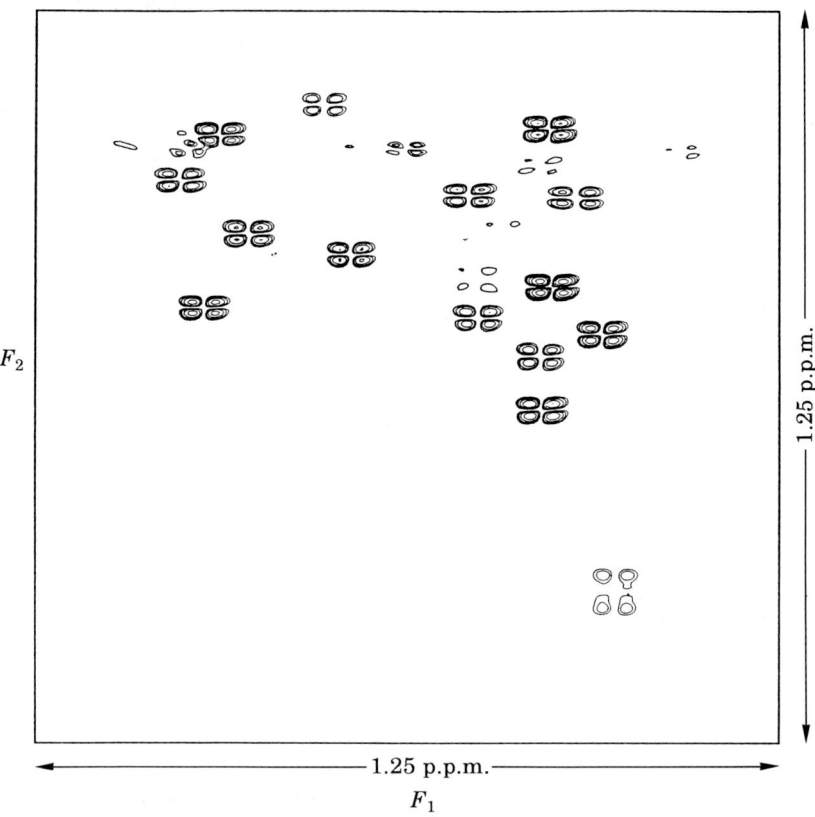

Fig. 7. – Detail of the COSY spectrum of ribonuclease A shown in fig. 6, examined using band-selective spin pinging for excitation and coherence transfer.

Multidimensional correlation spectroscopy places severe demands on data storage and instrument time. A three-dimensional correlation experiment with one thousand data points in each frequency dimension requires 10^9 words of data storage. Even without phase cycling, and with the maximum evolution and acquisition times restricted to 0.3 s each, the spectrometer would be tied up for a week on this single experiment, not an operation designed to win popularity with fellow spectroscopists. Band-selective pulses promise to alleviate such problems. By restricting the effective range of the radiofrequency pulses used to cover the first two («evolution») dimensions, we can greatly reduce the volume of frequency space under investigation and, if necessary, improve the fineness of digitization. This principle may be most easily illustrated by a two-dimensional example. The 400 MHz proton correlation spectrum of ribonuclease A is shown in fig. 6. A restricted 0.9 p.p.m. by 2 p.p.m. rectangular region has been selected in fig. 7 by using band-selective spin-pinging pulses. It is interesting to note that a spin-pinging sequence acts not only as an excitation pulse but also as a coherence transfer pulse since the hard 90° component provides the «mixing» while the soft 180° pulse filters out the desired band of frequencies. The selected region is examined under high digital resolution without entailing a prohibitively long experiment. This principle can clearly be extended to three- or four-dimensional spectroscopy.

9. – The Hartmann-Hahn experiment.

The double-resonance experiment described by HARTMANN and HAHN [12] was first conceived as a heteronuclear solid-state technique, designed to transfer polarization from an abundant spin species (I) to a rare spin species (S). In a solid, the dipole-dipole interactions ensure that the abundant I spins are strongly coupled among themselves, and it is then permissible to speak of an internal equilibrium and a «spin temperature». If the I spins are excited by a $90°(x)$ pulse followed by a continuous radiofrequency field B_1 applied along the y-axis, the I spins are said to be «spin locked». In a frame rotating at the transmitter frequency, the I spins see a static field B_1 but have the high polarization appropriate to the much more intense applied field B_0. If the Boltzmann equation is applied to this situation, the spin temperature in the rotating frame must be considered to be very low, reduced from ambient temperature by the ratio B_1/B_0. The S-spin temperature is unaffected. If the I and S spins can be brought into thermodynamic contact, the hot S spins will warm up the cold I spins and this transfer may be detected as a slight loss of I-spin polarization. The tightly coupled I-spin system constitutes a thermal reservoir, allowing multiple transfers to take place to boost the detected signal. Alternatively, an enhancement of the very weak S-spin signal may be observed directly.

The appropriate «contact» may be achieved by applying a second continuous radiofrequency field (B_2) to the S spins, satisfying the condition

$$\gamma_I B_1 = \gamma_S B_2 . \tag{1}$$

The S spin generates a local radiofrequency field at the site of the I spin that oscillates at the S-spin Larmor frequency in the rotating frame, and the I spin creates a similar field at the S site, providing a mechanism whereby mutual spin flips can occur. Thermodynamic mixing may then take place, heating up the I spins and cooling the S spins. If there is a significant mismatch of the Hartmann-Hahn condition this polarization transfer cannot take place.

In recent years attention has focused on the Hartmann-Hahn effect in liquids where spin temperature arguments are inappropriate. There are nevertheless two physical processes of polarization transfer in the liquid state, where the fast molecular reorientation averages the dipolar interactions to zero. They arise from the nuclear Overhauser effect and the scalar spin-spin coupling. Most of the experiments have involved homonuclear (proton-proton) spin systems.

The rotating-frame Overhauser effect is always positive and thus differs from the laboratory frame version which may reverse sign if the molecular-reorientation rate happens to fall in the intermediate regime. Consequently «ROESY» is preferred over «NOESY» for two-dimensional experiments designed to establish proximity of protons in molecules of biochemical interest. By contrast, the scalar coupling mechanism leads to an *oscillatory* interchange of polarization between the I and S spins at a rate determined by the coupling constant J_{IS}. This is analogous to the oscillatory exchange of energy between two pendulums that are coupled mechanically. This has been exploited in «total-correlation spectroscopy» (TOCSY)[13], sometimes known as the homonuclear Hartmann-Hahn experiment (HOHAHA)[14]. A single continuous field B_1 may be used for spin locking both the I and S spin species in this homonuclear case, but it is advantageous to increase the effective bandwidth over which the Hartmann-Hahn condition is satisfied by employing pulse sequences originally designed for broad-band heteronuclear decoupling, such as «MLEV»[15] or «WALTZ»[16] or new sequences specifically designed for Hartmann-Hahn mixing[17].

10. – Stepwise coherence transfer (DAISY).

In fact the Hartmann-Hahn experiment is inherently quite frequency selective when used in these circumstances. If a relatively weak radiofrequency field B_1 is used in a proton-proton system, it is only effective when set at the midpoint of the I and S chemical shifts. This is because the spins experience effective fields that are the resultants of B_1 and the appropriate offset ΔB_I or ΔB_S. Only when $|\Delta B_I| = |\Delta B_S|$ is the Hartmann-Hahn condition exactly satisfied. Alternatively we may perform the double-resonance analog of this experiment by applying a spin-lock field B_1 at the I-spin chemical shift and another spin-lock field B_2 at the S-spin chemical shift, with $B_1 = B_2$. The high selectivity then arises because any slight offset of one of these frequencies results in an increased effective field not matched at the other site. This method produces cleaner spectra.

Magnetization transfer by selective homonuclear Hartmann-Hahn mixing is then quite simple. A selective I-spin pulse generates I_y magnetization which is spin locked for a time $\tau = 1/J_{IS}$, chosen so that maximum transfer to site S takes place if a second spin-lock field is applied to the S spins. The two fields B_1 and B_2 are conveniently generated as interleaved DANTE sequences[5] each with a duty cycle of 50%. One DANTE sequence has all pulses of the same phase, while the other has the phase incremented in regular steps, mimicking a frequency shift. The new signal transferred to the S-spin site is in the pure absorption mode and shows very little intensity distortion with respect to the conventional high-resolution spectrum. Figure 8 illustrates such a transfer between two aromatic protons I and S in the 400 MHz spectrum of methyl salicylate[18].

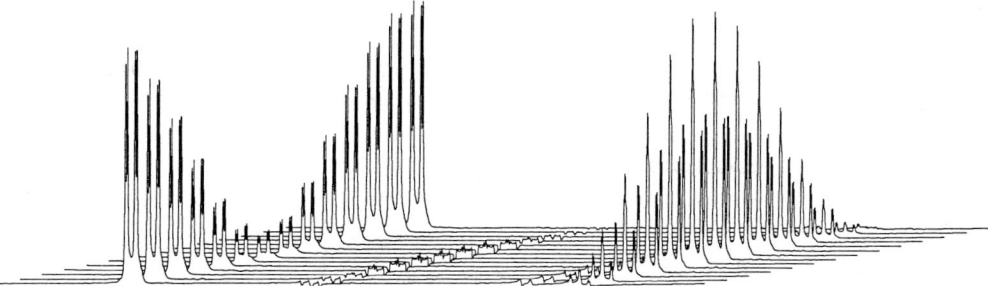

Fig. 8. – Selective Hartmann-Hahn coherence transfer between two of the four aromatic protons of methyl salicylate studied as a function of the spin-lock period. Site I loses signal intensity while site S gains intensity, reaching a maximum for $\tau = 1/2J_{IS}$ in this case.

This experiment was performed with a *single* spin-lock field at the mid-point of the chemical shifts, giving a maximum transfer when $\tau = 1/2J_{IS}$.

The beauty of the Hartmann-Hahn experiment is that the transfers can be cascaded [18]. Figure 9 shows a schematic diagram of the pulse sequence that would be used to propagate coherence in three consecutive steps, $1 \to 2 \to 3 \to$ $\to 4$. Each transfer can be optimized by setting spin-lock fields at the appropriate chemical-shift frequencies and by choosing the τ values according to the known spin-spin coupling constants. Each step is very selective so that no coherence

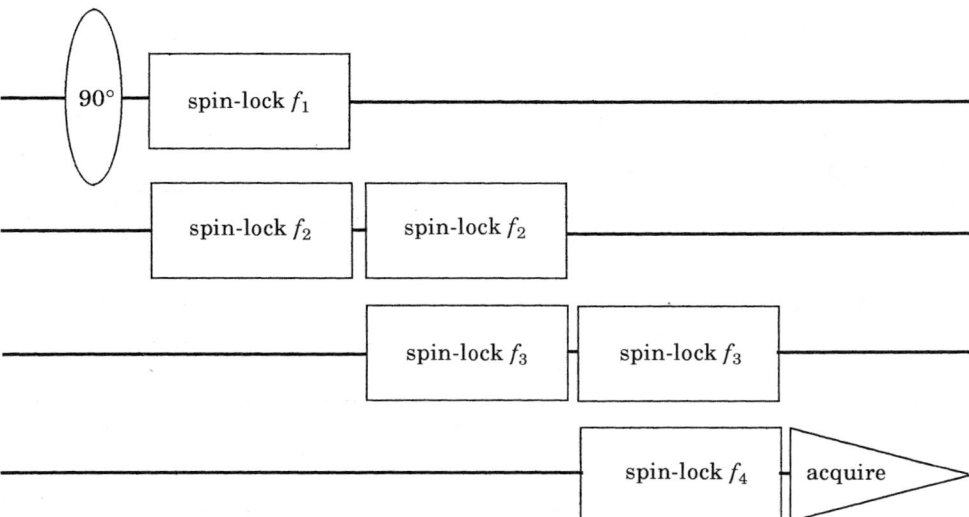

Fig. 9. – Pulse sequence for a three-stage Hartmann-Hahn coherence transfer. The first pulse is a soft $90°(x)$ excitation pulse; the spin-lock fields are applied along the $+ y$-axis of the rotating frame. The frequencies f_1 through f_4 are set at the respective chemical shifts. The three different spin-lock periods need not have the same durations.

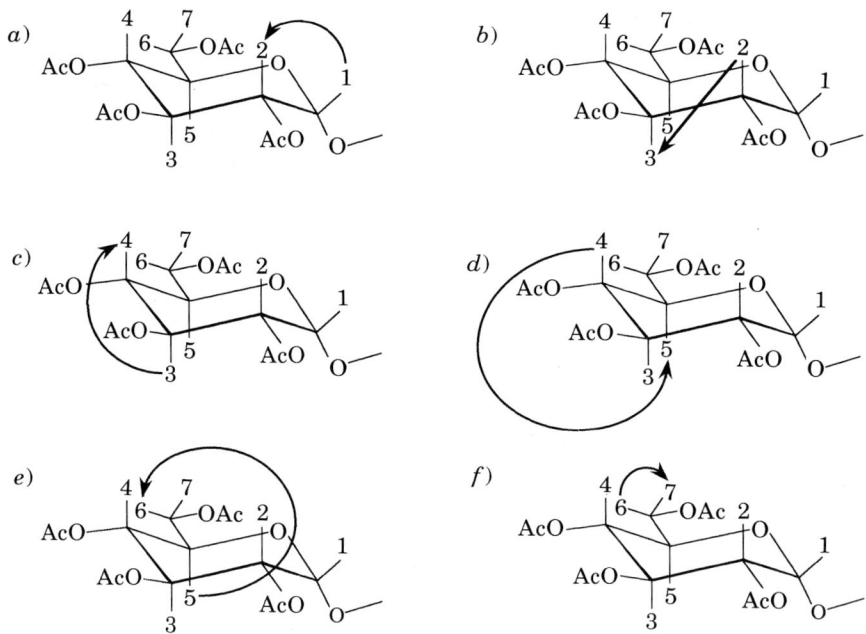

Fig. 10. – Schematic representation of the «DAISY» experiment on the protons of the glucose ring of sucrose octa-acetate. Six consecutive coherence transfer steps can be strung together, the final link being between the two inequivalent methylene protons.

can «go astray». The frequency selectivity is determined by setting the intensities $\gamma B_1/2\pi = \gamma B_2/2\pi$ comparable with the coupling constant. The practical limit on the number of consecutive transfers is determined by the sum of the individual spin-locking times compared with the spin-spin relaxation time. For example, six consecutive transfers can be implemented around the glucose ring of sucrose octa-acetate (fig. 10). The experiment is carefully planned in advance. The intensities, frequencies and durations of the spin-lock fields are adjusted for each stage of the transfer, taking into account chemical shifts and coupling constants observed in the conventional spectrum. An eight-step phase cycle alternates the phases of the soft pulse, the first and last spin-lock fields and the receiver. With these precautions, responses from other resonances are minimal and the spin multiplet from the final proton in the chain is faithfully reproduced in the pure absorption mode (fig. 11). Note that some of the glucose ring proton resonances in the conventional spectrum fall underneath responses from the fructose ring, but nevertheless give a clean response at the end of the coherence transfer chain. The method should, therefore, prove useful for examining spin multiplets buried under overlapping responses.

By analogy with the computer terminology «daisy chain», the experiment is called «DAISY» (direct assignment interconnection spectroscopy). It affords a

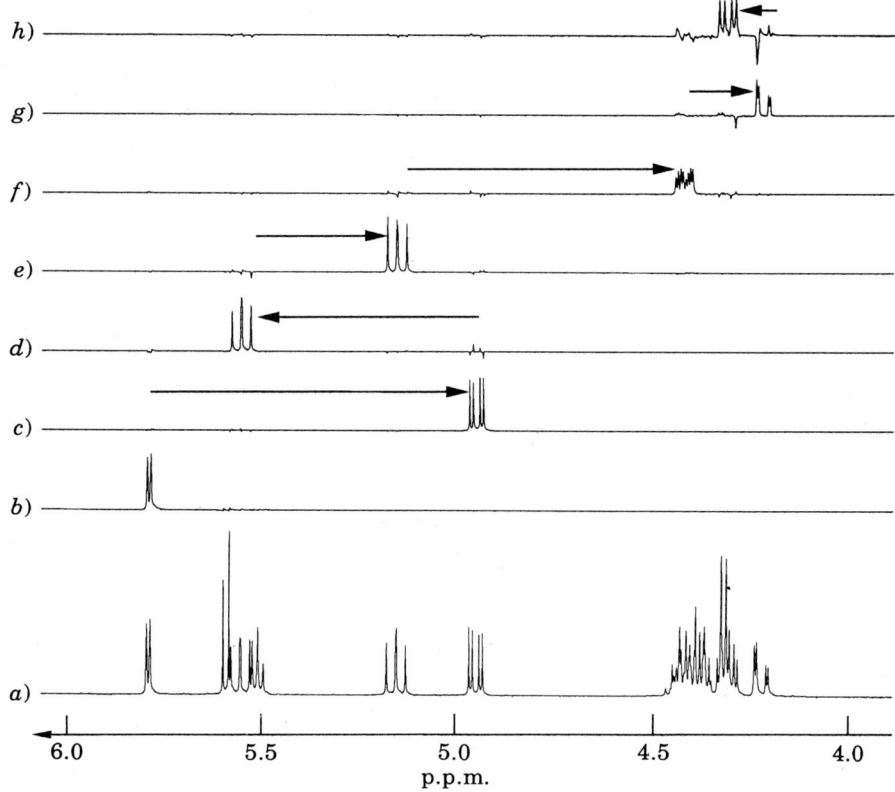

Fig. 11. – Experimental DAISY results for the protons in the glucose ring of sucrose octa-acetate. a) The conventional spectrum, exhibiting some regions of overlap with proton resonances from the fructose ring. b) Selective excitation of H_1. c) Coherence transfer $H_1 \to H_2$. d) $H_1 \to H_2 \to H_3$. e) $H_1 \to H_2 \to H_3 \to H_4$. f) $H_1 \to H_2 \to H_3 \to H_4 \to H_5$. g) $H_1 \to H_2 \to H_3 \to H_4 \to H_5 \to H_6$. h) $H_1 \to H_2 \to H_3 \to H_4 \to H_5 \to H_6 \to H_7$.

powerful method of establishing the connectivity of protons linked together through spin-spin coupling, even where intervening resonances are obscured by overlap. This approach differs radically from the indiscriminate and generalized diffusion of coherence that occurs in a TOCSY experiment. By carefully planning a suitable coherence transfer path to probe a particular structural feature, we can use DAISY as a precise diagnostic tool. In contrast, the TOCSY spectrum gives a global picture of the proton coupling network, and structural conclusions are usually drawn after the fact. One can visualize a continuous range of Hartmann-Hahn experiments with varying degrees of selectivity, from TOCSY at one extreme to DAISY at the other.

Several extensions of these experiments are possible. Simultaneous transfers can be implemented by interleaving several DANTE sequences during the

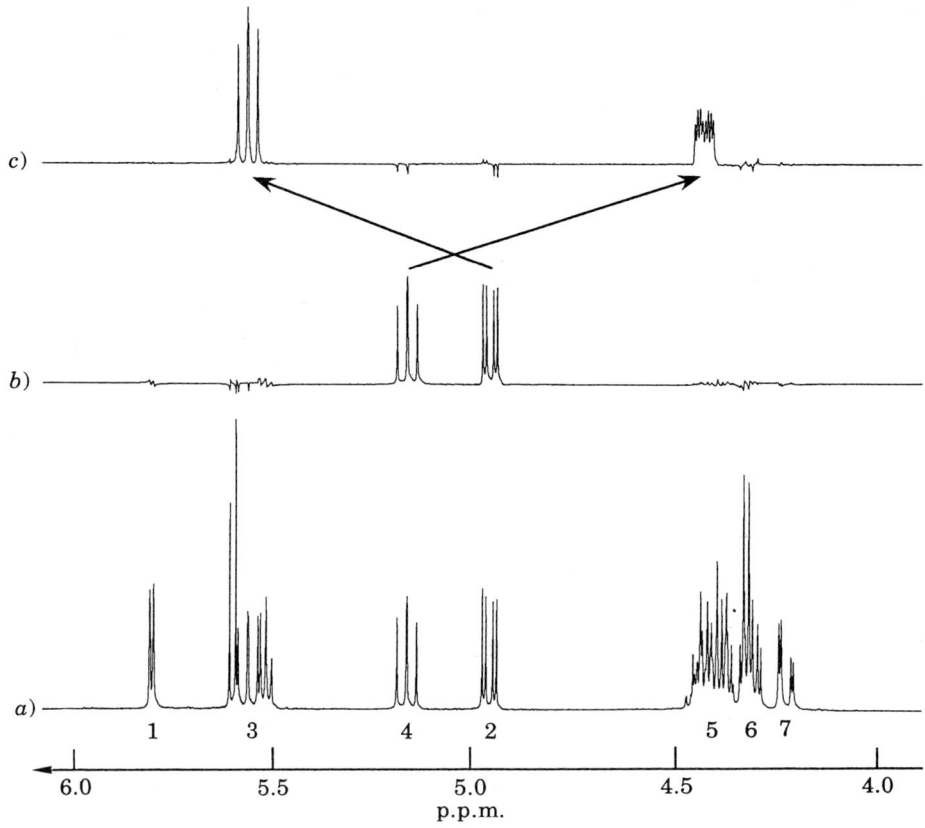

Fig. 12. – Two simultaneous Hartmann-Hahn transfers in the spectrum of sucrose octa-acetate. Protons H_2 and H_4 are excited by a band-selective pulse and coherence is then transferred $H_2 \to H_3$ and $H_4 \to H_5$ with four simultaneous spin-lock fields. To avoid cross-talk, the spin-lock level for $H_2 \to H_3$ transfer was deliberately made twice as intense as that for $H_4 \to H_5$ transfer.

spin-lock period. Two simultaneous transfers are illustrated in fig. 12, where protons H_2 and H_4 have been excited by a band-selective pulse that encompasses both chemical shifts. Then four spin-lock fields are applied at the appropriate chemical-shift frequencies to promote the simultaneous transfers $H_2 \to H_3$ and $H_4 \to H_5$. This highlights another valuable selectivity feature—the intensity of the spin-lock field. In principle, double excitation followed by four spin-lock fields could give rise to six possible coherence transfer paths. By setting two different levels of spin-lock field we can suppress this cross-talk and retain only the two desired links.

Another extension is to carbon-13 spectroscopy. For natural-abundance samples, carbon-carbon transfers are essentially limited to just a single step, but multistage proton transfers may be combined with carbon-13 detection by appending a nonselective heteronuclear polarization transfer step such as INEPT [19] at the end of the sequence. Thus the carbon-13 sites in a molecule may be identified one by one, using multistage proton transfers from a selected initial proton site. This could prove useful in situations where the INADEQUATE experiment [20] breaks down for lack of a direct carbon-carbon coupling in the chain.

This lecture has picked out only a few illustrative examples of selective excitation in order to illustrate the main principles. Incorporation of soft pulses into existing multidimensional experiments offers a fertile field of development [1] and it is clear we have not yet heard the last word on selective NMR spectroscopy.

* * *

The author is indebted to Dr. E̅. KUPČE for the spectra in fig. 11 and 12.

REFERENCES

[1] H. KESSLER, S. MRONGA and G. GEMMECKER: *Magn. Reson. Chem.*, **29**, 527 (1991).
[2] R. FREEMAN: *Chem. Rev.*, **91**, 1397 (1991).
[3] R. FREEMAN and S. WITTEKOEK: *J. Magn. Reson.*, **1**, 238 (1969).
[4] G. A. MORRIS and R. FREEMAN: *J. Magn. Reson.*, **29**, 433 (1978).
[5] H. GEEN, X. L. WU, P. XU, J. FRIEDRICH and R. FREEMAN: *J. Magn. Reson.*, **81**, 646 (1989).
[6] C. J. BAUER, R. FREEMAN, T. FRENKIEL, J. KEELER and A. J. SHAKA: *J. Magn. Reson.*, **58**, 442 (1984).
[7] W. S. WARREN: *J. Chem. Phys.*, **81**, 5437 (1984).
[8] H. GEEN and R. FREEMAN: *J. Magn. Reson.*, **93**, 93 (1991).
[9] X. L. WU, P. XU and R. FREEMAN: *J. Magn. Reson.*, **83**, 404 (1989).
[10] P. XU, X. L. WU and R. FREEMAN: *J. Magn. Reson.*, **99**, 308 (1992).
[11] A. BAX and R. FREEMAN: *J. Magn. Reson.*, **37**, 177 (1980).
[12] S. R. HARTMANN and E. L. HAHN: *Phys. Rev.*, **128**, 2042 (1962).
[13] L. BRAUNSCHWEILER and R. R. ERNST: *J. Magn. Reson.*, **53**, 521 (1983).
[14] A. BAX and D. G. DAVIS: *J. Magn. Reson.*, **63**, 207 (1985).
[15] M. H. LEVITT, R. FREEMAN and T. FRENKIEL: *J. Magn. Reson.*, **47**, 328 (1982).
[16] A. J. SHAKA, J. KEELER and R. FREEMAN: *J. Magn. Reson.*, **53**, 313 (1983).
[17] A. MOHEBBI and A. J. SHAKA: *J. Magn. Reson.*, **94**, 204 (1991).
[18] E̅. KUPČE and R. FREEMAN: *J. Magn. Reson.*, **100**, 208 (1992).
[19] G. A. MORRIS and R. FREEMAN: *J. Am. Chem. Soc.*, **101**, 760 (1979).
[20] A. BAX, R. FREEMAN and S. P. KEMPSELL: *J. Am. Chem. Soc.*, **102**, 4849 (1980).

Fine Structure in Two-Dimensional Spectra.

R. FREEMAN

Department of Chemistry, Cambridge University - Cambridge, U.K.

1. – Introduction.

While many applications of two-dimensional NMR look at the coarser features of the spectrum, for example the existence (or not) of a given cross-peak, it may occasionally prove very fruitful to examine fine structure in order to evaluate the coupling constants. This is particularly important for the more complex molecules where the one-dimensional spectrum may be far too crowded to permit a detailed examination. Two-dimensional spectra provide an inherently better access to fine-structure features by avoiding many of these problems of overlap. Nevertheless the practical restrictions on data storage and the total time available for the experiment usually impose some limits on the fineness of digitization attainable in multidimensional spectra. This lecture examines some of the procedures available for extracting accurate values of coupling constants from two-dimensional correlation spectra.

2. – The structure of COSY cross-peaks.

A cross-peak is defined by the frequency coordinates (δ_I, δ_S) of its centre, the active coupling J_{IS} and a set of passive couplings, J_{IR}, J_{SR}, J_{IQ}, J_{SQ}, etc. The active splitting J_{IS} is characterized by its appearance in both frequency dimensions and by the antiphase nature of the intensities. Assuming that we are dealing with the conventional COSY experiment where the second pulse is set at 90°, the cross-peak may be thought of as the basic square of side J_{IS}, split in F_1 and F_2 by the respective passive couplings. We may consider the cross-peak to be constructed in several stages, starting with a square pattern «stick spectrum» of side J_{IS}, with alternating intensities. This is then convoluted by suitable lineshape functions in each frequency dimension (these need not, of course, be the same). Then each passive spin is introduced in turn, splitting the pattern by a different amount in each frequency dimension (fig. 1).

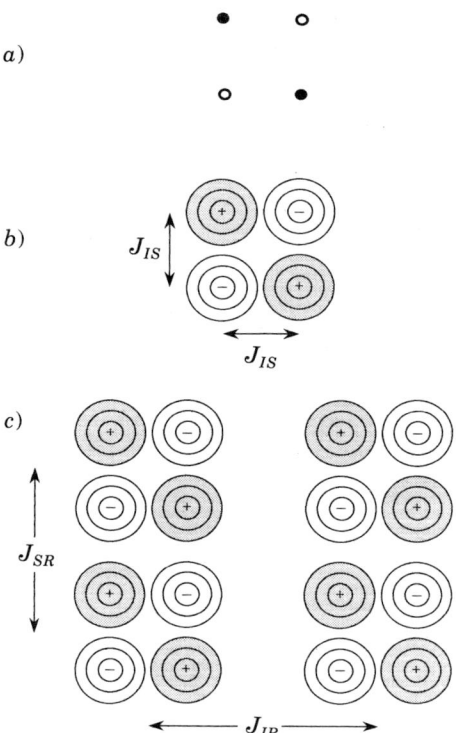

Fig. 1. – Structure of a COSY cross-peak representing coherence transfer from an I spin to an S spin, coupled to a passive spin R. a) Basic square pattern of alternating intensities, b) after convolution with a linewidth function, c) after introduction of the two passive splittings.

These symmetry elements are put to good use in the «pattern recognition» technique[1-3] for automatic analysis of COSY spectra by computer. For this purpose it is usual to employ a COSY-45 spectrum in which the passive spins do not change their spin states; this reduces the complexity of the cross-peak by a factor 2^n, where n is the number of passive spins. First a suitable region of the two-dimensional spectrum is located by cluster analysis and then it is examined by an algorithm designed to recognize the basic square patterns with alternating intensities. Adjacent identical square patterns are evidence for passive splittings. In principle all splittings can be removed, leaving a single response at the chemical-shift coordinates. The program can handle overlapping crosspeaks unless they are pathological cases. The processed COSY spectrum is greatly simplified, retaining only correlation information.

This characteristic geometry of a COSY cross-peak has an important and useful property that may not be immediately obvious—any two sections taken parallel to the F_1 and F_2 axes carry all the information necessary to reconstruct the entire cross-peak[4]. Naturally the sections should be chosen so as to have a

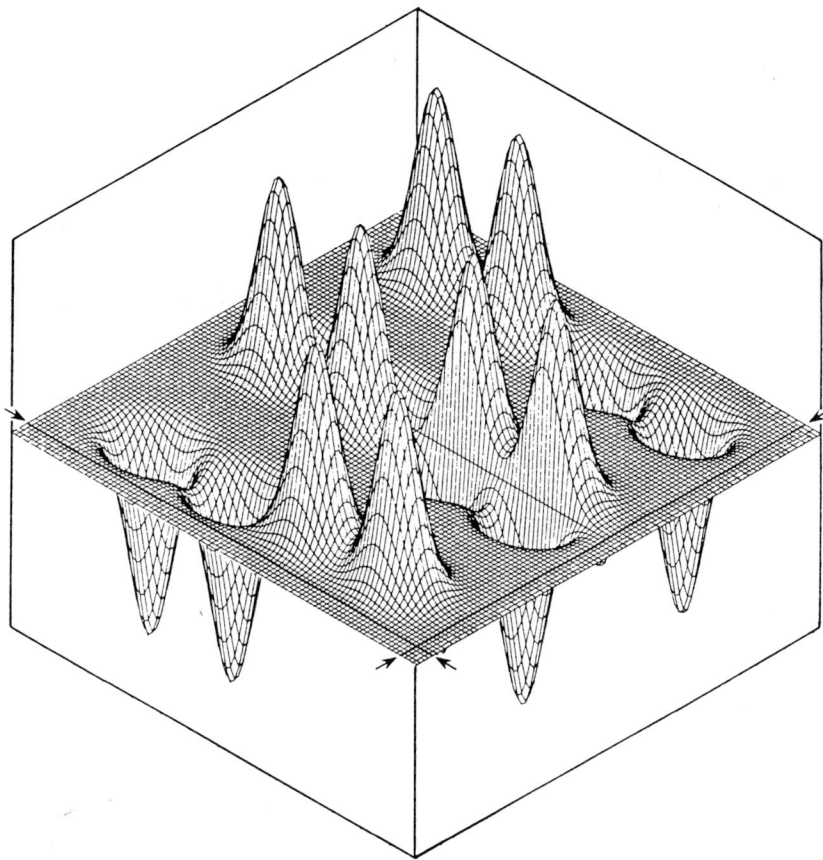

Fig. 2. – A simulated cross-peak shown as a net diagram. If the two orthogonal sections indicated by the arrows are multiplied together, the entire surface may be reconstructed.

good signal-to-noise ratio, but they need not intersect the peaks of the lines. The surface that defines the cross-peak is simply the product of these two orthogonal sections, divided by the intensity at the point where the two sections cross. Figure 2 shows the surface of a simulated cross-peak that has been reconstructed simply by multiplying the two sections indicated by the arrows. This reconstruction procedure only applies to a single *isolated* cross-peak, it is not, of course, appropriate when two or more cross-peaks overlap, unless the sections are carefully chosen to avoid all extraneous peaks.

This suggests a practical procedure for disentangling two overlapping cross-peaks by post-acquisition processing of the conventional COSY spectrum[4]. First we isolate the region of interest and then look for two suitable orthogonal sections that intersect lines from only one component cross-peak—they might, for example, skirt the extreme edges, where the tails of all extraneous lines are

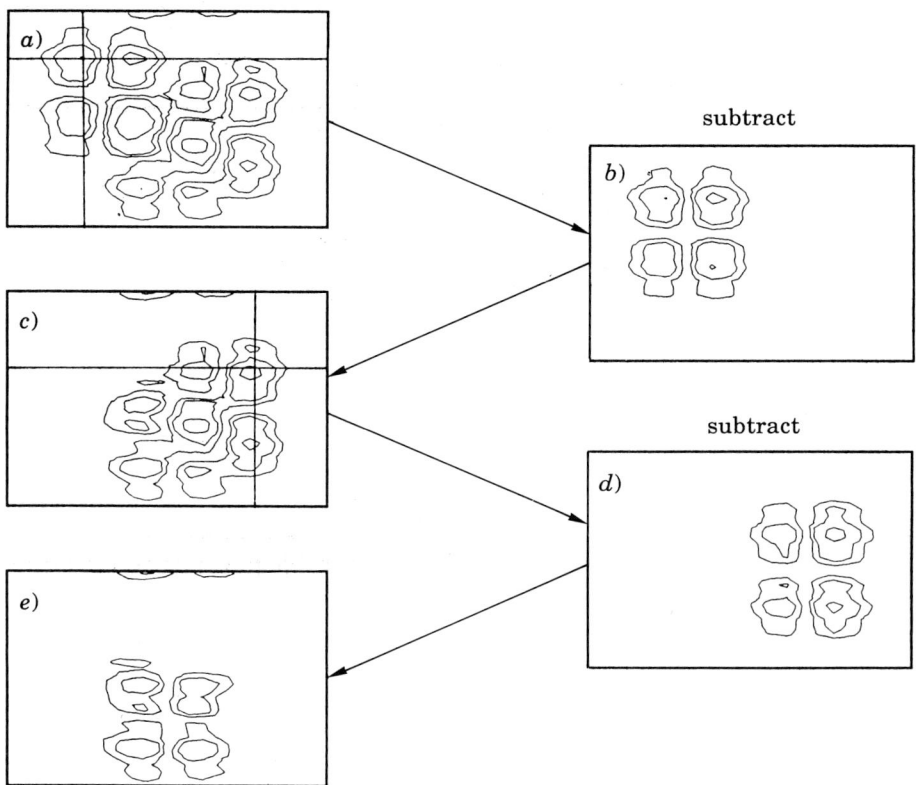

Fig. 3. – Separation of overlapping cross-peaks in the DQ-COSY spectrum of tolaasin. a) Two sections are found that intersect only the top left cross-peak. b) This is reconstructed and subtracted from the experimental data. c) Two further sections select the top right cross-peak. d) After reconstruction this is subtracted, leaving the lower cross-peak (e)).

of negligible intensity. These two sections are then multiplied together and the product normalized to the intensity at the crossing point. This reconstructed cross-peak is then subtracted from the experimental frequency domain matrix, leaving the residual cross-peaks to be further separated if necessary. An experimental example is provided by the three overlapping cross-peaks in the 400 MHz DQ-COSY spectrum of the natural product tolaasin obtained from *Pseudomonas tolaasii Paine*. The top left of the pattern shows two edges of a recognizable cross-peak. Two orthogonal sections are taken through these edges (fig. 3a)) to reconstruct the four-line pattern (fig. 3b)), which is then subtracted. This reveals another pattern at the top right (fig. 3c)) which is accessible through the two sections indicated. When this reconstructed cross-peak (fig. 3d)) has been subtracted, only a single four-line pattern remains (fig. 3e)).

A related method is to initiate the COSY sequence with a highly selective excitation pulse:

$$\text{soft } 90°\, (x)\text{-}t_1\text{-hard } 90°\text{-acquire } (t_2).$$

The frequency of the soft pulse is carefully chosen so as to slice through one of the overlapping cross-peaks without picking up significant intensity from the others [5]. The hard 90° pulse has the effect of spreading the intensity from this section into both frequency dimensions, giving a faithful reproduction of the chosen cross-peak. Normally this simplifies the residual pattern so that the process can be repeated, extracting the individual cross-peaks one at a time. As a practical demonstration, a selective spin-pinging pulse [6] has been used to separate three superimposed cross-peaks from the 400 MHz proton COSY spectrum of norambreinolide in $CDCl_3$ solution [5]. By identifying a suitable fre-

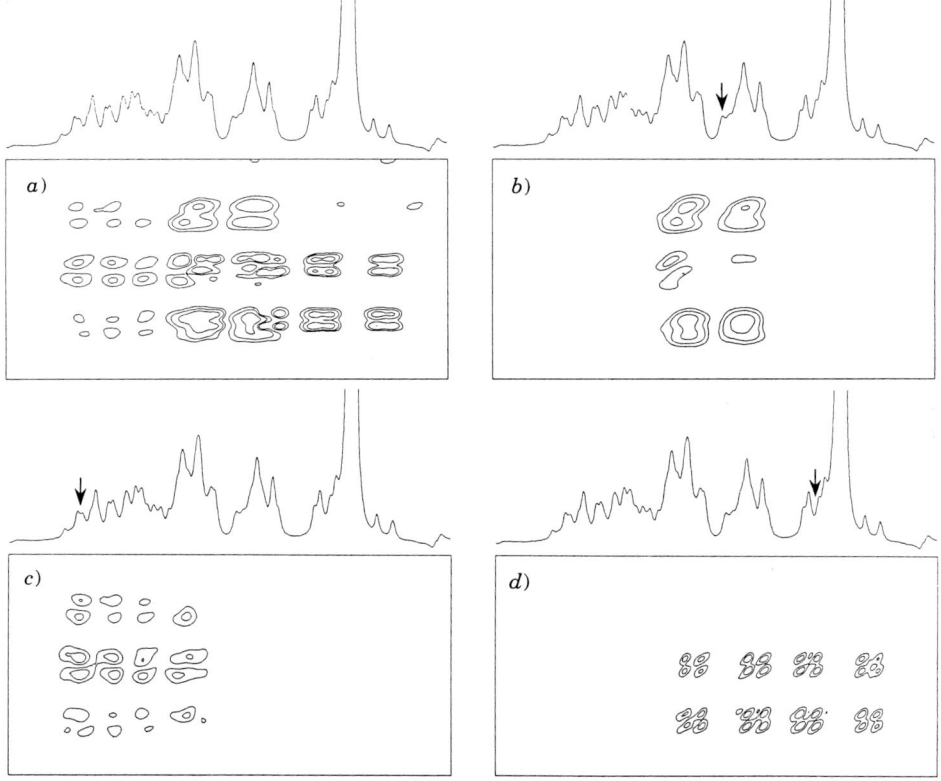

Fig. 4. – Separation of overlapping cross-peaks in the 400 MHz COSY spectrum of norambreinolide by spin pinging: a) the experimental cross-peak, b) through d) cross-peaks extracted by selective excitation at the frequencies indicated by the arrows. In both c) and d) the soft pulse was set to excite two adjacent transitions, thus doubling the intensities.

quency in the experimental spectrum (fig. 4a)) it is possible to set up a selective COSY experiment that excites only one cross-peak (fig. 4b)). Then a second soft pulse picks out the left-hand cross-peak (fig. 4c)). This experiment actually excites two close resonances of the same cross-peak, thus doubling the intensity. Finally a soft pulse at a third F_1 frequency serves to isolate the third component, again by excitation of two close resonances (fig. 4d)).

3. – Extraction of coupling constants.

In two-dimensional spectroscopy practical limits must be imposed on the number of increments in the evolution period (t_1) and on the total length of data gathering. This normally implies that the F_1 dimension is less finely digitized than the F_2 dimension since there is little penalty involved in increasing the maximum value of t_2. Consequently COSY cross-peaks may be better resolved in the F_2 dimension and any attempts to extract coupling constants are more likely to be successful with F_2 traces than with F_1 traces. Zero-filling can be used to ensure that there are sufficient data points per linewidth.

Since the active splitting in a COSY cross-peak generates an antiphase intensity pattern, it tends to represent an overestimate of the coupling constant if the linewidth is comparable with J_{IS}. In contrast, passive splittings have an in-phase intensity pattern and tend to underestimate the true coupling constant when the resolution is marginal. The usual combination of the two effects complicates the measurement of the coupling constants, suggesting that more sophisticated methods of data analysis would be advisable. The most straightforward approach is trial-and-error fitting in the frequency domain. This method is dogged by the familiar problem of false minima and becomes less reliable as the number of variable parameters (active and passive couplings) increases.

In some applications the active splitting of interest (J_{IS}) may occur in both in-phase and antiphase intensity patterns. For example, it might appear as an antiphase doublet in a cross-peak and as an in-phase doublet in the (rephased) diagonal peak. Provided that the two traces can be scaled to the same intensity, addition and subtraction of the two experimental traces removes the splitting, leaving one component line in the sum spectrum and the other in the difference spectrum. Since the splitting is now no longer partially resolved, an accurate value for J_{IS} can be measured. Passive splittings are now more readily evaluated since the complexity of the multiplet has been reduced. This procedure has come to be called the DISCO method[7].

Another approach has been taken by TITMAN and KEELER[8]. It also requires the two versions of the doublet. The recommended procedure is to record the cross-peak with an antiphase splitting from the COSY spectrum and the corresponding cross-peak with an in-phase doublet from the TOCSY spectrum[9]. This requires that the TOCSY spectrum be scaled to the same intensity as the COSY spectrum. A trial coupling constant J^* is introduced and an iterative search is set up to find the condition $J^* = J_{IS}$. In the time domain the in-phase version of doublet is represented by the term $\cos(\pi J_{IS} t)$, while the antiphase version is represented by $\sin(\pi J_{IS} t)$. The products

(1) $\qquad P_1 = a \cos(\pi J_{IS} t) \sin(\pi J^* t), \qquad P_2 = b \sin(\pi J_{IS} t) \cos(\pi J^* t)$

are then evaluated. The Fourier transforms of P_1 and P_2 generate two quite different spectra unless $a = b$ and $J^* = J_{IS}$. As a/b and $J^* - J_{IS}$ are varied in a two-dimensional iterative search program, the sum of the squares of the differences between corresponding points in the two spectra passes through a minimum. This gives $J^* = J_{IS}$.

4. – J-extension.

This is a scheme designed to remove the overlap of the two lines that define the coupling constant under investigation [10] so that the splitting can be measured by standard utility routines provided by the spectrometer manufacturers. In-phase and antiphase versions of the time domain signal are treated in a manner similar to that of Titman and Keeler, except that a large fixed coupling J_0 is employed,

(2) $\qquad Q_1 = a \cos(\pi J_{IS} t) \sin(\pi J_0 t), \qquad Q_2 = b \sin(\pi J_{IS} t) \cos(\pi J_0 t).$

If the scaling factor is correctly chosen so that $a = b$, addition of these two traces gives

(3) $\qquad\qquad\qquad Q_1 + Q_2 = a \sin[\pi(J_{IS} + J_0) t].$

Consequently, Fourier transformation of $Q_1 + Q_2$ gives a spectrum where $|J_{IS}|$ has been «extended» to $|J_{IS} + J_0|$, the intensities remaining in opposite phase. If J_0 is large compared with J_{IS}, the new splitting can be measured by standard methods and J_{IS} obtained by difference since J_0 is known exactly. Alternatively we might evaluate the two products

(4) $\qquad R_1 = a \cos(\pi J_{IS} t) \cos(\pi J_0 t), \qquad R_2 = b \sin(\pi J_{IS} t) \sin(\pi J_0 t).$

When $a = b$, subtraction of the two traces gives

(5) $\qquad\qquad\qquad R_1 - R_2 = a \cos[\pi(J_{IS} + J_0) t].$

Fourier transformation of $R_1 - R_2$ gives the in-phase version of the doublet $|J_{IS} + J_0|$. It may be more convenient to use existing frequency measurement routines when the two lines are in the same sense.

The J-extension method is particularly useful when there is a poorly resolved splitting to be measured. An example from the 400 MHz proton spectrum of metabromonitrobenzene is shown in fig. 5. The in-phase version of the

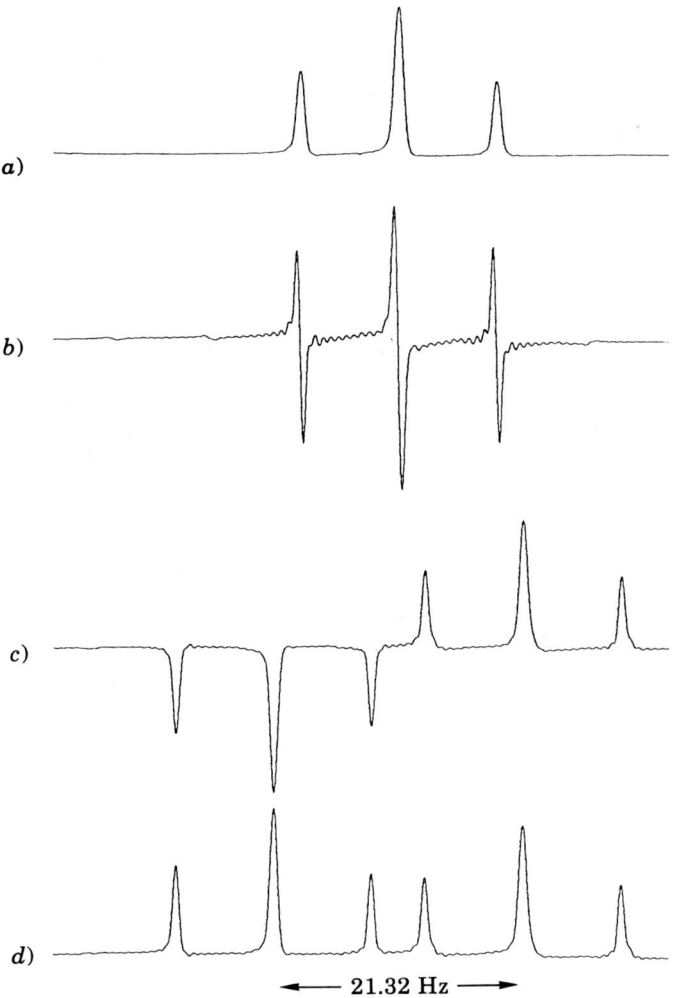

Fig. 5. – Measurement of a poorly resolved splitting from a cross-peak in the 400 MHz COSY spectrum of metabromonitrobenzene: *a)* the in-phase version, *b)* the antiphase version, *c)* extended by 21.00 Hz as an antiphase splitting, *d)* extended by 21.00 Hz as an in-phase splitting. The coupling constant is, therefore, 0.32 Hz.

multiplet appears to be a simple 1:2:1 triplet (fig. 5*a*)), but the antiphase version indicates that there is a small coupling comparable with the linewidth (fig. 5*b*)). However, the coupling cannot be measured from this trace since the observed splitting is an overestimate of the coupling constant. When processed according to eq. (3) with $J_0 = 21.00$ Hz, these traces give a pair of triplets with opposite intensities (fig. 5*c*)) and when eq. (5) is used the two triplets have the same sense (fig. 5*d*)). The new splitting $|J_{IS} + J_0|$ can be measured as 21.32 Hz, indicating that the small poorly resolved coupling was 0.32 Hz. Note that the lines in fig. 5c) and d) are perceptibly narrower than those in fig. 5a) where there is an unresolved splitting.

5. – J-deconvolution.

Consider a section through a COSY cross-peak from a two-spin system, containing an antiphase splitting J_{IS}. The Fourier transform is a time domain signal

(6) $$S(t) = \exp[-i2\pi\delta t]\exp[-\lambda t]\sin(\pi J_{IS} t),$$

where the modulation term $\sin(\pi J_{IS} t)$ reflects the active coupling. One approach to determining the coupling constant is to divide this time domain signal by $\sin(\pi J^* t)$, where J^* is a trial coupling constant [10, 11]. When $J^* = J_{IS}$, we would expect a sharp change in this processed signal. The Fourier transform would be a cross-peak section without the J_{IS} splitting. The method, originally proposed by BOTHNER-BY and DADOK [12], may, therefore, be called «J-deconvolution».

The section may be taken in either the F_1 or the F_2 dimension, but under normal circumstances the F_2 trace is more suitable since it has finer digitization. After zero-filling and transformation into the time domain this is divided by $\sin(\pi J^* t)$, where J^* is a trial coupling constant that is varied through the expected range of J_{IS}. However, there are numerical difficulties when we attempt to divide by zero or by a very small number, so it is essential to avoid the zero-crossings of $\sin(\pi J^* t)$. We may think of the process as multiplication by the corresponding cosecant function in which the singularities cause problems by interacting with noise in the experimental spectrum.

The trick is to sample the cosecant at regular intervals ($4a$ samples per cycle) calculated to avoid these singularities. The time domain signal is then

(7) $$S'(t) = \exp[-i2\pi\delta t]\exp[-\lambda t]\sin(\pi J_{IS} t)\operatorname{cosec}(\pi J^* t).$$

Now consider the summation

(8) $$\Sigma = 2\sin\theta \sum_{p=1}^{a} \sin[(2p-1)\theta] = 2\sin\theta\{\sin\theta + \sin 3\theta + \sin 5\theta + ...\}.$$

Each product of sines may be written in terms of the cosines of sums and differences,

(9) $$\Sigma = \cos 0 - \cos 2\theta + \cos 2\theta - \cos 4\theta + \cos 4\theta - ... - \cos(2a\theta) = 1 - \cos(2a\theta)$$

since the intermediate terms cancel. If θ is restricted to values such that

(10) $$\theta = (2k-1)\pi/4a,$$

where k is an integer, then the final cosine term is zero and Σ is unity. Consequently we may write

(11) $$\operatorname{cosec} \theta = 2 \sum_{p=1}^{a} \sin\left[(2p-1)\theta\right]$$

provided that the cosecant is only sampled at points that satisfy eq. (10). Thus the choice of sampling has had an important simplifying role. With the substitution $\theta = \pi J^* t$, the processed time domain signal may be written

(12) $$S'(t) = 2\exp[-i2\pi\delta t]\exp[-\lambda t]\sin(\pi J_{IS} t)\sum_{p=1}^{a}\sin[(2p-1)\pi J^* t].$$

This is an expression which is now readily Fourier transformed and which has no problems with zero-crossings. We call this Fourier transform the «test spectrum». If $a = 1$, then the time domain signal is just

(13) $$S'(t) = 2\exp[-i2\pi\delta t]\exp[-\lambda t]\sin(\pi J_{IS} t)\sin(\pi J^* t),$$

which may be rewritten

(14) $$S'(t) = \exp[-i2\pi\delta t]\exp[-\lambda t]\{\cos[\pi(J_{IS} - J^*)t] - \cos[\pi(J_{IS} + J^*)t]\}.$$

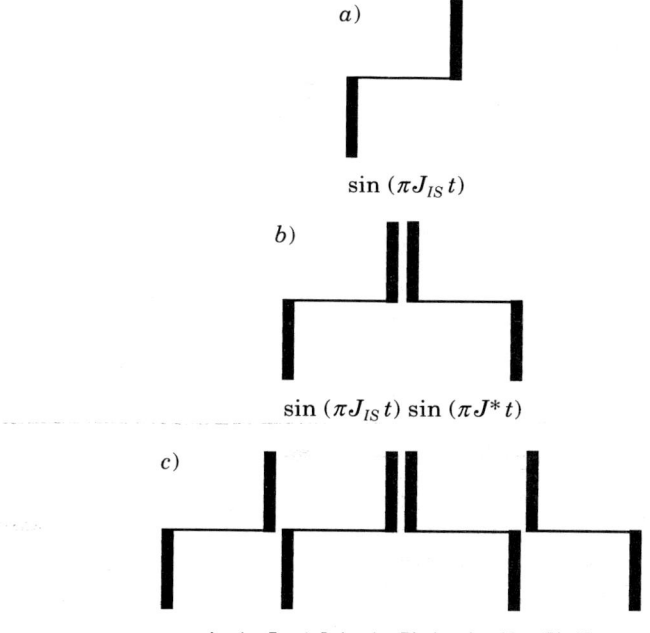

Fig. 6. – Simulated test spectra for J-deconvolution of a COSY cross-peak with an active coupling J_{IS}: a) the experimental doublet, b) processed by multiplication of the time domain signal by $\operatorname{cosec}(\pi J^* t)$ with four sample points per cycle, c) with eight points per cycle.

This is seen to be a spectrum with a small (positive) in-phase splitting $|J_{IS} - J^*|$ and a larger (negative) in-phase splitting $|J_{IS} + J^*|$. If the number of samples per cycle is increased to eight ($a = 2$), then

(15) $\quad S''(t) = 2\exp[-i2\pi\delta t]\exp[-\lambda t]\sin(\pi J_{IS} t)[\sin(\pi J^* t) + \sin(3\pi J^* t)]$.

Further increases in the rate of sampling, to $a = 3$ and then $a = 4$, result in addition of the terms $\sin(5\pi J^* t)$ and $\sin(7\pi J^* t)$ and so on. However, the general form of the test spectrum (fig. 6) can be described quite simply—a central doublet of splitting $|J_{IS} - J^*|$ flanked by an array of sidebands separated by multiples of J^*. Each sideband is an *antiphase* doublet of splitting $|J_{IS} - J^*|$. We thus have a remarkable result imposed by our particular choice of sampling—the number of antiphase sideband responses in the test spectrum is determined by the number of samples per cycle of the cosecant.

The antiphase nature of these doublets is the key to measuring J_{IS}, since it causes all the sidebands to disappear when $J^* = J_{IS}$. The integral of the absolute magnitude of the test spectrum goes through a sharp minimum at this condition. In practice there are also some other minima caused by interference between antiphase lines from adjacent and next-nearest neighbour sidebands, but the resulting minima are always less deep than the principal minimum at $J^* = J_{IS}$. What is important is that the critical condition can be located very accurately, even where the splittings are poorly resolved in the experimental spectrum. Analogous results can be achieved with passive splittings.

Experimental results on two passive splittings in metabromonitrobenzene bear out these expectations. A section was taken through one of the COSY cross-peaks and processed as described above, scanning the trial coupling over the range 5 to 11.5 Hz. The integral of the absolute magnitude of the test spectrum (fig. 7) shows two deep minima, at 7.98 Hz and 8.28 Hz. This illustrates how two couplings can be measured even though they differ

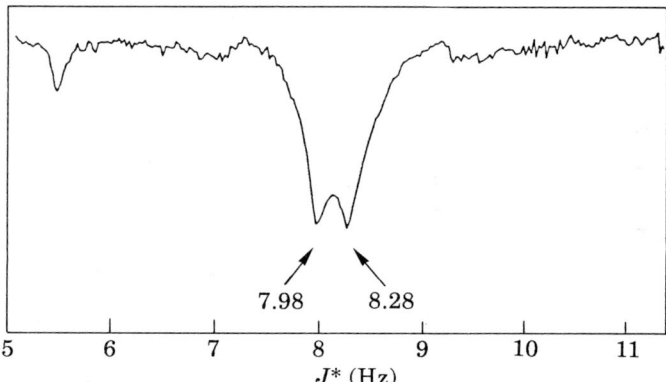

Fig. 7. – Two passive splittings from the COSY spectrum of metabromonitrobenzene measured by J-deconvolution. The integral plot shows two deep minima indicating couplings of 7.98 and 8.28 Hz.

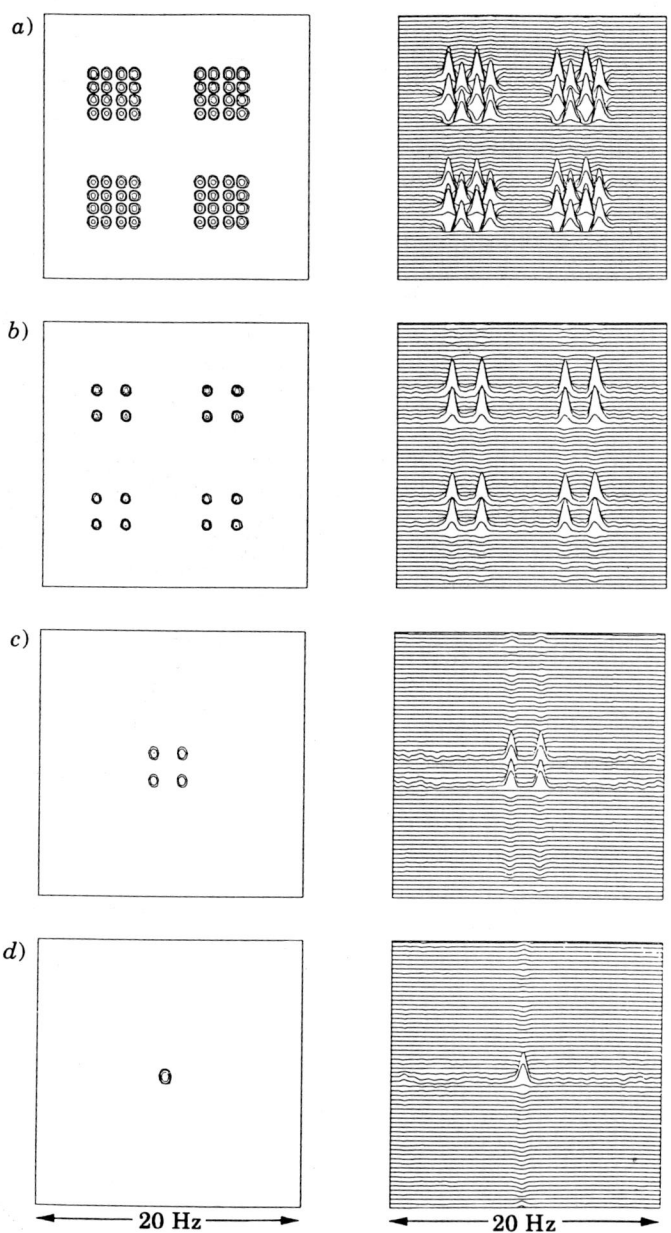

Fig. 8. – Progressive deconvolution of splittings from a cross-peak in the 400 MHz COSY spectrum of metabromonitrobenzene: a) experimental cross-peak, b) after removal of the active splitting (1.00 Hz), c) after removal of two passive splittings (7.98 Hz and 8.28 Hz), d) after removal of two further passive splittings (1.89 Hz and 2.18 Hz).

by only 0.30 Hz. The measurement accuracy has been estimated at better than ± 0.04 Hz.

The beauty of J-deconvolution is that, once the splitting has been measured, it is also removed from the processed spectrum. Thus a COSY cross-peak may be simplified step by step by deconvolution of the active and passive splittings until only a single peak is left at the chemical-shift coordinates. Figure 8 shows such a sequence of deconvolutions for a typical cross-peak from metabromonitrobenzene[11]. In principle a computer program could be written to measure all the splittings in a COSY spectrum and then reduce the entire correlation matrix to singlets instead of cross-peaks with fine structure.

6. – J-doubling.

This is an alternative to J-deconvolution which also invokes a trigonometrical identity, in this case the expression for doubling an angle, $\sin 2\theta = 2\sin\theta\cos\theta$. As before, the first step is to isolate the appropriate section through the COSY cross-peak, zero-fill and Fourier transform to give a time domain signal

$$(16) \qquad S(t) = \exp[-i2\pi\delta t]\exp[-\lambda t]\sin(\pi J_{IS} t).$$

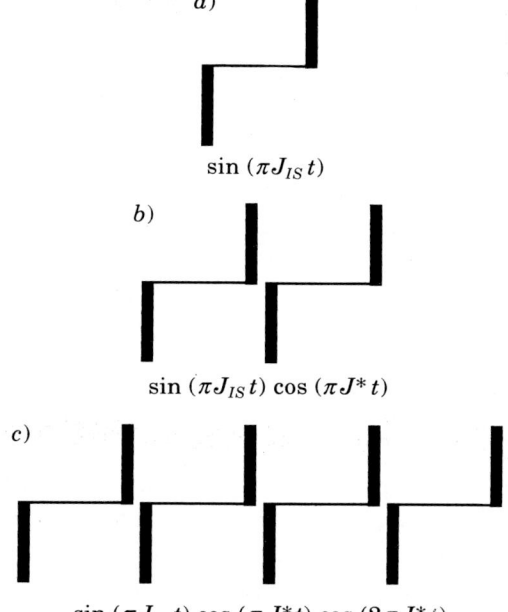

Fig. 9. – Simulated test spectra for J-doubling: a) the experimental active splitting, b) one stage of doubling, c) two stages of doubling.

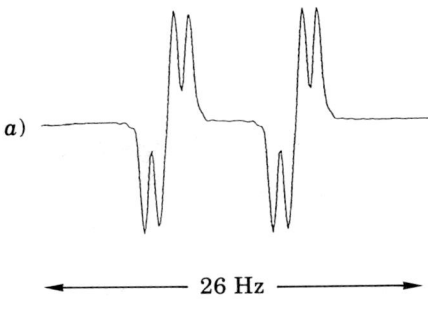

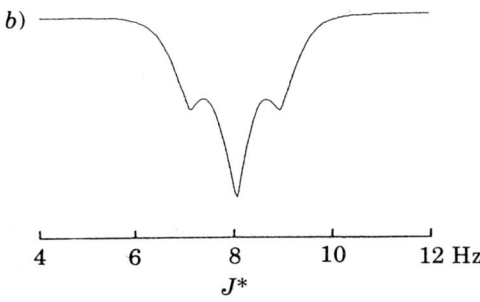

Fig. 10. – J-doubling of passive splittings in the 400 MHz COSY spectrum of metabromonitrobenzene: a) section through the experimental cross-peak, b) integral of the absolute magnitude of the test spectrum, indicating splittings of 7.99 Hz and 0.99 Hz.

This is then *multiplied* by a term involving a trial coupling constant J^* (equivalent to convolution in the frequency domain with a stick spectrum)

(17) $\qquad S_2(t) = \exp[-i2\pi\delta t]\exp[-\lambda t]\sin(\pi J_{IS} t)\cos(\pi J^* t)$.

This may be rewritten

(18) $\quad S_2(t) = 0.5\exp[-i2\pi\delta t]\exp[-\lambda t]\{\sin[\pi(J^* + J_{IS})t] - \sin[\pi(J^* - J_{IS})t]\}$.

This corresponds to a test spectrum (fig. 9b)) made up of a central antiphase doublet of splitting $|J^* - J_{IS}|$ flanked by an outer antiphase doublet $|J^* + J_{IS}|$. When $J^* = J_{IS}$, the inner doublet vanishes through destructive interference, while the outer doublet acquires the splitting $2|J_{IS}|$. As in the method of J-deconvolution, a good practical criterion for locating this condition is a minimum in the integral of the absolute magnitude of the test spectrum.

In practice it is useful to continue the doubling procedure, multiplying by

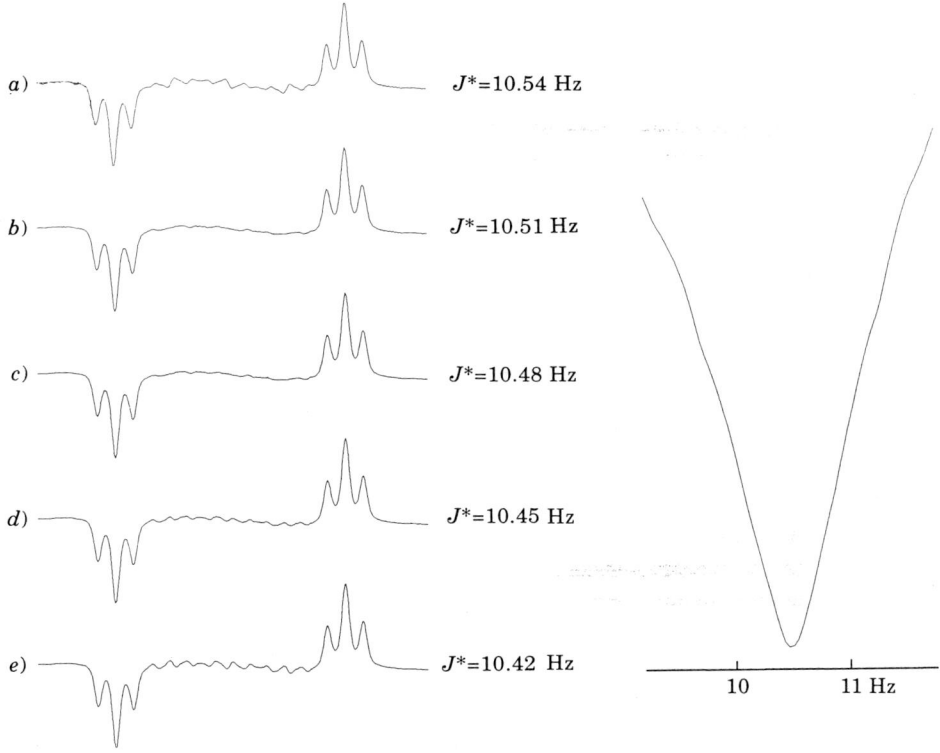

Fig. 11. – Two stages of J-doubling in the 400 MHz proton spectrum of strychnine. The trial coupling J^* is incremented in 0.03 Hz steps and the trace with the minimum residue of antiphase doublets is taken as a measure of $J_{IS} = 10.48$ Hz. The corresponding integral plot is shown on the right.

$\cos(2\pi J^* t)$, giving

(19) $\quad S_4(t) = \exp[-i2\pi\delta t]\exp[-\lambda t]\sin(\pi J_{IS} t)\cos(\pi J^* t)\cos(2\pi J^* t).$

This simply doubles the pattern in the test spectrum (fig. 9c)). Now three antiphase doublets vanish at the condition $J^* = J_{IS}$, leaving a single antiphase doublet of splitting $4|J_{IS}|$. In conditions of poor resolution, this has the advantage that any overlap is removed. Further doubling often proves useful, the only complication being that there are subsidiary minima in the integral plots owing to interference between nearest and next-nearest neighbour antiphase signals. Note the interesting parallel with test spectra by the J-deconvolution method. Passive splittings, which are treated in an analogous manner, show fewer of these interference effects.

A practical application of J-doubling is illustrated in fig. 10 for two passive splittings in the 400 MHz COSY spectrum of metabromonitrobenzene [13]. Figure 10a) shows a section through the experimental cross-peak, while fig. 10b) shows the 1:2:1 pattern of minima in the integral of the absolute magnitude of the test spectrum. The deep central minimum corresponds to a passive coupling of 7.99 Hz and the two satellites indicate a second passive coupling of 0.99 Hz. The accuracy is estimated as better than ± 0.04 Hz.

Perhaps a clearer appreciation of the precision with which the splitting can be measured is provided by direct observation of the test spectra as J^* is incremented in small steps. The highest-field proton in the 400 MHz COSY spectrum of strychnine was subjected to two stages of J-doubling (fig. 11). Weak residuals from the antiphase doublets can be seen in the central region between the two triplets. These pass through minimum intensity near the condition $J^* = 10.48$ Hz. A plot of the integral of the absolute-magnitude test spectrum is shown on the right.

In practical cases, J-doubling appears to function slightly more effectively than J-deconvolution, but it has the disadvantage that it does not remove the splittings from the spectrum. It might prove advantageous to use J-doubling to evaluate couplings, followed by J-deconvolution to simplify the cross-peak fine structure.

Many other methods have been applied to the problem of measurement of coupling constants in NMR spectra. The advantage claimed for J-deconvolution and J-doubling is the possibility of adaptation to automatic post-acquisition data processing by computer since there is a clear and simple criterion involved, the integral of the absolute magnitude of the test spectrum. Where there are «spurious» subharmonic conditions for minima, the program can readily check that they are subharmonics and rule them out. It may well be that, in future years, two-dimensional spectra will be obtained in such numbers that extraction of coupling constants by inspection will prove too tedious, suggesting the adoption of these automated schemes.

REFERENCES

[1] B. U. MEIER, G. BODENHAUSEN and R. R. ERNST: *J. Magn. Reson.*, **60**, 161 (1984).
[2] P. PFÄNDLER, G. BODENHAUSEN, B. U. MEIER and R. R. ERNST: *Anal. Chem.*, **57**, 2510 (1985).
[3] S. GLASER and H. K. KALBITZER: *J. Magn. Reson.*, **74**, 540 (1987).
[4] L. MCINTYRE and R. FREEMAN: *J. Magn. Reson.*, **89**, 632 (1990).
[5] P. XU, X. L. WU and R. FREEMAN: *J. Magn. Reson.*, **89**, 198 (1990).
[6] P. XU, X. L. WU and R. FREEMAN: *J. Magn. Reson.*, **99**, 308 (1992).
[7] H. OSCHKINAT and R. FREEMAN: *J. Magn. Reson.*, **60**, 164 (1984).
[8] J. TITMAN and H. KEELER: *J. Magn. Reson.*, **89**, 640 (1990).
[9] L. BRAUNSCHWEILER and R. R. ERNST: *J. Magn. Reson.*, **53**, 521 (1983).
[10] R. FREEMAN and L. MCINTYRE: *Israel J. Chem.*, **32**, 231 (1992).
[11] J. M. LE PARCO, L. MCINTYRE and R. FREEMAN: *J. Magn. Reson.*, **97**, 553 (1992).
[12] A. A. BOTHNER-BY and J. DADOK: *J. Magn. Reson.*, **72**, 540 (1987).
[13] L. MCINTYRE and R. FREEMAN: *J. Magn. Reson.*, **96**, 425 (1992).

Selective Double Resonance and Coherence Transfer.

C. ZWAHLEN, S. J. F. VINCENT and G. BODENHAUSEN

Section de Chimie, Université de Lausanne
Rue de la Barre 2, CH-1005 Lausanne, Switzerland

1. – Introduction.

Of the many ingenious inventions that have marked the history of magnetic resonance, the Hartmann-Hahn method is undoubtedly one of the most remarkable[1]. Much has been written about applications of this double-resonance technique to heteronuclear systems in the solid state, where two distinct spin species I and S (such as protons and carbons) are simultaneously subjected to two strong r.f. fields. If the magnetization of one species, say of the I spins, is initially prepared in a state I_x so that it is spin-locked by one of the r.f. fields, while the other r.f. field is applied in the vicinity of the S-spin resonance, one observes a transfer of magnetization from I_x to S_x. This transfer is most efficient if the amplitudes B_{1I} and B_{1S} of the two r.f. fields fulfil the Hartmann-Hahn condition $\gamma_I B_{1I} = \gamma_S B_{1S}$. The transfer may be mediated either by dipolar couplings, which are predominant in solids[1], or by scalar interactions as in isotropic solution[2]. It is only in the last decade or so that it has become apparent that a Hartmann-Hahn transfer can also occur between *like* spins, *i.e.* in homonuclear systems, where one may observe a transfer between I_x^A and I_x^X. In fact, the possibility of such a process was already implied in the thermodynamic description of heteronuclear experiments, since it was assumed that the magnetization could migrate between various I spins belonging to a common «reservoir», before being transferred to the S spin. But it is only with the advent of two-dimensional techniques such as «total-correlation spectroscopy» («TOCSY»), also known as «homonuclear Hartmann-Hahn spectroscopy» («HOHAHA»), that the transfer of transverse magnetization under spin-locked conditions could be exploited to study scalar coupled systems in liquids[3,4]. In its most popular form, the TOCSY method yields two-dimensional spectra where a cross-peak multiplet appears at frequency coordinates $\omega_1 = \Omega_A$ and $\omega_2 = \Omega_X$ whenever the spins A and X which have chemical shifts Ω_A and Ω_X *belong to a common coupling network*. Some complications arise in a TOCSY experiment when the r.f. field in the mixing time is not sufficiently strong, so that one must consider the effects of tilted effective spin-locking fields [5]. Various composite-pulse methods have been proposed to combat these effects [6]. Other complications arise when transfer through cross-relaxation (Overhauser effects) interferes with coherent transfer through scalar couplings. These complications can be brought under control by modified pulse sequences [7,8].

In this lecture, we describe a variant of the Hartmann-Hahn effect, which is *homonuclear*, in the sense that both radiofrequency fields are applied to like spin species (usually protons), and which is *selective*, in the sense that the r.f. fields are so weak that they only affect selected multiplets. In a high-resolution spectrum in isotropic solution, the Hartmann-Hahn effect can be used to probe for the existence of a scalar coupling between two selected spins A and X. Coherence is transferred while the magnetization vectors of the two selected spins are spin-locked simultaneously by two weak r.f. fields. It is the purpose of this lecture to provide a detailed «nuts-and-bolts» description of the transfer of coherence; the experimental aspects have been discussed elsewhere [9-14].

2. – Doubly selective irradiation.

A doubly selective radiofrequency field can be generated in practice [15, 16] by placing the carrier frequency midway between two selected chemical shifts at $\omega_0 = (1/2)(\Omega_A + \Omega_X)$, and by modulating the pulse with $\cos \omega_a t$, where the modulation frequency is $\omega_a = (1/2)(\Omega_A - \Omega_X)$. This generates two sidebands at frequencies $\omega_0 \pm \omega_a$, which coincide with the chemical shifts of A and X. Provided we have pure amplitude modulation (*i.e.* no admixture of phase modulation), the r.f. amplitudes of the two sidebands are exactly equal ($B_{1A} = B_{1X}$), so that the Hartmann-Hahn condition is fulfilled automatically, since $\gamma_A = \gamma_X$ in a homonuclear system. In this lecture we shall be specifically concerned with weak r.f. fields. The r.f. amplitudes are of the same order of magnitude as the scalar J couplings. This situation is usually not encountered in other applications of the Hartmann-Hahn effect and its implications deserve some discussion. In a two-spin system, the sideband frequencies $\omega_0 \pm \omega_a$ can easily be adjusted so that they precisely coincide with the two chemical shifts Ω_A and Ω_X, so that we need not be concerned about off-resonance effects and tilted effective fields. However, in systems with three or more spins, the scalar couplings between one of the «active» (*i.e.* irradiated) spins A and X and a «passive» spin M lead to multiplet structures that are more complex than the simple doublets encountered in a two-spin system. If the M spin is not affected by the two r.f. sidebands, these multiplets can be decomposed into A-X subspectra with effective chemical shifts $\Omega_A^{\text{eff}} = \Omega_A \pm \pi J_{AM}$ and $\Omega_X^{\text{eff}} = \Omega_X \pm \pi J_{XM}$. Under these circumstances, the r.f. sidebands cannot be resonant for all effective chemical shifts simultaneously, so that off-resonance effects must be taken into account.

To calculate the evolution of a spin system under the effect of doubly selective irradiation, one must solve the equation of motion of the density operator $\sigma(t)$ [17, 18]:

(1) $$\mathrm{d}\sigma(t)/\mathrm{d}t = -i[H, \sigma(t)].$$

If the Hamiltonian H is time independent, eq. (1) can be integrated to give

(2) $$\sigma(t) = U\sigma(0)U^{-1},$$

where U is the propagator defined by

(3) $$U = \exp[-iHt].$$

To be able to use this simple equation, it is necessary to find a suitable frame where the Hamiltonian is time independent [18]. Such a frame is usually known as «interaction representation». We begin by expressing the Hamiltonian describing two scalar coupled spins A and X in the presence of doubly selective irradiation (abbreviated DSI) in the laboratory frame:

(4) $$H_{\mathrm{DSI}}^{\mathrm{lab}} = \omega_0^A I_z^A + \omega_0^X I_z^X + 2\omega_1 \cos\omega_a t\{(I_x^A + I_x^X)\cos\omega_0 t +$$
$$+ (I_y^A + I_y^X)\sin\omega_0 t\} + \pi J 2 \boldsymbol{I}^A \cdot \boldsymbol{I}^X,$$

where ω_0^A is the Larmor frequency of spin A, $\omega_a = (1/2)(\omega_0^A - \omega_0^X)$ is the modulation frequency, $\omega_0 = (1/2)(\omega_0^A + \omega_0^X)$ is the carrier frequency, $2\omega_1 = -2\gamma B_1$ (expressed in rad s^{-1}) is the total amplitude of the r.f. field (ω_1 for each sideband), and J is the scalar coupling constant between A and X. As usual, we have neglected one of the two counterrotating parts of the r.f. field which appears at $-\omega_0$, so that the r.f. field can be described in terms of a circular rather than a linear polarization. The other counterrotating component, which in a spectrometer where the protons resonate at 300 MHz appears at -600 MHz from resonance, can be safely neglected. The Bloch-Siegert effects [19] associated with this component are too small to have any significant influence on the magnetization.

The Hamiltonian can be transformed from the laboratory to the rotating frame [18] by using the relation

(5) $$H_{\mathrm{DSI}}^{\mathrm{rot}} = \tilde{H} - H_0,$$

where

(6) $$\tilde{H} = U_0^{-1} H_{\mathrm{DSI}}^{\mathrm{lab}} U_0,$$

with the unitary propagator

(7) $$U_0 = \exp[-iH_0 t].$$

The aim is to choose H_0 in such a manner that the Hamiltonian is simplified as much as possible. We shall approach this goal in two consecutive steps. First we define

(8) $$H_0 = \omega_0 (I_z^A + I_z^X).$$

Hence

(9) $$U_0 = \exp[-i\omega_0 (I_z^A + I_z^X)t].$$

We obtain a Hamiltonian in the new reference frame:

(10) $$H_{\text{DSI}}^{\text{rot}} = \omega_a (I_z^A - I_z^X) + 2\omega_1 \cos \omega_a t (I_x^A + I_x^X) + \pi J\, 2\mathbf{I}^A \cdot \mathbf{I}^X.$$

This amounts to a transformation into a frame rotating at the average frequency $\omega_0 = (1/2)(\omega_0^A + \omega_0^X)$. Note that in this frame the chemical shifts are $\Omega_A = +\omega_a$ and $\Omega_X = -\omega_a$. The transformation of eq. (4) into eq. (10) can be verified either by using explicit matrix representations of the various operators or by using commutator rules. Obviously, the Hamiltonian of eq. (10) still contains time-dependent terms. Although it is not strictly possible to remove this time dependence, we can reformulate the problem by using a transformation into a doubly rotating frame (DR), where it will become more apparent that some terms can be neglected. We use the following unitary transformation:

(11) $$U^{\text{DR}} = \exp[-iH_\Delta t],$$

where

(12) $$H_\Delta = \omega_a (I_z^A - I_z^X).$$

The effective Hamiltonian in the resulting doubly rotating frame (also called «interaction frame») is then equal to

(13) $$H_{\text{DSI}}^{\text{DR}} = U^{\text{DR}-1} H_{\text{DSI}}^{\text{rot}} U^{\text{DR}} - H_\Delta =$$

$$= \omega_1 I_x^A + \omega_1 I_x^X + \omega_1 \{I_x^A \cos 2\omega_a t + I_y^A \sin 2\omega_a t\} +$$

$$+ \omega_1 \{I_x^X \cos 2\omega_a t - I_y^X \sin 2\omega_a t\} + \pi J\, 2I_z^A I_z^X +$$

$$+ \pi J\{2I_x^A I_x^X + 2I_y^A I_y^X\} \cos 2\omega_a t + \pi J\{2I_x^A I_y^X - 2I_y^A I_x^X\} \sin 2\omega_a t.$$

The Zeeman interactions have now been removed, since the doubly rotating frame behaves as if it were rotating on resonance for both spins A and X. The terms $\omega_1 I_x^A$ and $\omega_1 I_x^X$ represent the interaction of spin A with the (resonant) r.f. sideband located at Ω_A and the interaction of spin X with the sideband located at Ω_X. Both of these terms are time independent because we are in a doubly rotating frame. The time-dependent parts of eq. (13) proportional to ω_1 represent, for each spin, a field rotating at $\pm 2\omega_a$. Since we usually have $\omega_a > \omega_1 > \pi J$ (typically $\omega_a/2\pi > 100$ Hz, $\omega_1/2\pi \approx 25$ Hz and $J < 10$ Hz), the effect on spin A of the sideband located at Ω_X, and *vice versa*, is negligible. This allows one to neglect the oscillating time-dependent parts of eq. (13), much in the same way as we have neglected the counterrotating component in eq. (4) because the induced Bloch-Siegert shifts are small [20, 21]. Using the same hypothesis about the amplitudes of the frequencies, we can also say that the time-dependent parts of eq. (13) proportional to J can be neglected. The validity of this approximation is not obvious, and indeed there are situations where nonresonant effects must be taken into account [13]. With these simplifications, we obtain the following time-independent Hamiltonian:

$$H_{\text{DSI}}^{\text{DR}} = \omega_1 (I_x^A + I_x^X) + \pi J 2 I_z^A I_z^X, \tag{14}$$

which can be used to integrate eq. (1) analytically. In cases in which the approximations made between eqs. (13) and (14) appear doubtful, one can easily conduct numerical simulations using the full, untruncated Hamiltonian of eq. (13). In this way, we have shown elsewhere [13] that our approximations are quite safe.

3. – Matrix representations.

With the help of Pauli matrices (see below), one can readily construct a matrix representation of a Hamiltonian such as appears in eq. (14). To obtain an explicit matrix representation of the propagator $U = \exp[-iHt]$ of eq. (3), it is possible to use a series expansion of the exponential function. However, it is much simpler to diagonalize the matrix representation of the Hamiltonian:

$$U = \exp[-i\mathbf{H}t] = \mathbf{V} \exp[-i\mathbf{V}^{-1}\mathbf{H}\mathbf{V}t] \mathbf{V}^{-1} = \tag{15}$$

$$= \mathbf{V} \exp[-i\mathbf{H}^{\text{diag}} t] \mathbf{V}^{-1} = \mathbf{V} \mathbf{U}^{\text{diag}} \mathbf{V}^{-1},$$

where the bold symbols stand for matrix representations (which are of dimension 4×4 for a two-spin system). The only nonvanishing elements of $\mathbf{U}^{\text{diag}}$ are its diagonal elements:

$$(U^{\text{diag}})_{kk} = \exp[-i(H^{\text{diag}})_{kk} t]. \tag{16}$$

The matrix $\mathbf{V}$ of eigenvectors is defined so that $\mathbf{H}^{\text{diag}} = \mathbf{V}^{-1}\mathbf{H}\mathbf{V}$ is a diagonal matrix. This matrix can be found by solving the characteristic equation

$$|\mathbf{H}_{\text{DSI}}^{\text{DR}} - \mathbf{E}\lambda_k| = 0, \tag{17}$$

where the vertical bars denote the determinant of a matrix, $\mathbf{E}$ is a 4×4 unitary matrix, and λ_k ($k = 1, 2, 3, 4$) are the eigenvalues. The four corresponding eigenvectors $\mathbf{k}_k$ may be readily derived from the equation

(18) $$\mathbf{H}\mathbf{k}_k = \lambda_k \mathbf{k}_k .$$

To solve the characteristic equation (17), the time-independent Hamiltonian of eq. (14) must be expressed in matrix form. This can be achieved by forming direct products ($\otimes$) of the 2×2 Pauli matrices $\mathbf{I}_\alpha$ representing the operators I_α ($\alpha = x, y, z$) and the 2×2 unitary matrix $\mathbf{E}$. For two 2×2 matrices $\mathbf{M}$ and $\mathbf{N}$, the direct product is defined as

(19) $$\mathbf{M} \otimes \mathbf{N} = \begin{bmatrix} m_{11} & m_{12} \\ m_{21} & m_{22} \end{bmatrix} \otimes \begin{bmatrix} n_{11} & n_{12} \\ n_{21} & n_{22} \end{bmatrix} =$$

$$= \begin{bmatrix} m_{11}n_{11} & m_{11}n_{12} & m_{12}n_{11} & m_{12}n_{12} \\ m_{11}n_{21} & m_{11}n_{22} & m_{12}n_{21} & m_{12}n_{22} \\ m_{21}n_{11} & m_{21}n_{12} & m_{22}n_{11} & m_{22}n_{12} \\ m_{21}n_{21} & m_{21}n_{22} & m_{22}n_{21} & m_{22}n_{22} \end{bmatrix}.$$

To express our Hamiltonian, three different Cartesian operators must be represented by matrices:

(20) $$\mathbf{I}_x^A = \mathbf{I}_x \otimes \mathbf{E}_2 = \frac{1}{2}\begin{bmatrix} 0 & 1 \\ 1 & 0 \end{bmatrix} \otimes \begin{bmatrix} 1 & 0 \\ 0 & 1 \end{bmatrix} = \frac{1}{2}\begin{bmatrix} 0 & 0 & 1 & 0 \\ 0 & 0 & 0 & 1 \\ 1 & 0 & 0 & 0 \\ 0 & 1 & 0 & 0 \end{bmatrix},$$

(21) $$\mathbf{I}_x^X = \mathbf{E}_2 \otimes \mathbf{I}_x = \begin{bmatrix} 1 & 0 \\ 0 & 1 \end{bmatrix} \otimes \frac{1}{2}\begin{bmatrix} 0 & 1 \\ 1 & 0 \end{bmatrix} = \frac{1}{2}\begin{bmatrix} 0 & 1 & 0 & 0 \\ 1 & 0 & 0 & 0 \\ 0 & 0 & 0 & 1 \\ 0 & 0 & 1 & 0 \end{bmatrix},$$

(22) $$2\mathbf{I}_z^A \mathbf{I}_z^X = 2\mathbf{I}_z \otimes \mathbf{I}_z = 2\frac{1}{2}\begin{bmatrix} 1 & 0 \\ 0 & -1 \end{bmatrix} \otimes \frac{1}{2}\begin{bmatrix} 1 & 0 \\ 0 & -1 \end{bmatrix} = \frac{1}{2}\begin{bmatrix} 1 & 0 & 0 & 0 \\ 0 & -1 & 0 & 0 \\ 0 & 0 & -1 & 0 \\ 0 & 0 & 0 & 1 \end{bmatrix}.$$

We now express eq. (14) in terms of these matrices, so that we obtain a representation of the time-independent Hamiltonian of a two-spin system in the presence of doubly selective irradiation in the doubly rotating frame:

(23) $$\mathbf{H}_{\text{DSI}}^{\text{DR}} = \frac{1}{2}\begin{bmatrix} \pi J & \omega_1 & \omega_1 & 0 \\ \omega_1 & -\pi J & 0 & \omega_1 \\ \omega_1 & 0 & -\pi J & \omega_1 \\ 0 & \omega_1 & \omega_1 & \pi J \end{bmatrix}.$$

To express the results in a condensed way, we define the following frequencies:

(24) $$\omega_J = \pi J,$$

(25) $$\omega_{\text{eff}} = (4\omega_1^2 + \pi^2 J^2)^{1/2}.$$

The characteristic equation leads to an expression in fourth power in λ_k:

(26) $$\left(\lambda_k^2 - \frac{1}{4}\omega_J^2\right)\left(\lambda_k^2 - \frac{1}{4}\omega_J^2 - \omega_1^2\right) = 0.$$

The roots give the four eigenvalues λ_k:

(27) $$\lambda_1 = +\frac{1}{2}\omega_{\text{eff}}, \quad \lambda_2 = -\frac{1}{2}\omega_J, \quad \lambda_3 = +\frac{1}{2}\omega_J, \quad \lambda_4 = -\frac{1}{2}\omega_{\text{eff}}.$$

The corresponding eigenvectors $\boldsymbol{k}_k$ are

(28)
$$\boldsymbol{k}_1 = \frac{1}{2}\begin{pmatrix} c-s \\ c+s \\ c+s \\ c-s \end{pmatrix}, \quad \boldsymbol{k}_2 = \frac{1}{2}\begin{pmatrix} 0 \\ -\sqrt{2} \\ \sqrt{2} \\ 0 \end{pmatrix},$$

$$\boldsymbol{k}_3 = \frac{1}{2}\begin{pmatrix} \sqrt{2} \\ 0 \\ 0 \\ -\sqrt{2} \end{pmatrix}, \quad \boldsymbol{k}_4 = \frac{1}{2}\begin{pmatrix} c+s \\ -(c-s) \\ -(c-s) \\ c+s \end{pmatrix},$$

where $c = \cos\alpha$ and $s = \sin\alpha$, the angle α being defined by $\mathrm{tg}\,2\alpha = -\omega_J/2\omega_1$. By setting these eigenvectors in columns we obtain the unitary diagonalization matrix $\boldsymbol{V}$. The ordering of these eigenvectors is not of fundamental importance. The choice made here is consistent with earlier work [15, 16]:

(29) $$\boldsymbol{V} = \frac{1}{2}\begin{pmatrix} c-s & 0 & \sqrt{2} & c+s \\ c+s & -\sqrt{2} & 0 & -(c-s) \\ c+s & \sqrt{2} & 0 & -(c-s) \\ c-s & 0 & -\sqrt{2} & c+s \end{pmatrix}.$$

Since this matrix is orthogonal (both unitary and real), its inverse $\boldsymbol{V}^{-1}$ is obtained simply by transposition. We may easily verify that $\boldsymbol{H}_{\text{DSI}}^{\text{diag}} = \boldsymbol{V}^{-1}\boldsymbol{H}_{\text{DSI}}^{\text{DR}}\boldsymbol{V}$

(see eq. (15)) does indeed have a diagonal form:

$$
(30) \qquad H_{\text{DSI}}^{\text{diag}} = \frac{1}{2} \begin{pmatrix} \omega_{\text{eff}} & 0 & 0 & 0 \\ 0 & -\omega_J & 0 & 0 \\ 0 & 0 & \omega_J & 0 \\ 0 & 0 & 0 & -\omega_{\text{eff}} \end{pmatrix}.
$$

As expected, the diagonal elements correspond to the eigenvalues λ_k of eq. (27). This Hamiltonian can also be expressed in terms of products of Cartesian operators [22, 23]:

$$
(31) \qquad H_{\text{DSI}}^{\text{diag}} = \frac{1}{2} \left\{ [\omega_{\text{eff}} - \omega_J] I_z^A + [\omega_{\text{eff}} + \omega_J] I_z^X \right\}.
$$

It can be shown that this is consistent with earlier work [15, 16], where we have given a slightly different notation:

$$
(32) \qquad H_{\text{DSI}}^{\text{diag}} = (\pi J/2) \{ [(2\omega_1/\pi J) \cos 2\alpha - \sin 2\alpha - 1] I_z^A +
$$
$$
+ [(2\omega_1/\pi J) \cos 2\alpha - \sin 2\alpha + 1] I_z^X \}.
$$

We shall now turn our attention to the effect of this Hamiltonian on the density operator.

4. – Evolution, coherence transfer and detection.

To calculate the evolution of the density operator under the effect of the Hamiltonian of eq. (14), one has to calculate the propagator U defined in eq. (15). If we express the density operator $\sigma(\tau_{\text{DSI}} = 0)$ at the beginning of the doubly selective irradiation period in the doubly rotating frame, we obtain the following expression for the evolution:

$$
(33) \qquad \sigma(t) = U \sigma(0) U^{-1} = V U^{\text{diag}} V^{-1} \sigma(0) V U^{\text{diag}-1} V^{-1}.
$$

The expectation value of any operator Q can be calculated according to

$$
(34) \qquad \langle Q(t) \rangle = \text{Tr} \left\{ \boldsymbol{Q} \sigma(t) \right\},
$$

where Tr {} represents the trace (sum of the diagonal elements) of the matrix product in brackets. This expression is given in the usual Schrödinger representation in which the observable operators are time independent, whereas the system, described by the density operator $\sigma(t)$, evolves as a function of time [17].

If Q represents a product of Cartesian operators [22, 23], its evolution under the effect of a Hamiltonian H can be readily calculated by defining Q' as follows:

$$H = \omega Q' . \tag{35}$$

If the commutator $[Q', Q] = Q'Q - QQ' \neq 0$, one obtains

$$Q \xrightarrow{H\tau} Q \cos \omega\tau - i[Q', Q] \sin \omega\tau . \tag{36}$$

Some of the most frequently encountered commutators are

$$[I_\alpha^A, I_\beta^A] = iI_\gamma^A , \tag{37}$$

$$[I_\alpha^A, 2I_\beta^A I_{\alpha'}^X] = i2I_\gamma^A I_{\alpha'}^X , \tag{38}$$

$$[I_\alpha^A, I_{\alpha'}^X] = 0 , \tag{39}$$

with $\alpha, \beta, \gamma = x, y, z$ for spin A and $\alpha', \beta', \gamma' = x, y, z$ for spin X and cyclic permutations.

In a typical two-dimensional experiment, the doubly selective irradiation will not be applied to a spin system in thermal equilibrium, but only after excitation and partial evolution of transverse magnetization. If we start with selectively excited in-phase magnetization I_x^A, this evolves during t_1 under the effect of chemical shifts and scalar couplings. For a two-spin system in the frame rotating at the carrier frequency $\omega_0 = (1/2)(\Omega_A + \Omega_X)$, where the shifts are $\Omega_A = = +\omega_a$ and $\Omega_X = -\omega_a$, the free-precession Hamiltonian $H_{\text{FP}}^{\text{rot}}$ is

$$H_{\text{FP}}^{\text{rot}} = \omega_a I_z^A - \omega_a I_z^X + \pi J 2 I_z^A I_z^X . \tag{40}$$

Since all terms in this (diagonal) Hamiltonian commute, the transformations during the evolution period t_1 can be readily calculated with eqs. (35)-(39):

$$I_x^A \xrightarrow{H_{\text{FP}}^{\text{rot}} t_1} I_x^A \cos \omega_a t_1 \cos \pi J t_1 + 2 I_y^A I_z^X \cos \omega_a t_1 \sin \pi J t_1 + \tag{41}$$

$$+ I_y^A \sin \omega_a t_1 \cos \pi J t_1 - 2 I_x^A I_z^X \sin \omega_a t_1 \sin \pi J t_1 .$$

If a doubly selective irradiation is applied at the end of the evolution period, we may consider the fate of the four product operators of eq. (41) separately as initial conditions at $\tau_{\text{DSI}} = 0$ of the doubly selective irradiation period. Consider first the initial condition $\sigma(\tau_{\text{DSI}} = 0) = I_x^A$. We obtain during the doubly selective irradiation period

$$\langle I_x^A \rangle(\tau_{\text{DSI}}) = \frac{1}{2} \{ \cos \omega_J \tau_{\text{DSI}} + (\omega_J/\omega_{\text{eff}})^2 \cos \omega_{\text{eff}} \tau_{\text{DSI}} + (2\omega_1/\omega_{\text{eff}})^2 \} , \tag{42}$$

$$\langle I_x^X \rangle(\tau_{\text{DSI}}) = \frac{1}{2} \{ -\cos \omega_J \tau_{\text{DSI}} + (\omega_J/\omega_{\text{eff}})^2 \cos \omega_{\text{eff}} \tau_{\text{DSI}} + (2\omega_1/\omega_{\text{eff}})^2 \} , \tag{43}$$

(44) $\langle 2I_y^A I_z^X \rangle(\tau_{\text{DSI}}) = \frac{1}{2} \{ \sin \omega_J \tau_{\text{DSI}} + (\omega_J/\omega_{\text{eff}}) \sin \omega_{\text{eff}} \tau_{\text{DSI}} \}$,

(45) $\langle 2I_z^A I_y^X \rangle(\tau_{\text{DSI}}) = \frac{1}{2} \{ - \sin \omega_J \tau_{\text{DSI}} + (\omega_J/\omega_{\text{eff}}) \sin \omega_{\text{eff}} \tau_{\text{DSI}} \}$,

(46) $\langle 2I_z^A I_z^X \rangle(\tau_{\text{DSI}}) = (\omega_1 \omega_J/\omega_{\text{eff}}^2)(1 - \cos \omega_{\text{eff}} \tau_{\text{DSI}})$,

(47) $\langle 2I_y^A I_y^X \rangle(\tau_{\text{DSI}}) = -(\omega_1 \omega_J/\omega_{\text{eff}}^2)(1 - \cos \omega_{\text{eff}} \tau_{\text{DSI}})$,

all 10 other expectation values being zero. Thus, if we start with I_x^A, the coherence remains confined in a subspace spanned by only six operator products. The coefficients of the six product operator terms appearing in eqs. (42)-(47) are shown in fig. 1 as a function of the duration τ_{DSI} of the doubly selective irradiation period. All plots extend over a time scale from 0 to $1/J$, the upper limit corresponding to maximum transfer of in-phase magnetization [9, 13]. Because we have chosen $J = 5$ Hz, we have shown an interval $0 \leq \tau_{\text{DSI}} \leq 200$ ms. It is obvi-

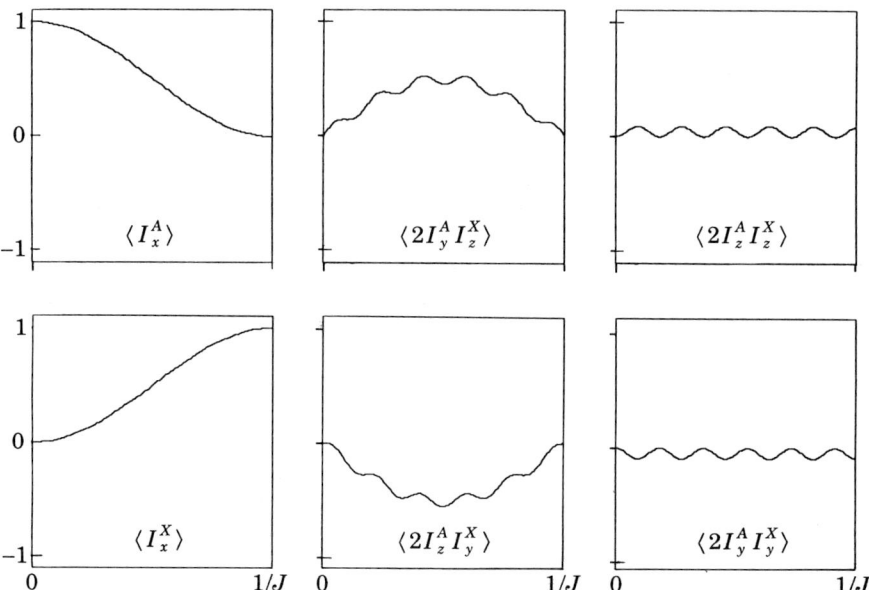

Fig. 1. – Evolution of the six nonvanishing terms in the course of a doubly selective irradiation interval τ_{DSI}, for an initial condition $\sigma(\tau_{\text{DSI}} = 0) = I_x^A$. The plots correspond to the analytical expressions of eqs. (42)-(47). The conditions used to generate these graphs were similar to experimental conditions, with a scalar coupling $J = 5$ Hz and an r.f. amplitude (for each sideband) $\nu_1 = 13.7$ Hz.

ous from fig. 1 that for $\tau_{\text{DSI}} = 1/J$ there is a complete conversion from $\langle I_x^A \rangle$ into $\langle I_x^X \rangle$. For $\tau_{\text{DSI}} < 1/J$, however, there is also a substantial amount of antiphase coherence $\langle 2I_y^A I_z^X \rangle$ and $\langle 2I_z^A I_y^X \rangle$. This has been shown many years ago by MÜLLER and ERNST [2].

We should note that we usually have an r.f. amplitude ω_1 that is strong compared to the scalar coupling ω_J, so that $(\omega_J/\omega_{\text{eff}})^2 < \omega_J/\omega_{\text{eff}} < 1$. If the r.f. field amplitude is much stronger than the scalar coupling (as in more conventional applications of the Hartmann-Hahn effect), we can also assume that $\omega_{\text{eff}} \approx 2\omega_1$, but in our homonuclear cases this approximation should not be made.

If we start with antiphase magnetization $2I_y^A I_z^X$ as initial state for $\tau_{\text{DSI}} = 0$, we obtain during of the doubly selective irradiation period

(48) $$\langle I_x^A \rangle(\tau_{\text{DSI}}) = \frac{1}{2} \left\{ -\sin \omega_J \tau_{\text{DSI}} - (\omega_J/\omega_{\text{eff}}) \sin \omega_{\text{eff}} \tau_{\text{DSI}} \right\},$$

(49) $$\langle I_x^X \rangle(\tau_{\text{DSI}}) = \frac{1}{2} \left\{ \sin \omega_J \tau_{\text{DSI}} - (\omega_J/\omega_{\text{eff}}) \sin \omega_{\text{eff}} \tau_{\text{DSI}} \right\},$$

(50) $$\langle 2I_y^A I_z^X \rangle(\tau_{\text{DSI}}) = \frac{1}{2} \left\{ \cos \omega_J \tau_{\text{DSI}} + \cos \omega_{\text{eff}} \tau_{\text{DSI}} \right\},$$

(51) $$\langle 2I_z^A I_y^X \rangle(\tau_{\text{DSI}}) = \frac{1}{2} \left\{ -\cos \omega_J \tau_{\text{DSI}} + \cos \omega_{\text{eff}} \tau_{\text{DSI}} \right\},$$

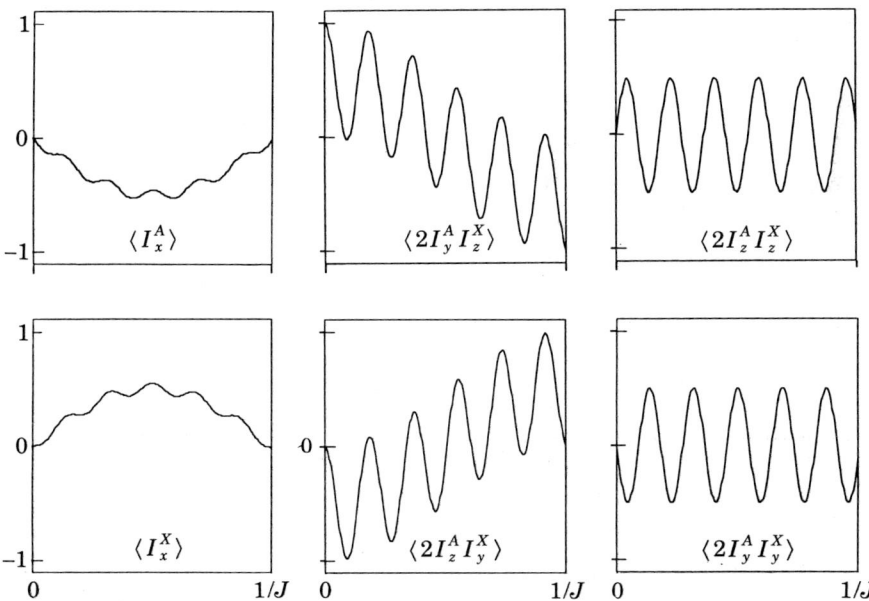

Fig. 2. – Same as fig. 1, but for an initial condition $\sigma(\tau_{\text{DSI}} = 0) = 2I_y^A I_z^X$, corresponding to eqs. (48)-(53).

(52) $\langle 2I_z^A I_z^X \rangle(\tau_{\text{DSI}}) = \frac{1}{2} \{(\omega_1/\omega_{\text{eff}}) \sin \omega_{\text{eff}} \tau_{\text{DSI}}\}$,

(53) $\langle 2I_y^A I_y^X \rangle(\tau_{\text{DSI}}) = \frac{1}{2} \{-(\omega_1/\omega_{\text{eff}}) \sin \omega_{\text{eff}} \tau_{\text{DSI}}\}$.

We note that we are confined to a subspace spanned by the *same* six operator products as in eqs. (42)-(47). All 10 other expectation values vanish for arbitrary values of τ_{DSI}. The τ_{DSI} dependence of the nonvanishing terms is shown in fig. 2.

If we start with an initial condition I_y^A at $\tau_{\text{DSI}} = 0$, we obtain during the doubly selective irradiation period

(54) $\langle I_y^A \rangle(\tau_{\text{DSI}}) =$

$= \cos \frac{1}{2} \omega_J \tau_{\text{DSI}} \cos \frac{1}{2} \omega_{\text{eff}} \tau_{\text{DSI}} - (\omega_J/\omega_{\text{eff}}) \sin \frac{1}{2} \omega_J \tau_{\text{DSI}} \sin \frac{1}{2} \omega_{\text{eff}} \tau_{\text{DSI}}$,

(55) $\langle I_z^A \rangle(\tau_{\text{DSI}}) = (2\omega_1/\omega_{\text{eff}}) \cos \frac{1}{2} \omega_J \tau_{\text{DSI}} \sin \frac{1}{2} \omega_{\text{eff}} \tau_{\text{DSI}}$,

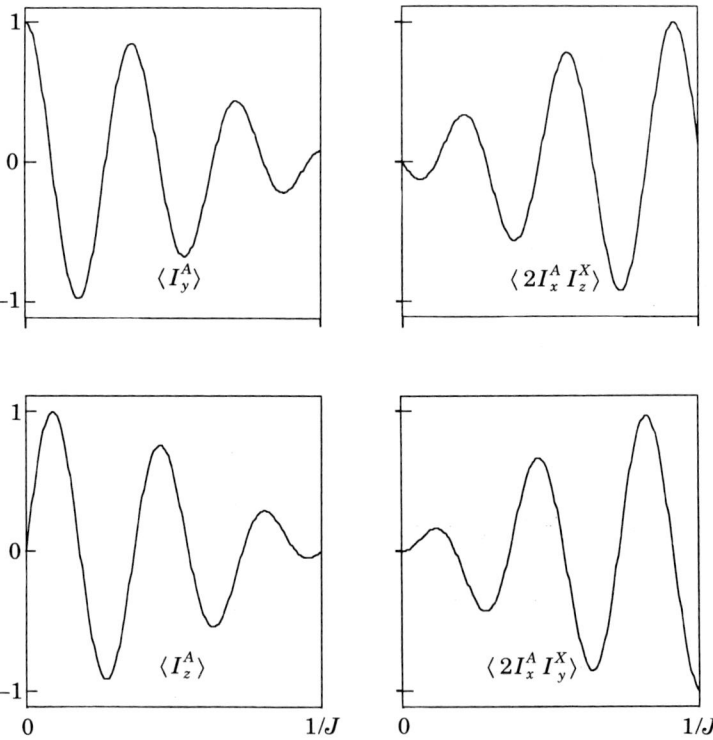

Fig. 3. – Same as fig. 1, but for an initial condition $\sigma(\tau_{\text{DSI}} = 0) = I_y^A$, corresponding to the four nonvanishing terms of eqs. (54)-(57).

(56) $\langle 2I_x^A I_z^X \rangle(\tau_{\text{DSI}}) =$

$= -\sin \frac{1}{2} \omega_J \tau_{\text{DSI}} \cos \frac{1}{2} \omega_{\text{eff}} \tau_{\text{DSI}} - (\omega_J/\omega_{\text{eff}}) \cos \frac{1}{2} \omega_J \tau_{\text{DSI}} \sin \frac{1}{2} \omega_{\text{eff}} \tau_{\text{DSI}},$

(57) $\langle 2I_x^A I_y^X \rangle(\tau_{\text{DSI}}) = (2\omega_1/\omega_{\text{eff}}) \sin \frac{1}{2} \omega_J \tau_{\text{DSI}} \sin \frac{1}{2} \omega_{\text{eff}} \tau_{\text{DSI}}.$

The transfer is now confined to a subspace comprising four operators. Their amplitudes are shown in fig. 3. All 12 other terms vanish for all values of τ_{DSI}.

Finally, if we start with antiphase magnetization $2I_x^A I_z^X$ at $\tau_{\text{DSI}} = 0$, we obtain

(58) $\langle I_y^A \rangle(\tau_{\text{DSI}}) =$

$= \sin \frac{1}{2} \omega_J \tau_{\text{DSI}} \cos \frac{1}{2} \omega_{\text{eff}} \tau_{\text{DSI}} + (\omega_J/\omega_{\text{eff}}) \cos \frac{1}{2} \omega_J \tau_{\text{DSI}} \sin \frac{1}{2} \omega_{\text{eff}} \tau_{\text{DSI}},$

(59) $\langle I_z^A \rangle(\tau_{\text{DSI}}) = (2\omega_1/\omega_{\text{eff}}) \sin \frac{1}{2} \omega_J \tau_{\text{DSI}} \sin \frac{1}{2} \omega_{\text{eff}} \tau_{\text{DSI}},$

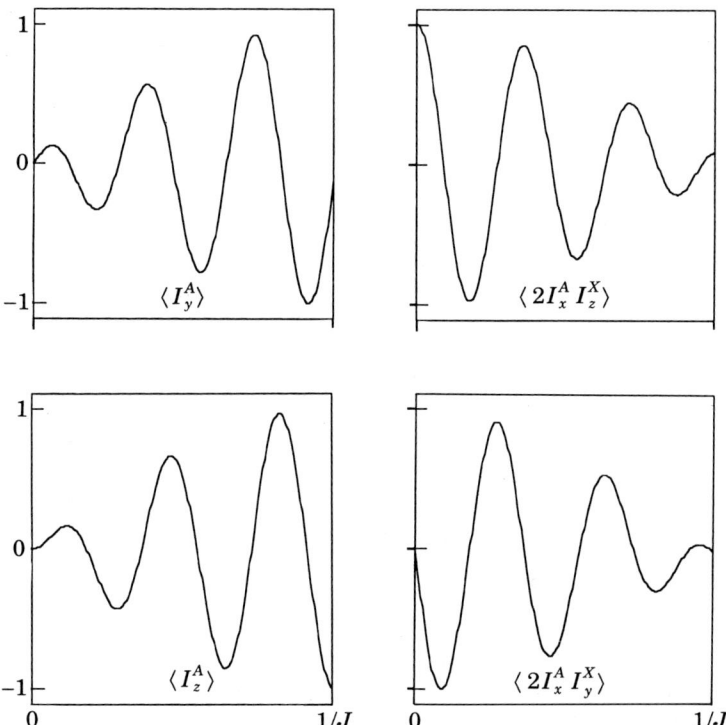

Fig. 4. – Same as fig. 1, but for an initial condition $\sigma(\tau_{\text{DSI}} = 0) = 2I_x^A I_z^X$, corresponding to eqs. (58)-(61).

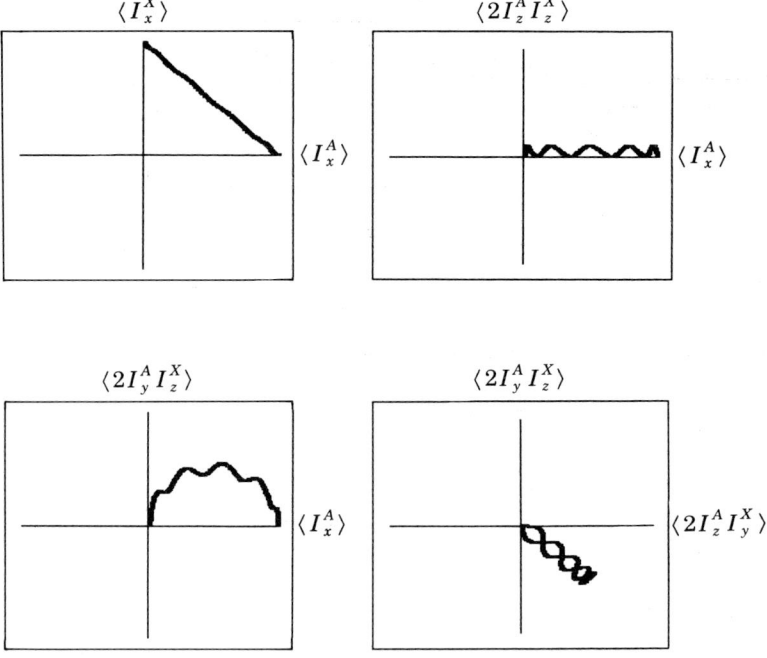

Fig. 5. – Trajectories in Liouville space corresponding to the transformations of eqs. (42)-(46).

(60) $\langle 2I_x^A I_z^X \rangle (\tau_{DSI}) =$

$= \cos \frac{1}{2} \omega_J \tau_{DSI} \cos \frac{1}{2} \omega_{eff} \tau_{DSI} - (\omega_J/\omega_{eff}) \sin \frac{1}{2} \omega_J \tau_{DSI} \sin \frac{1}{2} \omega_{eff} \tau_{DSI},$

(61) $\langle 2I_x^A I_y^X \rangle (\tau_{DSI}) = -(2\omega_1/\omega_{eff}) \cos \frac{1}{2} \omega_J \tau_{DSI} \sin \frac{1}{2} \omega_{eff} \tau_{DSI}.$

The transfer is again confined to a subspace comprising the *same* four operators of eqs. (54)-(57). Their amplitudes are shown in fig. 4.

It is instructive to think of these transformations as a walk in Liouville space [24]. Figure 5 illustrates how, according to eqs. (42)-(46), a density operator initially proportional to I_x^A is gradually transformed into I_x^X, and how some antiphase terms are created in a transient manner.

5. – Conclusions.

We have discussed the mechanism of coherence transfer between two scalar coupled spins irradiated by the sidebands of an audio-modulated radiofrequency field. This type of doubly selective coherence transfer can be used in a wide variety of experiments to investigate molecular systems in isotropic solution. In particular, one may use doubly selective coherence transfer to confirm the existence of scalar couplings [9], to separate overlapping signals in one- and two-dimensional spectra [11], to measure various relaxation parameters of resonances that are partly or completely hidden by overlapping signals [10, 14], to obtain correlation spectra with multiplets that are in pure positive absorption [12, 13], and to excite multiple-quantum coherences between selected pairs of nuclei [25]. All these applications demand that the mechanism of coherence transfer be perfectly understood, and the experimentalist should keep in mind all relevant components of the density operator.

* * *

We are indebted to Dr. N. D. KURUR for many stimulating discussions. This work was supported by the Fonds National Suisse de la Recherche Scientifique (FNRS) and by the Commission pour l'Encouragement de la Recherche Scientifique (CERS).

REFERENCES

[1] S. R. HARTMANN and E. L. HAHN: *Phys. Rev.*, **128**, 2042 (1962).
[2] L. MÜLLER and R. R. ERNST: *Mol. Phys.*, **38**, 963 (1979).
[3] L. BRAUNSCHWEILER and R. R. ERNST: *J. Magn. Reson.*, **53**, 521 (1983).
[4] A. BAX and D. G. DAVIES: *J. Magn. Reson.*, **65**, 355 (1985).
[5] K. ELBAYED and D. CANET: *Mol. Phys.*, **71**, 979 (1990).
[6] A. J. SHAKA, C. J. LEE and A. PINES: *J. Magn. Reson.*, **77**, 274 (1988).
[7] C. GRIESINGER, G. OTTING, K. WÜTHRICH and R. R. ERNST: *J. Am. Chem. Soc.*, **110**, 7870 (1988).
[8] J. BRIAND and R. R. ERNST: *Chem. Phys. Lett.*, **185**, 276 (1992).
[9] R. KONRAT, I. BURGHARDT and G. BODENHAUSEN: *J. Am. Chem. Soc.*, **113**, 9135 (1991).
[10] B. BOULAT, R. KONRAT, I. BURGHARDT and G. BODENHAUSEN: *J. Am. Chem. Soc.*, **114**, 5412 (1992).
[11] C. ZWAHLEN, S. J. F. VINCENT and G. BODENHAUSEN: *Angew. Chem.*, **104**, 1233 (1992); *Angew. Chem. Int. Ed. Engl.*, **31**, 1248 (1992).
[12] S. J. F. VINCENT, C. ZWAHLEN and G. BODENHAUSEN: *J. Am. Chem. Soc.*, **114**, 10989 (1992).
[13] S. J. F. VINCENT, C. ZWAHLEN and G. BODENHAUSEN: *J. Am. Chem. Soc.*, submitted.
[14] B. BOULAT and G. BODENHAUSEN: *J. Biomolec. NMR.* (in press).
[15] L. EMSLEY, I. BURGHARDT and G. BODENHAUSEN: *J. Magn. Reson.*, **90**, 214 (1990).
[16] Corrigendum: *J. Magn. Reson.*, **94**, 448 (1991).
[17] R. R. ERNST, G. BODENHAUSEN and A. WOKAUN: *Principles of Nuclear Magnetic Resonance in One and Two Dimensions* (Clarendon Press, Oxford, 1987).

[18] M. GOLDMAN: *Quantum Description of High-Resolution NMR in Liquids* (Clarendon Press, Oxford, 1988).
[19] F. BLOCH and A. SIEGERT: *Phys. Rev.*, **57**, 522 (1940).
[20] L. EMSLEY and G. BODENHAUSEN: *Chem. Phys. Lett.*, **168**, 297 (1990).
[21] M. A. MCCOY and L. MÜLLER: *J. Magn. Reson.*, **99**, 18 (1992).
[22] O. W. SØRENSEN, G. W. EICH, M. H. LEVITT, G. BODENHAUSEN and R. R. ERNST: *Prog. NMR Spectrosc.*, **16**, 163 (1983).
[23] U. EGGENBERGER and G. BODENHAUSEN: *Angew. Chem.*, **102**, 392 (1990); *Angew. Chem. Int. Ed. Engl.*, **29**, 374 (1990).
[24] M. H. LEVITT: *J. Magn. Reson.*, **99**, 1 (1992).
[25] S. NICULA and G. BODENHAUSEN: *J. Magn. Reson. A*, **101**, 209 (1993).

Double-Resonance J-Coupling Imaging (*).

R. CAMPANELLA, S. CAPUANI, F. DE LUCA, G. RAZA and B. MARAVIGLIA

Dipartimento di Fisica dell'Università degli Studi «La Sapienza»
00185 Roma, Italia

1. – Introduction.

A diagnostic magnetic-resonance image is usually obtained as a representation of the hydrogen nuclei belonging to the molecules of free water, which constitutes more than 70 % of the total body mass. Such a high concentration of resonant nuclei together with the high gyromagnetic ratio, characteristic of hydrogen, allows one to obtain high signal-to-noise ratio images. In spite of this, there is an increasing interest in the imaging of nuclei other than hydrogen, like ^{13}C, ^{10}B, ^{31}P, ^{15}N: this is justified from the roles they play in many physiological processes, which make them exploitable for functional studies, while conventional ^{1}H NMR images are usually important only for the morphological information they give rise to, and can be employed for functional studies only with particular procedures. Nuclei different from hydrogen might become important in MRI also for pharmaceutical studies: the capability to follow the uptake of some compound—in which a particular nucleus is used as a tracer—during its temporal evolution would make MRI an important tool.

2. – Double resonances to increase the signal-to-noise ratio.

The mentioned nuclei, like the others present *in vivo*, apart from hydrogen, are generally characterized by a low concentration, which can range from some mM down to a few μM, by a gyromagnetic ratio γ smaller and a spin-lattice relaxation time longer than those of hydrogen. These factors decrease the sensitivity, thus rendering imaging of these nuclei more difficult than for hydrogen, increasing the imaging time or diminishing the signal-to-noise ratio of the image for a fixed acquisition time.

(*) Work partially supported by INFN.

The signal-to-noise ratio in a NMR experiment depends on many factors which can be roughly divided into three groups[1,2]: some of them (*e.g.*, the noise figure of the preamplifier, the intensity of the static Zeeman field, the bandwidth of the receiver) take into account the performances and the operating conditions of the spectrometer; a second group can be cast out from the first, and is constituted by the parameters which more specifically describe the probe: its temperature, the quality factor, its volume with respect to the sample volume; finally come the parameters related to the sample: the volume and the temperature, the number of resonant spins per unit volume and the gyromagnetic ratio. We are not concerned here with the possibility of increasing the signal-to-noise ratio through a technical development of the various parts of the equipment which more strongly influence it, nor are we considering the gain in signal-to-noise ratio that can be obtained by raising the static magnetic field H_0, because there are obvious limitations, in particular in the case of human imaging, due to the technical impossibility of building whole-body magnets operating at fields above 4 T and, anyway, to patient protection limitations. Therefore, the only factors which can be acted on are the gyromagnetic ratio γ and the concentration of resonant spins. While it is evident that they are fixed by the constitution of the sample, double resonances can be effectively exploited to fictitiously increase one of these factors, or both: if some nuclei characterized by a low sensitivity, which will be denoted as S nuclei, interact with other I nuclei, which are instead characterized by a higher γ, or are present in a higher concentration, it is possible to exploit the interaction between them to increase the signal-to-noise ratio of the experiment performed on the S nuclei; basically this is the idea of the indirect detection: let us denote by I_c the nuclei belonging to the I species which interact with the S nuclei, and by I_u the I nuclei which do not interact with S. The image obtained as a representation of the I_c nuclei will be characterized by the high S/N of the I nuclei but, if the interaction is a short-range one, it will represent the spatial distribution of the S nuclei.

The detection of the I_c nuclei linked to the S nuclei instead of the latter, therefore, results in an increase—if the condition of higher γ_I with respect to γ_S, or I_c spin concentration greater than the concentration of the S nuclei is fulfilled—of, respectively, the factor γ or N in the S/N. The exact amount of the increase depends strongly on the characteristics of the molecule and on the method employed to exploit the interaction between the nuclei.

Another possibility which can be used to increase the signal-to-noise ratio of an image of rare nuclei is to increase the signal-to-noise ratio of the «rare» nuclei via a transfer of magnetization from the more sensitive nuclei; eventually this approach can be combined with an indirect detection to enhance further the performances of the experiment.

There are several ways to increase the signal from one nucleus by a magnetization transfer from a more sensitive one: historically the first experiment of polarization transfer has been the demonstration of the Overhauser effect[3,4], which consists in an enhancement of the polarization of nuclei in metals due to a saturation of the spin resonance of the electrons. Another method which transfers polarization from more sensitive nuclei to rare ones in the rotating frame has been developed by HARTMANN and HAHN [5]. The first of these methods has been demonstrated to be able to produce images of the distribution of free radicals *via* an enhancement of the proton signal [6]. Both these techniques suffer from some drawbacks, which are mainly the high power which must be delivered to the sample in the case of the Overhauser effect and the good excitation field homogeneity necessary to satisfy the Hartmann-Hahn condition.

More efficient techniques to transfer polarization from one nuclear species to another consist in employing sequences of pulses, which are in general selective with respect to the heteronuclear *J*-coupling interactions [7,8]. The first of these techniques to be introduced has been called SPI [9]. Its major limitations deal with the employment of selective pulses. To overcome these drawbacks the INEPT sequence [10] has been proposed; only hard pulses are used, so that the technique is effective over a broad range of proton frequencies and requires no *a priori* knowledge of the proton chemical shifts; in this technique the rare-nucleus signal is recorded while the proton frequency is irradiated. It is also possible to perform an indirect detection, where the proton signal is observed under a rare-nucleus excitation; this method is usually referred to as «reverse INEPT». A further progress of the polarization transfer techniques is DEPT [11], whose major advantages over INEPT are the lower distortions of relative intensities of the coupled spectra in comparison to those observed by a standard observation. Both techniques can be used for imaging purposes [12, 13], and produce a substantial increase in the signal-to-noise ratio. The last step in the development of polarization transfer techniques as a method to increase the signal-to-noise ratio is to perform first a transfer from the abundant nuclei to the rare ones, and second to transfer polarization back to the first; this results in an indirect observation of the rare nuclei, retaining the advantages of a double-magnetization transfer.

It is possible to give, at least qualitatively, the signal-to-noise ratio enhancement dependence in a double-resonance experiment, where two nuclear species I and S are involved, and in which any one of the two species is polarized or observed or both, reasoning as follows. The signal-to-noise ratio can be expressed as [14]

$$(1) \qquad \frac{S}{N} \propto \gamma_{exc} \gamma_{obs}^{3/2} \{ 1 - \exp[-T_R/T_1^{(exc)}] \},$$

where γ_{exc} and γ_{obs} represent, respectively, the gyromagnetic ratios of the nuclear species excited and observed, $T_1^{(exc)}$ is the spin-lattice relaxation time of the excited nuclei and T_R is the repetition time of the pulse sequence. The factor in braces takes into account the fact that, as hydrogen possesses the shortest T_1 among the nuclei naturally present *in vivo*, in an indirect observation a shorter recovery time can be employed without decreasing the polarization level of the observed species, thus increasing the signal-to-noise ratio per unit time.

Apart from the saturation factor, which can be examined separately, the signal-to-noise ratio enhancement that can be obtained employing the various techniques can be evaluated from eq. (1); a direct observation of the rare nucleus correponds to having both γ_{exc} and γ_{obs} equal to the γ of the rare nucleus, and the signal-to-noise ratio is proportional to $\gamma_S^{5/2}$. In INEPT or in DEPT the observed nucleus is the rare one as well, but polarization is transferred from the abundant nucleus, so that the factor related to γ_{exc} is given by the gyromagnetic ratio of the abundant nucleus, γ_I, and the signal-to-noise ratio is thus given by $\gamma_I \cdot \gamma_S^{3/2}$; the resulting enhancement is, therefore, given by γ_I/γ_S.

If the situation is reversed, *i.e.* the abundant nucleus is observed and polarization is transferred from the rare nucleus to the abundant one, then γ_{obs} is the gyromagnetic ratio of the abundant nucleus, while γ_{exc} is the rare-nucleus one, so that the enhancement of the signal-to-noise ratio is given by $(\gamma_I/\gamma_S)^{3/2}$. The last experiment that can be considered is to transfer polarization twice: the first time from I to S and successively from S back to I, so that both γ_{exc} and γ_{obs} are equal to the γ of the more sensitive I nucleus. We obtain, therefore, a total enhancement of the signal-to-noise ratio equal to $(\gamma_I/\gamma_S)^{5/2}$.

In the next section we shall describe an original method which allows one to obtain in a simpler way the same maximum enhancement of the signal-to-noise ratio.

3. – *J*-coupling imaging.

The method which is subsequently described[15] exploits the different behaviour of the spin-echo envelope[16] of the I spins of a J-coupled I-S system, which appears when also the S spins are excited with a 180° pulse applied at the same time of the 180° one on the I spins. The double-resonance sequence appears formally like the one employed in SEDOR[17], described by

$$I \quad \frac{\pi}{2}\text{-}\tau\text{-}\pi\text{-}\tau\text{-echo}, \quad S \quad 0\text{-}\tau\text{-}\pi,$$

and produces a cosine modulation of the spin-echo envelope at a frequency given by the J-coupling constant between the I and the S nuclei and with a modulation width which depends on the characteristics of the molecule. For τ values such that $\tau = \pi/J$ the echo amplitude exhibits a zero or at least a minimum; at that τ the signal contributed by the I spins coupled to the S spins, therefore, has a maximum difference with the signal originating from a conventional spin-echo experiment, which originates from all the I spins in the sample. A subtraction of the two signals, therefore, cancels the signal of the uncoupled I spins, which is the same for both sequences, and leaves only signal from the coupled I spins; the same consideration applies to the difference of two images, the first being obtained with a spin-echo excitation sequence, and the second with a SEDOR-like sequence.

For a two-spin system, which will be referred to as I and S, in which both a homonuclear and a heteronuclear J-coupling are present, the J-coupling interaction Hamiltonian can be written as

$$H_J \simeq J^{IS} \sum_{i,r} \boldsymbol{I}_i \cdot \boldsymbol{S}_r , \tag{2}$$

where $J^{IS} = (2\pi/3)\,\mathrm{Tr}\,\{\tilde{J}^{IS}\}$ is the trace of the coupling tensor and the sum is extended to all the J^{IS} coupled nuclei of the same molecule. The heteronuclear coupling has been here supposed to be greater than any homonuclear coupling. For such a system the transverse magnetization is

$$\langle I_+(t)\rangle \propto \sum_c \mathrm{Tr}\,\{\sigma_c(t) I_+^c\} + \sum_u \mathrm{Tr}\,\{\sigma_u(t) I_+^u\} , \tag{3}$$

where u denotes the uncoupled spins, c the J^{IS} coupled spins and $\sigma(t)$ the density matrix. The density matrices, at thermal equilibrium, will be, respectively, proportional to I_z^c and I_z^u, the Zeeman field being defined by $B_0\hat{z}$. At a time $t = 0$ immediately after a $\pi/2$ pulse applied to I spins, the transverse magnetizations will be given by $\sigma_c(0) = I_y^c$ and $\sigma_u(0) = I_y^u$ so that, after a time τ, the evolution of the two matrices can be described by

$$\begin{cases} \sigma_c(\tau) = \{\exp[i[(\gamma_I I_z^c + \gamma_S S_z)(\boldsymbol{G}\cdot\boldsymbol{r}_c)\tau - J^{IS} I_z^c S_z \tau]]\} I_y^c \{\ \}^{-1} , \\ \sigma_u(\tau) = \{\exp[i[\gamma_I I_z^u(\boldsymbol{G}\cdot\boldsymbol{r}_u)\tau]]\} I_y^u \{\ \}^{-1} , \end{cases} \tag{4}$$

where $\{\ \}^{-1}$ indicates the inverted left-hand side exponential. $\boldsymbol{G}$ is the applied magnetic-field gradient and $\boldsymbol{r}_c$ and $\boldsymbol{r}_u$ represent, respectively, the spatial position of the coupled and uncoupled I spins. The application of two π pulses at a time τ after the $\pi/2$ pulse, on both nuclear species, gives at a time $t_1 = t - \tau$

$$\begin{cases} \sigma_c(t_1) = \{\exp[i[(\gamma_I I_z^c + \gamma_S S_z)(\boldsymbol{G}\cdot\boldsymbol{r}_c) t_1 - J^{IS} I_z^c S_z t_1]]\} \cdot \\ \qquad\qquad \cdot R_{IS}(\pi)\,\sigma_c(\tau)\,R_{IS}^{-1}(\pi)\{\ \}^{-1} , \\ \sigma_u(t_1) = \{\exp[i[\gamma_I I_z^u(\boldsymbol{G}\cdot\boldsymbol{r}_u) t_1]]\}\, R_I(\pi)\,\sigma_u(\tau)\,R_I^{-1}(\pi)\{\ \}^{-1} , \end{cases} \tag{5}$$

where R_{IS} represents the rotation operator associated to the application of the two π pulses and R_I the rotation operator of the π pulse on spins I. At a time $t_2 = t - 2\tau$ we find

$$\begin{cases} \sigma_c(t_2) \simeq \{\exp[i[\gamma_I I_z^c(\boldsymbol{G}\cdot\boldsymbol{r}_c) t_2]]\}[-I_y^c \cos(J^{IS}\tau)]\{\ \}^{-1} , \\ \sigma_u(t_2) = \{\exp[i[\gamma_I I_z^u(\boldsymbol{G}\cdot\boldsymbol{r}_u) t_2]]\}[-I_y^u]\{\ \}^{-1} , \end{cases} \tag{6}$$

whilst, if only the I spins are irradiated, the density matrix is given just by

$$\sigma_{u,c}(t_2) = \{\exp[i[\gamma_I I_z^{u,c}(\boldsymbol{G}\cdot\boldsymbol{r}_{u,c})t_2]]\}[-I_y^{u,c}]\{\quad\}^{-1}. \tag{7}$$

If, as happens in practice, the spins are distributed with normalized density, the difference between the two spatially encoded spectra is given by

$$\langle I_+(\omega)\rangle_I - \langle I_+(\omega)\rangle_S = \tag{8}$$

$$= (1 - |\cos(J^{IS}\tau)|) \int_V \rho_c(\boldsymbol{r}_c) \exp[i[\gamma_I(\boldsymbol{G}\cdot\boldsymbol{r}_c)t + \omega t]]\,dv\,dt\,,$$

where $\langle I_+(\omega)\rangle_I$ is the Fourier transform of the signal eq. (3) obtained having irradiated only the I spins, i.e. with the density matrix given by eq. (7), and $\langle I_+(\omega)\rangle_S$ the Fourier transform of the signal eq. (3) after a double irradiation, i.e. with the density matrix given by eq. (6), V the sample volume and $\rho_c(\boldsymbol{r}_c)$ the J^{IS} coupled-spin normalized density.

The difference image given by eq. (8), although representative of the rare nuclei, is proportional to the density of the coupled abundant nuclei. In other words, the short-range nature of the J-coupling interaction constrains ρ_c to be representative of the density of the low-sensitivity and/or low-gyromagnetic-ratio S nuclei; conversely, from the signal-to-noise ratio point of view, the spatial distribution of the S spins is seen as if it were characterized by the coupled-I-nuclei density.

The method gives a substantial increase in the signal-to-noise ratio whenever the condition

$$\frac{I(I+1)}{S(S+1)}\left(\frac{\gamma_I}{\gamma_S}\right)^{11/4}\left(\frac{\rho_c}{\sqrt{2}}\right)(1-|\cos J^{IS}\tau|) \geq \rho_S \tag{9}$$

is fulfilled; eq. (9) has been derived rewriting the signal-to-noise ratio for a coil made of wire, in which the dominant source of noise is the coil resistance in the form of skin effect [2].

4. – Results and conclusions.

To test the method we have built a phantom formed with three tubes, the first one filled with a mixture of H_2O and D_2O, the second with a 0.2 M solution of $Li_2 B_{12} H_{12} \cdot 4H_2O$ in D_2O and the third with a 0.1 M solution of the same compound (fig. 2b)). The interest for such a molecule is due to the fact that it is an intermediate preparation compound for BSH, a ^{10}B compound employed in boron neutron capture therapy [18]; BNCT is a method for cancer therapy which consists in injecting boron-10 atoms into the tumour, and then irradiating them with thermal neutrons, with the final result of a destruction of the tumoral cells from the short-range products of the neutron capture process. This

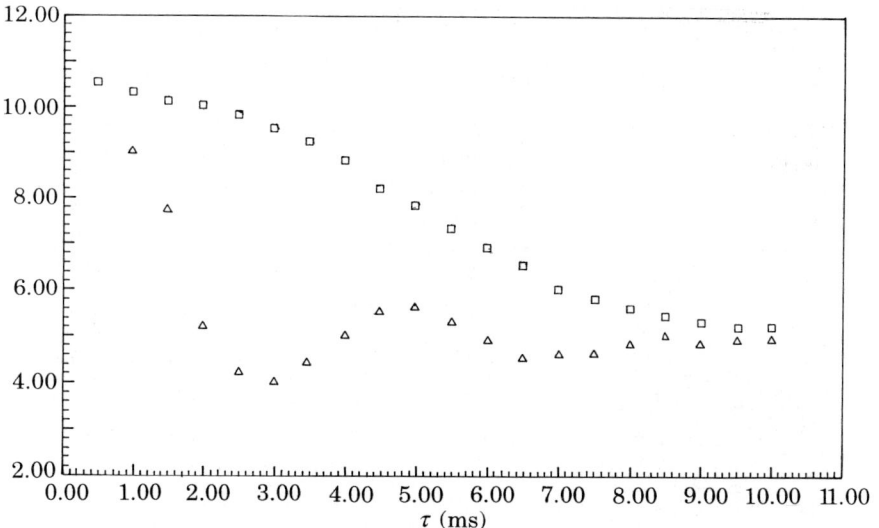

Fig. 1. – Spin-echo amplitude dependence (a.u.) vs. the interpulse delay for a spin-echo (squares) and a SEDOR-like (triangles) sequence. The modulation has a cosine behaviour and exhibits a minimum for $\tau = 2.6$ ms, and for such a value the difference between the two amplitudes is, therefore, maximum.

compound cannot be directly imaged for several reasons: a low gyromagnetic ratio, that is approximately nine times smaller than the hydrogen one, a strong quadrupole moment and, finally, the low concentration at which it is employed; on the other hand, it would be of great importance to be able to obtain a map of the spatial distribution of this compound after its injection into the body, to check where it concentrates and how it binds to the tumour.

First of all we obtained the echo amplitude modulation with a normal spin-echo sequence and with a SEDOR-like sequence: the results are shown in fig. 1. The SEDOR echo amplitude exhibits a minimum for $\tau = 2.6$ ms [19]. For this value of τ the images that are shown in fig. 2 were obtained: the two upper images are, respectively, a normal spin-echo image (left) and a J-modulated image obtained with a SEDOR-like sequence (right); the third image (bottom) is the difference of the two previous images; the two upper images are (75 × 75) bidimensional representations reconstructed from 20 projections, each one collected for 6 scans, with a repetition time of 6 s, a gradient strength of 0.022 T·m^{-1} and a Zeeman field intensity of 0.65 T.

The quality of the images is clearly affected by the low number of projections; nevertheless it is evident in the difference image that only the tubes where the boron compound is present are visible, while the water one disappears, showing that the result is a representation of the spatial distribution of the ^{10}B compound. It has not been possible to perform a quantitative evaluation of the signal-to-noise ratio enhancement obtained with the presented method,

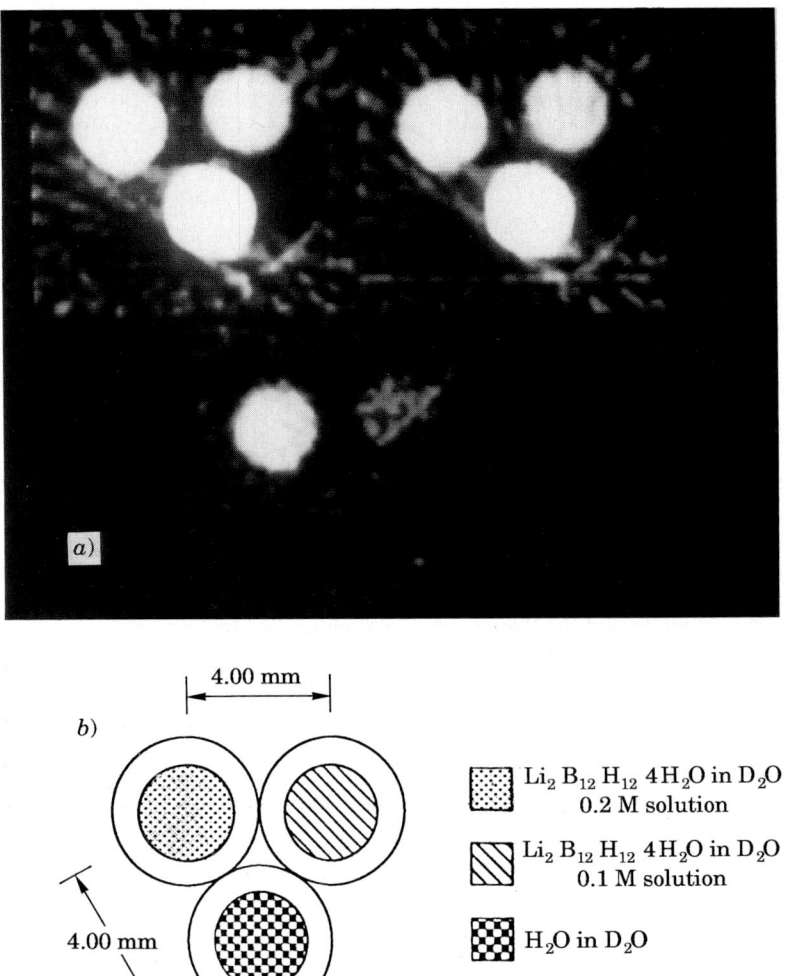

Fig. 2. – a) Top: spin-echo image (left) and J-modulated image obtained with a SEDOR-like sequence (right) of the phantom shown below; bottom: difference image of the two previous ones; the images are bidimensional representations of 75×75 pixels, reconstructed from 20 projections, collected from 6 scans each, with a repetition time of 6 s, an echo time of 5.2 ms; the gradient strength is $0.022\,\text{T}\cdot\text{m}^{-1}$ and the Zeeman field intensity 0.65 T. In the difference image only the tubes where the boron compound is present are visible, while the water one disappears; the result is, therefore, a representation of the spatial distribution of the ^{10}B compound, with a signal-to-noise ratio given by the J-coupled ^{1}H nuclei. b) Details of the construction and of the content of the three tubes that constitute the phantom employed for the images of a).

expressed by eq. (9), because the ^{10}B signal was not detectable, due to its low gyromagnetic ratio and to the strong quadrupole moment.

The grand result that we demonstrated was that, exploiting the intramolecular scalar coupling interaction, it is possible to reveal the spatial distribution of low-concentration or low-gyromagnetic-ratio nuclei, with a relevant signal-to-noise ratio enhancement, with a technique whose implementation is very easy on any tomograph, and whose efficiency does not depend on the main-field homogeneity; the method could be applied to many areas of interest, and seems to be very promising, for instance, for the imaging of ^{13}C.

REFERENCES

[1] A. ABRAGAM: *The Principles of Nuclear Magnetism* (Clarendon Press, Oxford, 1981).
[2] D. I. HOULT and R. E. RICHARDS: *J. Magn. Reson.*, **24**, 71 (1976).
[3] A. W. OVERHAUSER: *Phys. Rev.*, **92**, 411 (1953).
[4] T. R. CARVER and C. P. SLICHTER: *Phys. Rev.*, **92**, 212 (1953).
[5] S. R. HARTMANN and E. L. HAHN: *Phys. Rev.*, **128**, 2042 (1962).
[6] D. J. LURIE, D. M. BUSSEL, L. H. BELL and J. R. MALLARD: *J. Magn. Reson.*, **76**, 366 (1988).
[7] N. J. RAMSEY and E. M. PURCELL: *Phys. Rev.*, **85**, 143 (1951).
[8] E. L. HAHN and D. E. MAXWELL: *Phys. Rev.*, **88**, 1070 (1952).
[9] K. G. R. PACHLER and P. L. WESSELS: *J. Magn. Reson.*, **12**, 337 (1973).
[10] G. A. MORRIS and R. FREEMAN: *J. Am. Chem. Soc.*, **101**, 760 (1979).
[11] D. M. DODDRELL, D. T. PEGG and M. R. BENDALL: *J. Magn. Reson.*, **48**, 323 (1982).
[12] Y. VESHIMA, S. YAMAI, H. IKEIRA, T. HASHIMOTO, K. MORI, T. MAKI, H. FUKUDA and Y. TATENO: *Magn. Res. Med.*, **15**, 158 (1990).
[13] H. N. YEUNG and S. D. SWANSON: *J. Magn. Reson.*, **83**, 183 (1989).
[14] R. R. ERNST, G. BODENHAUSEN and A. WOKAUN: *Principles of Nuclear Magnetic Resonance in One and Two Dimensions* (Oxford University Press, Oxford, 1987).
[15] F. DE LUCA, R. CAMPANELLA, A. BIFONE and B. MARAVIGLIA: *J. Magn. Reson.*, **93**, 554 (1991).
[16] E. L. HAHN: *Phys. Rev.*, **80**, 580 (1950).
[17] C. P. SLICHTER: *Principles of Magnetic Resonance*, III edition (Springer-Verlag, Berlin, 1989).
[18] R. G. FAIRCHILD, V. P. BOND and A. D. WOODHEAD: *Clinical Aspects of Neutron Capture Therapy* (Plenum Press, New York, N.Y., 1989).
[19] F. DE LUCA, R. CAMPANELLA, A. BIFONE and B. MARAVIGLIA: *Chem. Phys. Lett.*, **186**, 303 (1991).

Some Double-Resonance Methods in Imaging Experiments.

E. W. RANDALL

Chemistry Department, Queen Mary and Westfield College, University of London Mile End Road, London E1 4NS, U.K.

Introduction.

Our Director has invited speakers to give historical introductions to their talks and also to bring along background material such as papers already published or those about to be. I have done both and venture to suggest some new experiments. I am also conscious that this is a school and that, therefore, some pedagogical eccentricity is not out of place. The introduction, therefore, takes on the aspect of a retrospective.

In my case I shall with your indulgence take the historical perspective back to 1959 when my own first NMR experiments were with both homonuclear and heteronuclear cases mainly in liquids. I had arrived at Harvard to work as postdoctoral Fellow with Prof. E. ROCHOW, an eminent inorganic chemist, but additionally I soon started a collaboration with J. BALDESCHWIELER (newly arrived as an Instructor) on nuclear magnetic double resonance. It was a topic he had determined to work on and which I had never heard of, but to which I was able to give some chemical insights. Later in 1963 we published the first review on the topic [1], at least so far as chemical applications were concerned.

In 1961 I went to a series of lectures in the Physics Department given by the Morris Loeb Lecturer of the day: one CH. SLICHTER. To a chemist just starting out on NMR they were a revelation. I have and cherish my notes to this day. I would always recommend his book which grew out of the lectures to any scientist interested in NMR. It is now in its third (enlarged) edition [2] and in my view has the widest perspective out of all the very good texts now available [3, 4].

1. – Some thoughts on definition.

It was apparent from the titles of the talks for the school that different authors had different operational definitions of «NM double resonance» which is the title of the school. Accordingly I offered some thoughts on the definition (or definitions) of this term. These I have modified twice: once from my submitted and distributed text for my presentation to the talk I actually presented; and now the second time in the light of the discussions and talks other authors gave.

In a sense it does not matter whether an interesting piece of work is classified as strictly being an NMDR experiment or not, but it seems to me to be intellectually unsatisfying not to have a good definition especially at the end of a very good advanced school.

Let us start with one approach which is based upon double irradiation (or excitation) of the sample. Double irradiation is not confined to NMR, but let us say that at least we conduct *one* NMR experiment (single resonance) with B_0 as the Zeeman field and B_1 as the oscillating field at frequency ν_1 and phase ϕ_1, and then repeat it with some second excitation of the sample. This excitation could involve electromagnetic radiation (B_2, ν_2 and ϕ_2) in any region of the electromagnetic spectrum, *e.g.* infrared, visible or ultraviolet radiation. These experiments which in some cases can lead to the CIDNIP effect (chemically induced nuclear polarization), for example, are double-*irradiation* experiments but obviously are not double-*resonance* (DR) experiments.

The DR experiments should involve at least two *resonances* by which we mean matching the static field B_0 both with the electromagnetic field (B_1, ν_1) *and* with (B_2, ν_2). The minimum number of levels involved is three (see the lecture of Mehring) but can involve more (see [2]). If the second resonance (in addition to the nuclear one) involves electron spins, then we come to ENDOR (electron-nuclear double resonance). Finally we come to *nuclear* magnetic *double* resonance (NMDR) (which also includes INDOR) where both resonances are nuclear(*).

A related alternative definition is to be found in Slichter's book and talks, namely the observation of one NMR transition while simultaneously irradiating another. But it turns out not to be the same thing as the DI version. This I accept is a good operational definition *if* one interprets simultaneously to be «in the same experiment» and not *instantaneously*.

Both definitions include trivial cases which can be accommodated in *another* additional classification, which I shall advance here, and both exclude cases which should, I suggest, be properly included.

(*) It is worth noting here that B_0 may not be constant nor weakly swept but may be changed drastically in a field-cycling experiment. ABRAGAM and PROCTOR, for example, looked at LIF for which they produced a nonequilibrium state at high field, allowed the magnetization to evolve at a much lower field and then observed resonances at high field again [5].

Take *case 1*: that of a «single spin», *i.e.* a system where the spins are equivalent and give only one line, say the ^{1}H signal from liquid water. In a single-resonance experiment we obtain one line. If, however, we apply two radiofrequency fields (ν_1, B_1) and (ν_2, B_2) and sweep B_0, we obtain *two* signals. Under the DI definition we have a double-resonance experiment. Under the unmodified CS definition we do *not*: we have *two single-resonance* experiments, since we are not exciting the resonances simultaneously even if we sweep the field slowly enough. If, however, we switched the field quickly (in relation to T_1), we would get a result, *e.g.*, if B_1 was very large and produced saturation. This situation should be classed as a double-resonance experiment. The word «simultaneous» can go. The modified CS definition seems fine.

Case 2 is where we have a single 90° pulse followed by Fourier transform of the free-induction delay of a system consisting of, say, four different (chemically shifted) lines from protons which are not J coupled and not exchanging in a liquid. The spectrum obtained is simply four lines (and the same result is obtained by a c.w. sweep method). Under the CS definition, even as modified, since all transitions are excited simultaneously this is a multiple-resonance experiment! With the DI definition, however, this is a single-resonance experiment, since only one irradiating field R_1 is used.

It seems only sensible to classify case 1 as *two single*-resonance experiments for slow-passage c.w. (or pulse)(*) and case 2 as *four single*-resonance experiments conducted simultaneously.

What is missing in each definition is the element of *interaction* between the two excitations. There are other examples where an interaction is missing.

One concerns nearly all experiments now conducted by (or for) chemists on liquid-state systems. Normally, despite the very high field stability of superconducting magnets, spectrometers use a «channel» (the locking channel) to stabilize the magnetic field. Almost exclusively this involves a second type of nucleus to that being investigated and it should be inert in the sense that it should not interfere with the main business of the experiment which in the simplest case involves only R_1 (either c.w. or pulsed). Clearly the locking does not increase the dimensionality of the experiment: it is a single-resonance experiment independent of the second experiment being conducted which could be in one or two (or more) frequency dimensions.

If we return to the simplest energy level system (*i.e.* composed of three levels) involving *two* transitions, then there *must* be some interaction. However, there are cases (the *original* cases) where, for example, we have *four* (or more) levels. The minimum number of *transitions* involved is still only *two*, but the

---

(*) Unless there is a Bloch-Siegert shift!

cardinal point is that one has an effect on the other. There are two possibilities: population (polarization) or coherence. In either case, therefore, one needs

  i) an interaction,
  ii) an excitation,
  iii) an observation.

The definition, therefore, is simply that double resonance occurs when excitation of one NMR transition produces an observed effect on another NMR transition.

2. – Classification.

An early attempt to classify NMDR experiments within my own experience is shown in table I. This classification is not as wide as any member attending this school would be expected to produce now.

TABLE I. – *Primary classification of* NM *double-resonance experiments* (*).

Mode	Zeeman field B_0	Observing radiation field, B_1	Irradiating radiation field, B_2	Notes
1) field sweep	vary (**)	fixed ν_1	fixed ν_2, B_2	ν_2 and B_2 incremented stepwise
2) frequency sweep	fixed	vary ν_1 (**)	fixed ν_2, B_2	ν_2 and B_2 incremented stepwise
3) INDOR	fixed	fixed ν_1	vary ν_2	ν_1, B_2 (and B_1) incremented
4) pulse + c.w.	fixed	pulse ν_1	fixed ν_2	ν_2 and B_2 incremented, vary α_1
5) multiple pulse + c.w.	fixed	multiple pulses	fixed ν_2	ν_2 and B_2 incremental combinations of various pulse angles α_1
6) multiple pulse	fixed	pulse ν_1	pulse ν_2	combinations of delays, pulse angles, α_1 and α_2

(*) Excludes field-cycling experiments.
(**) The sweep rate may be varied: slow passage and rapid passage.

In the early experiments, conducted by chemists at least, spectra were normally recorded by variation of B_0, whereas later (mode 2) B_0 was constant and ν_1 was swept. In each of these cases it was possible to increment ν_2 and B_2 and to record spectra for each pair of values. The results, of course, were multidimensional in frequency (and B_2) and were the forerunners of the 2fDFT experiments. Figure 1 shows an example taken from FREEMAN (one of the leading double resonators), which illustrates the point particularly well[7].

Table II shows some of the designations which have been used for experi-

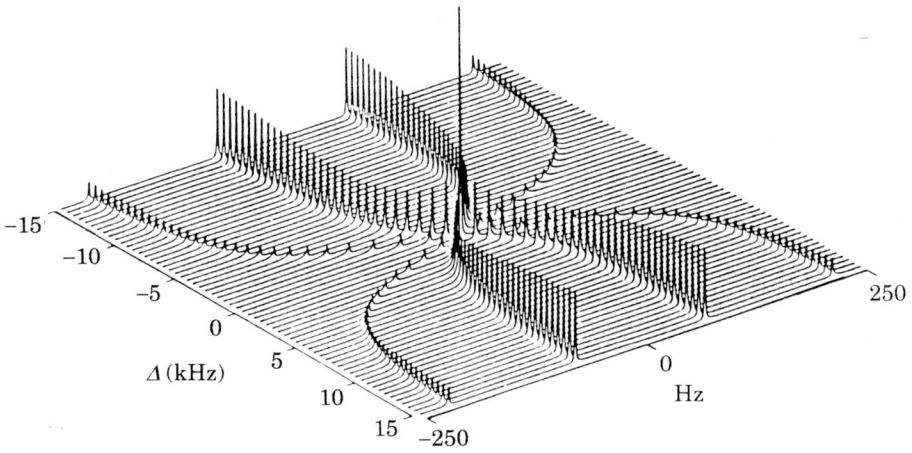

Fig. 1. – Simulated A-spin spectra of an AX_3 system with the decoupler applied to the X spins, as a function of decoupler resonance offset Δ. This simulation takes into account the spatial inhomogeneity of the B_2 field which broadens the lines and distorts the lineshapes.

TABLE II. – *Variations in B_2 and ν_2 (c.w. mode) liquids.*

Experiment		Notes
1) spin tickle (low B_2)	$\gamma B_2 \sim \Delta \nu_2$	ν_2 at a resonant line (L_1); incremented (through L_1, L_2, etc.)
2) decoupling (*) (high B_2)	$\gamma B_2 \gg J$	ν_2 at δ_2 (or δ_3, etc.)
3) intermediate	$\gamma B_2 \sim J$	ν_2 and B_2 incremented
4) off-centre double resonance (**)	$\gamma B_2 \gg J$	$\nu_2 \neq \delta_2$

Notes: $\gamma = \gamma/2\pi$; $\Delta\nu_2$ is a linewidth
J = scalar coupling constant
δ_2 = chemical shift of nucleus 2 (δ_3 of nucleus 3, etc.)

(*) The bandwidth over which decoupling occurs is increased by use of noise modulation or other techniques such as composite pulses and can cover δ_2, δ_3, δ_4, etc. in multinuclear systems, or for solutes dissolved in liquid crystals.
(**) Also known as «J scaling».

ments made possible in this way. Each one type of experiment is simply a subset of a whole multidimensional array. The full arrays were rarely mapped out completely since the experiments were time consuming and for most purposes

TABLE III. – *Some uses of double resonance: A{X}.*

	Notes
1) simplification of spectra of A	removal of *a)* quadrupole broadening *b)* coupling *c)* exchange broadening, through J (but not through δ)
2) improvement of S/N ratio for A	*a)* multiplet collapse *b)* nuclear Overhauser effect or magnetization transfer
3) INDIRECT observation of X	particularly if X has low receptivity, *e.g.* ^{15}N
4) determination of signs of couplings	relative signs only
5) assignments of A	if assignments for X are known
6) nuclear Overhauser effect	useful for studies of *a)* relaxation mechanisms, *b)* motion, *c)* internuclear distances
7) Measurement of T_1 values through gated decoupling experiments, and growth or decay of the Overhauser effect	

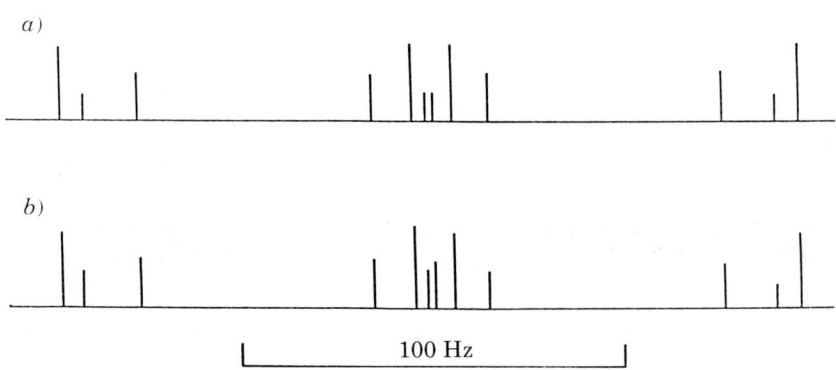

Fig. 2. – ^{1}H{^{15}N} spin tickling determination of the ^{15}N spectrum of formamide-^{15}N: *a)* theoretically calculated and *b)* experimentally determined.

only limited information was sought in any case. Table III shows some of the uses. The spin tickle experiment commonly was used to determine the relative signs of the various J's characterizing the system (an important fashion of the day), as could experiments of type 3. More importantly this type of experiment allowed *indirect* measurement of the whole spectrum for nucleus 2 (indirect mode). Figure 2 shows the ^{15}N spectrum of formamide-^{15}N (laboriously) determined in this way [8]. The advantage here is the sensitivity gained by observation of proton resonances (high γ_1) rather than by direct measurement of ^{15}N itself (low γ_2). This ratio of γ's amounts to about 10 for the ^{1}H$\{^{15}$N$\}$ case (see below). We shall see such experiments used for imaging ^{14}N and ^{15}N subsequently. The simple c.w. experiments of tables II and III are illustrated in fig. 3-6 to amuse the innocent.

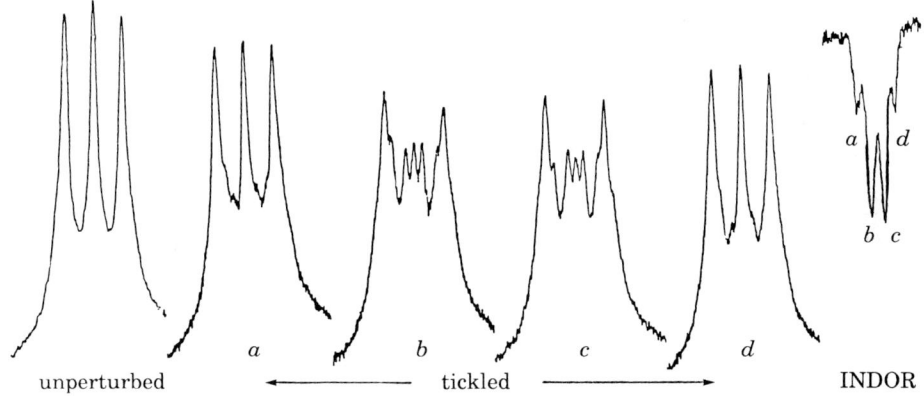

Fig. 3. – ^{1}H$\{^{14}$N$\}$ spin tickling and INDOR spectra of CH$_3$NC.

It is worth noting here that a useful symbolism was introduced (and used mainly by chemists) whereby for two nuclei the symbol AB indicated that they were of the same type ($\gamma_A = \gamma_B$; homonuclear case), whereas AX indicated that their resonance frequencies were far apart either because $\delta_1 - \delta_2$ was very large compared, say, to J_{AX} in the homonuclear case, or because $\gamma_A \neq \gamma_B$ (heteronuclear case). For double-resonance experiments the symbol for the second irradiated nucleus was enclosed in brackets [1]: initially $A\{X\}$ for field sweep, and later $A(X)$ for frequency sweep. In fact this subtlety concerning the use of the brackets has long gone since field sweep experiments in NMR are rarely used now unlike in ESR. The convention could, of course, be extended to include selective irradiations of separate lines (L) from the X nucleus simply by numbering them to produce, for example, $A(L_3)$.

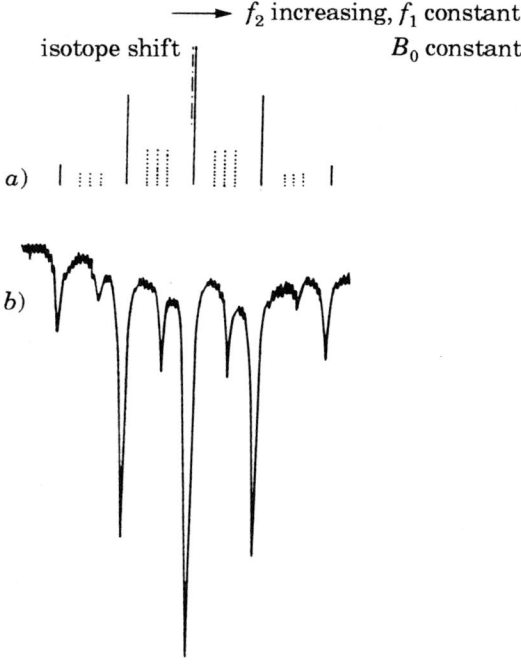

Fig. 4. – ^{15}N theoretical (a)) and experimental (b)) INDOR spectra of a mixture consisting mainly of $^{15}NH_4^+$ and $^{15}NH_3D^+$.

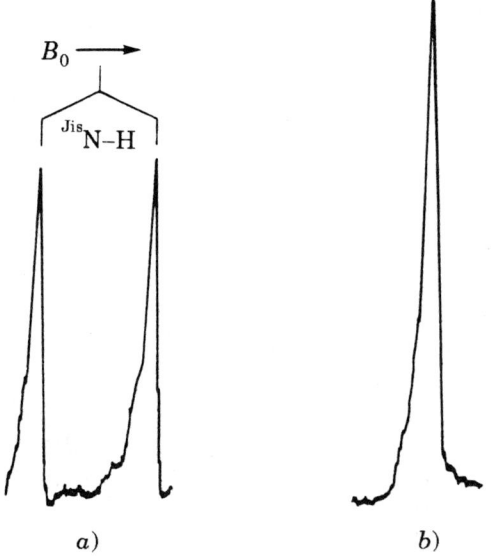

Fig. 5. – Spectra of the ^{15}NH proton in $(CH_3)_3Si^{15}NHC_6H_5$ at 40 MHz, a) without and b) with high-power irradiation at the ^{15}N resonance frequency.

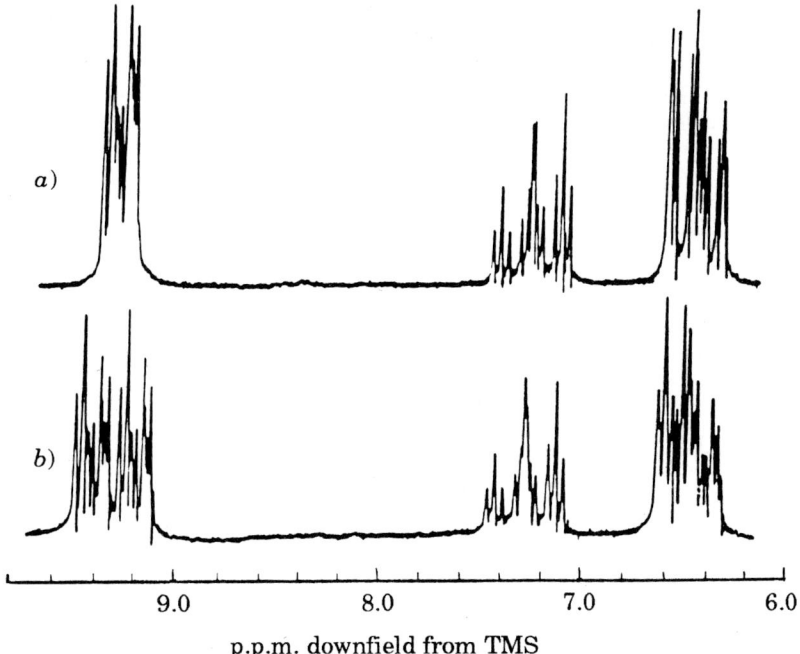

Fig. 6. – Spectra at 100 MHz of the α protons of pure pyridine-^{15}N: a) irradiated with high power at the ^{15}N resonance frequency of 10.13 MHz, and b) normal single-resonance spectrum.

Physicists (generally uninterested in the phenomenon of the chemical shift(*)) normally used the two symbols I and S for spins of different types.

With the widespread adoption of pulse techniques in the seventies by chemists (initially for R_1 only), the attention turned to direct observation of difficult nuclei («other» nuclei) such as ^{13}C and ^{15}N, since there was a considerable sensitivity gain over c.w. techniques [9, 10]. The emphasis now for double-resonance experiments was simply the decoupling of protons from the nucleus of interest in experiments such as ^{13}C{^{1}H}, ^{15}N{^{1}H} and ^{2}D{^{1}H} [9-11](**). The sensitivity gains stemming from the use of the pulse method could be augmented by X{^{1}H} nuclear Overhauser effects in favourable cases. Particular difficulties arose with ^{15}N for which nulling of the resonance would occur [12].

(*) It was a nuisance since it complicated the determination of the ratios of γ_s for different nucleotides. A similar bother occurs in imaging where dispacements of images produced by shift effects are referred to as «chemical-shift artefacts»: just another nuisance.

(**) The legacy of this era is that many authors, including some who are eminent, still refer to nearly all double-resonance experiments as decoupling experiments.

There was then consequently little interest in B_2 powers at the intermediate or tickling level. However, one significant use of low-power, noise, double resonance had emerged early [13]: it was to *broaden* the X resonance in cases where $J(X\text{-H})$ was large leaving the X resonances for which J was small as sharp lines.

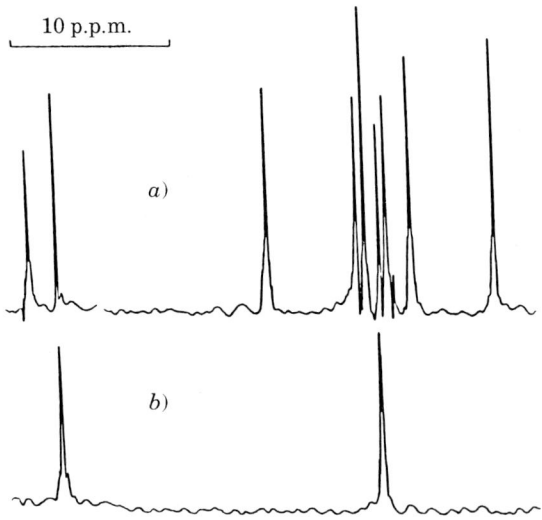

Fig. 7. – $^{13}C\{^1H\}$ spectrum of quinoline, *a)* with noise decoupling, *b)* with low-power noise.

This was one of the first editing techniques and enabled the assignment of resonances from quaternary ^{13}C centres [9, 13] (see fig. 7). These experiments would be referred to today as «nonquaternary suppression» techniques. Later we shall discuss the application of similar experiments in imaging. Despite this utility of intermediate B_2 values, there was almost no detailed attention paid to the $X\{^1H\}$ experiments in this regime. One of the exceptions was the experimental work done in my group, the results of which were interpreted with the aid of R. LYNDEN-BELL (and also A. BAIN). The basic approach proved interesting for its insights into the double-resonance Hamiltonian. The main contribution was the handling of the factorization of the population effect into two contributions arising from the tilt effect on the one hand and from relaxation effects on the other [14-16]. Some more recent applications are given in ref. [3, 17].

Some examples from the early FT era are reproduced in fig. 8 and 11. These are of the $^{15}N\{^1H\}$ type [18]. Additionally an example of the nonquaternary suppression technique in $^{13}C\{^1H\}$ double resonance with a pulsed ^{13}C field is given in fig. 7 [9]. To this is added an $^{15}N\{^1H\}$ example in the case of the $^{15}NH_4^+$ ion

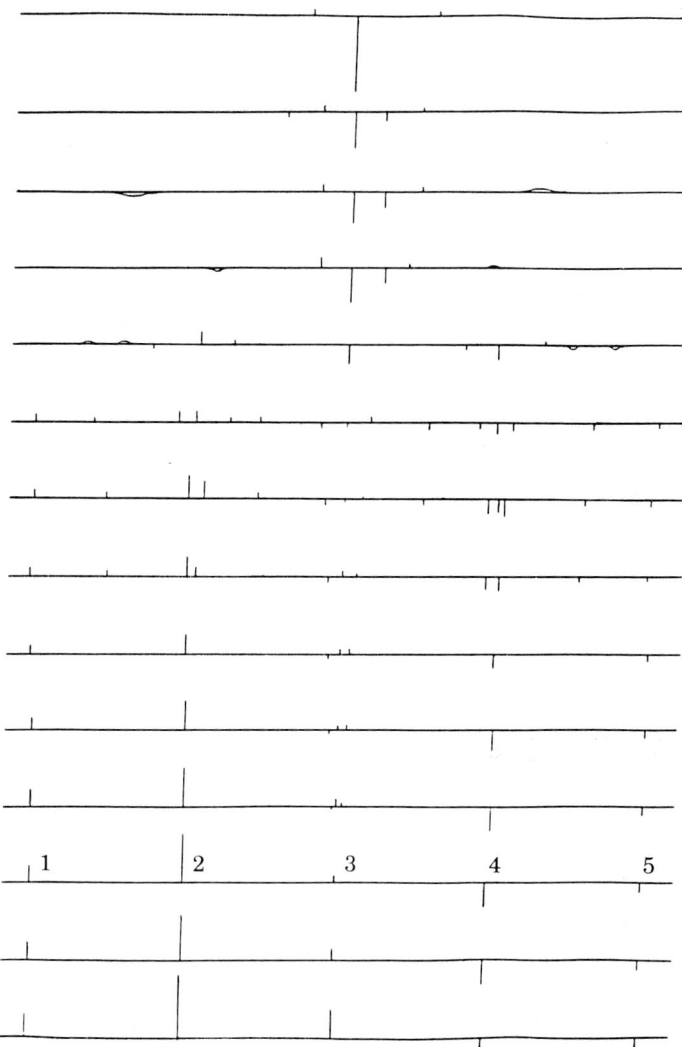

Fig. 8. – ^{15}N{^{1}H} double-resonance spectra of the ^{15}NH$_4^+$ ion (in 2M HNO$_3$). The power of the ^{1}H radiofrequency field increases from bottom to top. The numbers refer to the five lines of the ^{15}N quintet which in the single-resonance spectrum have the intensities in the ratio 1:4:6:4:1.

where B_2 is varied over a wide range (see fig. 8) [19]. The inversions of some lines which are seen occur also in cases where the magnetogyric ratios of the two spins concerned are of the *same* sign (*). The inversions, therefore, do not

---

(*) $^{15}\gamma$ and $^1\gamma$ have opposite signs.

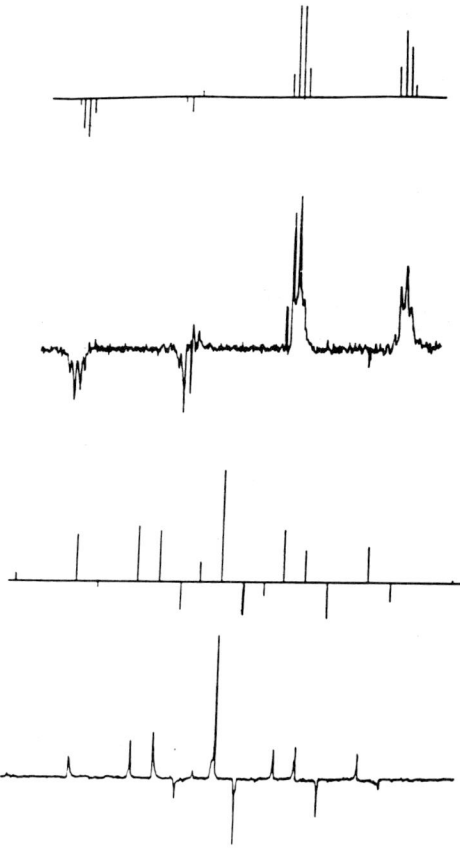

Fig. 9. – Observed and calculated ^{13}C{H$_3$} spectra of CH$_3$I. (Top) irradiation near a proton line, $\gamma B_2/2\pi J = 0.06$; (bottom) irradiation 20 Hz from centre: $\gamma B_2/2\pi J = 0.43$.

necessarily depend upon negative values for the nuclear Overhauser effects. Rather they occur because of the tilt effect on the population. This is shown in fig. 9, which illustrates some ^{13}C{^{1}H} results [14]. In general inversion of some lines can occur in an $A\{X\}$ experiment provided $\gamma_X > 2\gamma_A$ [14].

3. – Introduction to double-resonance imaging.

Most NMR imaging or localized spectroscopic studies are focussed on the ^{1}H and ^{31}P nucleotides, especially those investigations which are clinically oriented [4, 20]. Part of the reason is that these studies are relatively easy. With respect to nonclinical studies such as those declared to be material, biological or chemical studies, however, other nuclei come more easily into consideration. And indeed for medical studies it cannot be gainsaid that nuclei such as ^{13}C and

^{14}N (or ^{15}N) are not important, even if such studies are not yet prevalent in imaging.

3˙1. *Nitrogen.* – Thirty years ago (or more) the situation was not very different for spectroscopic studies of nitrogen or even carbon even though these were not driven by the health-care industry as the manufacturing market for imaging spectrometers was and to some extent still is. At that time I acquired, accidentally, an interest in what became referred to as «other nuclei». «Other» in this context means «other than protons» (*). The experimental means were then (because of the limitations of commercially available instrumentation) through double-resonance techniques and the «indirect» methods: ^{1}H$\{X\}$. Nitrogen assumed a personal (but not obsessive) importance. It was an attractive element since it had two magnetically active nuclear isotopes, one with $I = 1/2$, the other with $I = 1$. There was even a change of sign in the magnetogyric ratio to interest the spectroscopist. By contrast carbon, fluorine and phosphorus could boast only one magnetically active nucleotide each: ^{13}C, ^{19}F and ^{31}P, respectively. Boron had two but both were quadrupolar ($I = 3/2$ for ^{11}B and $I = 3$ for ^{10}B) and had the same sign for the magnetogyric ratios. The advent of higher-sensitivity FT techniques and their application to nitrogen studies [10, 18] laid bare the possibilities of direct-detection methods but even these appeared useless to some. The only response to the question «why?» was a facetious «because it is there». The first FT studies of ^{15}N at natural abundance appeared in 1971 [10] and thereafter they rapidly increased in number. Nowadays spectroscopic studies involving ^{15}N at natural abundance have become routine, even those involving 2fDFT work on large molecules (so I remember when colleagues thought such ^{15}N spectroscopic studies seemed rather bizarre!).

It was against this background that it seemed to me a year or two ago only sensible to start imaging nitrogen. It was bizarre enough! Not only «is it there», that is to say intrinsically interesting, but nitrogen is of considerable (not to say fashionable) interest in, for example, metabolic processes occurring in many subjects (human, animal, plant and microbial), and also in environmental transformations not least those concerning fertilizers which intimately involve nitrogen. This, however, is only by way of justification for as CH. SLICHTER has written [2] concerning the sensitivity problem in NMR: «sooner or later all resonators want to observe a resonance which is too weak to be seen». The next question which I have added is simply «can it be imaged?».

The basic problems concerning the nitrogen nucleotides are the same for imaging as for spectroscopy. ^{14}N is abundant, has $I = 1$, and hence gives broad

---

(*) This phrase is almost as tasteless as the use of «pseudocontact» interactions to denote anything other than Fermi contact effects.

lines (hundreds of Hz or more in liquids) because of the normally rapid electric quadrupolar relaxation mechanism. ^{15}N, on the other hand, has $I = 1/2$, gives narrow lines, but has a very low natural abundance, 0.37%. Each nucleus has a low value for its magnetogyric ratio, and, therefore, a low receptivity. Either one deters the average chemist especially for studies in the solid state despite the early demonstration of the plausibility of such studies[21].

3˙2. *Imaging* ^{14}N. – In our imaging experiments we nevertheless started with ^{14}N because of its high abundance. The main question was: how broad a resonance could be imaged on our instrument(*) with such a modest maximum field gradient of 2 G/cm? The answer was not known since most studies involved ^{1}H or ^{31}P nuclei giving linewidths less than about 20 Hz even *in vivo* and no one had tried anything broader on our installation. The phantom we used was simple: a single vial but containing compounds (ammonium nitrate and sodium azide in aqueous solution) with the ^{14}N nucleus in four environments giving four different linewidths, *viz.*, 13 Hz, 130 Hz, 42 Hz and 24 Hz, arising from the ^{14}NH$_4^+$ ion, the terminal resonances of the N$_3^-$ ion, the central resonance of the N$_3^-$ ion and the NO$_3^-$ signal, respectively (given in order of chemical shift from low to high frequency). The linewidth limit, even with special windowing techniques, was found to be about 150 Hz and even then there were distortions of the image in the read-out dimension[22].

Increasing the field gradient makes broader lines accessible(**). At 10 G/cm, for example, we could reach about 700 Hz. These distortions have been overcome lately[24] by using only phase-encoding gradients and increasing the dimensionality of the experiment, so that the so-called read-out dimension, to which the distortions are confined, is not used.

At this point the utility of ^{1}H{^{14}N} in indirect methods became evident. Not only would the sensitivity be increased, but also the maximum width (at any gradient field strength) which could be imaged was extended simply because the nucleus being imaged (^{1}H) was less broadened by the ^{14}N electric quadrupolar relaxation than is the ^{14}N resonance itself. Of course, the method relies on the presence of a scalar interaction (J) between ^{1}H and ^{14}N. It is necessary though that the relaxation is not so fast that these nuclei are completely decoupled.

The method involves a simple *difference* technique: the difference spectrum between the ^{1}H single-resonance experiment and the spectrum for the ^{1}H{^{14}N} experiment. The difference comes only from ^{1}H nuclei J-coupled to ^{14}N. Since J

(*) A Sisco-200 operating at 4.7 T.
(**) We have shown this recently for lines of widths of the order of kilohertz by the use of gradients as large as 5 kG cm^{-1} in the stray field imaging (STRAFI technique), but for ^{1}H and not ^{14}N so far[23].

interactions are intramolecular, this difference comes only from molecules which contain ^{14}N. These spectra can be used, with appropriate application of gradients, to produce successively ^{1}H images, ^{1}H$\{^{14}$N$\}$ images and the difference image, which is effectively the indirect ^{14}N image.

There are more modern double-resonance techniques of the 2fDFT type which utilize only pulses in both R_1 and R_2. These are elegant and subtle and have been applied recently and very effectively in the case of ^{1}H$\{^{13}$C$\}$ by KNÜTTEL and KIMMICH and co-workers [25, 26].

The problem with imaging ^{13}C, ^{14}N, or ^{15}N indirectly through hydrogen by all methods is that proton resonances are far more ubiquitous and, therefore, additional techniques are required in order to filter out (or *edit*) the resonances of protons attached to ^{13}C, ^{14}N, or ^{15}N from other proton resonances. The situation is reminiscent of the «hidden proton» problem which was solved by d.r. methods for the detection of select proton resonances overlaid by others years ago [1]. Multiple-pulse and multiquantum techniques are the new and very powerful methods for accomplishing this editing of the proton signals. In addition to multiple-quantum filtering, these workers [25, 26] used cyclic polarization transfer and split-pathway compensation for the ^{1}H$\{^{13}$C$\}$ case.

None of these methods unfortunately is appropriate to the ^{1}H-^{14}N situation, however, since both the ^{1}H and ^{14}N resonances are generally broad because of the rapid ^{14}N relaxation (due to ^{14}N's electric-quadrupole moment). Broad lines render 2fDFT methods difficult at best and even impossible. This is simply because interpulse delays must be introduced to allow J evolution. This will not occur where partial decoupling is occurring because of either relaxation or exchange.

For the ^{1}H-^{14}N case the use of ^{1}H$\{^{14}$N$\}$ double resonance with ν_2 applied at the ^{14}N frequency *in the continuous-wave or equivalent pulse mode*, however, can be used successfully, in the following way [27].

Suppose we have a complex proton spectrum containing sharp lines (S), and lines (B_N) broadened by ^{1}H-^{14}N coupling and rapid ^{14}N relaxation, such that no splitting is observed but the decoupling is not complete. Completion of the decoupling process is then accomplished by a ^{1}H$\{^{14}$N$\}$ experiment which makes the B_N lines sharpen to S_N. Subtraction of the two spectra will yield only the difference, $B_N - S_N$. Repetition of this sequence in an imaging mode will yield only the difference image arising from $B_N - S_N$. This image is formed by protons coupled to ^{14}N and is equivalent to a direct ^{14}N image. The image will be intensified the more protons are coupled [25] but not in a simple way since the effect depends upon the extent of the broadening which will vary with the J value.

An example is shown in fig. 10. The phantom consisted of two concentric vials with formamide in the inner vial (o.d. 1.7 cm) and methanol in the annular space (o.d. 2.3 cm). The peaks in fig. 10 ii) *a*) from low to high field are: CHO, 8.2 p.p.m.; NH (cis), 7.5 p.p.m.; NH (trans), 7.3 p.p.m. The ^{14}N-^{1}H coupling

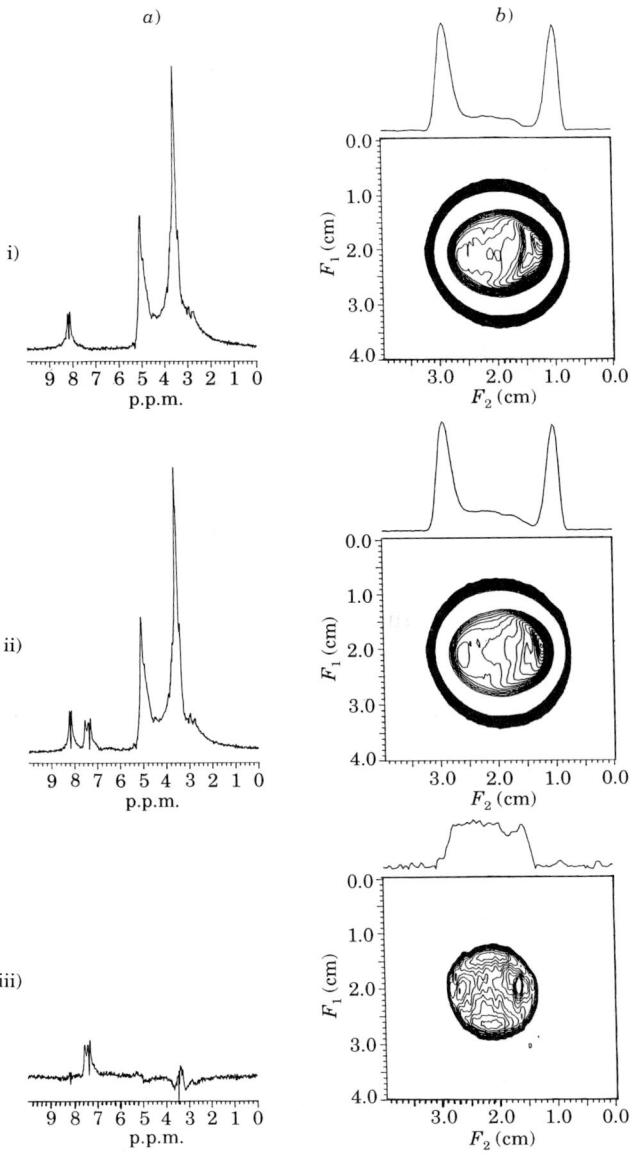

Fig. 10. – a) Magnitude reconstructed ^{1}H spectra (single acquisition) of a phantom consisting of two concentric vials (o.d. 1.7 cm and 2.3 cm) with formamide in the inner vial and methanol in the annular space. i) Single resonance. ii) With ^{14}N d.r. iii) The difference spectrum, ii) − i). b) Fourier 2D spin echo ^{1}H images without slice selection. A trace through the centre of each image is also shown. Echo time (TE)=20 ms, $\pi/2$ pulse width (PW)=240 μs. Number of transients per phase encoding step (NT) = 4. Number of phase-enconding steps (NPE) = 64. Number of data points collected in F_2 (NP) = 256. Repetition time (TR) = 3.023 s. Read gradient (G_R) = 1.5 G cm^{-1}. In-plane spatial resolution = = 1.250 mm × 0.313 mm. i) Single resonance, total imaging time = 12.9 min. ii) With ^{14}N d.r., total imaging time = 12.0 min. iii) The difference image, ii) − i).

constants, which do not appear in fig. 10 i) a) since the quadrupole relaxation rate is too fast, may be deduced from the ^{15}N-^{1}H values in the ^{15}N isotopomer. They are: ^{2}J(CH-N) = 10.3 Hz, ^{1}J(NH$_t$) = 62.0 Hz and ^{1}J(NH$_c$) = 65.1 Hz (*). These three lines are broadened by the rapid ^{14}N relaxation. The broadenings, however, are in the ratio of the J values, and the effect on the NH signals is that they are broadened beyond detection in the single-resonance case of fig. 10 i) a), whereas the CH signals can still be easily detected. There is, therefore, an effective T_2 weighting in the experiment.

It may be thought that in an ^{1}H{^{14}N} experiment there would be some gain from the nuclear Overhauser effect (nOe). Two factors mitigate against this: firstly the maximum value that can be expected is only 1.04 if the dipole-dipole mechanism dominates, and secondly this mechanism *never* dominates since the electric quadrupolar mechanism is always the most important [8].

3'3. Imaging ^{15}N.

3'3.1. Direct method. The first experiments we conducted concentrated on samples at *natural abundance* and used no enhancement techniques [28]. They were undertaken as basic «benchmark» experiments to which enhancement techniques could be added later. Three aids to reduce the sensitivity problem were tried initially, however: i) the samples were pure liquids, ii) the steady-state free-procession (SSFP) technique was used, and iii) in one case (aniline) the T_1 of the ^{15}N was reduced by addition of chromium acetylacetate as a relaxation reagent to the sample (**). Such reagents have been renamed in the imaging context as contrast agents (and a considerable industry has grown around them).

Long accumulation times of several hours were required as expected. This fact makes such simple experiments impossible in clinical situations and undesirable in general. Enhancement techniques, therefore, should be brought to bear.

3'3.2. ^{15}N{^{1}H} Overhauser method. In the direct mode the simplest enhancement is available with ^{15}N{^{1}H} d.r. experiments in which the Overhauser effect is used to advantage. The maximum value of the effect is − 3.93 for ^{1}H{^{15}N}. This is obtained in the extreme narrowing regime when the ^{15}N relaxation mechanism is purely dipolar. Nulling of the resonance can occur, however, in a number of circumstances: i) when the dipole-dipole contribution to the total relaxation is only 20%, ii) under certain off-centre d.r. conditions (see fig. 11), iii) if the correlation time is (un)suitably long such as happens for large molecules [12].

(*) ^{1}J indicates a coupling through *one* bond, ^{2}J through *two*.
(**) One characteristic of ^{15}N resonances is the prevalence of long T_1 values.

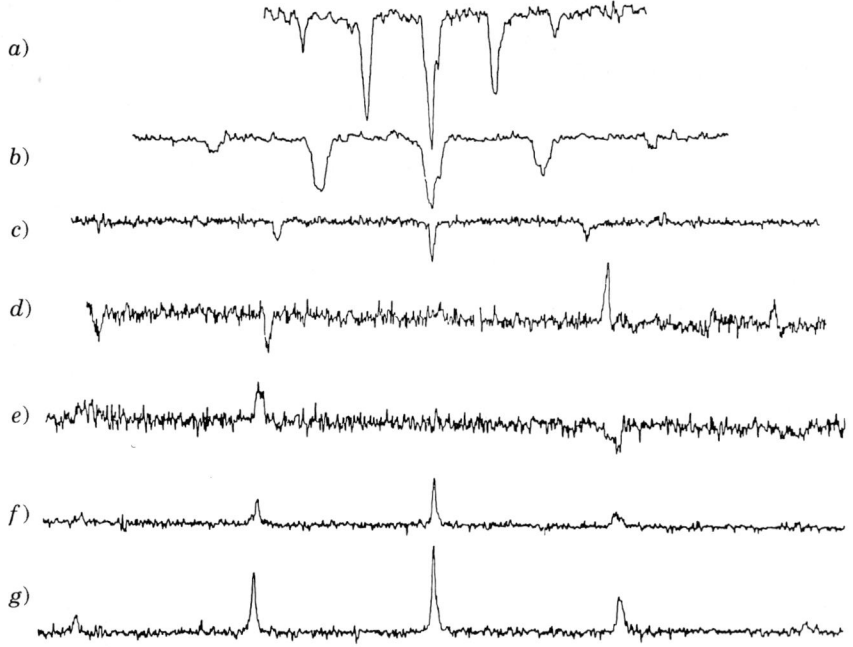

Fig. 11. – ^{15}N{^{1}H} «off-centre» double-resonance spectra of the ^{15}NH$_4^+$ ion (in 2M HNO$_3$). The ^{1}H radiofrequency power decreases from a) to g). In a) the signals are inverted by the nuclear Overhauser effect and exhibit a reduced separation compared to g), the single-resonance spectrum.

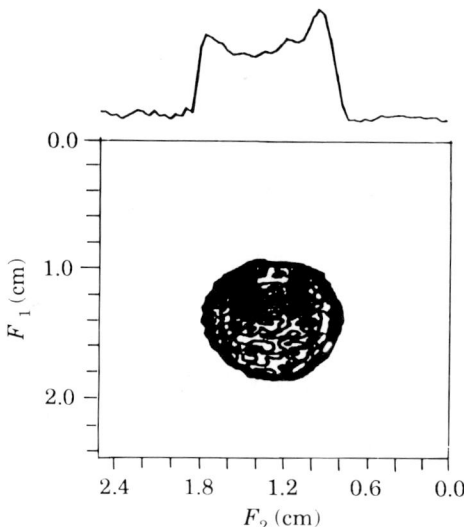

Fig. 12. – ^{15}N image and 1D profile of an ampoule containing aniline-^{15}N.

The first example which we reported, aniline-^{15}N, is interesting because in our sample the exchange processes cause partial collapse of the ^{1}H and the ^{15}N multiplets to broad lines[29]. These multiplets are observed only in very dry conditions[30] and exhibit the $^1J(^{15}$N-^{1}H) coupling of about 80.0 Hz, a value which varies slightly with conditions such as solvent and temperature. A single but still broadened ^{15}N line ($\Delta\nu$ = 20 Hz) was obtained for this sample (a 99.5% enriched sealed aniline-^{15}N sample)(*). Irradiation at the ^{1}H resonance frequency of the nearly collapsed NH$_2$ proton doublet ($\Delta\nu_{1/2}$ = 42 Hz) at sufficient power causes sharpening of the ^{15}N resonance to 18 Hz, but more importantly results in its enhancement by a factor of -3.7. IsD profiles and 2sD images were obtained in the magnitude mode (see fig. 12 and 13). The total acquisition time was 16 min which resulted in an S/N ratio for the profile of approximately 10:1.

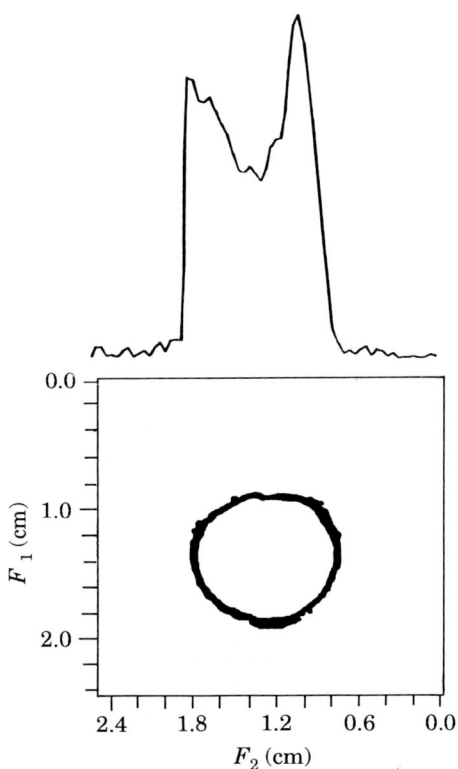

Fig. 13. – ^{15}N image and 1D profile of the same ampoule as in fig. 12 obtained with ^{15}N{^{1}H} imaging exhibiting the Overhauser gain.

(*) Yes! At last we resorted to enrichment.

The sample was contained in a cylindrical ampoule (o.d. 10 mm) inserted into a solenoid tuned to 20.28 MHz wound around a shortened 13 mm NMR tube. The proton B_2 field was produced by a standard (8 cm internal diameter (Alderman-Grant type)) imaging coil operating at 200 MHz on a Sisco 200 spectrometer (4.7 T, 33 cm bore). The ^{1}H and ^{15}N circuits contained suitable high- and low-pass frequency filters, respectively. The B_2 field was applied either continuously or only during the ^{15}N acquisition period (gated mode). The basic imaging sequence employed the steady-state free-precession method (SSFP) which has the merit of allowing more rapid accumulation than would otherwise be dictated by the relatively long ^{15}N T_1 value. This was measured to be 9.7 s (± 0.1 s) by a conventional inversion-recovery experiment. Spin echo images were also obtained using a 10 s delay.

The nOe enhancement of -3.7 allows a reduction in the spectral or imaging time by a factor of $(-3.7)^2$, *i.e.* approximately 14.

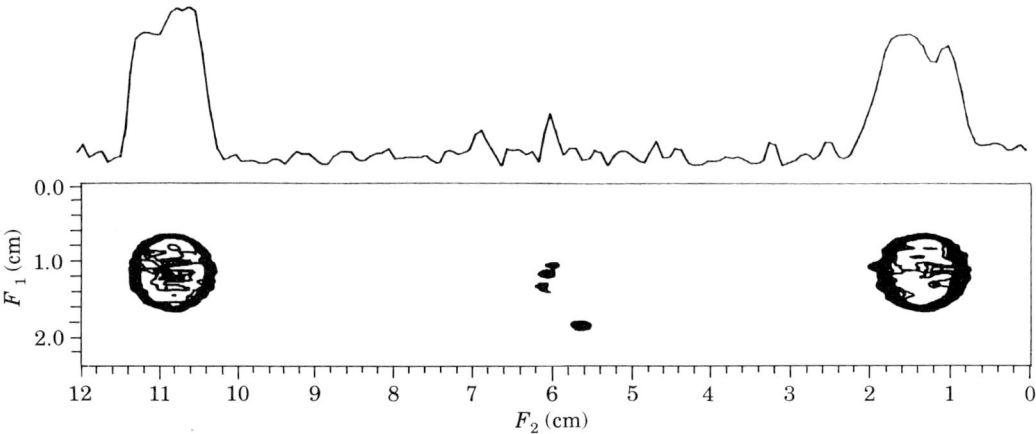

Fig. 14. – ^{15}N image and 1D profile of a 5M NH_4NO_3-$^{15}N_2$ sample in 2M HNO_3 contained in an ampoule. The two images arise because of the large ^{15}N shift (~ 354 p.p.m.) between the NO_3^- ion (left-hand image) and the $^{15}NH_4^+$ ion.

In the case of a 5M NH_4NO_3-$^{15}N_2$ sample in 2M HNO_3, the T_1 values were measured to be (50.6 ± 1.6) s for NH_4^+ and (105.6 ± 1.6) s for NO_3^-. Figures 14 and 15 show the normal ^{15}N single-resonance and the ^{15}N$\{^1$H$\}$ double-resonance profiles and images in which the nOe of -2.4 is apparent for the $^{15}NH_4^+$ signal. Here the factor for the reduction in time for the experiment is $(-2.4)^2$, *i.e.* ~ 5.8.

The sample used for the study at natural abundance was phenyl hydrazine for which the dynamics are more complex. As in the case of aniline, proton ex-

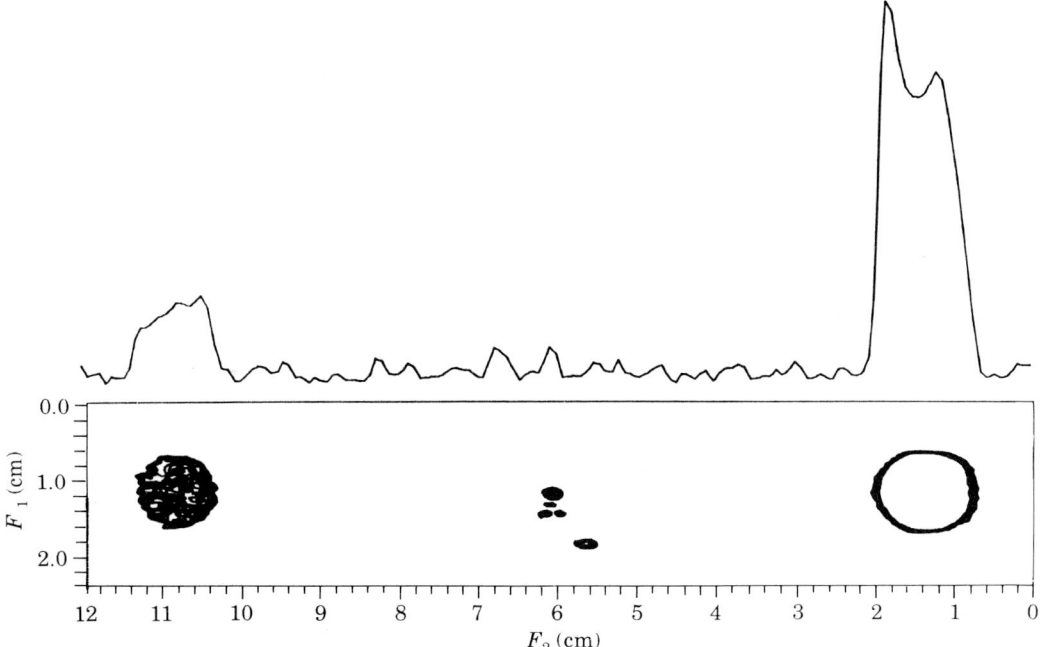

Fig. 15. – ^{15}N image and 1D profile of the same ampoule as in fig. 14 obtained with ^{15}N{^{1}H} imaging exhibiting the Overhauser effect on the right-hand image of the ^{15}NH$_4^+$ species.

change leads to *nearly complete collapse* of the ^{15}N-H couplings for both the NH and NH$_2$ groups. The two cases are different, however, in that the NH$_2$ process is more rapid than the NH process and so leads to a narrower ^{15}N line (16 Hz in our sample at 20 °C) compared with the NH$_2$ value of 80 Hz in the single-resonance experiment. The nOe values are nearly equal[10] at a value of about -2.9, and we have measured the ^{15}N T_1 values as (8.40 ± 0.1)s (for NH) and (3.91 ± 0.01)s (for NH$_2$).

Nevertheless from our instrument the spectral profiles and the images show a smaller intensity in the left-hand (NH) resonance (see fig. 16). We conclude that the B_2 level in our apparatus is insufficient to give the full nOe for the NH line but sufficient for NH$_2$ which is more nearly completely decoupled by its faster exchange. It was probably not caused by differential offset of the ν_2 frequency since no change was noted with noise modulation of ν_2 nor with the employment of the Waltz sequence. The linewidths in ^{15}N{^{1}H} spectra are 8.5 Hz and 9.4 Hz, respectively. The time gain in this case is about 8.4.

So it may be concluded that imaging of ^{15}N by direct methods is feasible and can be improved by ^{15}N{^{1}H} experiments even when the ^{15}N resonance is broadened by rapid exchange of attached protons.

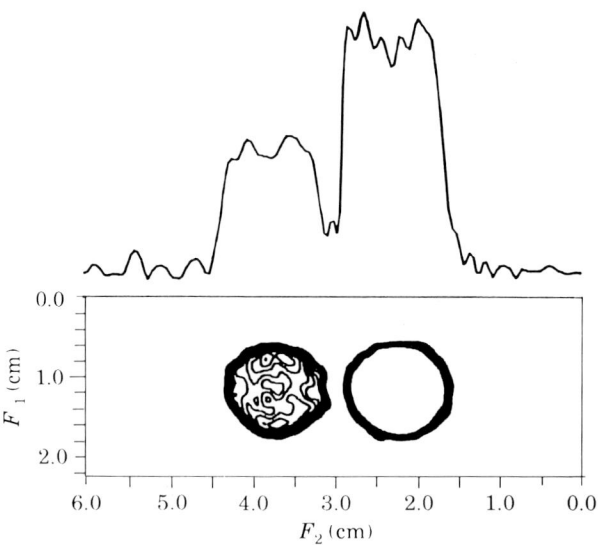

Fig. 16. – ^{15}N image and 1D profile of phenylhydrazine ($C_6H_5NHNH_2$) obtained at natural abundance with ^{15}N{^{1}H} imaging.

Despite this success, however, the fact is that if the compound of interest has ^{15}N-^{1}H J interactions, then *indirect* detection by a ^{1}H{^{15}N} experiment offers much greater sensitivity. What are the options? And when would experiments involving the ^{14}N isotopomers be preferable?

3˙3.3. Indirect methods, ^{1}H{^{15}N}. The best enhancements of poor ^{15}N signals which arise from its low magnetogyric ratio come from the so-called indirect methods at least when the ^{15}N nuclei are spin coupled to ^{1}H, by use of ^{1}H{^{15}N} experiments. The ratio of $\gamma(^1H)/\gamma(^{15}N)$ is nearly -10 (actually -9.858). Additionally the ^{15}N nucleus generally has respectably long relaxation times (perhaps *too* long) unlike its ^{14}N isotope, such that in single-resonance experiments the couplings $J(^{15}N\text{-}^1H)$ are resolvable, except in cases where the hydrogen atoms are exchanging. The whole panoply of spectroscopic indirect methods, especially the multipulse methods, can be invoked, therefore, in imaging modes to improve the sensitivity of the imaging experiment.

The methods of choice, therefore, involve such techniques as INEPT and the multiple-quantum techniques which KNÜTTEL and co-workers have used for the imaging of ^{13}C at natural abundance. The gains to be expected for ^{15}N are greater than for ^{13}C because the γ for ^{15}N is lower (approximately 10 times relative to ^{1}H) than the ^{13}C γ-value (approximately 4). The exact evaluation of these gains is, however, not straightforward, and vary by a square-root factor in the literature. The enhancement factors are given in table IV which has been adapted from [31].

TABLE IV. – *Comparison of single and $^1H\{^{15}N\}$ indirect methods for ^{15}N detection* (*).

			Gain, $|G|$ (**)
direct $^{15}N\{^1H\}$	1) single resonance	1	1
	2) $^{15}N\{^1H\}$ with nOe	$R/2$ (max)	4(***)
	3) INEPT	R(****)	9.9
indirect (polarization transfer) $^1H\{^{15}N\}$	4) inverse INEPT	$R3/2$	31
	5) multiple quantum	$R5/2$	306

(*) There is in each case the possibility of an additional saturation factor [32].
(**) The gain $|G|$ is expressed as a ratio of signal relative to mode 1.
(***) The nOe gain η is -5, but the observed signal is $1+\eta$. $1+\eta$ may vary between -4 and 1 (and includes 0).
(****) $R = \gamma(^1H)/\gamma(^{15}N)$.

So far we have not investigated the multiple-pulse techniques for ^{15}N, but we have looked at the simpler $^1H\{^{15}N\}$ (c.w.) experiments in which ν_2 is applied continuously as in the ^{14}N case reported above.

Here we propose the use of a simple trick to add to the subtraction technique which worked well for ^{14}N. There the $^1H\{^{14}N\}$ experiment was used to *sharpen* ^1H resonances arising from protons coupled to ^{14}N. With ^{15}N we may expect that the ^1H resonances will be already sharp if there is very slow ^1H exchange.

The idea here then is simply that application of ν_2 rather than being used to *sharpen* the ^1H resonance (if exchange *is* present) is to *broaden* the ^1H resonance. This broadening can be induced by suitable choice of the decoupling power B_2 (with offset of ν_2 if necessary), such as has been successful in the $^{15}N\{^1H\}$ mode with low-power noise decoupling (see fig. 11).

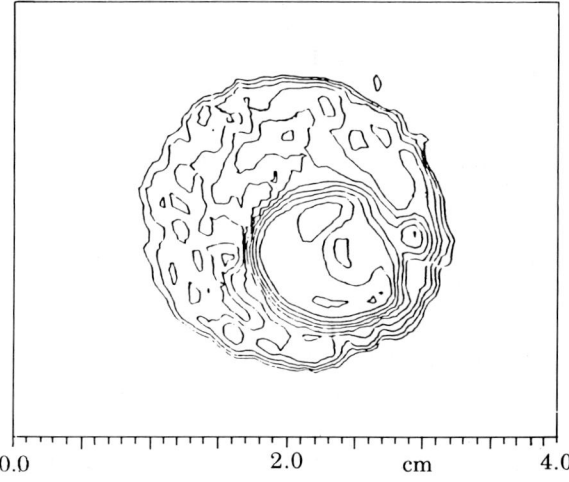

Fig. 17. – ^1H cross-sectional image of two tubes. The inner tube contains $^{15}NH_4^+$ and the annular space contains methanol.

The subtraction is again required in order to filter out those ^{1}H contributions to the image arising from protons not coupled to ^{15}N (just as in the ^{14}N case).

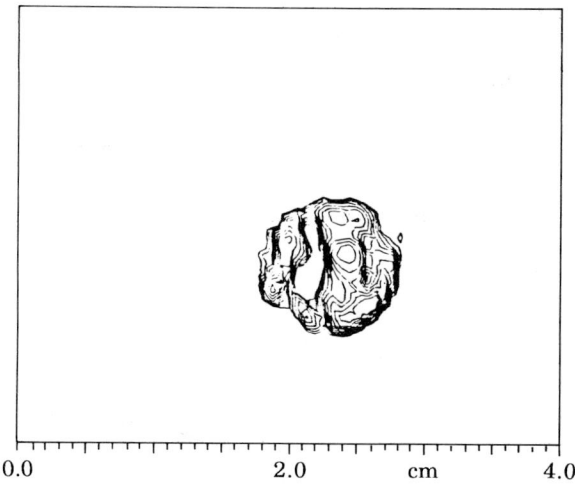

Fig. 18. – The difference ^{1}H image obtained by subtracting the image of fig. 17 from an ^{1}H{^{15}N} image.

Our attempts so far are shown in fig. 17 and 18. The experiments are successful, but the gains are not spectacular (perhaps we have not broadened the ^{1}H resonances sufficiently). It could be that we have simply detected the Overhauser difference in the experiment. This, of course, is very small since $(1/2)[\gamma(^{15}N)/\gamma(^1H)]$ is only 1/20 (approximately) in this mode. The experiments continue.

In this type of situation the preferred methods will surely involve ^{1}H{^{15}N} INEPT methods or those of the multiple-quantum type (see table IV).

Suppose, however, that the ^{1}H resonances become broadened because of ^{1}H exchange, which is frequently the case with NH groups. Here the INEPT and multiple-quantum techniques may be inapplicable, and the best approach may be to mimic the ^{1}H{^{14}N} experience.

For the sample of aniline-^{15}N (already studied in the ^{15}N{^{1}H} mode, see above), in which there is proton exchange occurring, application of B_2 did *not* sharpen the ^{1}H resonance of the NH_2 protons, however. We conclude here that the broadening which they exhibit is due to chemical-shift broadening and that the ^{15}N and ^{1}H resonances are completely decoupled by the exchange process. This proposition we shall test.

* * *

I am most grateful to all my collaborators named in the references, past and present. For the achievement of the imaging experiments I should like particularly to thank Dr. P. KINCHESH and Dr. S. C. R. WILLIAMS for their indulgence and many new insights.

I am grateful to Longman Scientific and Technical for permission to reproduce fig. 1 from *A Handbook of Nuclear Magnetic Resonance* by R. FREEMAN (1987), to Pergamon Press for fig. 2-6, to Academic Press for fig. 9 and 10.

This work was supported in part by Grant LRG 68 269 from the Agriculture and Food Research Council.

Note added in proofs.
An alternative double-resonance experiment in cases where there is J modulation has been used by the Rome group [33, 34]. They employ a version of the SEDOR experiment which has been described elsewhere in this volume by SLICHTER and MEHRING.

REFERENCES

[1] J. D. BALDESCHIELER and E. W. RANDALL: *Chem. Rev.*, **63**, 81 (1963).
[2] C. P. SLICHTER: *Principles of Magnetic Resonance*, 3rd edition (Springer-Verlag, Berlin, 1989).
[3] R. R. ERNST, G. BODENHAUSEN and A. WOKAUN: *Principles of Nuclear Magnetic Resonance in One and Two Dimensions* (Oxford University Press, Oxford, 1987).
[4] P. T. CALLAGHAN: *Principles of Nuclear Magnetic Resonance Microscopy* (Oxford University Press, Oxford, 1991).
[5] A. ABRAGAM and W. G. PROCTOR: *Phys. Rev.*, **106**, 160 (1957); see also ref. [2], p. 271.
[6] Z. WU, W. HAPPER and J. M. DANIELS: *Phys. Rev. Lett.*, **59**, 1480 (1987).
[7] R. FREEMAN: *A Handbook of Nuclear Magnetic Resonance* (Longman, Harlow, 1988), p. 41.
[8] R. J. CHUCK, D. G. GILLIES and E. W. RANDALL: *Mol. Phys.*, **16**, 121 (1969).
[9] E. W. RANDALL: *Chem. Br.*, **7**, 371 (1971).
[10] J. M. BRIGGS, L. F. FARNELL and E. W. RANDALL: *Chem. Commun.*, 680 (1971).
[11] J. M. BRIGGS, L. F. FARNELL and E. W. RANDALL: *Chem. Commun.*, 70 (1973).
[12] G. E. HAWKES, W. M. LITCHMAN and E. W. RANDALL: *J. Magn. Reson.*, **19**, 255 (1975).
[13] E. WENKERT, A. O. CLOUSE, D. W. COCHRAN and D. DODDRELL: *J. Am. Chem. Soc.*, **91**, 879 (1969).
[14] A. D. BAIN, R. M. LYNDEN-BELL, W. M. LITCHMAN and E. W. RANDALL: *J. Magn. Reson.*, **25**, 315 (1977).
[15] G. E. HAWKES, S. PAGET and E. W. RANDALL: *J. Magn. Reson.*, **30**, 393 (1978).

[16] P. S. ALBRAND, E. W. RANDALL and R. M. LYNDEN-BELL: *J. Magn. Reson.*, **37**, 61 (1980).
[17] R. R. ERNST, G. BODENHAUSEN and A. WOKAUN: *Principles of Nuclear Magnetic Resonance in One and Two Dimensions* (Oxford University Press, Oxford, 1987), p. 220.
[18] E. W. RANDALL: in *Nitrogen NMR* (Plenum Press, London, 1973), p. 41.
[19] W. M. LITCHMAN and E. W. RANDALL: unpublished results (1975).
[20] P. MANSFIELD and P. G. MORRIS: *Advances in Magnetic Resonance*, Supplement 2, (Academic Press, San Diego, Cal., 1982), p. 65.
[21] M. G. GIBBY, R. G. GRIFFIN, A. PINES and J. S. WAUGH: *Chem. Phys. Lett.*, **17**, 80 (1972).
[22] P. KINCHESH, D. S. POWLSON, E. W. RANDALL and S. C. R. WILLIAMS: *J. Magn. Reson.*, **97**, 208 (1992).
[23] P. KINCHESH, E. W. RANDALL and K. ZICK: *J. Magn. Reson.*, **100**, 411 (1992).
[24] P. KINCHESH, E. W. RANDALL and S. C. R. WILLIAMS: *Magn. Reson. in Chem.*, **31**, 495 (1993).
[25] A. KNÜTTEL, R. KIMMICH and K.-H. SPOHN: *J. Magn. Reson.*, **86**, 526 (1990).
[26] A. KNÜTTEL, K.-H. SPOHN and R. KIMMICH: *J. Magn. Reson.*, **86**, 542 (1990).
[27] P. KINCHESH, E. W. RANDALL and S. C. R. WILLIAMS: *J. Magn. Reson.*, **100**, 625 (1992).
[28] P. KINCHESH, E. W. RANDALL and S. C. R. WILLIAMS: *J. Magn. Reson.*, **98**, 458 (1992).
[29] P. KINCHESH, E. W. RANDALL and S. C. R. WILLIAMS: *J. Magn. Reson. A*, **103**, 234 (1993).
[30] E. W. RANDALL and J. J. ZUCKERMAN: *J. Am. Chem. Soc.*, **90**, 3167 (1968).
[31] W. BERMEL: Bruker Spectrospin Report.
[32] R. R. ERNST, G. BODENHAUSEN and A. WOKAUN: *Principles of Nuclear Magnetic Resonance in One and Two Dimensions* (Oxford University Press, Oxford, 1987), p. 468.
[33] F. DE LUCA, R. CAMPANELLA, A. BIFONE and B. MARAVIGLIA: *Chem. Phys. Lett.*, **189**, 303 (1991).
[34] F. DE LUCA, R. CAMPANELLA, A. BIFONE and B. MARAVIGLIA: *J. Magn. Reson.*, **93**, 554 (1991).

Rotating-Frame Spectroscopy and Imaging under Radio- and Audio-Frequency Excitation.

N. LUGERI, F. DE LUCA, B. C. DE SIMONE and B. MARAVIGLIA

Dipartimento di Fisica dell'Università «La Sapienza»
piazzale Aldo Moro 5, 00185 Roma, Italia

1. – Introduction.

There is no need, by now, to emphasize the power of the wide world of techniques related to NMR, as a tool of investigation in the equally wide fields of study, both in fundamental research and in practical applications. Whenever a new «promising» technique starts to maintain its promises, new extensions and expansions of the basic idea are soon pursued towards wider ranges of physical systems to be studied by means of the new method, and improvements both on the quality of the results and on the related technical problems are seeked.

In this lecture we shall be dealing with a branch of the more general ensemble of rotating-frame experiments, namely the direct-excitation and detection ones, where a resonant double excitation with radio- (RF) and audio-frequency (AF) oscillating magnetic fields is used to perform pulsed NMR in the rotating frame.

It should be immediately pointed out that this kind of double excitation is somehow out of the «classical» definition of double-resonance experiments, since the same spin species undergoes both resonant irradiations, according to the different static fields it experiences in the laboratory (LCF) and in the rotating frame (RCF).

A general review on the various aspects of RCF experiments will be given, starting from a theoretical basis, with particular emphasis to the transformation properties of the spin-interaction Hamiltonians in the rotating frames, or interaction representations, to be introduced in order to get a description of the system as simple as possible.

Applications of this technique to relaxation studies and imaging of solids will be presented together with both preliminary and conclusive experimental results on the various applicative aspects of the method, obtained in our laboratory.

2. – General features of RCF experiments.

The concept of rotating reference frame appears in the first pages of some NMR monographs as a geometrical tool applied to the description of NMR, in the same way as in many other physical phenomena. In magnetic-resonance experiments, the role of the rotating frame has soon been developed much further than this pictorial significance gaining a deep physical and not only mathematical meaning. In fact it should be pointed out that, while in the classical interpretation of the RCF as a «freeze frame» for the precession of a vector around an axis (magnetization in MR, angular momentum in rigid-body dynamics and so on) the rotation frequency is given by quantities inherent in the physical system, a more interesting character is gained by the RCF in MR experiments due to its relation to the frequency of the externally imposed resonant magnetic field.

After REDFIELD with his fundamental paper on NMR saturation in solids and the first rotary saturation experiments[1], the rotating-frame spectroscopy obtained its own place among the various classes of NMR experiments, especially thanks to the wide applications to relaxation study of slow molecular dynamics by relaxation dispersion in the RCF [2,3].

Later on the possibility of modulation of the effects of the spin interactions on the RCF spectrum started to be exploited to perform line-narrowing experiments typically on dipolar broadened solids. After LEE and GOLDBURG[4], who first proposed this technique with an indirect detection procedure, to be described later, and a rather long period of quiescence in the 70's, the magic angle in the RCF (MARF) narrowing technique underwent a sort of renaissance due to the proposal of the direct-detection approach both by c.w.[5] and pulsed excitation[6] and with the recent application to imaging of solids [7-9].

3. – Theoretical backgrounds.

The nuclear-spin Hamiltonian can be usefully expressed in tensorial form as

$$(1) \qquad H = \sum_\lambda C_\lambda \sum_{jk=1}^{N} I^j \tilde{A}_\lambda^{jk} S_\lambda^k \,,$$

where I is always the nuclear-spin vector, while both the vector S_λ and the 2nd-rank interaction tensor $\tilde{A}_\lambda$ change according to the interaction λ, as listed in table I for the most important components. S can then represent the same spin species as I or another one (either nuclear or electronic) or else a magnetic field

TABLE I. – *List of the entries in* eq. (1), *relative to the internal and external Hamiltonians considered in the text.*

Hamiltonians	λ	C_λ	S_λ	$\tilde{A}_\lambda$
Zeeman	z, RF, AF	$-\gamma_I$	B_0, B_1, B_2	1
chemical shift	CS	$-\gamma_I$	B_0	σ
dipolar	D	$-2\gamma_I\gamma_S\hbar$	S	D
J-coupling	J	l	S	J
quadrupolar	Q	$\dfrac{eQ}{6\hbar I(2I-1)}$	I	V
field gradient	G	γ_I	r	G

and $\tilde{A}_\lambda$, which depends only on spatial co-ordinates, represents the kind of coupling intervening between I and S.

If we now define the second-rank, spin-dependent, tensors $\tilde{T}_\lambda$ as $S_\lambda^+ I$ and if we consider that spherical co-ordinates rather than Cartesian ones are better suitable to describe transformations of functions under rotation, we can express eq. (1) as the scalar product of $\tilde{A}_\lambda$ and $\tilde{T}_\lambda$ in spherical-tensor notation [10]:

$$(2) \qquad H = \sum_\lambda C_\lambda \sum_{jk=1}^{N} \sum_{L=0}^{2} \sum_{M=-L}^{L} (-1)^M A_{L,-M}^{\lambda,jk} T_{LM}^{\lambda,jk}.$$

Although this comprehensive expression is useful in describing the general features of the nuclear-spin Hamiltonian, a distinction based on the entries of table I should be borne in mind because of the rank-dependent form of the transformation properties of eq. (2) under rotation in spin space. In fact, while $\tilde{A}_\lambda$ is of rank 2 in real space (rank 0 in spin space) whichever the interaction λ described is, $\tilde{T}_\lambda$ is of rank 2 in spin space only for homonuclear two-spin interactions ($\lambda = D, J, Q$) and for their heteronuclear counterparts, in the case of double resonant excitation of I and S spin systems.

That is why the various Hamiltonians undergo rank-selective transformations when rotations are brought about by RF resonant pulses or continuous irradiation. In the case of cyclic behaviour of the perturbation on the spin co-ordinates, selective averaging of the anisotropic part of $\tilde{T}_\lambda$ can be obtained [11-13], thus providing a very useful tool for the investigation of particular aspects of the spin interactions or of spin dynamics which are usually masked by the presence of dominant line-broadening interactions (of course, this consideration does not apply to liquidlike systems where the anisotropic character of H_λ is lost). By contrast transformations imposed by rotations of spatial co-ordinates [13-15] cannot provide such selectivity due to the common rank shared by all the $\tilde{A}_\lambda$ tensors in real space.

In ordinary initial condition for the NMR experiment the leading term in (1) is provided by the Zeeman interaction with the main static field $\boldsymbol{B}_0 = B_0 \hat{Z}$, i.e. $\|H_z\| \gg \|H_\lambda\|$. In such cases an interaction representation which effectively removes this dominant part of the Hamiltonian is required, in order to get a more direct insight into the dependence of the evolution of the system on the interaction terms.

The interaction representation is formally defined by the transformation of the Schrödinger equation (in angular frequency units)

$$(3) \qquad i\frac{\partial \psi}{\partial t} = H\psi \xrightarrow{R} i\frac{\partial \psi'}{\partial t} = H'\psi',$$

where $H = H_0 + H_\varepsilon$ with $\|H_0\| \gg \|H_\varepsilon\|$, and

$$(3') \qquad \begin{cases} R = \exp[-iH_0 t], \\ \psi' = R^{-1}\psi, \\ H' = R^{-1}HR - iR^{-1}\frac{\partial R}{\partial t} = R^{-1}H_\varepsilon R. \end{cases}$$

If we apply this definition to NMR systems, we can see that the main Zeeman Hamiltonian is removed by the operator

$$(4) \qquad R_{z0} = \exp[-iH_{z0} t] = \exp[-i\omega_0 t I_Z].$$

As already pointed out, this applies regardless of the presence of an external oscillating magnetic field $\boldsymbol{B}_1 = 2B_1 \cos\omega t \hat{X}$ (H_{RF} in table I). Since it is always useful to transform in such a way that H_{RF} turns out to be static [16], transformation (4) can define a «useful» interaction representation only in the case of on-resonance RF perturbation. In order to keep the formal validity of the identity between RCF and interaction representation, we can split the static Zeeman term as follows:

$$(5) \qquad H_{z0} = H_z + H_{\mathrm{off}} = -\gamma(B_0 - \Delta B_0)I_Z - \gamma\Delta B_0 I_Z = \omega I_Z + \Delta\omega I_Z,$$

where ω is the «physical» angular velocity of the RF field, and define the interaction representation by replacing H_{z0} with H_z in (4). Since definition (4) represents a rotation by $\omega_0 t$ around I_Z, in the following we shall refer to transformations brought about by such considerations as rotating frames.

Since rotations introduced by operators like eq. (4) are performed in $\boldsymbol{I}$ spin space, the only terms in the Hamiltonians which are affected are the $\tilde{T}_\lambda$ spin-dependent tensor. We can then use the general expression of the transformation of an irreducible spherical tensor of rank L in terms of the irreducible representation of the three-dimensional rotation group [10]:

$$(6) \qquad R^{-1}(\alpha\beta\gamma) T_{LM} R(\alpha\beta\gamma) = \sum_{M'} D^l_{MM'}(\alpha\beta\gamma) T_{LM'},$$

where (α, β, γ) are the Euler angles which define the rotation described by

R [10, 17] and where also the possibility of having two different ranks of $\tilde{T}_\lambda$ and of the Wigner matrix $\tilde{D}^l$ is considered.

How T_{LM}^λ are obtained in terms of Cartesian components derives directly by the definition $\tilde{T}^\lambda = S_\lambda^+ \boldsymbol{I}$ and table I. If we restrict ourselves to spin-1/2 systems, we obtain for the single-spin (linear) Hamiltonians the following identities [13, 14]:

(7) $$\begin{cases} T_{00} = \tilde{1}, \\ T_{10} = I_Z, \\ T_{1\pm 1} = \frac{\mp 1}{\sqrt{2}} (I_X \pm i I_Y) = \mp \frac{I_\pm}{\sqrt{2}}, \end{cases}$$

while for two-spin (bilinear) interactions, the second-rank tensor of components $T_{LM}^{(12)}$ is given by [17]

$$T_{LM}^{(12)} = \sum_{M_1} (L_1 L_2 M_1, M - M_1 / LM) \, T_{L_1 M_1}^{(1)} T_{L_2, M-M_2}^{(2)}$$

with the well-known Clebsch-Gordan coefficients in parentheses and where (12) denotes the two intervening single-spin operators of rank L_1 and L_2, respectively.

If we neglect $L = 1$ components, whose effect on NMR spectra is proven to be negligible [14], we obtain

(7') $$\begin{cases} T_{00}^{(IS)} = -\frac{1}{\sqrt{3}} \boldsymbol{I} \cdot \boldsymbol{S}, & T_{20}^{(IS)} = \sqrt{\frac{1}{6}} (3 I_Z S_Z - \boldsymbol{I} \cdot \boldsymbol{S}), \\ T_{2\pm 1}^{(IS)} = \mp \frac{1}{2} (I_\pm S_Z + I_Z S_\pm), & T_{2\pm 2}^{(IS)} = \frac{1}{2} I_\pm S_\pm. \end{cases}$$

Correspondingly, the irreducible spherical-tensor components of the $\tilde{A}_\lambda$ tensors can be expressed in terms of Cartesian co-ordinates as [14]

(8) $$\begin{cases} A_{00}^\lambda = -\frac{1}{\sqrt{3}} \, \text{tr} \{A^\lambda\}, \\ A_{20}^\lambda = \frac{1}{\sqrt{6}} [3 A_{ZZ}^\lambda - \text{tr} \{A^\lambda\}], \\ A_{2\pm 1}^\lambda = \mp \frac{1}{2} [A_{XZ}^\lambda + A_{ZX}^\lambda \pm i(A_{YZ}^\lambda + A_{ZY}^\lambda)], \\ A_{2\pm 2}^\lambda = \frac{1}{2} [A_{XX}^\lambda - A_{YY}^\lambda \pm i(A_{XY}^\lambda + A_{YX}^\lambda)]. \end{cases}$$

The formal tools to study the dynamics of the spin system in terms of the time evolution of observables are completed by the well-known density matrix formalism and by the average Hamiltonian theory which is of great advantage when periodically time-dependent perturbations are imposed on the spin system. We are not going into a detailed description of such theories, well established and exhaustively explained in many textbooks and monographs on NMR (*e.g.*, [13, 14, 18]). What we need for the presentation of RCF experiments are just the basic formulae which moreover provide a descriptive connection between RCF NMR and traditional experiments.

Instead of solving the Schrödinger equation (3), the introduction of the reduced spin density matrix ρ allows one to solve for the time evolution of the interesting observables, O, of the system, as follows:

(9) $$\langle O \rangle(t) = \text{tr}\{\rho(t)\, O\}$$

with $\rho(t)$ obeying the Liouville-von Neumann equation

(10) $$\frac{d\rho}{dt} = -i[H, \rho].$$

With the interaction representation(s) introduced by terms like eq. (5), we obtain the time evolution of O in the chosen rotating frame by

(9') $$\langle O \rangle_R(t) = \text{tr}\{\rho_R(t)\, O\}$$

and

(10') $$\frac{d\rho_R(t)}{dt} = -i[H_R^{\text{int}}, \rho_R]$$

with

$$\rho_R = R^{-1} \rho R, \qquad H^{\text{int}} = R^{-1} H_{\text{int}} R.$$

R is now defined in a more general way than in (4) in the case of an explicitly time-dependent external Hamiltonian $H = H_{\text{ext}}(t) + H_{\text{int}}$:

(11) $$R(t) = \hat{T} \exp\left[-i \int_0^t dt'\, H_{\text{ext}}(t')\right]$$

with $\hat{T}$ the Dyson time-ordering operator.

The solution of the Liouville-von Neumann equation for ρ_R is then given by the propagator

(12) $$L(t) = \hat{T} \exp\left[-i \int_0^t dt'\, H_{\text{int}}(t')\right]$$

in the relation

(13) $$\rho_R(t) = L\rho_R(0)L^{-1}.$$

An approximate description is obtained through an average defined on the Hamiltonian in the propagator (12). Two main procedures have been defined: coherent averaging [11-14], based on the definition of a cyclic behaviour of the perturbation ($R(t + t_c) = R(t)$) with t_c the so-called cycle time of the Hamiltonian, and secular averaging [14, 19, 20], founded on the separation of both the Hamiltonian and the density matrix in terms of secular and nonsecular contributions and in the commutation relations obtained by substitution into the Liouville-von Neumann equation.

As an example we consider the case of coherent averaging where the time domain for the observation of the evolution of the system is restricted to integer multiples, nt_c, of the cycle time (stroboscopical observation). The propagator is then $L(t_c) = \hat{T} \exp[-i\overline{H}_{\text{int}} \cdot t_c]$ and $\overline{H}_{\text{int}}$ is expressed by the Magnus expansion [13] stopped at the desired order of approximation. The lowest one is the zeroth-order average Hamiltonian which is nothing but the time average over one cycle of the transformed internal Hamiltonian:

(14) $$\overline{H}_{\text{int}}^{(0)} = \frac{1}{t_c} \exp\left[-i \int_0^{t_c} dt\, H_R^{\text{int}}(t)\right].$$

If we wish to consider the relative contribution to the total «size» of the interaction Hamiltonian from the n-th-order terms, it can be shown that, roughly speaking in terms of the unperturbed linewidth of the system, $\delta\omega$, they are of the order of $(\delta\omega \cdot t_c)^n$ with respect to the 0-th-order contribution [13].

The case of interest in RCF experiments is that of a time-independent external perturbation or, better, of effectively time-independent external perturbations in some interaction representation. As we shall soon see, we are dealing with Zeeman Hamiltonians in the successive rotating frames and the cyclic character they generate with the precession of the spins [13, 18]. This cyclic behaviour allows one to define approximate average Hamiltonians in the «RCF's», where they are transformed according to the transformation properties of the spin-dependent tensors under rotations. The statements of the average Hamiltonian theory thus lead the experimenter in designing the experimental way to average out the undesired part of the interaction Hamiltonians.

4. – Magic-angle rotating-frame experiments.

4`1. *Interaction representation.* – All of the line-narrowing techniques share the common feature that a peculiar frame of reference is sought where the evolution of the system is ruled by a suitably manipulated interaction Hamiltonian.

Our version of the MARF method can be easily explained in terms of a transposition of the NMR experiments onto a lower Larmor frequency, under a «tunable» internal Hamiltonian. In the typical arrangement of MARF experiments a new set of Zeeman levels, whose distribution is given by the transformed Hamiltonian, can be defined in the desired frame. Among them transition can be induced by resonant excitation.

Let us now state this in terms of interaction representation [5,6,8]:

$$(15) \quad H_{\text{tot}} = H_{\text{ext}}(t) + H_{\text{int}}(t), \qquad H_{\text{ext}} = H_z + H_{\text{off}} + H_{\text{RF}}(t) + H_{\text{AF}}(t),$$

is the Hamiltonian relative to direct MARF experiments, with

$$H_{\text{AF}} = 2\gamma B_2(t) \cos \omega_e t I_Z,$$

while the other entries have already been defined. H_{AF} represents the audiofrequency excitation on the RCF Zeeman levels.

It is worth pointing out that the time dependences in the total Hamiltonian (15) arise from totally different sources. The one on H_{ext} is introduced by, and under control of, the experimenter, the other one arises from «Heaven sent» fluctuations on the spatial parameters of the internal interactions.

Referring to eq. (11) we can divide the operator $R(t)$ into a cascade of interaction representations as

$$(16) \quad R(t) = R_Z(t) R_E(t) R_A(t)$$

to define the full interaction representation used for MARF experiments. Each entry defines a frame rotating around the direction of the static field which is obtained in each step of the transformation, after an average Hamiltonian approximation rejecting fast oscillating contributions in the transformed external perturbations.

According to ref. [21] and eq. (15), we obtain

$$(17) \quad \begin{cases} R_z = \exp[-i\omega t I_Z], \\ R_E = \exp[-i\theta I_y] \exp[-i\overline{H}_0' t] \exp[i\theta I_y], \\ R_A = \exp[-i(\pi/2) I_{\eta'}] \exp[-i\overline{H}'' t] \exp[i(\pi/2) I_{\eta'}]. \end{cases}$$

The first RCF $(x, y, z \equiv Z)$ does not need to be described. As long as the (quasi-)resonant RF perturbation is applied, we are led to the definition of the effective field $\boldsymbol{B}_{\text{eff}} = (\omega_0 - \omega)\hat{z} + B_1 \hat{x}$ by

$$(18) \quad \overline{H}'(t) = \frac{1}{t_c} \int_0^{t_c} [R_z^{-1}(H_{\text{ext}} - H_z) R_z] \, dt = \Delta\omega(t) I_z + \omega_1(t) I_x + H_{\text{AF}}(t) =$$

$$= \gamma \boldsymbol{B}_{\text{eff}}(t) \cdot \boldsymbol{I} + H_{\text{AF}}(t)$$

with $t_c = 2\pi/\omega$. Provided that $\|\gamma \boldsymbol{B}_{\text{eff}} \cdot \boldsymbol{I}\| \gg \|H_{\text{int}}\|$, $\boldsymbol{B}_{\text{eff}}$ defines a «new» Zeeman quantization axis, thus a new frame with the «Z»-axis pointing along $\boldsymbol{B}_{\text{eff}}$ is introduced by a rotation by $\theta = \boldsymbol{B}_{\text{eff}} z = \text{tg}^{-1}(\gamma B_1/\Delta\omega)$ around y. In this tilted rotating frame or TRCF ($\xi, \eta \equiv y, \zeta$)

(19) $$\overline{H}'_\theta = \omega_e(t) I_\zeta + 2\omega_2(t) \cos\omega_e t (I_\zeta \cos\theta - I_\xi \sin\theta)$$

with $\omega_e = \gamma B_{\text{eff}}$ and $\omega_2 = \gamma B_2$.

The whole of R_E thus defines a doubly rotating tilted frame, DRTF ($\xi', \eta', \zeta' \equiv \zeta$), where the field along ζ' is cancelled, since we have chosen ω_e as the frequency of the transverse perturbing field (on-resonance MARF experiments). The three steps introduced thus far can be viewed geometrically in fig. 1.

We are then left with an external average Hamiltonian

(20) $$H''(t) = \frac{1}{t'_c} \int_0^{t'_c} R_E^{-1}(\overline{H}' - I_\zeta \omega_e) R_E \, dt = \omega_2(t) \sin\theta I_{\xi'}$$

with $t'_c = 2\pi/\omega_e$, i.e. a further static Zeeman term which can be removed (if it results dominant with respect to the transformed H_{int} in the DRTF) by a third rotating frame, expressed by R_A in complete analogy with the arguments which led us to R_E. In this latter passage only, the definition of the rotating frame is not linked to a physical rotating field of excitation,

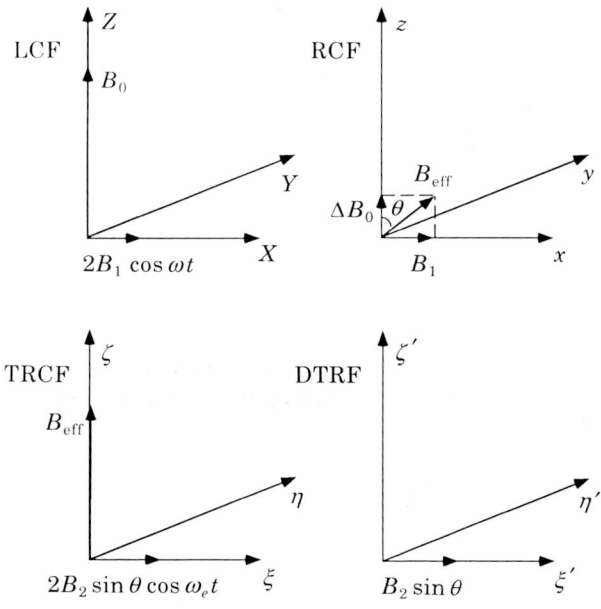

Fig. 1. – Pictorial representation of the frames of reference introduced by the transformations described in the text.

but it is evident that it is formally possible to introduce more and more rotating frames, if necessary.

In each step, a time dependence on the intensity of the external fields is indicated, describing the possibility of pulsed excitation of any timing and shaping, compatible with the MARF scheme. In the case of pulses in the TRCF, which is the one here considered, we can see that by means of the introduction of the third rotating frame, through $R_A(t)$, one can describe the AF pulses $B_2(t)$ as rotation operators in the DRTF, thus providing a direct resemblance to the classic pulsed experiments in the laboratory frame. Any kind of pulse sequence can be in principle performed in the TRCF, thus allowing the preparation of the transformed density matrix in a favourable nonequilibrium state; its evolution, effectively free from external perturbations, under a «manipulated» internal Hamiltonian, is then observed. The modality of the detection and the expression of the internal Hamiltonians will be presented in the next section.

In conclusion of this section, it is worth noting that we have introduced approximations in the definition of R_E and R_A since the only 0-th order of the Magnus expansion of H' and H'' has been retained. This means that terms oscillating at 2ω in H' and at ω_e and $2\omega_e$ in H'' have been ignored. The entity of this approximation can be estimated by the evaluation of

$$(21) \quad \overline{H}^{(1)}(t_c) = -\frac{i}{2t_c} \int_0^{t_c} dt_2 \int_0^{t_2} dt_1 \, [H(t_1), H(t_2)]$$

in the various cases

$$\overline{H}_\theta'^{(1)}(t_c) = \frac{1}{4} \frac{\omega_1^2}{\omega} I_z - \frac{1}{2} \frac{\omega_1 - \Delta\omega}{\omega} I_x.$$

This term reduces to the so-called Bloch-Siegert shift, when on-resonance irradiation is performed. With typical experimental values it weights less than 1‰ and decreases in importance with increasing B_0. The correction on θ is correspondingly negligible.

If we accept to neglect $\overline{H}_\theta'^{(1)}$, the same result holds true for $\overline{H}''^{(1)}$ in the DRTF Zeeman term, with the suitable change of indices. In our case of on-resonance $H_{AF}(t)$, we get

$$\overline{H}''^{(1)}(t_c') = \frac{1}{4} \frac{\omega_2^2}{\omega_e} I_{\zeta'}.$$

This term is usually one order of magnitude «heavier» than the corresponding one in the RCF and it can lead to nonnegligible effects in the case of strong AF irradiation [22].

For present purposes we can disregard also this correction term and use definition (17) for the interaction representation to find the expression for the transformed internal Hamiltonians.

4'2. *TRCF Hamiltonians.* – Let $\boldsymbol{M}$ be aligned along $\boldsymbol{B}_{\text{eff}}$ by a «traditional» spin-lock sequence [8, 16]. Then, in the simplest direct RCF experiment a 90° AF pulse ($\omega_2 t_p = \pi/2$) creates the transverse magnetization in the DRTF $M_{r'} = |\boldsymbol{M}| \hat{r}'$, which then evolves towards equilibrium according to eq. (10') with $R = R_z \cdot R_E$ and the corresponding propagator eq. (12).

$M_z(t) = M_{r'}(t) \sin \theta$ will be detected by means of a very sensitive AF coil placed along Z and tuned to ω_e, thus providing the DRTF FID once the signal collected at the coil is phase-sensitively demodulated with the AF carrier as a reference. The block diagram which describes the general features of the apparatus for the direct MARF method is presented in fig. 2 [23].

The reason why the component along B_0 of the magnetization is detected relies on the trivial argument that the detection apparatus is fixed in the laboratory frame, and only along Z we find no time dependences in the passage from DRTF to LCF.

The Fourier transform of the signal provides the TRCF spectrum, determined by

$$(22) \qquad H_R^{\text{int}} = R_E^{-1} R_z^{-1} H_{\text{int}} R_z R_E .$$

It can soon be realized that the expression of this transformation in terms of eq. (6) requires a set of Euler angles $\Omega = (-\omega t, \theta, -\omega_e t)$, so that, recalling that

$$(23) \qquad D_{MM'}^l(\alpha \beta \gamma) = \exp[iM\alpha] d_{MM'}^l(\beta) \exp[iM'\gamma]$$

with $\tilde{d}^l(\beta)$ the reduced Wigner matrix of rank l, M and M' involve the introduction of periodical time dependence on H_R^{int}, unless $M = M' = 0$. The time-independent term in the final expression of H_R thus represents the secular contribution to the spectrum and is given by that part of the «usual» LCF secular Hamiltonian (time-independent in the RCF) which stays secular in the transformation to the TRCF (time-independent in the DTRF).

Let us now present the expressions of some of the most important interaction Hamiltonians in the DTRF. The ω-dependent contributions are neglected, while the nonsecular contributions in the TRCF (depending on ω_e) are indicated; this means that the only LCF secular term has been transformed to the DTRF, an approximation sufficiently valid indeed.

As usual we must distinguish between linear and bilinear interaction; starting then from first-rank transformations in I-spin space, we have:

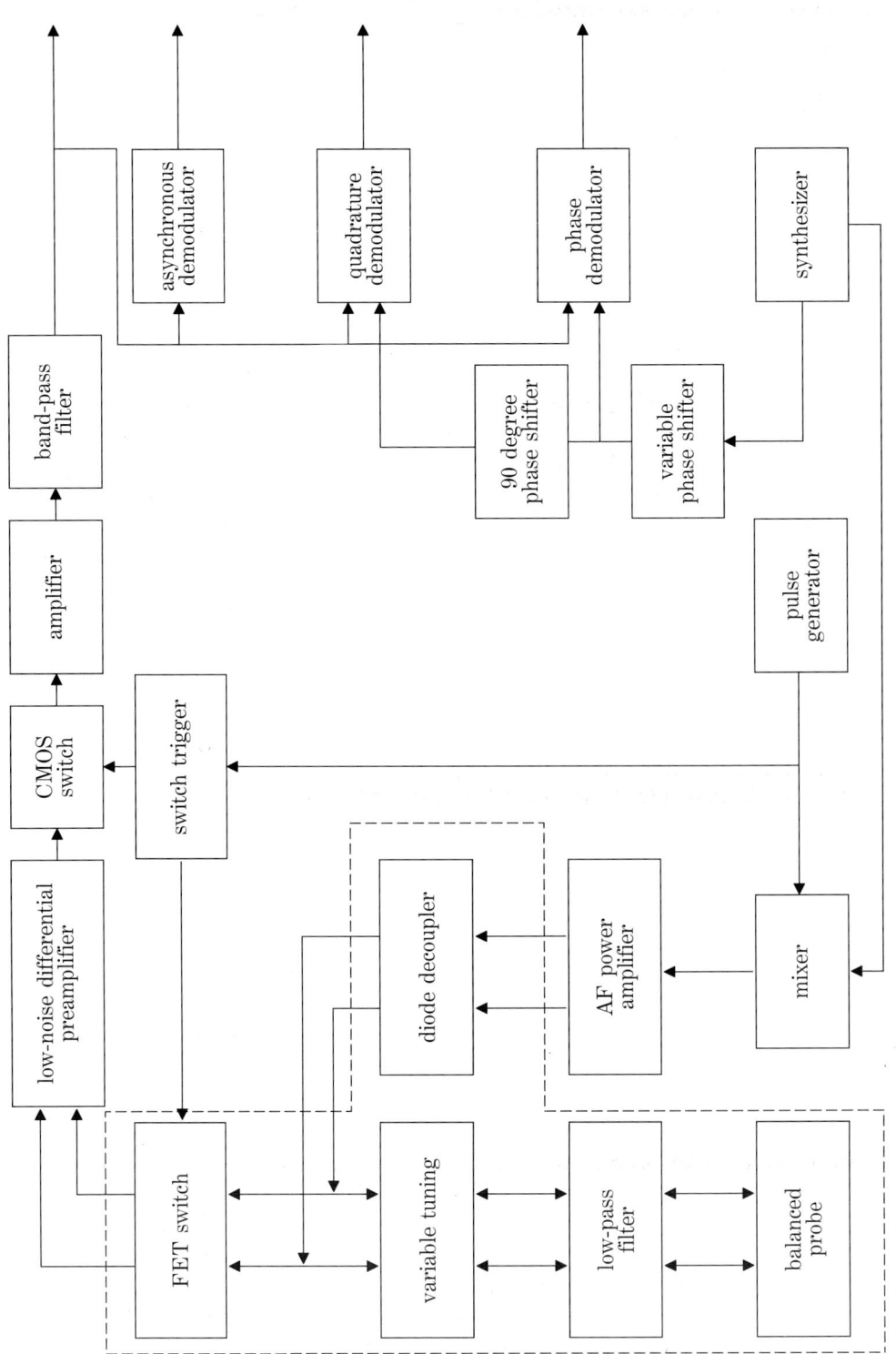

Fig. 2. – Schematic block diagram of the AF part of the radio- audio-frequency spectrometer used in MARF direct-detection experiments (from ref. [23]).

chemical shift:

(24) $H_{R,\text{sec}}^{\text{CS}} =$

$$= \left(-\frac{\sigma_{00}}{\sqrt{3}} + \sqrt{\frac{2}{3}}\,\sigma_{20}\right) B_0 \left[I_0 \cos\theta + \frac{\sin\theta}{\sqrt{2}} (I_1 \exp[i\omega_e t] - I_{-1} \exp[-i\omega_e t])\right] =$$

$$= \sigma_{ZZ} B_0 [I_{\zeta'} \cos\theta + (I_{\xi'} \cos\omega_e t - I_{\eta'} \sin\omega_e t)\sin\theta]$$

and

heteronuclear dipolar coupling:

(25) $H_{R,\text{sec}}^{J_{IS}} =$

$$= \left(-\frac{J_{00}}{\sqrt{3}} + \sqrt{\frac{2}{3}}\,J_{20}\right) S_0 \left[I_0 \cos\theta + \frac{\sin\theta}{\sqrt{2}} (I_1 \exp[i\omega_e t] - I_{-1} \exp[-i\omega_e t])\right] =$$

$$= J_{ZZ} S_Z I_{\zeta'} \cos\theta + J_{ZZ} S_Z (I_{\xi'} \cos\omega_e t - I_{\eta'} \sin\omega_e t)\sin\theta.$$

Here, also direct dipolar interactions can be read, by considering the tensor $\tilde{D}$ instead of $\tilde{J}$. It should be noted that the final expression is the same for both interactions despite the absence of the term D_{00} in the direct interaction (tr$\{\tilde{D}\} = 0$), since the first term in parentheses is identically equal to J_{ZZ}, as obtained by definition (8). This is a consequence of the fact that, in the case of heteronuclear couplings, the usually called «scalar» component ($A_{00} T_{00}$) is not invariant under rotation in I-spin space, showing instead rank 1 in the transformation, so that both σ_{00} and J_{00} also acquire an angular dependence [13].

Bilinear interactions claim for second-rank Wigner matrix, so that we get:

homonuclear dipolar:

(26) $H_{R,\text{sec}}^{J_{jk}} = -\dfrac{J_{00}}{\sqrt{3}} \mathbf{I}^j \cdot \mathbf{I}^k + J_{20} \Bigg\{ \dfrac{1}{2}(3\cos^2\theta - 1)\dfrac{1}{\sqrt{6}}(3I_{\zeta'}^j I_{\zeta'}^k - \mathbf{I}^j \cdot \mathbf{I}^k) -$

$- \sqrt{\dfrac{3}{8}} \dfrac{\sin 2\theta}{2} [(I_{\zeta'}^j I_+^k + I_+^j I_{\zeta'}^k) \exp[i\omega_e t] + (I_{\zeta'}^j I_-^k + I_-^i I_{\zeta'}^k) \exp[-i\omega_e t]] +$

$+ \sqrt{\dfrac{3}{8}} \dfrac{\sin^2\theta}{2} [I_+^j I_+^k \exp[2i\omega_e t] + I_-^j I_-^k \exp[-2i\omega_e t]] \Bigg\} =$

$= \dfrac{1}{3} \text{tr}\{\tilde{J}\} \mathbf{I}^j \cdot \mathbf{I}^k + \dfrac{1}{\sqrt{6}} (3J_{ZZ} - \text{tr}\{\tilde{J}\}) \Bigg\{ \dfrac{1}{2}(3\cos^2\theta - 1)\dfrac{1}{\sqrt{6}}(3I_{\zeta'}^j I_{\zeta'}^k - \mathbf{I}^j \cdot \mathbf{I}^k) +$

$$+ \sqrt{\frac{3}{2}} \sin\theta \cos\theta \left[(I^j_{\eta'} I^k_{\zeta} + I^j_{\zeta'} I^k_{\eta}) \sin\omega_e t - (I^j_{\xi'} I^k_{\zeta} + I^j_{\zeta'} I^k_{\xi}) \cos\omega_e t \right] +$$

$$+ \sqrt{\frac{3}{8}} \sin^2\theta \left[(I^j_{\xi'} I^k_{\xi} - I^j_{\eta'} I^k_{\eta}) \cos 2\omega_e t - (I^j_{\xi'} I^k_{\eta} + I^j_{\eta'} I^k_{\xi}) \sin 2\omega_e t \right] \bigg\},$$

where the scalar part is clearly invariant, reducing to zero in the case of direct interaction, and, similarly,

quadrupolar:

$$(27) \quad H^Q_{R,\text{sec}} = \frac{eQ}{2I(2I-1)\hbar} V_{ZZ} \left\{ \frac{1}{2}(3\cos^2\theta - 1) \frac{1}{\sqrt{6}} [3I_{\zeta'} I_{\zeta'} - I(I+1)] - \right.$$

$$- \sqrt{\frac{3}{8}} \frac{\sin 2\theta}{2} [(I_{\zeta'} I_+ + I_+ I_{\zeta'}) \exp[i\omega_e t] + (I_{\zeta'} I_- + I_- I_{\zeta'}) \exp[-i\omega_e t]] +$$

$$+ \sqrt{\frac{3}{8}} \frac{\sin^2\theta}{2} [I_+ I_+ \exp[2i\omega_e t] + I_- I_- \exp[-2i\omega_e t]] \bigg\},$$

where the last passage as in (26) has been omitted since it looks very similar to that of $H^J_{R,\text{sec}}$.

The selective modulation of the size of the interaction Hamiltonians then arises from the rank-sensitive angular dependence here written. The reason for the name of the technique derives from the above-written Hamiltonians. It is in fact clear that two different «magic» values of θ, $\theta^{(1)}_M = \cos^{-1} 0 = 90°$ and $\theta^{(2)}_M = \cos^{-1}(1/\sqrt{3}) = 54°44'$ reduce to zero the secular part of the linear and bilinear Hamiltonians, respectively. When second-rank Hamiltonians are reduced to second-order effects by $\theta^{(2)}_M$, first-rank ones are scaled by $\cos\theta^{(2)}_M = 1/\sqrt{3}$. On the other hand, when, say, chemical shifts are to be nulled, bilinear interactions are halved and changed in sign $((3/2)\cos^2\theta^{(1)}_M - 1/2 = -1/2)$.

Several applications of those features have been devised, especially concerning line-narrowing capabilities both for homogeneous and inhomogeneous broadened solidlike systems [4-6, 24-27]. In the following section we are going to describe the ones upon which experimental research has been developed in our laboratory during the last few years.

5. – Applications of MARF experiments.

5˙1. *MARF imaging*. – Feasibility of NMR imaging is ruled by the resolution condition [28]

$$(28) \quad \gamma G \Delta r \geq \delta\omega,$$

which states that the interaction with the magnetic-field gradient G has to be dominant with respect to the internal ones described by the linewidth $\delta\omega$, if a resolution Δr is to be obtained.

When, as in solidlike systems of great interest in materials science and in biomedicine, $\delta\omega$ becomes as wide as several tens of kHz, mostly because of homonuclear dipolar interactions, alternative methods with respect to the traditional ones are needed if one wants to avoid the use of large gradients, which nevertheless are used in some solid-state imaging (SSI) techniques [29-31]. Most of them rely on the combination of standard imaging methods with a line-narrowing technique [32-34] and our one [7, 34] can be included in this class of experiments. Although the concept is valid for any kind of broadening interaction, via the proper choice of the «rank» of the magic angle, we shall only deal with dipolar broadened solids, intending as 55°44' the value of the magic angle which is here considered.

The basic idea is to detect the evolution of the system in the TRCF under a magic-angle-oriented effective field whose magnitude is a linear function of the spatial co-ordinates (X, Y, Z). In this situation the MARF narrowed linewidth, $\delta\omega_R$, of the system should replace the unperturbed one in eq. (28) as well as an effective-field gradient G_{eff} has to be intended entering the left-hand side of the imaging condition. Such a gradient must accomplish both features of changing the intensity of B_{eff} and keeping fixed its magic orientation in the RCF, in all planes perpendicular to its direction. Such a condition requires the use of both a static gradient acting on ΔB_0, $G_0 = (\partial B_0/\partial X, \partial B_0/\partial Y, \partial B_0/\partial Z)$ and a RF one acting on B_1, $G_1 = ((\partial B_1/\partial X)\cos\omega t, (\partial B_1/\partial Y)\cos\omega t, (\partial B_1/\partial Z)\cos\omega t)$. Their composition creates the required static-effective-field gradient $G_{\text{eff}} = (\partial B_\zeta/\partial X, \partial B_\zeta/\partial Y, \partial B_\zeta/\partial Z)$ in the TRCF, where imaging can then be performed on a narrow-band system, provided that $G_1/G_0 = \text{tg}\,\theta_M$, to maintain the magic-angle condition all over the sample.

For fixed resolution, a much smaller intensity with respect to the traditional methods is required for the field gradient, depending on the degree of narrowing obtained, which is usually limited by the B_1 inhomogeneity. In fact this effect, not only introduces an inhomogeneous broadening of the TRCF line due to the distribution in ω_e, but also determines a valuable retaining of the homogeneous one relative to the incomplete nulling of the secular dipolar Hamiltonian in the TRCF, because of a distribution in the values of the angle between B_{eff} and B_0.

Two main methods have been implemented in our laboratory, their basic difference relying on the modality of detection of the TRCF evolution of the magnetization.

The first one is the full rotating-frame method, based on the direct detection technique which we have been dealing with throughout this lecture. In principle, anyone of the traditional imaging methods can be implemented, although in practice severe limitations are imposed by the probe architecture. The main problems are due to the coupling between the RF homogeneous and gradient coils, which prevents them from being controlled independently and reduces the degrees of freedom at disposal of the experimenter. The first images of

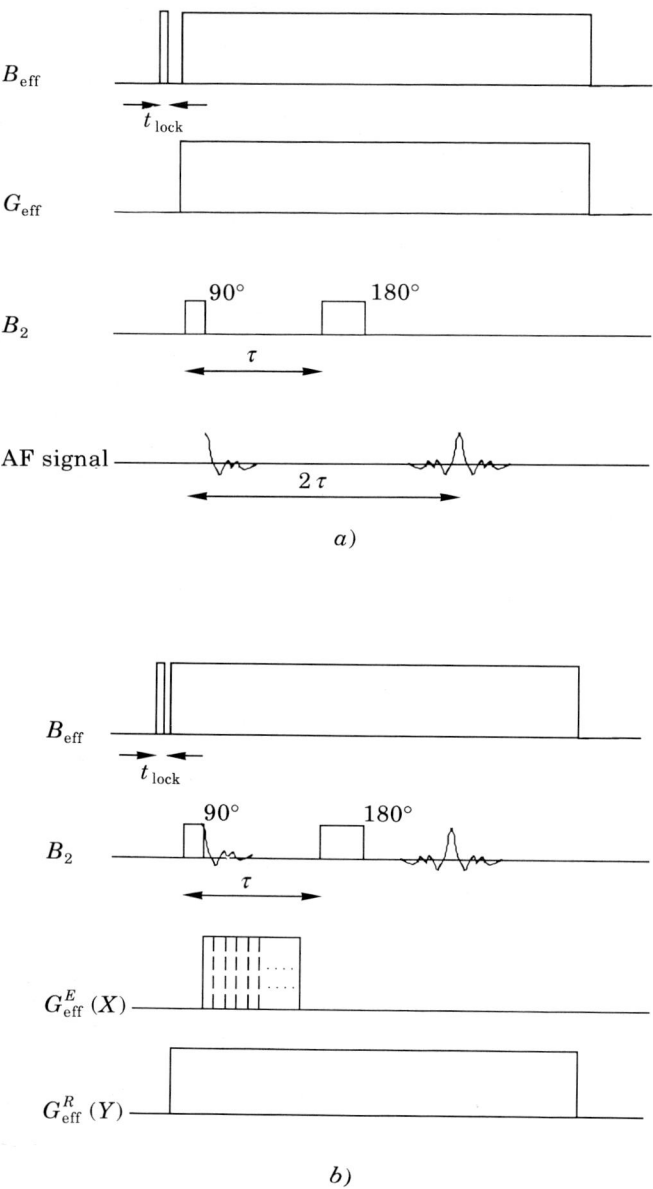

Fig. 3. – Timing diagram of the sequences proposed for the MARF solid-state imaging method. *a)* Monodimensional projection, with a fixed effective-field gradient direction; *b)* 2DFT scheme, with the readout $G_{\text{eff}}^E(X)$ and the phase-encoding $G_{\text{eff}}^R(Y)$ effective-field gradients along two orthogonal directions.

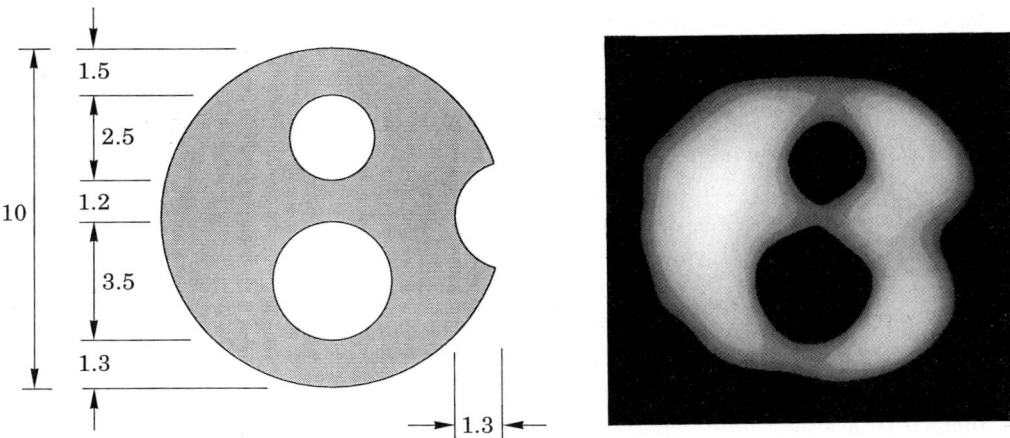

Fig. 6. – 2D projection-reconstruction fluorine image of a teflon cylinder drilled as sketched beside the image, performed with the L-G MARF method. 18 projections have been collected, in this case, too, by rotating the sample. The field gradient intensity was $G_{\text{eff}} \approx 1.4 \, \text{G/cm}$, to obtain a resolution of about 1 mm on a solid system with a LCF bandwidth of about 10 kHz, reduced to about 500 Hz by the magic-angle condition (from ref. [35]).

For this version of SSI experiments, too, we have started performing projective imaging, as that of fig. 6 [35], while the feasibility of faster and more versatile schemes, mostly based on 2DFT methods, is under investigation. Also here the problem related to the coupling of the RF coils is an important one, although, due to the use of separate coils to produce the inhomogeneous RF field and the homogeneous 90° pulse, coil decoupling procedures of the type proposed in some double-coil detection schemes [36] can be exploited.

The two approaches to detection, of course, bring to different sensitivity and different ranges of application. The direct one is chacterized by a potentially wide applicability and versatility, but also shows an inherently low sensitivity, due to the low frequency of the TRCF precession, only in part recovered by the increase in the induction on the AF coil with the usually high number of windings (about one thousand) and by the possibility of signal averaging. The Lee-Goldburg one is, of course, more sensitive, being based on LCF signal detection and moreover has a very easy experimental set-up, needing only a dedicated design of the probe, but shows a reduced field of applicability especially concerning the relaxation time sensitivity which can be introduced in the images so obtained. A more detailed discussion of the features offered by our SSI methods and their sensitivity comparison can be found in [35].

5˙2. *Relaxation studies.* – It is well known how NMR relaxation times are sensitive to the spectrum of the molecular motions, in relation to the Zeeman solids obtained with this technique have been performed by the projection-reconstruction method [28], using a single RF coil to provide both B_1 and G_1 and manually rotating the sample, following the simple sequence of fig. 3a), and the result of ref. [8] is here presented in fig. 4. 2DFT procedures have been implemented with the sequence of fig. 3b), but the above-described technical problems and the low B_0 field we had to use in the first version of the experiments allowed us to get images only in particularly favourable cases. At present a new version whith a higher value for B_0 and a new architecture of the gradient coils is under test. In each case the system is perturbed by a spin echo sequence in the TRCF, generated by $90°$-t-$180°$ B_2 pulses, which allows the detection of a signal not affected by the AF coil ringing [23] and can be exploited for T_{2_c} contrast on the image. Other pulse sequences can be introduced in order to introduce a relaxation time sensitivity in the results of the imaging experiment, as will be described in the next point.

The second MARF SSI method is based on the well-known indirect detection method by LEE and GOLDBURG [4], where the TRCF evolution of the magnetization is recorded by the observation of the LCF FID's in a set of magic-angle experiments where the period of the evolution in the presence of $\boldsymbol{B}_{\text{eff}}(\boldsymbol{r}) = \boldsymbol{B}_{\text{eff}}(0) + \boldsymbol{G}_{\text{eff}} \cdot \boldsymbol{r}$ is incremented in steps of duration Δt. In fact, referring to the first line of fig. 5, the first point of the LCF signal is proportional to M_Z, which precesses around ζ when the TRCF is «on», and the reconstructed time evolution of the magnetization in the TRCF is thus described by

$$(29) \qquad M_Z^R(n\Delta t, r) = M_0(r)[\sin^2 \theta_M \exp[-i\gamma B_{\text{eff}}(r) n\Delta t] + \cos^2 \theta_M],$$

with $M_0(r)$ the equilibrium magnetization of the voxel at position r.

Δt can thus be considered as the sampling time of the TRCF FID, so that the corresponding spectrum will appear in a window $1/2\Delta t$ Hz wide and, since the distribution of the spatially encoded precession frequencies in the sample is centred at $\omega_{\text{eff}}(r = 0) \gtrsim \delta\omega$, the time step has to be accurately chosen in order to avoid aliasing, but also exceedingly long acquisition times.

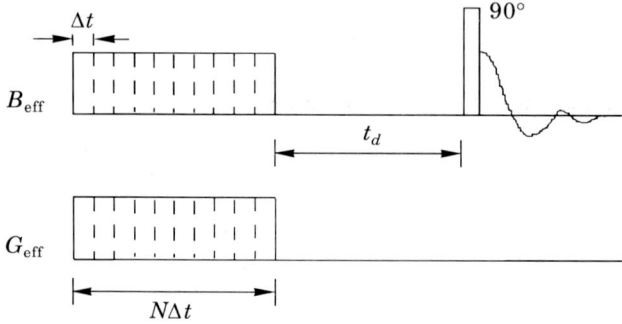

Fig. 5. – Timing diagram of the indirect-detection version (Lee-Goldburg) of the MARF SSI experiments.

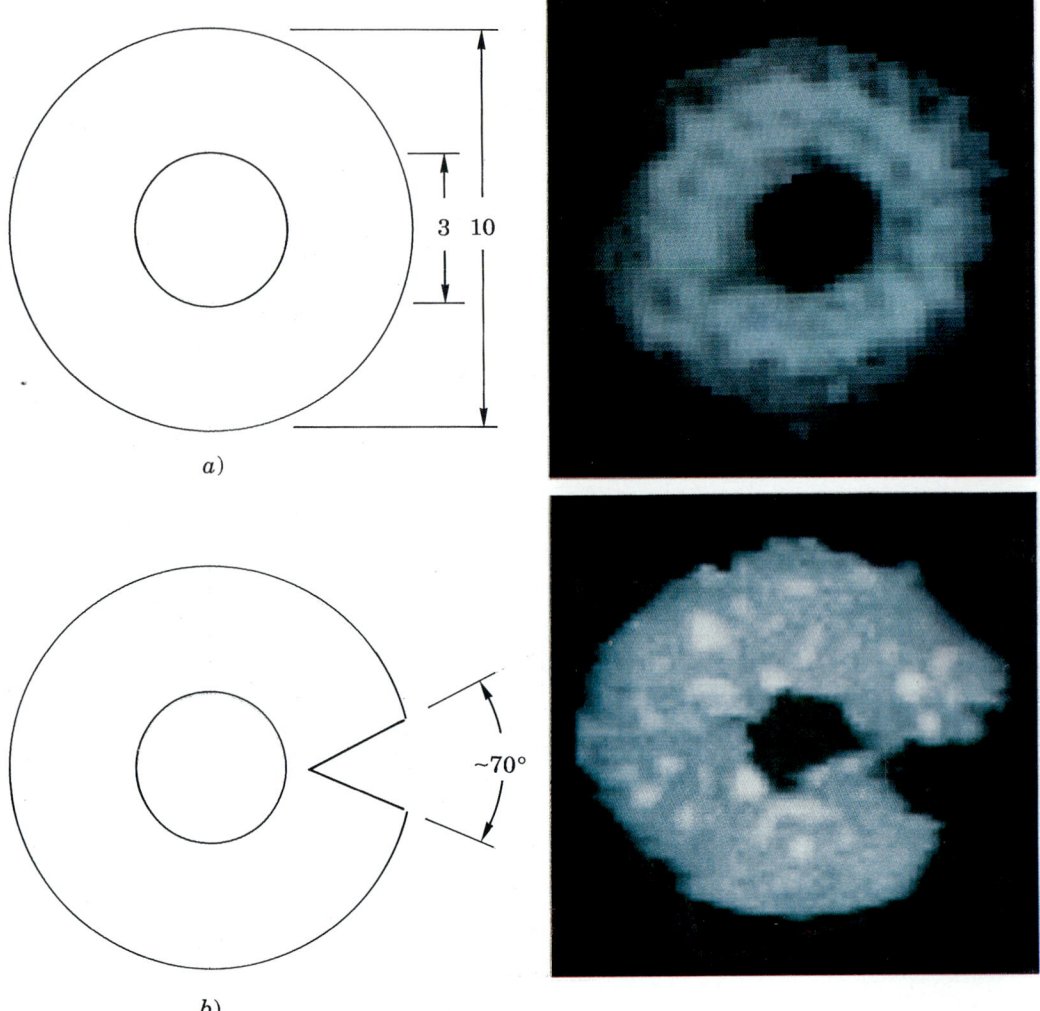

Fig. 4. – Bidimensional projection-reconstruction proton image of two phantoms composed by compressed powdered adamantane, as obtained with the sequence of fig. 3a), repeated to obtain 16 independent projections, by sample rotation. The LCF bandwidth of adamantane is about 18 kHz, while the spectral width of each image is about 8 kHz, due to a gradient $G_{eff} \approx 1.7$ G/cm. The resolution is about 1 mm (from ref. [8]).

field intensity, and that relaxation dispersion provides valuable information on the dynamics of the system.

As reviewed by GOLDMAN in this volume [20], one can get easy relationships between $T_1(\omega)$ or $T_2(\omega)$ and τ_c, the correlation time of the molecular motion considered, only within rather strict constraints. The most important one is the distinction between the weak ($\tau_c \ll T_{2,\text{r.l.}}$) and strong ($\tau_c \gg T_{2,\text{r.l.}}$) collision regimes. $T_{2,\text{r.l.}}$ here denotes the transverse relaxation time of the rigid-lattice state of the system.

What is more commonly studied is the longitudinal relaxation dispersion, which gets the well-known form (BPP theory [37])

$$\text{(30)} \qquad T_1^{-1} \approx \frac{B_\text{L}^2 \tau_c}{1 + \omega^2 \tau_c^2}$$

only in the weak-collision limit. B_L is the average dipolar local field. Moreover, a limit is imposed on the size of the static field involved, with respect to the local field, to preserve the perturbative approach which leads to eq. (30), namely $B_0 \gg B_\text{L}$. This is the reason why full LCF relaxometry cannot investigate exceedingly slow motions, still keeping (30) as a starting point.

By transferring the problem to the above-defined interaction representations, one can extend the limits of feasibility of simple relaxation studies towards slower and slower dynamics. Dipolar relaxation in the TRCF is in fact described by the following approximate relations (weak collision, $B_\text{eff} \gg B_\text{L}$, $\omega \gg \omega_e$) [38, 39]:

$$\text{(31)} \qquad T_{1\rho}^{-1} \approx \tau_c \frac{[d_{02}(\theta)]^2}{1 + 4\omega_e^2 \tau_c^2},$$

$$\text{(32)} \quad T_{2\rho}^{-1} \approx \tau_c \left\{ [d_{00}(\theta)]^2 + \frac{[d_{01}(\theta)]^2 + [d_{0-1}(\theta)]^2}{1 + \omega_e^2 \tau_c^2} + \frac{[d_{02}(\theta)]^2 + [d_{0-2}(\theta)]^2}{1 + 4\omega_e^2 \tau_c^2} \right\}.$$

Due to the ω_e-dependence, the range of τ_c's detectable is remarkably shifted towards larger values. Since the spin system is initially polarized along B_0, the sensitivity of the experiments is usually satisfactory both in the traditional $T_{1\rho}$ measurements by spin-lock sequences [40], with LCF signals detected, and in the new direct versions [25, 41-43], where the classic pulse sequences can be performed and $T_{2\rho}$ is easily evaluable from the spectrum.

A very interesting feature arises from the θ-dependence in (31), (32): when $\theta = \theta_\text{M}$ first-order dipolar interaction is nulled so that the size of the dipolar local field is reduced by something like two orders of magnitude, depending only on 2nd-order terms. As a consequence of the line narrowing obtained, $T_{2\rho}$ is lengthened, and moreover it acquires a frequency dependence similar to that of $T_{1\rho}$ showing a minimum for $\tau_c \approx \omega_e^{-1}$. The frequency-independent term in (32), which usually damps the transverse relaxation time dispersion, is replaced by a much weaker zero-frequency contribution from 2nd-order perturbation theory [4, 26], and becomes important for $\tau_c \gg \omega_e^{-1}$, thus allowing the use of $T_{2\rho}$ dispersion for slow-motion studies.

However, since the condition $B_{\text{eff}} \gg B_\text{L}$ must be fulfilled in order to reach damping of dipolar interaction, it is not possible to extend the RCF relaxation time dispersion towards those frequencies which are too low to keep the RCF high-field approximation still holding.

Such an extension is instead obtained by taking advantage of the result that in the DRTF the high-field condition is determined by $B_2 \gg B_{\text{L}\rho} \approx 10^{-2} B_\text{L}$.

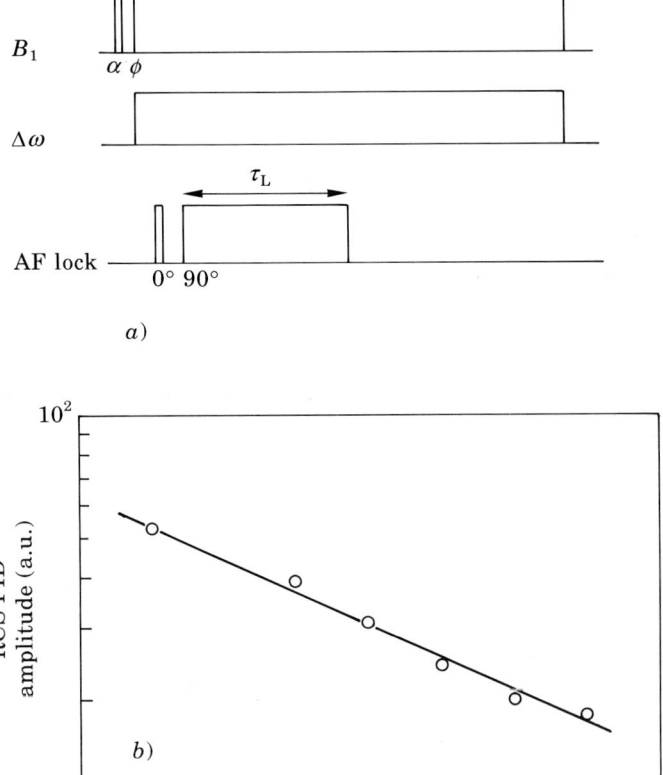

Fig. 7. – Measurements of the longitudinal relaxation time in the DTRF, $T_{1\rho\rho}$. a) Timing diagram of the spin-lock sequence in the DTRF, b) $T_{1\rho\rho}$ of doped water, as measured by the above-described method. A value $T_{1\rho\rho} \approx 10$ ms is evaluated from the graph (from ref. [41]).

A BPP-like expression for the longitudinal relaxation rate in the DRTF

$$T_{1\rho\rho}^{-1} \approx \frac{B_{L\rho}^2 \tau_c}{1 + (\gamma B_2 \tau_c)^2} \tag{33}$$

is thus obtained, relying on a «reduced» high-field approximation and an «extended» weak-collision case. $T_{1\rho\rho}$ then shows a sensitivity on B_2 (maximum relaxation rate for $\tau_c \approx \omega_2^{-1}$) which can be exploited down to frequency values of the order of hundreds of Hz.

Measurements of $T_{1\rho\rho}$ can be performed by a spin-lock sequence in the RCF (a direct transposition of the «classic» one [40]), thus observing the amplitude of the DRTF FID with increasing locking time, as in fig. 7, where a preliminary result showing a measurement of the $T_{1\rho\rho}$ of doped water as obtained by this technique is presented. The experiment has been performed with both RF and AF excitations set on resonance, since the sample chosen to prove the feasibility of the direct method did not need a particular action on dipolar couplings. Nevertheless, the direction of B_2 in the TRCF or, better, of $B_{2,\text{eff}}$, if the AF field is applied off resonance, does affect the transformation of the residual dipolar terms into the third rotating frame, and a new magic direction can allow a further damping of the dipolar interaction [25]. This consideration can in principle be iterated to find sensitivity on very slow dynamic processes.

Several studies which prove the validity of the DRTF relaxometry approach to slow-motion studies have already been presented [42,43]. The new method which we propose is currently being developed with a particular interest about the application to relaxation time contrasted imaging of systems where the slow dynamics components are the interesting ones. In fact a lot of physical systems, other than the «true» rigid solids, show both liquidlike and solidlike behaviours and often several interesting features are linked to the latter one.

As an example we can quote dense colloidal systems such as gels and pastes, biological macromolecular aggregates, liquid crystals, ceramics during some phases of preparation and so on. Especially as far as biomedicine is concerned, it is likely that study of the slow and ultraslow dynamics contributions can provide valuable information on intramolecular interactions, on hydration dynamics and for a model study of the biostructures. As a further example, this kind of approach can be exploited in solids with various degrees of rigidity to study the response of the system to an applied stress [41,44].

$T_{1\rho}$ contrasted imaging methods have already been presented [45,46]; since they are based on the LCF high-field approximation, localized slow-motion studies can be performed only on those components for which the «standard» weak-collision approximation holds. Also with our imaging method described in the preceding paragraph, $T_{1\rho}$ and $T_{2\rho}$ can be performed in both detection versions: in the direct one the usual inversion recovery and spin echo sequences can be straightforwardly implemented; in the Lee-Goldburg one, spin-locking along

B_{eff} and refocusing of M_z by B_{eff} inversion should be performed, for $T_{1\rho}$ and $T_{2\rho}$ imaging, respectively.

Finally, if we want to apply to imaging the above-presented considerations about the extension of the weak-collision and high-field approaches, one needs to implement a $T_{1\rho}$-sensitive imaging sequence. This can be obtained only with the direct-detection method, by combining the spin-lock sequence of fig. 7 with a MARF imaging one, inserted at the end of the AF locking pulse [41].

This kind of approach may then provide the possibility of the ultraslow-dynamics characterization of heterogeneous systems or of homogeneous ones under the action of spatially localized perturbations.

REFERENCES

[1] a) A. G. REDFIELD: *Phys. Rev.*, **98**, 1787 (1955); b) J. R. FRANZ and C. P. SLICHTER: *Phys. Rev.*, **148**, 287 (1966).
[2] A. ABRAGAM: *Principles of Nuclear Magnetism* (Clarendon Press, Oxford, 1961).
[3] a) M. GOLDMAN: *Spin Temperature and Nuclear Magnetic Resonance in Solids* (Clarendon Press, Oxford, 1970); b) J. F. JAQUINOD and M. GOLDMAN: *Phys. Rev. B*, **8**, 1944 (1973).
[4] M. LEE and W. I. GOLDBURG: *Phys. Rev.*, **140**, 1261 (1965).
[5] A. E. MEFED and V. A. ATSARKIN: *Sov. Phys. JETP*, **47**, 378 (1978).
[6] F. DE LUCA, B. C. DE SIMONE, B. MARAVIGLIA and C. NUCCETELLI: *Solid State Commun.*, **70**, 797 (1989).
[7] F. DE LUCA and B. MARAVIGLIA: *J. Magn. Reson.*, **67**, 169 (1986).
[8] F. DE LUCA, B. C. DE SIMONE, N. LUGERI, B. MARAVIGLIA and C. NUCCETELLI: *J. Magn. Reson.*, **90**, 124 (1990).
[9] F. DE LUCA, B. C. DE SIMONE, N. LUGERI and B. MARAVIGLIA: *Solid State Commun.*, **82**, 151 (1992).
[10] M. E. ROSE: *Elementary Theory of Angular Momentum* (Wiley, New York, N.Y., 1967).
[11] U. HAEBERLEN and J. S. WAUGH: *Phys. Rev.*, **185**, 420 (1969).
[12] J. S. WAUGH, L. M. HUBER and U. HAEBERLEN: *Phys. Rev. Lett.*, **20**, 180 (1968).
[13] U. HAEBERLEN: *High Resolution NMR in Solids—Selective Averaging, Adv. Magn. Reson.*, Supplement 1 (Academic Press, New York, N.Y., 1976).
[14] M. MEHRING: *Principles of High Resolution NMR in Solids*, 2nd edition (Springer-Verlag, Berlin, 1983).
[15] E. R. ANDREW, A. BRADBURY and G. EADES: *Nature (London)*, **182**, 165 (1958).
[16] C. P. SLICHTER: *Principles of Magnetic Resonance*, II edition (Springer-Verlag, Berlin, 1990).
[17] M. TINKHAM: *Group Theory and Quantum Mechanics* (McGraw-Hill, New York, N.Y., 1964).
[18] R. ERNST, G. BODENHAUSEN and A. WOKAUN: *Principles of One and Two Dimensional NMR Spectroscopy* (Springer-Verlag, Berlin, 1987).

[19] L. L. BUISHVILI and M. G. MENABDE: *Sov. Phys. JETP*, **50**, 1176 (1979).
[20] M. GOLDMAN: this volume, p. 1.
[21] M. MEHRING and J. S. WAUGH: *Phys. Rev. B*, **5**, 3459 (1972).
[22] V. G. POKAZAN'EV and L. I. YAKUB: *Sov. Phys. JETP*, **46**, 114 (1977).
[23] F. DE LUCA, P. FATTIBENE, N. LUGERI, R. CAMPANELLA and B. MARAVIGLIA: *Appl. Magn. Reson.*, **2**, 93 (1991).
[24] D. BAARNAL and I. J. LOWE: *Phys. Rev. Lett.*, **11**, 258 (1963).
[25] V. A. ATSARKIN, A. E. MEFED and M. I. RODAK: *Sov. Phys. JETP*, **21**, 1537 (1979).
[26] Y. N. IVANOV, B. N. PROVOTOROV and E. B. FEL'DMAN: *Sov. Phys. JETP*, **48**, 930 (1978).
[27] W. K. RHIM, A. PINES and J. S. WAUGH: *Phys. Rev. B*, **3**, 684 (1971).
[28] P. MANSFIELD and P. G. MORRIS: NMR *Imaging in Biomedicine, Adv. Magn. Reson.*, Supplement 2 (Academic Press, New York, N.Y., 1982).
[29] S. EMID and J. H. N. CREIGHTON: *Physica B*, **128**, 81 (1985).
[30] P. J. MCDONALD, J. J. ATTARD and D. G. TAYLOR: *J. Magn. Reson.*, **72**, 224 (1987).
[31] A. A. SAMOILENKO and K. ZICK: *Bruker Report*, **1**, 40 (1990).
[32] D. G. CORY, J. W. M. VAN OS and W. S. VEEMAN: *J. Magn. Reson.*, **76**, 543 (1988).
[33] G. C. CHINGAS, J. B. MILLER and A. N. GARROWAY: *J. Magn. Reson.*, **66**, 530 (1986).
[34] N. LUGERI, F. DE LUCA and B. MARAVIGLIA: in NMR *Microscopy*, edited by R. KIMMICH and W. KUHN (VCH Verlagsgesellschaft, Weinheim, 1991) p. 103.
[35] F. DE LUCA, B. C. DE SIMONE, N. LUGERI and B. MARAVIGLIA: *J. Magn. Reson. A*, **102**, 287 (1993).
[36] A. HAASE: *J. Magn. Reson.*, **61**, 130 (1985).
[37] N. BLOEMBERGEN, E. M. PURCELL and R. V. POUND: *Phys. Rev.*, **73**, 679 (1948).
[38] G. P. JONES: *Phys. Rev.*, **148**, 332 (1966).
[39] J. S. BLICHARSKI: *Acta Phys. Pol. A*, **41**, 223 (1973).
[40] D. LOOK and I. J. LOWE: *J. Chem. Phys.*, **144**, 2995 (1966).
[41] F. DE LUCA, N. LUGERI, B. C. DE SIMONE and B. MARAVIGLIA: *Colloids and Surfaces A*, **72**, 63 (1993).
[42] A. E. MEFED and V. A. ATSARKIN: *Phys. Status Solidi A*, **93**, K21 (1986).
[43] A. E. MEFED, V. A. ATSARKIN and M. E. ZHABOTINSKII: *Sov. Phys. JETP*, **64**, 397 (1986).
[44] B. BLÜMICH, P. BLÜMLER, E. GÜNTHER, J. JANSEN, G. SCHAUSS and H. W. SPIESS: in NMR *Microscopy*, edited by R. KIMMICH and W. KUHN (VCH Verlagsgesellschaft, Weinheim, 1991), p. 167.
[45] R. E. SEPPONEN, J. A. POHJONEN, J. T. SIPPONEN and J. I. TANTLU: *J. Comput. Ass. Tomography*, **9**, 1007 (1985).
[46] E. ROMMEL, R. KIMMICH, H. KOPERICH, C. KURSE and K. GERSONDE: *Magn. Reson. Med.*, **24**, 199 (1992).

A Proposed Stern-Gerlach Experiment on Individual Spins in Solids.

M. BLOOM

Canadian Institute for Advanced Research and Department of Physics
University of British Columbia
6224 Agricultural Road, Vancouver, B.C., Canada, V6T 1Z1

1. – Introduction.

One of the oft-cited disadvantages of NMR in relation to other spectroscopic methods is its relatively low sensitivity. The minimum number of spins required to produce an observable NMR signal depends on many parameters such as gyromagnetic ratio, magnetic field, the relaxation times T_1 and T_2, temperature, etc., so that one cannot specify a number applicable to all systems, even for optimal signal/noise. Still, I suppose that most people would accept a statement that there are very few situations in which fewer than 10^{14} spins are detectable. Consequently, the proposal in a recent paper by SIDLES[1] of a novel method for detecting a type of magnetic-resonance signal of *a single spin in a solid* is sensational. Though no successful experiment has been reported as yet, some detailed design specifications have been given[2, 3]. The proposed experiment involves a type of resonant Stern-Gerlach coupling between the precession of spins at their Larmor frequency and the oscillations of a mechanical oscillator, *i.e.* an unusual type of double resonance, which makes it appropriate for discussion at this school. I have the impression that, though the claims for sensitivity look preposterous at first glance, the basic idea is sound and may lead to some new approaches to the detection of nuclear and/or electron spins. In any case, with the help of the published papers[1-3], I shall attempt to explain, in simple terms, the broad principles of the experiment and to examine quantitative aspects of its sensitivity. I shall also discuss briefly some related experiments that might be carried out.

For reasons to be explained later, SIDLES has used the term *folded* Stern-Gerlach (FS-G) experiment to characterize his proposed technique. It is closely related to a resonant modification of the Stern-Gerlach (S-G) experiment developed in our laboratory more than 25 years ago[4, 5] that we called the transverse Stern-Gerlach (TS-G) experiment. The TS-G experiment makes use of an oscillating or rotating *inhomogeneous* magnetic field at a frequency close to the Larmor frequency of the spins. The plan of my seminar is to first review briefly some essential features of the S-G, TS-G and FS-G experiments and the connections between the TS-G and FS-G; I will then address quantitative aspects of the FS-G experiment in order to assess its experimental feasibility.

2. – Review of the Stern-Gerlach experiment.

2`1. *Classical description.* – In the S-G experiment, a beam of neutral particles of angular momentum $I\hbar$ and magnetic moment $\boldsymbol{\mu} = \gamma \boldsymbol{I}\hbar$ is subjected to a magnetic field having a homogeneous part $\boldsymbol{H}_0$ and an inhomogeneous part $\boldsymbol{h}_1(\boldsymbol{r})$ satisfying the conditions $\boldsymbol{\nabla} \cdot \boldsymbol{h}_1(\boldsymbol{r}) = \boldsymbol{\nabla} \times \boldsymbol{h}_1(\boldsymbol{r}) = 0$. For simplicity, and with no loss of generality for our purpose, we restrict ourselves to the simple case of a large homogeneous field ($h_1 \ll H_0$) oriented along the z-axis, so that a classical magnetic dipole precesses around the direction of $\boldsymbol{H}_0$ at an angular Larmor frequency close to $\omega_0 = -\gamma H_0$. Then, μ_z is a constant of the motion while μ_x and μ_y are proportional to $\cos\{\omega_0 t + \phi\}$ and $\sin\{\omega_0 t + \phi\}$, respectively. As a result, the S-G force, given by

$$(1) \quad \boldsymbol{F}_{\text{S-G}} = -\boldsymbol{\nabla}(\boldsymbol{\mu} \cdot \boldsymbol{h}_1) = -(\boldsymbol{\mu} \cdot \boldsymbol{\nabla})\boldsymbol{h}_1 = -\left\{\mu_x \frac{\partial \boldsymbol{h}_1}{\partial x} + \mu_y \frac{\partial \boldsymbol{h}_1}{\partial y}\right\} - \mu_z \frac{\partial \boldsymbol{h}_1}{\partial z},$$

consists of two distinct parts, one of which is proportional to μ_z and is independent of time, thus giving rise to cumulative momentum changes in momentum and another, given in the $\{\;\}$ brackets of eq. (1), which oscillates at the Larmor frequency and contributes only oscillatory momentum changes. This has the consequence that the change of momentum due to $\boldsymbol{F}_{\text{S-G}}$ after a time τ corresponding to many Larmor periods, *i.e.* $\omega_0 \tau \gg 1$, is

$$(2) \quad \Delta \boldsymbol{p}_{\text{S-G}} = \int_0^\tau \boldsymbol{F}_{\text{S-G}} \, dt = -\gamma \hbar \tau \boldsymbol{G} I_z + O\left\{\frac{1}{\omega_0 \tau}\right\},$$

where $\boldsymbol{G} = \partial \boldsymbol{h}_1/\partial z$ is the effective (vector) gradient of the inhomogeneous magnetic field. The well-known result of this analysis is that *the displacement of the particle after it leaves the* S-G *deflecting region provides a measurement of a specific component of angular momentum, I_z in this case.* Such measurements always show that I_z is quantized and takes on only a discrete number of values. This quantization is detectable even if the displacement of the particles is adequately described classically, as is the case in almost all of the atomic- and molecular-beam experiments carried out since the early experiments of Stern and Gerlach. The same is likely to be the case for the FS-G experiment to be described later. It is, however, worth reminding ourselves that there is no great difficulty in setting up a quantum-mechanical description of the S-G experiment.

2˙2. *Quantum description* [4,6]. – Consider a beam of spin-1/2 neutral particles in which there is no correlation between the z-coordinate and spin I of the particles, so that the initial-state vector of a particle, $|\psi(z,t)\rangle$ at $t = 0$, can be written as a product of a spatial and spin parts

$$(3) \qquad |\psi(z,0)\rangle = \psi(z)\{C_+(0)|+\rangle + C_-(0)|-\rangle\},$$

where

$$(4) \qquad I_z|\pm\rangle = \pm\left(\frac{1}{2}\right)|\pm\rangle$$

and the initial, one-dimensional wave packet is expressed in terms of the momentum ($p_z = \hbar k$) eigenfunctions ($\exp[ikz]$) and their probability amplitude ($\phi(k)$) as

$$(5) \qquad \psi(z) = \int dk\, \phi(k) \exp[ikz].$$

The solution to the time-dependent Schrödinger equation for an inhomogeneous magnetic field having quadrupole symmetry, $\mathbf{h}_1(\mathbf{r}) = G\{\mathbf{x} - \mathbf{z}\}$, and $h_1 \ll H_0$, is simple. In first-order perturbation theory, the term $\gamma\hbar Gx I_x$ in the Zeeman Hamiltonian may be neglected and the solution is

$$(6) \qquad C_\pm(\tau) = C_\pm(0) \exp[-(\pm)[\omega_0 + \gamma Gz]\tau/2]$$

giving

$$(7) \qquad |\psi(z,\tau)\rangle \approx \int dk \left\{C_+(0)\phi\left(k - \frac{\gamma G\tau}{2}\right)|+\rangle + C_-(0)\phi\left(k + \frac{\gamma G\tau}{2}\right)|-\rangle\right\}.$$

Thus the initial wave packet is split into two wave packets in momentum space corresponding to the values of $\Delta p_{\text{S-G}}$ given by eq. (2) with the quantum values $I_z = \pm 1/2$. In a typical beam experiment, this splitting of the wave packet in momentum space is translated into a spatial separation by inertial flight of the particles over a longer period of time.

3. – The transverse Stern-Gerlach experiment.

In the TS-G experiment [4, 5], the *time-independent*, inhomogeneous magnetic field $\mathbf{h}_1(\mathbf{r})$ is replaced by an *oscillating*, inhomogeneous field $2\mathbf{h}_1(\mathbf{r})\cos\omega t$. For $h_1 \ll H_0$, it is convenient to replace the oscillating field by a superposition of a rotating and a counterrotating field, as in the usual description of magnetic resonance and keep only the component rotating in the same sense as the spin precession [7].

3'1. *Resonance in the* TS-G *experiment—quantization of* I_{x_e}. – When the oscillating field satisfies the resonance condition, $\omega \approx \omega_0$, the two terms in eq. (1) for $\boldsymbol{F}_{\text{S-G}}$ interchange their roles. Cumulative changes in momentum are produced by μ_x rather than μ_z, so that the separation of particles deflected by the S-G force results in quantization of I_{x_e}, *i.e.* the momentum change of the particle is proportional to I_x in the reference frame rotating about the z-axis at an angular frequency ω, instead of being proportional to I_z as in the normal S-G experiment. More generally, for $\omega \neq \omega_0$, the quantized component of angular momentum is I_{e_e}, the component of $\boldsymbol{I}$ along the *effective field in the rotating frame* [7], given by $\boldsymbol{H}_e = (H_0 - \omega_0/\gamma)\boldsymbol{k} + h_{1x}\boldsymbol{i}$. As appreciated by students of NMR, $\boldsymbol{H}_e$ plays a crucial conceptual role in the understanding of many magnetic-resonance phenomena. Experimental demonstration [5] of the quantization of I_{x_e} exhibited the operational significance of $\boldsymbol{H}_e$ as a direction of quantization for angular momentum.

Before turning to the *folded Stern-Gerlach experiment*, I emphasize three aspects of the TS-G experiment that provide insight into some features of the FS-G experiment:

i) A necessary condition for achieving quantization of the component of $\boldsymbol{I}$ in the direction of $\boldsymbol{H}_e$ is that the spins must precess many times in $\boldsymbol{H}_e$ rather than in $\boldsymbol{H}_0$, *i.e.* the condition that must be satisfied on resonance is that $\gamma h_1 \tau \gg 1$ rather than $\omega_0 \tau \gg 1$ as in the normal S-G experiment.

ii) In the TS-G experiment, as in the normal S-G experiment, cumulative changes in momentum are always associated with a nonzero time-averaged force. By contrast, the FS-G experiment achieves quantization of angular momentum via the resonant coupling to a mechanical oscillator of a S-G force oscillating at the Larmor frequency of the spins!

iii) Although the TS-G experiment is a resonance technique, it is *not* a high-resolution experiment in the frequency domain, though SIDLES and his collaborators [1-3] have suggested that it is capable, on its own, of providing high *spatial* resolution. The width of the TS-G resonance, $\approx \gamma h_1$, cannot be made arbitrarily small because of the requirement of a large gradient. The TS-G experiment is a state selector and can be combined with other procedures, as in the Rabi modification of the S-G experiment [8] to make precision measurements. The same is true of the FS-G experiment.

4. – The folded Stern-Gerlach experiment.

The idea of Sidles [1] is to use a *homogeneous, r.f. field* $2\boldsymbol{H}_1 \cos \omega t$, with $\boldsymbol{H}_1 = H_1 \boldsymbol{i}$ oriented along the x-axis, and an angular frequency ω close to that of the Larmor frequency $\omega_0 = -\gamma H_0$ in a *large, homogeneous, static field* $\boldsymbol{H}_0 = H_0 \boldsymbol{k}$ applied along the z-axis. These fields are combined with a *static, inhomoge-*

neous field $h_1(r)$. Furthermore, *either* the spin *or* the source of $h_1(r)$ is attached to a mechanical oscillator of mass m which oscillates along the x-axis at a natural frequency $(\omega + \Delta)$ that is also close to the Larmor frequency. The basic idea behind the FS-G experiment is that at resonance ($\Delta = 0$) the S-G force, arising from the interaction between the precessing spin and the inhomogeneous magnetic field, excites the mechanical oscillator in its resonant mode to an amplitude proportional to I_{x_c}. The term *folded* is used to convey the fact that the particle trajectory associated with the S-G force is spatially confined by the mechanical oscillations.

4`1. *Analysis of the* FS-G *experiment for a frictionless oscillator.* – A geometry leading to a simple analysis of the oscillator motion, and realizable in practice, is the quadrupole inhomogeneous field given by $h_1(r) = G(z - x)$, which leads to the following equation of motion for the oscillator at $\omega = \omega_0$ in the absence of oscillator or spin damping:

$$\text{(8)} \qquad \frac{d^2 x}{dt^2} + (\omega + \Delta)^2 x = \left[\frac{F_{\text{S-G}}}{m}\right] \cos \omega t + \text{nonresonant force terms},$$

where

$$\text{(9)} \qquad F_{\text{S-G}} = \gamma \hbar G I_{x_c}$$

and we have retained only the force term that is resonant with the mechanical oscillator.

Equation (8) is easily integrated and it is instructive to compare the solution at mechanical resonance ($\Delta = 0$) with the S-G experiment. For initial values of position and velocity given by $x(0) = x_0$ and $v(0) = v_0$, the solution is

$$\text{(10)} \qquad x(t) = x_0 \cos \omega t + \left\{\frac{v_0}{\omega} + \frac{F_{\text{S-G}} t}{2m\omega}\right\} \sin \omega t.$$

The term involving $F_{\text{S-G}}$ represents the perturbation of the oscillator from its original state. The resulting perturbation of the momentum, obtained from the contribution of $F_{\text{S-G}}$ to $m(dx/dt)$, for $\omega t \gg 1$, is given by

$$\text{(11)} \qquad (\Delta p)_{\text{S-G}} = \frac{1}{2} \gamma \hbar G I_{x_c} \cos \omega t,$$

a result similar to the normal S-G result of eq. (2) and to the corresponding TS-G result [5], but with a trajectory «folded» into that of an oscillator. If $(\Delta p)_{\text{S-G}}$ can be measured for a system under experimental conditions that its magnitude is proportional to I_{x_c}, as predicted for the FS-G experiment at resonance, then repetition of the experiment should result in the system being found in various *oscillator states* exhibiting *space quantization* in a manner analogous to that observed for particles in beams.

Under some conditions, it may be possible to decouple the mechanical oscillator thermally from its surroundings for a long enough time, τ_d, to measure the quantum state of the oscillator and to detect the FS-G resonance via a change in the quantum state. Then I propose, as an intuitive *semi-classical criterion for detectability*, that the perturbation to the amplitude of oscillation due to $F_{\text{S-G}}$ be greater than the characteristic quantum size of the oscillator $\lambda_0 = \sqrt{\hbar/m\omega}$. Using eqs. (9) and (10), this gives

$$\text{(12)} \qquad \frac{\gamma \hbar G I_{x\rho} \tau_{\text{d}}}{2m\omega} \geq \lambda_0 \qquad \text{(semi-classical detectability criterion)}.$$

Under conditions that the averaging time for measurement of the oscillator amplitude is greater than the thermal damping time of the oscillator τ_{d}, it is necessary to take into account the intrinsic thermal size of the oscillator $\lambda(T) = \sqrt{kT/m\omega^2}$. As described below, the *thermal criterion for detectability* is the same as eq. (12) with λ_0 replaced by $\lambda(T)$.

4'2. *Inclusion of oscillator damping.* – We now examine the criterion for detectability of the FS-G resonance under the more conservative condition that oscillator damping is included [3]. The damping time constant τ_{d} and the «quality factor» Q of an oscillator are related via $Q = \omega \tau_{\text{d}}$. Equivalently, for an oscillator having $Q \gg 1$, a fraction $1/Q$ of its energy is dissipated per cycle due to oscillator friction. Thus the steady-state amplitude $L_{\max}$ of an oscillator under the influence of $F_{\text{S-G}}$ for a time $\gg \tau_{\text{d}}$ in a FS-G experiment is given by the condition that the energy dissipated per cycle is balanced by the work done per cycle by $F_{\text{S-G}}$

$$\text{(13)} \qquad \frac{m\omega^2 L_{\max}^2}{Q} \approx F_{\text{S-G}} L_{\max} \approx \gamma \hbar G I_{x\rho} L_{\max},$$

which gives the following result expressed in several different, but equivalent, formulae:

$$\text{(14)} \qquad L_{\max} \approx \frac{\gamma \hbar G Q}{m\omega^2} I_{x\rho} \approx \frac{\gamma \hbar G \tau_{\text{d}}}{m\omega} I_{x\rho} \approx \gamma G \tau_{\text{d}} \lambda_0^2 I_{x\rho}.$$

As mentioned earlier in this section the resulting *detectability criterion* given by

$$\text{(15)} \qquad L_{\max} \geq \lambda(T) \qquad \text{(thermal detectability criterion)}$$

is equivalent to eq. (12) with the quantum size of the oscillator λ_0 replaced by its thermal size $\lambda(T)$, where $\lambda(T)/\lambda_0 = \sqrt{kT/\hbar\omega}$.

4'3. *Some practical considerations concerning mechanical oscillators.* – The detectability criteria given by eqs. (12) and (15) for the FS-G resonance suggest that improved sensitivity is obtained for oscillators of small mass and low frequency. Although low frequency in a mechanical oscillator is normally associated with large mass, it is possible that some unconventional solutions remain to be found in the optimization of the sensitivity of FS-G detectors. Here, I will only examine the detectability criteria for a very conventional quartz oscillator of a single-arm cantilever design following the discussion by SIDLES et al. [3], to which the reader is referred for details.

A single cantilever of width w and length b has an effective mass of $m = \rho b w^2 / 4$, where the density of quartz is $\rho = 2.65$ g·cm^{-3}. A quartz oscillator of dimensions $w = 57$ nm and $b = 2.86$ μm has a frequency of $\omega/2\pi = 7.17 \cdot 10^6$ Hz and an effective mass of $m = 6.2 \cdot 10^{-15}$ g. The characteristic effective oscillator sizes are then calculated to be

$$\lambda_0 = 5.7 \cdot 10^{-11} \text{ cm} \quad \text{(quantum size)}, \quad \lambda(T) = 3.43 \cdot 10^{-9} \sqrt{T} \text{ cm} \quad \text{(thermal size)},$$

while the oscillator damping time for $Q = 2 \cdot 10^6$ is $\tau_d = 0.044$ s.

4'4. *Criterion for detectability of a single proton spin by* FS-G *resonance.* – We now ask whether the FS-G resonance of a single nuclear spin can be detected, say a proton for which $|I_{x_\rho}| = 1/2$ and $\gamma = 2.67 \cdot 10^4$ e.m.u. (s^{-1}G^{-1}) or, equivalently, $2.67 \cdot 10^8$ SI units (s^{-1}T^{-1}). Using eqs. (14) and (15), we ask what value of the magnetic-field gradient G is required for detection and obtain

(16) $$G \geqslant \frac{m\omega\lambda(T)}{\gamma\hbar\tau_d I_{x_\rho}} \approx 1.75 \cdot 10^9 \sqrt{T} \text{ G/cm} = 17.5 \sqrt{T} \text{ G/Å}.$$

Upon regarding this number, most NMR spectroscopists would feel that the detection of the FS-G resonance for a single nuclear spin is completely impractical, in view of the fact that magnetic-field gradients $G > 10^3$ G/cm are difficult to attain under normal laboratory conditions. SIDLES et al. [3] acknowledge that such is the case, but argue that the experimental conditions involved in detecting the FS-G resonance are so different from those of ordinary NMR that one should keep an open mind and try to take advantage of the experimental flexibility associated with recently developed techniques in *scanning tunneling microscopy* and *atomic-force microscopy*, to which the FS-G experiment should really be compared. They argue that there remains much scope for increasing the sensitivity by scaling the detector to lower sizes.

Until someone actually carries out the experiment or demonstrates conclusively its impossibility, the possibilities of the FS-G experiment will remain a matter of opinion. My feeling is that, though technically difficult to do, there is no fundamental reason why the FS-G experiment cannot be carried out successfully. There may well be some improvements in sensitivity inherent in the length scale of a single-spin experiment. On the question of attainable gradients, for example, the field gradient at a distance r from a single spin is

(17) $$G \approx \frac{d}{dr}\left(\frac{\gamma\hbar}{2r^3}\right) \approx \frac{3\gamma\hbar}{2r^4},$$

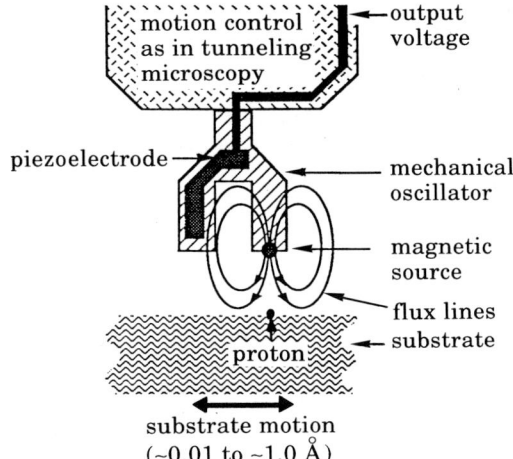

Fig. 1. – Schematic diagram illustrating a proposed experimental approach [2] to the *folded Stern-Gerlach experiment* on single proton located on the surface of a solid substrate. In this proposal, which is discussed in the text, the source of the inhomogeneous magnetic field is fixed to the quartz oscillator and the oscillating magnetic fields are generated by scanning the substrate in a manner analogous to *scanning tunneling microscopy*, while superimposing an oscillatory motion on the substrate.

which gives a value of $G \approx 3.6 \cdot 10^9$ G/cm at a distance of ≈ 1 Å from a single proton and $G \approx 2.4 \cdot 10^{12}$ G/cm for a single unpaired electron spin! This is much larger than the field gradient required for detectability of a single proton at $T = 1$ K, according to eq. (16).

I have to add a cautionary remark on the *specific* experimental design suggested by SIDLES [2], and reproduced here as fig. 1. His proposal is to generate the oscillating magnetic field in the FS-G experiment by jiggling the substrate containing the (single) proton on its surface. As I see it, the *inhomogeneous field* $h_1(r)$ would be modulated in this geometry along with the *homogeneous field* H_1, so that important parts of the resulting $F_{\text{S-G}}$ would then be constant in time rather than tuned to the frequency of the mechanical oscillator as is desired—*i.e.* this design includes characteristics of the TS-G in addition to the FS-G experiment. It would be important to arrange the geometry such that the time-averaged part of the inhomogeneous field were dominant while still producing an effective oscillating magnetic field; alternatively, it should be possible to generate a large, homogeneous r.f. field using a large coil and to scan the surface without r.f. modulation.

5. – Basic differences between Stern-Gerlach experiments and NMR.

The Stern-Gerlach experiment and its resonance counterparts (TS-G and FS-G) differ from magnetic-resonance experiments in a fundamental way. In order to carry out a normal NMR (or ESR) experiment, it is first necessary to wait a time comparable with the spin-lattice relaxation time T_1 to allow the spin system to become magnetized and/or to establish population differences among spin states via the exchange of energy between the spin system and a reservoir characterized by a temperature T. All measurements in NMR start from the ordered state associated with such a magnetization and/or population difference.

By contrast, S-G experiments can (and usually are!) carried out successfully without any initial order at all. In a beam experiment, the beam could be completely unpolarized initially. For example, an unpolarized beam in the initial state given by eq. (3) would correspond to $C_+(0) = (1/\sqrt{2})\exp[i\alpha]$ and $C_-(0) = (1/\sqrt{2})\exp[i\beta]$, where α and β are random phase factors over which the final results must be averaged. The general nature of the outcome of an experiment for such an initial state would be similar to that for an initially polarized state in the sense that the displacements of individual particles would depend only on the properties of the states $|+\rangle$ and $|-\rangle$ and not on the initial populations $|C_\pm|^2$ or phases, α and β, of the states that are affected by the order. After a sufficient change in momentum to separate the spin states has been induced by $F_{\text{S-G}}$, an individual particle must have «declared itself» to be in one of these states via its new trajectory, *i.e.* that particle's state vector will have been modified by the measurement process to correspond to one of the pure states specified by the particular measurement being made. Thus, in the S-G experiment, the final state of the particle is selected by the measurement process itself.

These remarks should hold for the FS-G experiment as well so that the process of measurement of a particular resonant excitation of a mechanical oscillator proportional to I_{x_ρ} for $I = 1/2$ will give one of two outcomes: *either* the measurement will result in a change in the state of the oscillator corresponding to the spin state $|+\rangle_x = (1/\sqrt{2})(|+\rangle + |-\rangle)$, having $I_{x_\rho} = +1/2$; *or* to the state $|-\rangle_x = (1/\sqrt{2})(|+\rangle - |-\rangle)$, having $I_{x_\rho} = -1/2$. These outcomes are predictable only in a probabilistic sense. Since the feasibility of the FS-G experiment depends only on being able to detect signals for *different* values of I_{x_ρ}, it would be possible to carry out FS-G resonance measurements on some systems in which NMR measurements would be prohibitively difficult because of extremely long T_1 values.

6. – Stern-Gerlach experiments on macroscopic spin systems.

One of the intriguing possibilities opened up by the FS-G experiment is that of doing Stern-Gerlach experiments on macroscopic spin systems. One might wish to do this in order to increase the FS-G signal amplitude through the use of a spin system composed of N spins, where $N \gg 1$, *and/or* to explore questions on the role of quantum mechanics in characterizing the physical properties of macroscopic systems. There is considerable interest in questions concerning the manifestations of quantum mechanics at the macroscopic level at the present time, particularly with respect to magnetic systems [9]. I will confine myself to a few brief remarks and leave further exploration of this subject as an exercise for the student.

6`1. *Polarized samples.* – An obvious method of increasing the FS-G signal strength would be to wait a time $\gg T_1$ for the nuclear-spin system to establish its equilibrium magnetization $\boldsymbol{M}_0$ and then carry out a *spin-locking* experiment to re-orient $\boldsymbol{M}_0$ along $\boldsymbol{H}_1$ in the rotating frame before beginning the FS-G experiment. For $\boldsymbol{M}_0$ given by Curie's law [7], this would increase the value of $|I_{x_\rho}| = = 1/2$ (for $N = 1$) by a factor of $A \approx (\hbar\omega/2kT)N$. An experiment with this polarization could only be carried out for a time of order the spin-lattice relaxation time in the rotating frame, $T_{1\varrho}$ [7], which is, however, very long for many low-temperature solids. For the oscillator frequency of ≈ 7.17 MHz used earlier and $T = 1$ K, $\hbar\omega/2kT \approx 2 \cdot 10^{-4}$, while a mass of (*e.g.*) H_2O solid, $m(H_2O) \approx 10^{-15}$ g (chosen to be somewhat less than the mass of the quartz oscillator in our earlier discussion), corresponds to $N \approx 4 \cdot 10^8$ protons. The apparent gain of $A \approx 8 \cdot 10^4$ is only illusory, however, since the average gradient G per spin, being proportiuonal to $\langle r^{-4} \rangle$, would decrease drastically with sample size if the microscopic method leading to eq. (17) were used. From a practical standpoint, it would seem appropriate, therefore, to combine a macroscopic sample with a macroscopic source of magnetic-field gradient, which experienced NMR spectroscopists tell us limits the gradient to $G \leq 10^3$ G cm^{-1}. Some consideration would have to be given to the efficiency with which relatively large masses could be coupled to mechanical oscillators of not too low frequency in order to increase N substantially.

6`2. *Unpolarized samples.* – In the previous subsection, some consideration was given to a possible gain in signal for the FS-G resonance by using large numbers of spins, $N \gg 1$, and by waiting for a macroscopic magnetization in the external magnetic field to be attained via spin-lattice relaxation. As discussed in the previous section, the FS-G experiment does not require that the system be polarized initially. It is well known that an unpolarized sample may be represented by an ensemble in which the probability that the difference ($m = n_+ - n_-$, *for* $N = n_+ + n_-$) between the number of spins pointing up and down with respect to any direction for $N \gg 1$ is given by

$$(18) \qquad P(m) = \left\{ \frac{1}{2\pi N} \right\}^{1/2} \exp\left[-\frac{m^2}{2N} \right],$$

so that a value of $|I_{x_\rho}| \approx \sqrt{N}\,(1/2)$, or even larger, would have a fairly high probability at any time. As an aside, note that, for values of $N \leq 10^8$, polarization of the sample does not give a large advantage over an unpolarized sample insofar as typical signal amplitudes are concerned. More importantly, I would like to emphasize that a FS-G experiment performed *successfully* on an unpolarized sample would force the system into one of its possible values of I_{x_c} or into a statistical distribution of width much smaller than given by the initial ensemble of eq. (18). Just as in the case of a S-G experiment performed on a single atom in a beam experiment, an initially unpolarized system would be left in a polarized state following the completion of the experiment. If the experiment were repeated on the same system at a different time, a different experimental outcome would not be surprising for an initially unpolarized sample.

The remarks made above on a «gedanken experiment» involving a macroscopic sample would only apply to a system having value of T_{1_c} much longer than the time of the experiment. During a time longer than T_{1_c}, the nuclear-spin system would exchange angular momentum with the other degrees of freedom of the solid in which the spins are located and this would presumably manifest itself via a corresponding change in the oscillator state. Another influence that has been ignored in our discussion is that of dipolar coupling between the spins. It seems to me that there is potentially interesting physics, and possibly some surprises, to be found in the investigation of these phenomena as manifested in the FS-G experiment on macroscopic systems.

Note added (January, 1993).

Since my seminar was presented, a successful FS-G experiment has been reported by D. RUGAR, C. S. YANNONI and J. A. SIDLES: *Nature (London)*, **360**, 563 (1992). The system studied was a *macroscopic* sample of < 30 ng of diphenylpicrilhydrazyl, which exhibits electron spin paramagnetism. Therefore, the experiment should be classified as EPR rather than NMR.

In the language of my seminar, the experimental conditions corresponded to a short spin-lattice relaxation time $T_1 \ll \tau_s$ and field gradient G of magnitude between approximately 10^3 G/cm and $6 \cdot 10^3$ G/cm. The mechanical oscillator was not tuned to the EPR Larmor frequency of 220 MHz, but to the second harmonic of a sinusoidal modulation of H_0, a frequency in the vicinity of 10 kHz. It was estimated that a spatial resolution of $\approx 19\,\mu m$ in one dimension was achieved.

The authors concluded from their experiment that there was reason to be optimistic about prospects for a successful performance of the FS-G experiment on a single nuclear spin, in agreement with the opinion expressed in my seminar.

REFERENCES

[1] J. A. SIDLES: *Phys. Rev. Lett.*, **68**, 1124 (1992).
[2] J. A. SIDLES: *App. Phys. Lett.*, **58**, 2584 (1991).
[3] J. A. SIDLES, J. L. GARBINI and G. P. DROBNY: *Rev. Sci. Instrum.*, **63**, 3881 (1992).

[4] M. BLOOM and K. L. ERDMAN: *Can. J. Phys.*, **40**, 179 (1962).
[5] M. BLOOM, E. ENGA and H. LEW: *Can. J. Phys.*, **45**, 1481 (1967).
[6] D. BOHM: *Quantum Theory* (Prentice-Hall, Reading, Mass., 1951), p. 593.
[7] C. P. SLICHTER: *Principles of Magnetic Resonance*, third edition (Springer-Verlag, Berlin, 1990).
[8] I. I. RABI, J. R. ZACHARIAS, S. MILLMAN and P. KUSCH: *Phys. Rev.*, **53**, 318 (1938).
[9] P. C. E. STAMP, E. M. CHUDNOVSKY and B. BARBARA: *Int. J. Mod. Phys. B*, **6**, 1355 (1992).

Proton-Electron Double-Resonance Imaging of Exogenous and Endogenous Free Radicals *in Vivo*.

D. J. LURIE and I. NICHOLSON

Department of Bio-Medical Physics and Bio-Engineering, University of Aberdeen Foresterhill, Aberdeen AB9 2ZD, U.K.

1. – Introduction.

Free radicals are defined as molecules which have one or more unpaired electrons in their outer orbitals. Changes in the concentrations of naturally occurring free radicals have been associated with the pathogenesis of many diseases, including cancer, inflammatory disease, heart disease and many more [1,2]. Stable nitroxide free radicals, already investigated for use as contrast agents in clinical NMR imaging [3], might be imaged directly and used as «tracers» when labelled chemically to another molecule of interest. The ability to image the distribution of free radicals in the bodies of animals or humans would, therefore, have profound implications for medicine and biology.

Several research groups worldwide are investigating electron paramagnetic-resonance (EPR) imaging for various applications, including the imaging of free-radical distributions *in vivo* [4]. The principles behind EPR imaging are identical to those used in NMR imaging, namely the use of magnetic-field gradients to provide spatial discrimination, but a major difficulty lies in the breadth of EPR resonance lines, necessitating the use of magnetic-field gradient strengths at least an order of magnitude larger than those currently used in NMR imaging. Consequently, EPR imaging is currently limited to samples approximately 10 cm in size. The short electron relaxation times of most free radicals means that the majority of EPR imaging experiments are continuous wave, rather than pulsed, so that slice selection is not possible, and neither can the highly developed pulse sequences used in NMR imaging be applied directly in EPR imaging.

Over the last five years, our group has been developing a technique for imaging free radicals using nuclear magnetic double resonance. Since the

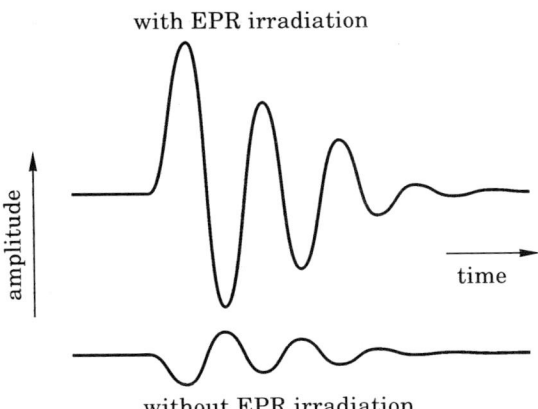

Fig. 1. – Diagram of NMR FID signals (simulated), showing the effect of EPR irradiation in a DNP experiment, with an enhancement factor of $E = -5$.

method is based on conventional proton NMR imaging, it does not suffer from the drawbacks currently associated with EPR imaging. We call the technique proton-electron double-resonance imaging (PEDRI); it is a combination of proton NMR imaging with dynamic nuclear polarization (DNP) (also known as the Overhauser effect or as nuclear-electron double resonance). In DNP, an NMR resonance of the sample is observed in the usual manner while an EPR resonance of a free-radical solute is irradiated. Under favourable conditions, irradiation of the solute's EPR can cause a large increase in the amplitude of the observed NMR signal (fig. 1). In PEDRI, then, the EPR irradiation is carried out while a proton NMR image is collected: the NMR signal is enhanced in regions of the sample containing free radical, and these regions appear with correspondingly greater intensity in the final NMR image[5]. Subtraction of NMR images obtained with and without EPR irradiation yields an image showing only the distribution of the free radical in the sample.

2. – Summary of relevant results from dynamic nuclear polarization theory.

Since the main topic of this lecture is the application of DNP to the imaging of free radicals, no attempt will be made here to explore the theoretical background, nor to derive the relevant expressions describing the enhancement of the NMR signal. The background to the double-resonance experiment has in any case been covered elsewhere in the school, and the reader is also directed to three excellent reviews of the use of DNP in the study of free-radical solutions[6-8]. It is useful, however, to summarize the main results of DNP theory.

2˙1. *The DNP enhancement.* – The enhancement of the NMR signal in a DNP experiment may be written empirically as

$$E = \frac{A_z}{A_0}, \tag{1}$$

where A_z and A_0 are, respectively, the NMR signals obtained with and without EPR irradiation. The observed enhancement factor depends on a number of factors through the relationship

$$E = 1 - \rho f s \frac{|\gamma_S|}{\gamma_P}, \tag{2}$$

where ρ is called the coupling factor, f the leakage factor, and s the saturation factor, with γ_S and γ_P being the electron and proton gyromagnetic ratios.

2˙2. *The coupling factor.* – The coupling factor, ρ, depends on the nature and time dependence of the electron-proton interactions: it takes a value of -1 if the interactions are purely scalar, while $\rho = +0.5$ when dipole-dipole interactions are dominant, as is the case in the experiments which will be discussed here.

2˙3. *The leakage factor.* – The leakage factor, f, measures the fraction of the proton longitudinal relaxation arising from proton-electron interactions, and can be written as

$$f = \frac{R_1 - R_1^0}{R_1}, \tag{3}$$

where R_1 is the proton longitudinal relaxation rate of the free-radical solution and R_1^0 is the relaxation rate of the pure solvent. ($R_1 = 1/T_1$, $R_1^0 = 1/T_1^0$, where T_1 and T_1^0 are the longitudinal, or spin-lattice, relaxation times of the free-radical solution and pure solvent, respectively.) Since $R_1 = R_1^0 + kc$, where k is the relaxivity of the free radical and c is its concentration, one may rewrite eq. (3) as

$$f = \frac{kc}{R_1^0 + kc} \tag{4}$$

in order to reveal the dependence of the observed enhancement on the concentration of the free-radical solute. For small values of kc (*i.e.* for low concentrations), $1 - E$ is proportional to c.

2˙4. *The saturation factor.* – The saturation factor, s, contains the dependence of the enhancement on the strength of the EPR irradiation and the electron relaxation times. With EPR irradiation assumed to be at the electron Larmor frequency, the saturation factor may be written as

$$s = \frac{1}{n}\left[\frac{\gamma_S^2 B_2^2 \tau_1 \tau_2}{1 + \gamma_S^2 B_2^2 \tau_1 \tau_2}\right], \tag{5}$$

where B_2 is the EPR irradiation r.f. magnetic-field strength in the rotating frame, and τ_1 and τ_2 are the electron longitudinal and transverse relaxation times. n is approximately equal to the number of lines in the free radical's EPR spectrum (e.g., the stable nitroxide free radicals have three hyperfine lines, so $n \simeq 3$).

2'5. *The maximum theoretical enhancement.* – Equations (2)-(5) can be used to estimate the maximum achievable enhancement value: taking $\rho = +0.5, f = 1, s = 1/3, |\gamma_S/\gamma_P| = 658$, we obtain the value $E^{\max} \simeq -109$ for a proton DNP experiment. (A negative enhancement factor simply means that the phase of the NMR signal is changed by 180° upon EPR irradiation, as shown in fig. 1.) While the maximum possible enhancement cannot be observed in practice, due to the impossibility of achieving complete saturation of the EPR line, large enhancement values ($E \geqslant 50$) can be obtained without too much difficulty.

2'6. EPR *irradiation power.* – Again assuming $\rho = +0.5$, we can rewrite eq. (2) as

$$\frac{1}{1-E} = \frac{n}{329}\left(1 + \frac{1}{\alpha P}\right)\left(1 + \frac{1}{kcT_1^0}\right), \tag{6}$$

where α is a constant and P is the power of the applied EPR irradiation. Here, we have used the fact that $P \propto B_2^2$ in order to illustrate the dependence of the enhancement on P.

Of course, the observed enhancement factor is a function of B_2, and the actual applied power required to achieve a given value of B_2 (or, in turn, E) with a given sample depends on the EPR irradiation frequency ω_2 and on the volume V_c of the tuned coil or resonator used to apply the EPR irradiation to the sample:

$$P \propto \omega_2^2 V_c \tag{7}$$

or

$$P \propto B_0^2 V_c, \tag{8}$$

where B_0 is the strength of the applied static magnetic field. It will be seen that eqs. (7) and (8) have important implications for the practical implementation of PEDRI.

3. – Implementation of PEDRI.

3'1. *Magnetic-field strength considerations.* – As already mentioned above, in a PEDRI experiment one applies an EPR irradiation to the sample while collecting a proton NMR image. When designing apparatus from scratch, one must first decide the value of the static magnetic field B_0. From the point of view of the NMR imaging experiment, one would clearly like to operate at high field in order to optimize the signal-to-noise ratio (SNR). In practice, however, the magnetic-field strength in a PEDRI experiment is limited by the nonresonant absorption of the EPR irradiation. In a given magnetic-field strength, the EPR frequency is approximately 660 times the proton NMR frequency. Since the dielectric losses in conducting samples (*e.g.*, animals) increase rapidly with frequency, and since the applied EPR irradiation power necessary to achieve a given enhancement increases quadratically with the EPR frequency (eq. (7)), it is desirable to use an EPR irradiation frequency of less than 300 MHz, corresponding to magnetic-field strengths of less than or equal to 0.01 T.

This rationale has been confirmed by our experimental results. Initial PEDRI experiments were carried out at a field strength of 0.04 T [5], with an EPR irradiation frequency of 1.123 GHz, where an applied power equivalent to 400 W/kg of sample was needed to achieve an enhancement of −5, the sample being a 2 mM aqueous solution of TEMPOL nitroxide free radical. In later experiments performed at 0.01 T, with an EPR irradiation frequency of 288 MHz, only 23 W/kg irradiation power was needed in order to achieve the same enhancement [9].

It should be noted that not all of the applied r.f. power is actually absorbed by the lossy sample, the exact proportion depending on the frequency as well as

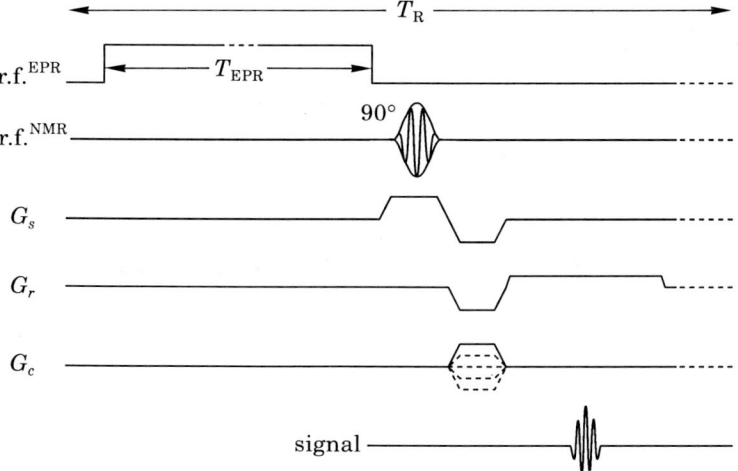

Fig. 2. – A basic PEDRI pulse sequence, based on a spin-warp, gradient-echo, partial-saturation NMR imaging sequence.

the geometry and nature of the sample. As a rough guide, one might expect one quarter to one half of the applied power to be absorbed in an animal if the EPR irradiation is in the (200 ÷ 300) MHz range.

3'2. *Basic PEDRI pulse sequence.* – The basic PEDRI experiment involves irradiating the free radical's EPR while collecting a proton NMR image. Figure 2 shows an example of a PEDRI pulse sequence, here based on a simple gradient-echo, partial-saturation, spin-warp sequence. It can be seen that the EPR irradiation is not applied continuously, but is switched on for a time T_{EPR} before every NMR detection pulse. The increase in proton polarization upon EPR irradiation is not instantaneous, but occurs with a time constant equal to the proton longitudinal relaxation time T_1, as does the decay of proton polarization after switching off the EPR irradiation. Thus the value of T_{EPR} should be of the order of or longer than T_1 in order to achieve a high enhancement value. An advantage of pulsed (rather than continuous) EPR irradiation is that the time-averaged applied EPR irradiation power is reduced in proportion with the duty cycle (*i.e.* T_{EPR}/T_R, where T_R is the pulse sequence repetition time).

3'3. *Interleaved PEDRI pulse sequence.* – In order to produce an image representing the distribution of the free radical under study, it is necessary to subtract images obtained with and without EPR irradiation. While this procedure could be accomplished by collecting and processing with- and without-EPR images separately, then performing the subtraction, it is preferable to use an interleaved pulse sequence, as shown schematically in fig. 3, mainly in order to alleviate the effects of sample movement during data acquisition. In the interleaved PEDRI pulse sequence, each phase-encoding gradient is applied twice, and each alternate NMR detection pulse is preceded by a period of EPR irradiation. This results in three images, namely «with-EPR», «without-EPR» and «difference», the latter exhibiting nonzero intensity only in regions of the sample containing free radical. It should be noted that, since the phase of the NMR signal is reversed upon EPR

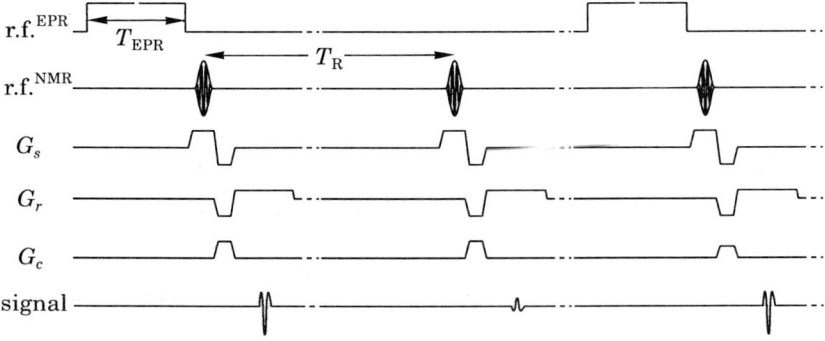

Fig. 3. – An interleaved PEDRI pulse sequence.

irradiation, it is necessary to subtract the complex image data (*i.e.* real and imaginary), rather than simply the pixel intensities.

3'4. *Field-cycled* PEDRI. – When performing *in vivo* PEDRI experiments, it is important to ensure that the applied EPR irradiation power is sufficiently low that excessive heating of the sample by nonresonant r.f. absorption does not occur. While this is important for experiments on animals, where hyperthermia would interfere with the normal metabolism and would probably affect the result of the experiment, it is vital if PEDRI imaging of human subjects is ever to be considered, for obvious reasons! Current guidelines on r.f. exposure in clinical NMR imaging recommend that the r.f. power absorption be limited to between 1 and 20 W/kg, depending on the time scale of the examination and on the region of the body being imaged.

Reducing the EPR frequency by suitable choice of B_0 can certainly reduce the applied r.f. power to manageable levels, as discussed in subsect. 3'1, but this is inevitably at the expense of the SNR of the NMR experiment, which in turn compromises sensitivity and, of course, image quality. The use of magnetic-field cycling with PEDRI offers the possibility of lowering the r.f. power requirement without reducing the SNR.

A field-cycled pulse sequence has three periods, during each of which the «static» magnetic field B_0 takes a different value: the polarization period at B_0^P (high field), during which the spin polarization builds up; the evolution period at B_0^E (low field) in which the spin system evolves; and the detection period at B_0^D (intermediate field), during which the NMR signal is detected [10]. In field-cycled PEDRI (FC-PEDRI), the EPR irradiation is applied during the evolu-

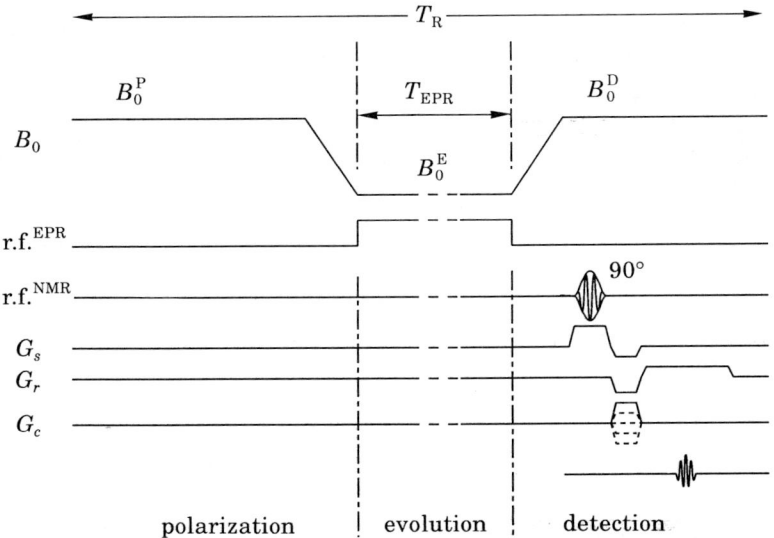

Fig. 4. – A field-cycled PEDRI pulse sequence.

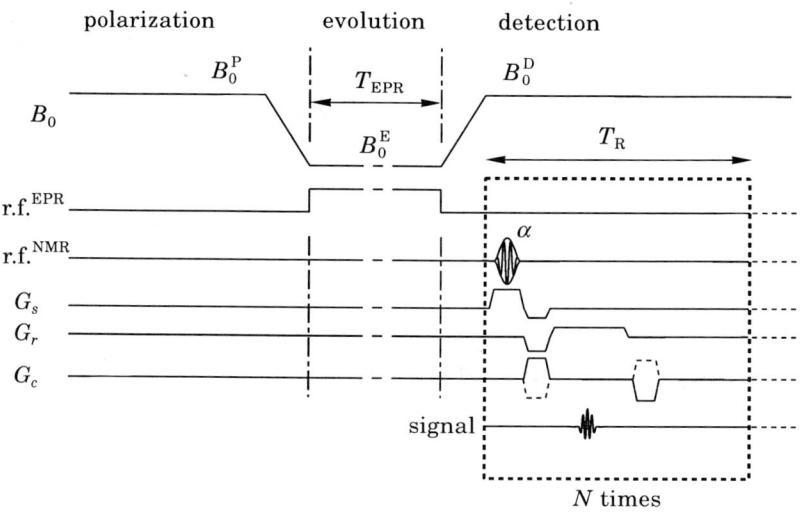

Fig. 5. – A snapshot FC-PEDRI pulse sequence.

tion period, at a frequency (and power) determined by B_0^E. The values of B_0^P and B_0^D are chosen in order to give an acceptable SNR [9]. Figure 4 shows a typical FC-PEDRI pulse sequence (with $B_0^P = B_0^D$, for simplicity). In order to preserve the enhanced proton polarization arising from the EPR irradiation, the time taken to switch between B_0^E and B_0^D must be short compared to the T_1 of the sample (or the minimum T_1 in an *in vivo* experiment).

3˙5. *Snapshot* PEDRI. – Another way of lowering the total energy absorbed in the sample from the EPR irradiation is to reduce the number of periods of EPR irradiation per image. In the pulse sequences described in the previous sections, an image with matrix size $N \times N$ requires N periods of EPR irradiation to be applied, one for each phase-encoding gradient step, each of which contributes to sample heating. In snapshot PEDRI, only a single period of EPR irradiation is applied, and the image data are then collected as rapidly as possible (before the enhanced proton polarization has decayed), using any conventional «snapshot» or «single-shot» NMR imaging pulse sequence [11]. Figure 5 shows a snapshot field-cycled PEDRI pulse sequence, in which the snapshot technique has been combined with magnetic-field cycling [12]. In this case, a snapshot FLASH fast imaging pulse sequence is illustrated [13].

4. – Hardware for PEDRI.

4˙1. *The magnet.* – Since PEDRI is based on proton NMR imaging, there are no special requirements for the magnet itself. As PEDRI is exclusively a low-field technique, it is convenient to use resistive or permanent magnets. Our experiments to date have been performed on a resistive, vertical-field, whole-body magnet, manufactured by Oxford Instruments in 1979. Originally designed to operate at 0.04 T, the magnet was run at 0.01 T for our PEDRI experiments, giving NMR and EPR frequencies of 425 kHz and 288 MHz, respectively (the latter corresponding to the middle line in a nitroxide free radical's triplet).

Field-cycling experiments, of course, require that the magnetic field be changed rapidly between levels, and, therefore, need special-purpose magnets and/or power supplies. There are two ways in which field cycling can be accomplished. Perhaps the most obvious method is to change rapidly the current flowing in a resistive magnet. The difficulty in this approach occurs because of the inductance of the magnet: rapid changes in current cause a back-e.m.f. to be generated which might compromise the magnet windings' insulation. With a suitably designed magnet, this technique can be successful, however, and the use of a «pre-emphasized» voltage wave form (of the sort frequently used in magnetic-field gradient drivers) would allow trapezoidal field-cycling wave forms to be used, as illustrated in fig. 4.

The stability of the magnetic field during the detection period is of crucial importance: any instability will introduce unacceptable image artifacts (primarily ghosting). Thus, in the method described above, the magnet power supply must clamp accurately the current during the detection period. Another solution is to use a technique called «field compensation», in which a high-homogeneity magnet provides the detection magnetic field, which is offset during the evolution period by the field generated by a smaller, coaxial magnet[10]. In FC-PEDRI, the magnetic field applied during the evolution period only needs to be sufficiently homogeneous to irradiate the EPR line of interest throughout the sample. Nitroxide free radicals, for example, have a linewidth of approximately 4 MHz at an EPR frequency of 200 MHz, so that a variation of B_0^E of $\pm 1\%$ over the sample volume can be tolerated. Field stability during the detection period is provided by switching off the current in the secondary magnet; this is easier to achieve than the accurate clamping to a nonzero current required in the former method of field cycling described above.

We have used the field compensation technique in our FC-PEDRI experiments[9]; figure 6 shows the arrangement of the magnet coils. The previously described whole-body magnet generated the polarization and detection magnetic fields, each equal to 0.01 T, and a 22 cm diameter Helmholtz pair was used as the secondary magnet. The secondary coil had an inductance (L) of 24 mH, with a resistance (R) of 1 Ω, giving a time constant (L/R) of 24 ms. By using pre-emphasis of the driving voltage, the field at the sample could be switched from 0.01 T to 0.005 T in less than 5 ms. At the lower field strength the EPR frequency was reduced to 160 MHz, and the applied r.f. power required to achieve $E = -5$ with a 2 mM TEMPOL sample was lowered to 5.6 W/kg, in broad agreement with eq. (7)[9].

The field compensation approach has two disadvantages: firstly, it requires

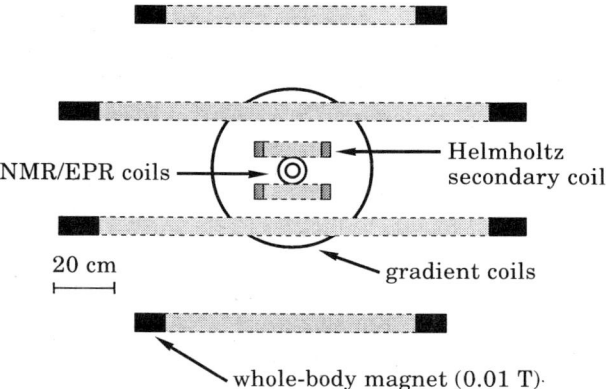

Fig. 6. – Cross-sectional view of coils used for FC-PEDRI experiments.

duplication of hardware (namely two separate magnets and two power supplies); secondly, the mutual inductance between the coaxial magnet coils can result in instability of the primary magnet's field after switching of the secondary magnet. In our apparatus, this effect limited the polarization and detection magnetic fields to 0.01 T, instead of the 0.04 T which the whole-body magnet was capable of providing. One solution to this problem might be to use a permanent magnet to generate the primary field.

4'2. *The double-resonance r.f. coil assembly.* – In PEDRI, as in DNP, one needs to excite and detect the nuclear resonance, and also to excite the EPR resonance. Since the NMR and EPR frequencies are at least two orders of magnitude apart, even when field cycling is used, it is necessary to use separate, coaxial coils or resonators for the two purposes. The requirements for the double-resonance coil assembly are as follows:

a) The B_1 field produced by the NMR coil and the B_2 field produced by the EPR resonator must be as homogeneous as possible (as in any imaging experiment).

b) The EPR resonator should be efficient, producing a large B_2 magnetic field per unit applied r.f. power. In other words, the quality factor (Q) of the resonator should be large.

c) The sensitivity of the NMR coil (in detection mode) should be high.

d) The filling factor of both the NMR coil and the EPR resonator should be high.

e) The NMR coil and EPR resonator should be electrically isolated. Assembling the double-resonance coil should not reduce the Q-factor of either coil.

It is obvious that some of these requirements are contradictory. For example, with two coaxial coils, it is impossible to optimize the filling factor of both. Thus the design of any double-resonance assembly will necessarily be a compromise, depending to a large extent on the nature of the experiment to be performed.

The most fundamental design choice concerns the relative positions of the NMR coil and EPR resonator. Placing the NMR coil on the inside, nearer the sample, will optimize its filling factor, improving the SNR. Placing the EPR resonator on the inside will result in a higher B_2 field per unit applied power, resulting in greater EPR saturation and hence greater enhancement values. In *in vivo* applications, it is preferable to adopt the former configuration, with the EPR resonator outside the NMR coil, in order to maximize the SNR of the NMR experiment. Although the efficiency of the EPR resonator is reduced, one can simply apply a higher power (assuming an appropriate r.f. amplifier exists) in order to recover the required enhancement. Since both the B_1 and B_2 r.f. magnetic fields must be orthogonal to B_0, the choice of coil or resonator type depends on whether the static field is horizontal or vertical.

As an example, one double-resonance assembly which we have found to be useful in our vertical-field imager is the combination of a solenoidal NMR transmit/receive coil with an Alderman-Grant, slotted-tube EPR resonator. The Alderman-Grant resonator has a high Q, and the electric field is mainly confined to the gaps in the «H»s, making it suitable for use with aqueous samples [14]. Figure 7 shows a schematic diagram of the coil/resonator assembly. The resonator was constructed from 0.25 mm copper sheet; the diameter of the resonator was 11 cm and its length was 8 cm. For the sake of clarity, neither the guard rings, situated between the resonator and NMR coil, nor the external, cylindrical,

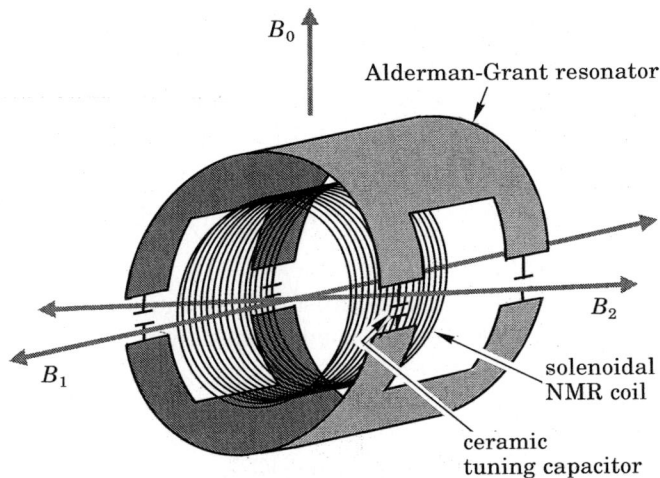

Fig. 7. – Diagram showing double-resonance coil/resonator assembly based on Alderman-Grant EPR resonator and solenoidal NMR coil. B_0, B_1 and B_2 are mutually orthogonal.

copper r.f. shield are shown here. Coupling was achieved by a single-turn loop situated between the r.f. shield and the resonator, a series trimming capacitor being used to match the resonator to 50 Ω. The resonator was tuned to 237 MHz by four 2.2 pF ceramic capacitors, situated at the gaps between the «H» pieces. One of the capacitors was an adjustable trimmer to allow fine tuning. The solenoidal NMR coil, diameter 7.5 cm, length 7 cm, comprised 70 turns of 0.4 mm copper wire, centre-tapped, and tuned to 425 kHz with ceramic capacitors. In this configuration, B_1 and B_2 are orthogonal, and the interaction between the NMR coil and EPR resonator is minimal.

5. – Applications of PEDRI.

5˙1. *Imaging exogenous free radicals in vivo.* – The purpose of this type of experiment is to inject an animal with a solution of «stable» free radical, and then to use PEDRI to follow the progress of the free radical through the body, the concept being analogous to the use of radioactive tracers in nuclear medicine.

5˙1.1. Desirable features of a PEDRI contrast agent. The properties of an ideal free-radical «contrast agent» or «tracer» for use with PEDRI are as follows:

a) The compound should be as stable as possible *in vivo*, in order to provide adequate time for the imaging experiment.

b) The toxicity of the compound should be low.

c) The EPR spectrum should contain as few lines as possible (n in eqs. (5) and (6)), in order to maximize the enhancement.

d) The EPR lines should be as narrow as possible, *i.e.* the electron spin-spin relaxation time τ_2 should be long (eq. (5)), in order to maximize the saturation of the EPR line, and hence increase the enhancement.

e) The NMR relaxivity of the compound should be high (k in eqs. (4) and (6)), in order to increase the leakage factor, f, again increasing the enhancement factor.

Not surprisingly, no free radicals satisfy all of the above requirements simultaneously, and the choice of compound is inevitably a compromise.

5˙1.2. Nitroxide free radicals. At present, probably the most convenient compounds to use are members of the nitroxide family of free radicals. Nitroxides have been known for many years to EPR spectroscopists, and were explored as contrast agents for clinical NMR imaging in the early 1980s [3]. The nitroxides are chemically versatile compounds, the common feature of which is the N—O group on which the unpaired electron is located. Many different nitroxide free radicals exist, either based on a piperidine, six-membered ring, or on a pyrrolidine, five-membered ring. Figure 8 shows the structures of one piperidine (4-hydroxy-2,2,6,6-tetra-methyl-piperidine-1-oxyl, or TEMPOL) and one pyrrolidine (3-carboxy-2,2,5,5-tetra-methyl-pyrrolidine-1-oxyl, or pyrrolidine carboxylic acid, or PCA).

Various nitroxide free radicals go some way to satisfying the requirements listed in paragraph 5˙1.1. The stability of the compounds in biological systems has been studied by a number of workers [15-17]. Nitroxides are reduced to diamagnetic hydroxylamine compounds, with, in general, pyrrolidine nitroxides being reduced more slowly than piperidine free radicals. Negatively charged groups also make the compounds less vulnerable to reduction, and PCA has been found to be one of the most stable [3].

Despite being free radicals, nitroxides are not especially toxic. For example, the acute toxicity of PCA has been investigated [18], and its LD50 in the rat was found to be 15.1 mmol/kg, of the same order of magnitude as that quoted for Gd-DTPA, a compound currently used as a paramagnetic contrast agent in clinical NMR imaging.

The number of EPR lines exhibited by a nitroxide is determined by the hyperfine interaction between the unpaired electron and the adjacent nitrogen nucleus. Since ^{14}N has a spin of 1, the usual nitroxide EPR spectrum comprises three principal lines. However, if ^{15}N is substituted into the molecule, the number of lines is reduced to two, since the nuclear spin of this nitrogen isotope is 1/2. It is also possible to reduce the width of each line in the spectrum by substituting ^2H for ^1H, since the linewidth is partly due to interactions between the methyl hydrogen nuclei and the unpaired electron. Perdeuterated, ^{15}N-substituted nitroxides have been used in EPR imaging to increase sensitivity [19], and would undoubtedly be of great use in PEDRI for the reasons outlined above, but unfortunately they are rather expensive (^{15}N, perdeuterated PCA is currently 150 times more

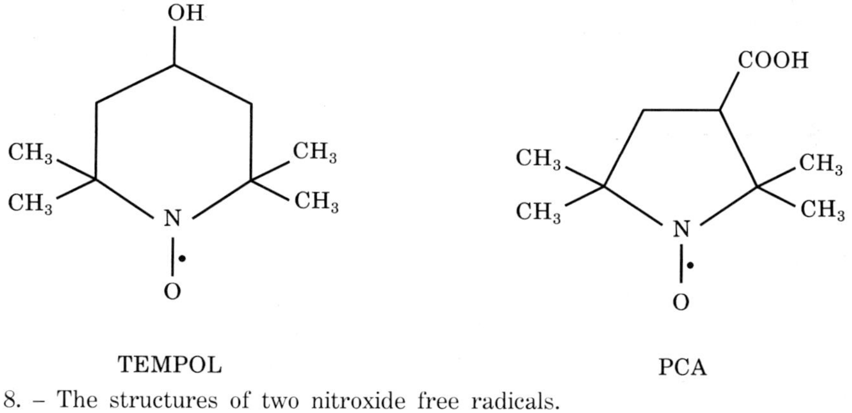

Fig. 8. – The structures of two nitroxide free radicals.

expensive than the unsubstituted compound, at roughly $1000 per 100 mg!).

The influence of nitroxides on water proton relaxation has been studied by BENNETT et al. [20] among others, who found that the relaxivity of nitroxides is an order of magnitude less than that of Gd-DTPA. (This is why the latter was developed as a clinical contrast agent in favour of the nitroxides.) It is possible to increase the relaxivity of nitroxides by binding them to larger molecules such as proteins [20], but this may be at the expense of broadening the EPR linewidth, providing no net improvement in the compound's suitability for PEDRI experiments.

5`1.3. *In vivo* PEDRI studies. The first use of PEDRI *in vivo* was performed by GRUCKER [21], who administered an intraperitoneal injection of a 10 mM aqueous solution of Fremy's salt (nitroso disulphonate) to an anaesthetised rat, and imaged its distribution in the peritoneal cavity. The field strength used was 6.8 mT, with an EPR frequency of 197.4 MHz. Although Fremy's salt has a rather narrow EPR linewidth (26 µT in a deoxygenated 2 mM aqueous solution, as compared to 122 µT for PCA [22]), giving a large enhancement factor per unit applied EPR irradiation power, it is too unstable and toxic for intravenous injection, and was used only to demonstrate the feasibility of *in vivo* PEDRI.

In vivo PEDRI experiments have also been carried out using intravenously administered PCA in the rat [23]. Adult Sprague-Dawley rats ((300 ÷ 350) g) were anaesthetised and injected with 1 ml of bicarbonate-buffered 150 mM PCA solution, via a cannula placed in the external jugular vein. A series of images was then obtained over a period of approximately 30 min, using an interleaved PEDRI pulse sequence (subsect. 3`3). The field strength used was 0.01 T, with an EPR irradiation frequency of 238 MHz, corresponding to the low-frequency line in the nitroxide's hyperfine triplet. Figure 9 shows a set of «difference» images obtained after injection of the free radical. Image 1 was obtained before injection, and the subsequent images were collected over the course of the next 21 min. The images are transaxial, showing the highest free-radical concentration in the animal's kidneys. Analysis of the time dependence of the difference image intensities gave a value for the half-life of the free-radical clearance of (11 ± 2) min, identical to the value observed by other workers using EPR imaging [24].

5`2. *Imaging endogenous free radicals in vitro and in vivo.* – The second category of biological applications of PEDRI involves the detection of endogenous, or naturally occurring free radicals. Of these, the so-called oxygen-derived free radicals (ODFR) hydroxyl (•OH) and superoxide (O_2^- •) are perhaps the most important in terms of their roles in the pathogenesis of many diseases [25, 26].

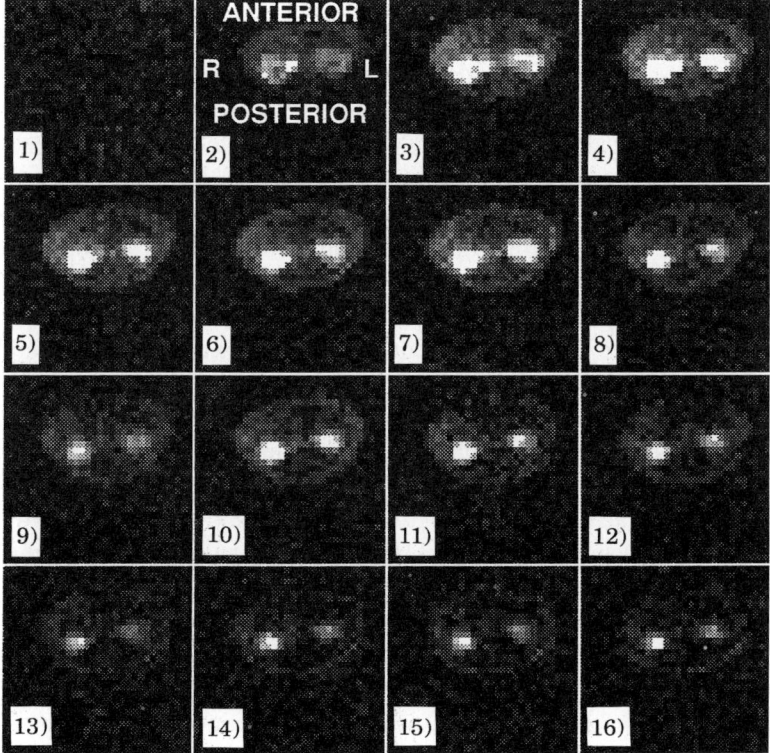

Fig. 9. – Consecutive series of PEDRI «difference» images showing the clearance of PCA nitroxide through the kidneys of a rat. Images are 32×32, transaxial, with a field of view of 10 cm $\times$ 10 cm and a slice thickness of 2 cm. The pulse sequence was as fig. 3, with $T_{EPR} = 300$ ms and $T_R = 500$ ms; two averages were collected.

One condition in which ODFR are thought to be involved is so-called reperfusion injury. When a tissue is deprived of oxygen for some time (such as the heart muscle during a «heart attack»), and oxygen is then re-introduced, much of the resulting tissue damage is thought to arise because of the generation of hydroxyl free radicals[27].

5˙2.1. Spin trapping. Hydroxyl free radicals are extremely reactive, and consequently are also very short lived (submicrosecond). It is, therefore, not possible to observe these free radicals directly, either by EPR spectroscopy or by PEDRI. Fortunately, however, a technique called spin trapping can be employed[28]. In this technique, a spin trap molecule is first dissolved in the system under study (the spin trap is not a free radical). When short-lived radicals are formed in the vicinity of a spin trap molecule, there is a high probability that they will interact to form a longer-lived free radical called a spin adduct, which can be detected by EPR (or PEDRI).

Spin trapping has been used to detect hydroxyl radicals formed during the reperfusion of an isolated, ischaemic rat heart[29]; here, the perfusate (containing spin trap) was passed through the excised heart and into the cavity of an X-band EPR spectrometer, where the spin adducts were detected.

Spin trapping has also been used *in vivo;* a review of some of these experiments is given in ref.[30]. Basically, the studies involve the administration of a spin trap to a living animal, which is then exposed to an insult (drug, toxin or regional ischaemia) in order to stimulate free-radical production. Spin adducts are then sought using EPR spectroscopy, either on extracted body fluids, or on tissue samples after sacrifice of the animal.

5˙2.2. PEDRI with spin trapping. In combination with spin trapping, PEDRI offers the possibility of imaging the production of endogenous free radicals in intact animals. As a first step towards this goal, we have recently demonstrated the feasibility of the combined technique by imaging the production of hydroxyl free radicals generated by the action of ultraviolet (UV) light on a solution of hydrogen peroxide[31].

The spin trap DMPO (5,5-dimethyl-1-pyrroline-N-oxide) was used in this work, and the first step was to measure the EPR spectrum of the DMPO/•OH spin adduct at low magnetic field, in order to determine the optimum frequency at which to irradiate the EPR in a PEDRI experiment. An aqueous solution containing 50 mM DMPO and 25 mM H_2O_2 was exposed to UV light and was then transferred to the FC-PEDRI apparatus, where its EPR spectrum was measured using field-cycled DNP (FC-DNP)[32]. The FC-DNP pulse sequence is the same as that used in FC-PEDRI (fig. 4), except that no magnetic-field

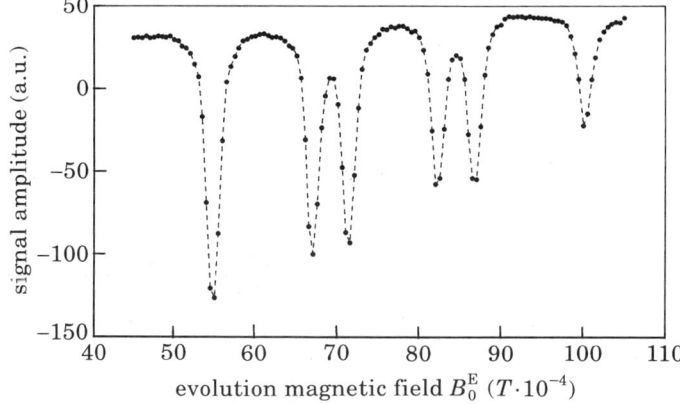

Fig. 10. – FC-DNP spectrum of the DMPO/•OH spin adduct (reproduced from ref.[31] with permission). The EPR irradiation frequency in this case was 223 MHz.

gradients are applied. Instead, the EPR irradiation frequency is kept constant and the sequence is repeated with different values of B_0^E, a plot of NMR signal amplitude vs. B_0^E revealing the positions of the EPR lines. The FC-DNP spectrum of DMPO/•OH, shown in fig. 10, revealed six separate lines[33]. At a field strength of 0.01 T, the lowest-frequency line occurred at 223 MHz, and this was used as the irradiation frequency in subsequent PEDRI experiments to image the production of the spin adduct in a UV-transparent quartz container situated in the PEDRI apparatus[31].

Work is currently underway in our laboratory to apply this technique to the detection of endogenous free radicals in biological systems.

6. – Summary and conclusions.

PEDRI is proving to be a useful technique for the study of free radicals in biological systems. In comparison with EPR imaging, which has a higher inherent sensitivity, and can image free radicals in solids as well as liquids, PEDRI has the advantage of using standard (albeit low-field) NMR imaging hardware and software, making it very flexible. PEDRI could, in principle, be used to image free radicals in humans, but clearly these applications would require the development of suitable contrast agents and/or spin traps.

Further work must be performed in order to increase the sensitivity of PEDRI. This will involve the design of better stable free radicals for use as contrast agents, in particular with regard to their EPR characteristics. Although the potential of FC-PEDRI has been demonstrated, further development work must be carried out in order to explore its full potential. In particular, the use of considerably higher polarization and detection magnetic fields in order to increase the SNR and sensitivity would be valuable.

* * *

This work was supported by the Medical Research Council and by the Leverhulme Trust.

REFERENCES

[1] T. F. SLATER: *Biochem. J.*, **222**, 1 (1984).
[2] B. HALLIWELL and M. GROOTVELD: *FEBS Lett.*, **213**, 9 (1987).
[3] R. L. EHMAN, G. E. WESBEY, K. L. MOON, R. D. WILLIAMS, M. T. MCNAMARA, W. R. COUET, T. N. TOZER and R. C. BRASCH: *Magn. Reson. Imaging*, **3**, 89 (1985).

[4] G. R. EATON, S. R. EATON and K. OHNO, Editors: EPR *Imaging and in vivo* EPR (CRC Press, Boca Raton, Fla., 1991).
[5] D. J. LURIE, D. M. BUSSELL, L. H. BELL and J. R. MALLARD: *J. Magn. Reson.*, **76**, 366 (1988).
[6] K. H. HAUSSER and D. STEHLIK: in *Advances in Magnetic Resonance*, edited by J. S. WAUGH, Vol. 3 (Academic Press, New York, N.Y., 1968), p. 79.
[7] R. A. DWEK, R. E. RICHARDS and D. TAYLOR: *Annu. Rev.* NMR *Spectrosc.*, **2**, 293 (1969).
[8] W. MULLER-WARMUTH and K. MEISE-GRESCH: in *Advances in Magnetic Resonance*, edited by J. S. WAUGH, Vol. **11** (Academic Press, New York, N.Y., 1983), p. 1.
[9] D. J. LURIE, J. M. S. HUTCHISON, L. H. BELL, I. NICHOLSON, D. M. BUSSELL and J. R. MALLARD: *J. Magn. Reson.*, **84**, 431 (1989).
[10] F. NOACK: *Prog.* NMR *Spectrosc.*, **18**, 171 (1986).
[11] G. J. EHNHOLM, R. JOENSUU, A. JARVI, S. PETERSSON, O. SALO and S. VAHASALO: in *Proceedings of the Society of Magnetic Resonance in Medicine*, 10th Annual Meeting, San Francisco, Works-in-Progress (1991), p. 1247.
[12] D. J. LURIE, I. NICHOLSON and J. R. MALLARD: in *Proceedings of the Society of Magnetic Resonance in Medicine*, 11th Annual Meeting, Berlin (1992), p. 72.
[13] A. HAASE: *Magn. Reson. Med.*, **13**, 77 (1990).
[14] D. W. ALDERMAN and D. M. GRANT: *J. Magn. Reson.*, **36**, 447 (1979).
[15] W. R. COUET, R. C. BRASCH, G. SOSNOVSKY and T. N. TOZER: *Magn. Reson. Imaging*, **3**, 83 (1985).
[16] J. F. W. KEANA, S. POU and G. M. ROSEN: *Magn. Reson. Med.*, **5**, 525 (1987).
[17] P. D. MORSE, J. M. PETRUSZAK and L. REMINGER: in *Proceedings of the 13th International EPR Symposium*, 32nd Rocky Mountain Conference, Denver, Colorado, Abstract No. 164 (1990).
[18] V. AFZAL, R. C. BRASCH, D. E. NITECKI and S. WOLFF: *Invest. Radiol.*, **19**, 549 (1984).
[19] G. BACIC, F. DEMSAR, Z. ZOLNAI and H. M. SWARTZ: *Magn. Reson. Med. Biol.*, **1**, 55 (1988).
[20] H. F. BENNETT, R. D. BROWN, S. H. KOENIG and H. M. SWARTZ: *Magn. Reson. Med.*, **4**, 93 (1987).
[21] D. GRUCKER: *Magn. Reson. Med.*, **14**, 140 (1990).
[22] D. GRUCKER and J. CHAMBRON: *Phys. Med.*, **5**, 329 (1989).
[23] D. J. LURIE, I. NICHOLSON, M. A. FOSTER and J. R. MALLARD: *Philos. Trans. R. Soc. London, Ser. A*, **333**, 453 (1990).
[24] V. QUARESIMA, M. ALECCI, M. FERRARI and A. SOTGIU: *Biochem. Biophys. Res. Commun.*, **183**, 829 (1992).
[25] B. HALLIWELL and J. M. C. GUTTERIDGE: *Free Radicals in Biology and Medicine*, 2nd edition (Clarendon Press, Oxford, 1989), Chapt. 2.
[26] P. A. SOUTHERN and G. POWIS: *Mayo Clinic Proc.*, **63**, 390 (1988).
[27] L. H. OPIE: *Circulation*, **80**, 1049 (1989).
[28] G. R. BUETTNER: in *Superoxide Dismutase*, edited by L. W. OBERLEY, Vol. 2 (CRC Press, Boca Raton, Fla., 1982), Chapt. 4.
[29] S. PIETRI, M. CULCASI and P. J. COZZONE: *Eur. J. Biochem.*, **186**, 163 (1989).
[30] C. H. KENNEDY, K. R. MAPLES and R. P. MASON: *Pure Appl. Chem.*, **62**, 295 (1990).
[31] D. J. LURIE, J. MCLAY, I. NICHOLSON and J. R. MALLARD: *J. Magn. Reson.*, **95**, 191 (1991).

[32] D. J. LURIE, I. NICHOLSON and J. R. MALLARD: *J. Magn. Reson.*, **95**, 405 (1991).
[33] D. J. LURIE, I. NICHOLSON, J. S. McLAY and J. R. MALLARD: *Appl. Magn. Reson.*, **3**, 917 (1992).

PROCEEDINGS OF THE INTERNATIONAL SCHOOL OF PHYSICS «ENRICO FERMI»

Course I
Questioni relative alla rivelazione delle particelle elementari, con particolare riguardo alla radiazione cosmica
edited by G. PUPPI

Course II
Questioni relative alla rivelazione delle particelle elementari, e alle loro interazioni con particolare riguardo alle particelle artificialmente prodotte ed accelerate
edited by G. PUPPI

Course III
Questioni di struttura nucleare e dei processi nucleari alle basse energie
edited by C. SALVETTI

Course IV
Proprietà magnetiche della materia
edited by L. GIULOTTO

Course V
Fisica dello stato solido
edited by F. FUMI

Course VI
Fisica del plasma e relative applicazioni astrofisiche
edited by G. RIGHINI

Course VII
Teoria della informazione
edited by E. R. CAIANIELLO

Course VIII
Problemi matematici della teoria quantistica delle particelle e dei campi
edited by A. BORSELLINO

Course IX
Fisica dei pioni
edited by B. TOUSCHEK

Course X
Thermodynamics of Irreversible Processes
edited by S. R. DE GROOT

Course XI
Weak Interactions
edited by L. A. RADICATI

Course XII
Solar Radioastronomy
edited by G. RIGHINI

Course XIII
Physics of Plasma: Experiments and Techniques
edited by H. ALFVÉN

Course XIV
Ergodic Theories
edited by P. CALDIROLA

Course XV
Nuclear Spectroscopy
edited by G. RACAH

Course XVI
Physicomathematical Aspects of Biology
edited by N. RASHEVSKY

Course XVII
Topics of Radiofrequency Spectroscopy
edited by A. GOZZINI

Course XVIII
Physics of Solids (Radiation Damage in Solids)
edited by D. S. BILLINGTON

Course XIX
Cosmic Rays, Solar Particles and Space Research
edited by B. PETERS

Course XX
Evidence for Gravitational Theories
edited by C. MØLLER

Course XXI
Liquid Helium
edited by G. CARERI

Course XXII
Semiconductors
edited by R. A. SMITH

Course XXIII
Nuclear Physics
edited by V. F. WEISSKOPF

Course XXIV
Space Exploration and the Solar System
edited by B. ROSSI

Course XXV
Advanced Plasma Theory
edited by M. N. ROSENBLUTH

Course XXVI
Selected Topics on Elementary Particle Physics
edited by M. CONVERSI

Course XXVII
Dispersion and Absorption of Sound by Molecular Processes
edited by D. SETTE

Course XXVIII
Star Evolution
edited by L. GRATTON

Course XXIX
Dispersion Relations and their Connection with Causality
edited by E. P. WIGNER

Course XXX
Radiation Dosimetry
edited by F. W. SPIERS and G. W. REED

Course XXXI
Quantum Electronics and Coherent Light
edited by C. H. TOWNES and P. A. MILES

Course XXXII
Weak Interactions and High-Energy Neutrino Physics
edited by T. D. LEE

Course XXXIII
Strong Interactions
edited by L. W. ALVAREZ

Course XXXIV
The Optical Properties of Solids
edited by J. TAUC

Course XXXV
High-Energy Astrophysics
edited by L. GRATTON

Course XXXVI
Many-Body Description of Nuclear Structure and Reactions
edited by C. BLOCH

Course XXXVII
Theory of Magnetism in Transition Metals
edited by W. MARSHALL

Course XXXVIII
Interaction of High-Energy Particles with Nuclei
edited by T. E. O. ERICSON

Course XXXIX
Plasma Astrophysics
edited by P. A. STURROCK

Course XL
Nuclear Structure and Nuclear Reactions
edited by M. JEAN and R. A. RICCI

Course XLI
Selected Topics in Particle Physics
edited by J. STEINBERGER

Course XLII
Quantum Optics
edited by R. J. GLAUBER

Course XLIII
Processing of Optical Data by Organisms and by Machines
edited by W. REICHARDT

Course XLIV
Molecular Beams and Reaction Kinetics
edited by CH. SCHLIER

Course XLV
Local Quantum Theory
edited by R. JOST

Course XLVI
Physics with Intersecting Storage Rings
edited by B. TOUSCHEK

Course XLVII
General Relativity and Cosmology
edited by R. K. SACHS

Course XLVIII
Physics of High Energy Density
edited by P. CALDIROLA and H. KNOEPFEL

Course IL
Foundations of Quantum Mechanics
edited by B. D'ESPAGNAT

Course L
Mantle and Core in Planetary Physics
edited by J. COULOMB and M. CAPUTO

Course LI
Critical Phenomena
edited by M. S. GREEN

Course LII
Atomic Structure and Properties of Solids
edited by E. BURSTEIN

Course LIII
Developments and Bonderlines of Nuclear Physics
edited by H. MORINAGA

Course LIV
Developments in High-Energy Physics
edited by R. R. GATTO

Course LV
Lattice Dynamics and Intermolecular Forces
edited by S. CALIFANO

Course LVI
Experimental Gravitation
edited by B. BERTOTTI

Course LVII
History of 20th Century Physics
edited by C. WEINER

Course LVIII
Dynamics Aspects of Surface Physics
edited by F. O. GOODMAN

Course LIX
Local Properties at Phase Transitions
edited by K. A. MÜLLER and A. RIGAMONTI

Course LX
C*-Algebras and their Applications to Statistical Mechanics and Quantum Field Theory
edited by D. KASTLER

Course LXI
Atomic Structure and Mechanical Properties of Metals
edited by G. CAGLIOTI

Course LXII
Nuclear Spectroscopy and Nuclear Reactions with Heavy Ions
edited by H. FARAGGI and R. A. RICCI

Course LXIII
New Directions in Physical Acoustics
edited by D. SETTE

Course LXIV
Nonlinear Spectroscopy
edited by N. BLOEMBERGEN

Course LXV
Physics and Astrophysics of Neutron Stars and Black Holes
edited by R. GIACCONI and R. RUFFINI

Course LXVI
Health and Medical Physics
edited by J. BAARLI

Course LXVII
Isolated Gravitating Systems in General Relativity
edited by J. EHLERS

Course LXVIII
Metrology and Fundamental Constants
edited by A. FERRO MILONE, P. GIACOMO and S. LESCHIUTTA

Course LXIX
Elementary Modes of Excitation in Nuclei
edited by A. BOHR and R. A. BROGLIA

Course LXX
Physics of Magnetic Garnets
edited by A. PAOLETTI

Course LXXI
Weak Interactions
edited by M. BALDO CEOLIN

Course LXXII
Problems in the Foundations of Physics
edited by G. TORALDO DI FRANCIA

Course LXXIII
Early Solar System Processes and the Present Solar System
edited by D. LAL

Course LXXIV
Development of High-Pozer Lasers and their Applications
edited by C. PELLEGRINI

Course LXXV
Intermolecular Spectroscopy and Dynamical Properties of Dense Systems
edited by J. VAN KRANENDONK

Course LXXVI
Medical Physics
edited by J. R. GREENING

Course LXXVII
Nuclear Structure and Heavy-Ion Collisions
edited by R. A. BROGLIA, R. A. RICCI and C. H. DASSO

Course LXXVIII
Physics of the Earth's Interior
edited by A. M. DZIEWONSKI and E. BOSCHI

Course LXXIX
From Nuclei to Particles
edited by A. MOLINARI

Course LXXX
Topics in Ocean Physics
edited by A. R. OSBORNE and P. MALANOTTE RIZZOLI

Course LXXXI
Theory of Fundamental Interactions
edited by G. COSTA and R. R. GATTO

Course LXXXII
Mechanical and Thermal Behaviour of Metallic Materials
edited by G. CAGLIOTI and A. FERRO MILONE

Course LXXXIII
Positrons in Solids
edited by W. BRANDT and A. DUPASQUIER

Course LXXXIV
Data Acquisition in High-Energy Physics
edited by G. BOLOGNA and M. VINCELLI

Course LXXXV
Earhquakes: Observation, Theory and Interpretation
edited by H. KANAMORI and E. BOSCHI

Course LXXXVI
Gamow Cosmology
edited by F. MELCHIORRI and R. RUFFINI

Course LXXXVII
Nuclear Structure and Heavy-Ion Dynamics
edited by L. MORETTO and R. A. RICCI

Course LXXXVIII
Turbulence and Predictability in Geophysical Fluid Dynamics and Climate Dynamics
edited by M. GHIL, R. BENZI and G. PARISI

Course LXXXIX
Highlights of Condensed-Matter Theory
edited by F. BASSANI, F. FUMI and M. P. TOSI

Course XC
Physics of Amphiphiles: Micelles, Vesicles and Microemulsions
edited by V. DEGIORGIO and M. CORTI

Course XCI
From Nuclei to Stars
edited by A. MOLINARI and R. A. RICCI

Course XCII
Elementary Particles
edited by N. CABIBBO

Course XCIII
Frontiers in Physical Acoustics
edited by D. SETTE

Course XCIV
Theory of Reliability
edited by A. SERRA and R. E. BARLOW

Course XCV
Solar-Terrestrial Relationships and the Earth Environment in the Last Millennia
edited by G. CINI CASTAGNOLI

Course XCVI
Excited-State Spectroscopy in Solids
edited by U. M. GRASSANO and N. TERZI

Course XCVII
Molecular-Dynamics Simulations of Statistical-Mechanical Systems
edited by G. CICCOTTI and W. G. HOOVER

Course XCVIII
The Evolution of Small Bodies in the Solar System
edited by M. FULCHIGNONI and Ľ. KRESÁK

Course XCIX
Synergetics and Dynamic Instabilities
edited by G. CAGLIOTI and H. HAKEN

Course C
The Physics of NMR Spectroscopy in Biology and Medicine
edited by B. MARAVIGLIA

Course CI
Evolution of Interstellar Dust and Related Topics
edited by A. BONETTI and J. M. GREENBERG

Course CII
Accelerated Life Testing and Experts Opinions in Reliability
edited by C. A. CLAROTTI

Course CIII
Trends in Nuclear Physics
edited by P. KIENLE, R. A. RICCI and A. RUBBINO

Course CIV
Frontiers and Borderlines in Many-Particle Physics
edited by R. A. BROGLIA and J. R. SCHRIEFFER

Course CV
Confrontation between Theories and Observations in Cosmology: Present Status and Future Programmes
edited by J. AUDOUZE and F. MELCHIORRI

Course CVI
Current Trends in the Physics of Materials
edited by G. F. CHIAROTTI, F. FUMI and M. TOSI

Course CVII
The Chemical Physics of Atomic and Molecular Clusters
edited by G. SCOLES

Course CVIII
Photoemission and Absorption Spectroscopy of Solids and Interfaces with Synchrotron Radiation
edited by M. CAMPAGNA and R. ROSEI

Course CIX
Nonlinear Topics in Ocean Physics
edited by A. R. OSBORNE

Course CX
Metrology at the Frontiers of Physics and Technology
edited by L. CROVINI and T. J. QUINN

Course CXI
Solid-State Astrophysics
edited by E. BUSSOLETTI and G. STRAZZULLA

Course CXII
Nuclear Collisions from the Mean-Field into the Fragmentation Regime
edited by C. DETRAZ and P. KIENLE

Course CXIII
High-pressure Equation of State: Theory and Applications
edited by S. ELIEZER and R. A. RICCI

Course CXIV
Industrial and Technological Applications of Neutrons
edited by M. FONTANA and F. RUSTICHELLI

Course CXV
The Use of EOS for Studies of Atmospheric Physics
edited by J. C. GILLE and G. VISCONTI

Course CXVI
Status and Perspectives of Nuclear Energy: Fission and Fusion
edited by R. A. RICCI, C. SALVETTI and E. SINDONI

Course CXVII
Semiconductor Superlattices and Interfaces
edited by A. STELLA

Course CXVIII
Laser Manipolation of Atoms and Ions
edited by E. ARIMONDO, W. D. PHILLIPS and F. STRUMIA

Course CXIX
Quantum Chaos
edited by G. CASATI, I. GUARNERI and U. SMILANSKY

Course CXX
Frontiers in Laser Spectroscopy
edited by T. W. HÄNSCH and M. INGUSCIO

Course CXXI
Perspectives in Many-Particle Physics
edited by R. A. BROGLIA and J. R. SCHRIEFFER

Course CXXII
Galaxy Formation
edited by J. SILK and N. VITTORIO

Nuova Tipografia Compositori, Bologna

State College at Framingham

10-68-948217